PHYSICS of STRENGTH and
FRACTURE CONTROL
Adaptation of Engineering
Materials and Structures

PHYSICS of STRENGTH
and
FRACTURE CONTROL
Adaptation of Engineering Materials and Structures

Author
Anatoly A. Komarovsky
Scientific Editor
Viktor P. Astakhov

CRC PRESS

Boca Raton   London   New York   Washington, D.C.

Library of Congress Cataloging-in-Publication Data

Komarovsky, Anatoly A.
   Physics of strength and fracture control : adaptation of engineering materials and
structures / Anatoly A. Komarovsky ; scientific editor, Viktor P. Astakhov.
      p. cm.
   Includes bibliographical references and index.
   ISBN 0-8493-1151-9 (alk. paper)
   1.Strength of materials. 2. Fracture mechanics. I. Astakhov, Viktor P. II. Title.

TA405 .K555 2002
620.1'12—dc21

TA
405
K555
2003

2002066454

Visit the CRC Press Web site at www.crcpress.com

© 2003 by CRC Press LLC

No claim to original U.S. Government works
International Standard Book Number 0-8493-1151-9
Library of Congress Card Number 2002066454
Printed in the United States of America 1 2 3 4 5 6 7 8 9 0
Printed on acid-free paper

Preface

Modern engineering materials and structures operate on the ground, under the water, and in space, at normal, high and cryogenic temperatures, in aggressive environments, and under conditions of intensive radiation. Requirements for their strength, reliability, and durability continuously increase.[1,2] Engineering and applied sciences try to solve multiple problems associated with these requirements using mechanical–mathematical methods rather than physical methods.[3,4] This practice imposes substantial limitations on the level of technogenic safety of engineering objects already achieved and does not lead to improvements in many technological processes associated with engineering materials.

Although the prospects for developing engineering materials are correlated to solid-state physics,[5] the related aspects of this science are insufficiently elaborated.[6,7] As a result, the gap grows between improving the durability of engineering materials and structures and increasing requirements for their safety and reliability, and the corresponding scientific and technical support to meet these challenges. Developed at a time when only invariable or slowly varying force fields were considered, many existing concepts of the physical nature of resistance of solids to different combinations of external effects are obsolete.[3]

According to data presented by Kluev,[3] technogenic accidents and catastrophes resulting in economic losses amounting to $2 billion occur every 10 to 15 years on average; those with losses of up to $100 million occur every 14 to 15 days. Technical progress and inadequacy of traditional methods for ensuring reliable service conditions lead to a 10 to 30% increase in these losses annually. Moreover, for critical objects, they are aggravated by environmental, moral, and social consequences.

From today's standpoint, this means that standard methods of ensuring strength, reliability, and durability have already exhausted their potential. The development of these methods follows a curve with negative first and second derivatives. The time when a qualitatively new stage in the development of design methods should be introduced to support technical progress has passed. Because existing methods and concepts have no reserves left, qualitatively new principles of ensuring technogenic safety need to be introduced.

This book demonstrates that the advances in modern physics that became evident in the mid-20th century form a reliable and sufficient ground for revising common notions of the nature of resistance of solids to diverse external fields (force, thermal, radiation, etc.) and aggressive environments. The book formulates, and then theoretically and experimentally proves, new concepts to control deformation and fracture. It offers methods for inhibition of fracture and reconditioning of damaged structures; a number of nontraditional methods are developed and applied to solve typical practical problems.

This book introduces a new physical concept in the development of the science of resistance of materials to external effects. At its core, the proposed approach has the thermodynamic state of solids equation derived by the author. The book demonstrates that the system of ensuring reliability and durability of engineering structures commonly used today is at an embryonic stage of its development and thus still passive and uncontrollable. The current system is not able to provide corrections or replenishments to the used part of the service life of engineering materials and structures. As a result, failures of engineering materials and structures can be neither predicted nor avoided, thus leading to "unpredictable" accidents and catastrophes. In contrast to this existing system, the concept suggested in the book allows controllability of the stressed–deformed state of materials and structures by activating or preventing undesirable deformations and fractures.

This work develops a new stage in the science of materials strength and provides an introduction to the theory of adapting materials and structures to operating conditions based on physical principles of leading-edge technologies. It introduces new avenues for industries dealing with advanced technologies and products for the improvement of technogenic safety of engineering objects, as well as for the reduction of power consumption of special technological processes. It also discusses practical, but nontraditional, methods of solving many typical problems.

The book is intended for a wide range of readers specializing in the fields of solid-state physics, statistical physics, thermodynamics, materials science, manufacturing technology and processing of structural materials, technology for production and processing of mineral resources, resistance of materials to various external loads, quality of parts and structures, reliability and durability of machines and mechanisms, etc. Thus, all who are directly or indirectly involved in the activation (as in metal cutting) or elimination (engineering structures) of deformation and fracture will find the information useful. It is also very helpful for students because it covers the fundamental aspects of the physics of solids and their resistance to external energy fields and aggressive environments.

Anatoly A. Komarovsky

Preface of the Scientific Editor

Background

To be practical and efficient, materials simulations should be based on proper understanding of the physics of materials; moreover, a way to "convert" such understanding into a mathematical model should be clearly indicated. Unfortunately, this has not yet occurred. Currently, the approach to this problem is to create new materials research centers and laboratories supported by industry and by the National Science Foundation (NSF). The results obtained thus far are not encouraging.

This book explains why solution of actual problems in physics and engineering of materials within the scope of traditional ideas is not possible, regardless of the amount of money granted by NSF or invested by industry. Using numerous examples, this book demonstrates the drawbacks of existing approaches to the mechanics of material and mechanical metallurgy. Attention is drawn to the fact that well-known books on the subject pay little attention to the physics of materials resistance to various external effects (external forces, fields, etc.). Although existing books consider a number of microlevel phenomena, including the property of AM bonds, dislocations, etc., the relation between the microphysics and macrophysics of materials, which defines their actual behavior, is explained qualitatively and thus cannot be used in practice.

Handbooks, reference books, engineering manuals, and standards on the engineering calculations of the strength of parts and structures do not follow advances in materials science. Not one essential property has been added in the last 50 years to the known properties of materials available to a designer. (One can see this on the most popular Web site for materials properties, www.matweb.com.) As a result, design methodology based on failure criteria (largely obtained in the 19th century) and an enormous "safety factor" (that costs billions and "covers" lack of knowledge of materials) prevails in practice.

This book pioneers a new direction in materials science. For the first time, a physical explanation of the strength of materials is offered. The book is multidisciplinary and should be of great interest to all specialists concerned with materials and their properties, design of parts and structures, durability, and reliability.

The Aim of This Book

The ultimate goal of this book is to achieve full understanding of the physics of solid matter through the derived equation of the state of a solid. Using this equation as the basis, this book aims to describe the interaction of a solid with external energy fields. Another essential goal is to suggest new methods to control failure of solids under a full diversity of service conditions.

This book demonstrates that physical–mechanical properties can be controlled from the point of design of a material to the point of fulfilling specific consumer functions, at the stage of solidification and processing, and during service periods of machines and mechanisms. At the first two stages the control functions are performed by passive or materials science methods, whereas at the last stage they are achieved by using active energy methods. Fundamentals of technology for making materials with preset properties in this book will be of interest to materials scientists involved in the development of advanced materials. In addition to solving a direct problem, i.e., control of the structure-formation or destructive processes, the suggested approach allows an inverse problem, i.e., prediction, to be handled successfully.

The book clarifies physical principles of operation and advantages and disadvantages of existing methods of technical diagnostics and nondestructive testing, as well as ways of expanding their capabilities. It also indicates guidelines for development of new, advanced methods for prediction of the technical state of materials and structures. This will be particularly interesting to specialists involved in the development and application of methods of technical diagnostics and nondestructive testing, reliability, and durability of engineering objects and their components. In general, knowing the trends in a specific science and technology area widens the horizons for those working in the area and is very helpful to other specialists.

Why One Needs This Book

This book is essential for understanding how structural materials behave in reality in various external fields and aggressive environments, for understanding how to control the processes of deformation and fracture of solids, and, finally, to build high-reliability engineering objects whose structural components can be adapted to operating conditions. It will not answer all questions about materials, but it will supply knowledge about the physical nature and behavior of materials. Using this knowledge can provide the answers to many theoretical and practical problems.

Viktor P. Astakhov

Acknowledgments

I was extremely fortunate that Viktor P. Astakhov kindly agreed to be the scientific editor of this book. As initiator of its publication and a great advisor during the course of its preparation, he made many useful comments and suggestions regarding content, structure, and presentation of major ideas. His incredible engineering sense and broad interdisciplinary knowledge about materials and their technology and about technical physics and applied mathematics helped me enormously in clarifying content. I am deeply grateful to Dr. Zelnichenko for his coordination of the translation of this book into English and to Mrs. T.K. Vassilenko and Ms. Kutianova for their translation. Special thanks are extended to T.Yu. Snegireva and I.S. Batasheva for the clear illustrations.

The Author

Anatoly A. Komarovsky, Ph.D., Dr.Sci., is currently the chief of the Laboratory of Physics of Strength, Scientific and Engineering Center for Non-Traditional Technologies (SALUTA), Kyiv. He received his B.S. and M.E. Mech.E. degrees from Kiev Aviation University (Ukraine) in 1964, his B.S. and M.S. degrees in physics from Kiev National State University (Ukraine) in 1969, and his Ph.D. from the Highest Scientific-Attestation Committee of the U.S.S.R., Moscow, in 1973. Dr. Komarovsky was awarded a Dr.Sci. (Ukraine) designation in 1992 for his outstanding performance and profound impact on science and technology.

After a career in industry with Kiev Civil Engineering Research Institute, where he eventually became the chief of the civil engineering and multidisciplinary research laboratory, Dr. Komarovsky became managing director and professor of Kiev Industrial Technical College. There he taught a number of materials-related courses and continued his study on the physics of materials. After achieving great theoretical results and experimental confirmations, he joined SALUTA in 1991 to further research and implement his results in practice. Dr. Komarovsky has written four books and more than 60 scholarly articles; he holds seven patents.

The Scientific Editor

Viktor P. Astakhov, Ph.D., Dr.Sci., received his B.Eng. in manufacturing from Odessa College of Industrial Automation (U.S.S.R.) in 1972, his M.Eng. from Odessa National Polytechnic University (U.S.S.R.) in 1978, his B.Sc. and M.Sc. in applied mathematics from Mechnicov State University (Odessa, U.S.S.R.) in 1990, and his Ph.D. from Tula Polytechnic University (U.S.S.R.) in 1983. Dr. Astakhov was awarded a Dr.Sci. designation (U.S.S.R.) in 1991 for his outstanding performance and profound impact on science and technology.

Dr. Astakhov's first teaching appointment was in the department of metal cutting and cutting tools at Odessa National Polytechnic University in 1984, where he became a full professor and head of the deep-hole machining industrial center. Dr. Astakhov has served on a number of national scientific and planning committees and has also been a consultant to the machine tool building, and aerospace, nuclear, and gas turbine industries. Currently, he is R&D director of Hypertool Co., a consultant to Ford Motor Co., and adjunct professor at Concordia University (Montreal) and the University of Manitoba (Winnipeg).

Dr. Astakhov's principal research interest is in manufacturing, including the theory of metal cutting, mechanical metallurgy, and physics of fracture. Active in fundamental and industrial research, he has published and edited several books and more than 100 papers, and he holds more than 40 patents. He serves as a board member and frequent reviewer for several international scientific journals.

Introduction: History and Overview, Objectives and Problems

From the remote past to the present, the problem of strength has been the focus of considerable attention; multiple intensive investigations on the matter have been carried out. Interest has always been stimulated by practical needs. The consideration of a solid as an ideal, defect-free, and continuous medium has been a dominant idea for a long time. Based on this simplification, a number of theories have been developed. In modern studies of materials resistance to external mechanical loads they are called classical mechanical theories of strength.[3]

In fracture mechanics a solid is described in more realistic terms. This has allowed explanation of the observed substantial differences between theoretical concepts and experimental data. However, more and more facts in poor agreement with the concepts of fracture mechanics have accumulated with time.[4]

Some of these discrepancies were explained using the statistical approach. Using methods of probability theory, this approach leads to the development of a procedure for prediction of mechanical properties of materials and to understanding the level of their susceptibility to damage in a given force field. The prospects of the design methods, as the supporters of this approach view them, lie in design characteristics not thought of as limits (as at present), but rather as mean values, variances, or distribution laws. The majority of researchers consider this approach very general and promising;[3] however, this opinion has no objective basis because it does not consider the physical nature of processes occurring in a structure of materials under the effect of various external fields and aggressive environments.

Any structural process results in a certain outward appearance and thus any observed outward appearance has some internal reason. Even the most occasional event has its causes and obeys some laws peculiar to it alone. The occasional or random nature of fracture appears so only because its causes and the laws that it obeys are unclear, unknown, and therefore obscure to us.[9] Only a deep insight into the nature of a phenomenon, investigation of the governing mechanism, and cognition of cause–consequence relationships make this event nonrandom — regular, and thus, predictable. Only such an approach can substantially decrease the role of chance in the behavior of materials and structures and thus create conditions for development of the theory of design and reliability having a physical rather than a probability background.

Formalized approaches described solids from the mechanical-mathematical positions regarding destruction processes as fatally inevitable. Transfer of the science of strength to a physical domain should be considered the most important stage in the development of such approaches. It took place when solids began to be regarded not as hypothetical viscoelastic continua but as atomic–molecular (AM) systems, which were perceived as peculiar structures of atoms related to each other by adhesion forces. This allowed construction of a purely mechanical picture of loading: an external force is distributed via interatomic bonds stressing the bonds. Thus, the stability of a solid was judged by the relationship between external and internal forces. For example, if tensile forces are smaller than adhesion forces, the solid will be deformed; in contrast, if tensile forces are greater than adhesion forces, the solid will fracture or be irreversibly deformed. However, as more and more experimental data were accumulated, it became apparent that the mechanical model was limited and incomplete.

As was established at the beginning of the 1950s, the thermal motion of atoms causes a fundamental change in the purely "mechanical" model. It was found that an external force does not actually interact with a structure of rigidly connected atoms but rather with a dynamic system. In such a system, each particle is in thermal motion, which changes local tensions of interatomic bonds. This dramatically changes the nature of the material reaction to any external effects. This approach to the problem of strength became known as "kinetic."[10]

Results of systematic investigations of strength using the kinetic approach led to a radical alteration of the entire system of views on the phenomenon of mechanical fracture at the beginning of the 1970s. Extensive experimental studies conducted by the authors of this approach showed that fracture should not be regarded as a certain critical event but rather as a physical process developing with time in the AM system. Even in the absence of other external fields, the course of this process depends on the value and velocity of application of the external pressure and temperature of the environment. This conclusion is of fundamental importance, changing the seemingly evident notions of the role of external factors that cause deformation and fracture.

At the same time, the evolution of the science of strength was not very consistent. Having accepted and widely employed the notions of the AM structure of solids, researchers nevertheless ignored the most important and integral part of the AM concept, namely that behavior of a bounded system of atoms can be described in any situation by the laws of statistical physics and thermodynamics.[11] The atomistic concept is completed if and only if this fact is adopted, i.e., not merely atomic–mechanical (as was thought in the 1920s) or atomic–kinetic (in the 1970s),[10] but thermodynamic (in the 1990s).[12]

The modern science of strength studies, observes, systematizes, and develops practical recommendations, and describes and explains phenomena and effects that accompany the processes of deformation and fracture. At the present stage, it appears as a mountain of experimental facts poorly correlated by so-called phenomenological approaches.[3,4] A considerable gap exists between physical and engineering notions accepted and used in physics[6,7] and those in the strength of materials and materials science[2,13] because the latter use empirical postulates based on the previously discussed mechanical–mathematical concept.

The problem of strength is the focus in solid-state physics, materials science, technology, manufacture and processing of structural materials, production, and processing of mineral resources and thus related to the resistance of materials to various external effects, quality of parts and structures, reliability and durability of machines and mechanisms, and other branches of science and technology. Therefore, the price of a mistake in solving the related problems is extremely high.[8,14]

Frequent failures, accidents, and catastrophes of engineering objects are often results of imperfections in the existing design and calculation methods. Costs of prevention and resolution of their consequences increase.[14] Costs of production and processing of mineral resources and technological processes associated with forming structural materials[16] increase annually. Traditional approaches based entirely on phenomenological data fail to offer efficient methods for handling typical technical problems; thus, continued movement in this direction leads the science of strength to a dead end.[17]

No section or chapter dedicated to the physics of resistance of materials to various external effects can be found in the courses of studies of the physics of structural materials and in corresponding textbooks.[6,7] No universal concept has been developed to allow all processes to be considered from common positions and to indicate prospects of development of science of strength and its practical applications.[3,18]

In the modern periodical,[1,8] encyclopedic,[19] and reference[3] literature a possible solution of the problem of technogenic safety of engineering objects is directly related to advances in materials science. Technical progress is said to be entirely dependent on developments of structural materials to enable their performance in diverse (constant and alternating) external fields (force, thermal, radiation, electromagnetic, etc.), aggressive environments, and vacuum. In our opinion, however, the major obstacles to overcome are traditional stereotypes, such as "the ultimate goal of materials science is production of materials with any desirable properties" and "the guarantee of failure-free operation of an object designed is a successful match of properties of materials with anticipated service conditions."[3] This implies that the materials science approach to solution of the problem of strength, reliability, and durability is the only one possible, leaving no room for any alternative. The use of this approach, however, leads to two distinctive trends.

According to the first trend, the attempt is made to reduce the physics of structural materials to physical metallurgy because, despite great practical importance, until recently theoretical physicists have studied this problem to an intolerably small degree.[5,20] Therefore, it is no wonder that modern solid-state physics fails to answer many of the practical questions raised by developing technology. Unfortunately, it is for the same reason that the physics of strength, reliability, and durability is still at an initial stage of development.[21] It appears that a great deal of time is needed to find comprehensive answers to many practical questions and thus to address practical problems. Although specialists in the field always indicate the complexity of the problem,[3,22] in our opinion, its solution should be pursued because of the enormous significance of the expected results.

The second trend is historical. The need to solve everyday practical problems at low costs prompted specialists to abandon physical considerations and to replace them with much simpler mechanical considerations. The overwhelming majority of researchers in the field use this approach, known as the phenomenological approach.[3,4,22] Using this phenomenological approach, they often develop logically faultless and mathematically beautiful models, but pay little attention to the physical nature of the phenomenon under study. As a result, many such models are too schematic, valid only for specific conditions, and serve as the first (and sometimes, very rough) approximation to reality.

The achieved level of theoretical concepts, experimental procedures, and instrumentation defines entirely any result obtained. Attempts to use this level to gain new knowledge or to develop a novel approach usually result in the accumulation of senseless, randomly varying experimental data. This explains why some researchers use probability theory to deal with these data, hoping to solve practical problems.[3]

Considering the fracture process to be fatally inevitable, supporters of the mechanistic direction maintain that the physics of strength can originate only from solid-state mechanics. Accepting the existing limitations, they hope to solve the problem of technogenic safety using methods of quantum fracture mechanics and they expect to achieve this in 2010.[1]

Advocates of the practical approach maintain that there are no general problems in the physics of strength. Instead, specific problems of everyday engineering practice are dictated by practical needs. Among the variety of approaches, they recognize only the statistical approach and believe that it is a great help in solving a problem under consideration.[1,3]

In other words, everything that does not fit in the familiar frames established by existing concepts and thus does not promise immediate, or at least rapid, application, is no longer regarded as a "problem."

In this book the science of strength, reliability, and durability is considered to be an evolving discipline, rather than a field established within narrow historical limits. This immediately raises a question: Are the known materials science methods of ensuring technogenic safety the only possibility or does the technology of the future offer other alternatives? In this way, the "older" and more elaborated parts of physics[11,23] may turn out to be very useful compared with the methods of quantum fracture mechanics popular today.[1]

For clear psychological reasons and inertia, both extremes give rise to a number of uncomfortable feelings associated with revising traditional views on the processes of deformation and fracture.[17,24] The outlooks are still obstructed by current needs of the industry, as well as by delusions and myths introduced into our consciousness by previous generations of researchers.

The purpose of the book is to substantiate a novel approach to the problem of strength, deformation, and fracture which provides clear explanations for a wide variety of experimental facts from the common positions and behavior of materials and structures in diverse external fields and aggressive environments. The hope is to offer a working tool to develop efficient methods to control the stressed–deformed state of materials and structures.

Achievement of this purpose requires that multiple problems that are subjects of investigation in different areas of knowledge, such as materials science, statistical physics, thermodynamics, solid-state physics, thermal physics, theory of design, strength, reliability, etc., be considered from a single point of view. Such an approach covers the complete succession of situations in which a material passes from the moment of its solidification to the end of the service life of parts and structures. The necessity for this consideration stems from the intention to reduce the excessive safety factors, reduce the weight of engineering objects, and utilize more rationally scarce structural materials, power, and labor, providing that reliable performance of a material or structure within the assigned period of time is assured. The intention is also to increase the efficiency of resource-saving technologies associated with the fracture of solids (for example, metal cutting).

In the development of the subject, the author did not follow the known approaches and deliberately departed from traditional methods. He tried to originate his considerations from the most general standpoint possible and to show, maintaining clarity and consistency, that the ideas of modern physics form a reliable basis sufficient for explanation, description, and prediction of the behavior of materials under various external conditions. In doing so, the author had to rule out some existing concepts, clarify others, and introduce totally new ones. This concerns primarily the thermodynamic ambiguity of interatomic bonds, the possibility of their phase transformation, and, particularly, the role of the Debye temperature in destruction processes. The concept suggested is based entirely on a single fundamental idea, namely, on the mobility of electrically charged structure-forming particles (electrons and polarized ions).

The concept developed in this book is based on well-studied parts of modern physics and on the analysis of comprehensive factorial material accumulated over many years by researchers from different countries. Therefore, the conclusions are quite realistic. Their perception will require attention from readers and compulsory correction of habitual notions.

This book is interdisciplinary in intent, subject, and content. It should be helpful to physicists and to application-related specialists (materials engineers and strength scientists, production engineers and designers, etc.) because it solves fundamental problems pertinent to these and allied disciplines by the methods of solid-state physics. The logic of the book is as follows: it begins with principles and physics of AM complexes, in which physical–mechanical properties of solids are initiated and where the fluctuations of local parameters (temperature, density, etc.) are inevitable. Then it continues through the microstructure to a macrolevel and engineering practice that deals with mean parameters of these complexes (strength, temperature, deformation, durability, etc.).

List of Symbols

A	constant, external parameter; mechanical working plane; normalization factor
B	constant; the mean length of one crack
B_o	constant corresponding to mean-statistical service temperature T_o
C	constant
c	velocity of light
C_v	heat capacity at constant volume
C_p	heat capacity at constant pressure
D	electromagnetic dipole; electrical induction; dynamic effect (increment in the resistance of the AM structure on the high-rate deformation)
d	diameter; plastic indirect hardness indicator
$\mathbf{E}$	intensity of the electric field
E	magnitude of the intercity of the electric field; elasticity modulus
e	mass of an electron
$\mathbf{F}$	free energy or F-potential
F	Lorentz force (Equation 1.16); force
F_a	magnetic attractive force (Ampere force)
F_k	electric repulsive force (Coulomb force)
F_o	Coriolis force
f	function; free energy per unit volume; density of electric charges; degree of absorption expressed in percents
$f(\tau/v)$	rate of failures of similar type elements under conditions v
G	Gibbs free energy or G-potential
g	eccentricity of elliptical orbit; electrical conductivity of dielectric; ultrasonic pulse released in separation of atoms of the rotos
$g_{1,2}$	internal thermal flows
H	enthalpy; magnitude of the intensity of uniform magnetic field; heat content of material
$\hbar$	Planck's constant; elastic indirect hardness indicator

I	configuration integral (Chapter 2); electric current
i	circumferential electric current (Chapter 1); component
j	inner quantum number
K	polarization coefficient; coefficient of crushed stone enrichment of concrete
K_a	asymmetry factor
k	Boltzmann constant; coefficient accounting for the influence of the surface shape on the intensity of aging processes
l	azimuthal quantum number; length of a microcrack
L	length; specimen thickness
L_o	length of oriented part of crack
L_i	length of isometric part of crack
M	mechanical momentum of atomic bond
M_e	total electron momentum
M_l	angular momentum
M_x	projection of the momentum of an electron
M_s	projection of spin (intrinsic electron momentum)
m	magnetic quantum number; mass
m_e	charge of an electron
m_n	mass of the nucleus
N	number of atoms or rotoses in volume V
n	principal quantum number; total quantity of photons; coefficient of diffusion; number of cycles; cement-to-sand ratio
$\mathbf{P}$	pressure (vector parameter); internal resistance
$\mathbf{P_c}$	compressive (compresson) part of the internal resistance
$\mathbf{P_d}$	tensile (dilaton) part of the internal resistance
P	pressure (magnitude)
P_c	resistance under static loading
P_d	resistance under dynamic loading
$\mathbf{p}$	tensor of microtension; polarization vector
p	polarization vector
Q	macroscopic heat
Q_σ	time of ultrasound propagation at the last stage of loading
q	phonon radiation of rotos; internal heat flow; concentration of protons
$q\,(n)$	oscillation energy quantum

$q\,(l)$	rotation energy quantum
q_T	kinetic energy impulse
r	radius of an electron circular orbit; distance (Chapter 1)
R	distance radius; interaction radius; universal gas constant; electrical resistance
r_a	parameter of the elliptical orbit (Equation 1.37.II)
$\dot{r} = dr/d\tau$	radial velocity of a given mass of charges of the rotos
S	macro area; statistical variation; standard deviation
s	spin quantum number; microsurface; resource
$s(\tau)$	residual part of the resource
$\mathbf{s}$	entropy (vector parameter)
T	absolute temperature; intercity of a thermal field
T_c	temperature of brittle fracture
T_e	temperature of evaporation
T_m	melting temperature
T_n	preferential temperature
t	submicro local temperature
t_c	submicro local phase transformation temperature
U	macroscopic energy; electric voltage; acoustic emission power
U_o	activation energy
u	overall energy of rotos
u_r	potential part of energy
u_t	kinetic part of energy
V	volume
V_d	dilaton part of the volume
V_c	compresson part of the volume
v	velocity; rate; rate of application of an external load; relative radio-polarization coefficient of mechanical stress
W	parameter accounting for the change of moisture content in a porous structure; water–cement ratio of concrete
W_{max}	maximal water saturation
w	area of orbit
$\dot{w}$	sector velocity
Z	number of allowable states or function of state
z	the valence of a chemical element

α	angle; configuration factor; thermal expansion coefficient
$\alpha_p = \sigma/\sigma_p$	level of stress in compression
α_s	level of stress state in tension
β	Madelung constant; angle; degree of orientation (Equation 3.66) isothermal compressibility (Chapter 9)
β_r	characteristic of the shape of a single pulse
Γ	parameter determined from the normality condition; accounts for the physical nature of a phenomenon
γ	angle
Δ	rigidity of a solid; actual temperature deviation from the average level, T_o
Λ	thermal resistance
λ	thermal conductivity; electromagnetic wavelength; electrodynamic potential of interaction
δ	Maxwell-Boltzmann factor or function
θ	macroscopic Debye temperature
θ_t	macroscopic Debye temperature of the first kind
θ_s	macroscopic Debye temperature of the second kind
ε	longitudinal strain
$\varepsilon = \varepsilon_p + \varepsilon_o$	the sum of inherent (ε_p) and force field-induced (ε_o) dielectric permittivity
$\dot{\varepsilon}$	strain rate
ε_p	dielectric permittivity
ε_o	dielectric permittivity of vacuum
ε_τ	deformability (strain) at σ_τ
μ	magnetic susceptibility; strain rate sensitivity
σ	stress; effective stress
σ_c	mean level of the applied stress
σ_p	compressive strength
σ_s	tensile stress
σ_M	dynamic strength
σ_τ	sustained (long-time) strength
ξ	electrodynamic potential of interaction of the atoms following from Equation 1.17
υ	frequency of revolutions in Equation 1.50; frequency of oscillations of atoms of the rotos at the θ-temperature

ν	microscopic Debye temperature; frequency of oscillations; frequency of the IR spectrum expressed in cm^{-1}
ν_f	microscopic Debye temperature of the first kind
ν_s	microscopic Debye temperature of the second kind
η	transverse strain
φ	angular displacement; tensile prestrain; neutron flow density; shear angle (Chapter 9)
$\dot{\varphi} = d\varphi/d\tau$	angular velocity of a given mass of charges of the rotos
$\varphi\tau$	radiation dose equal to the product of neutron flow density, $\varphi\, n/cm^2 \cdot s$, and time τ, s.
Φ	critical subset in region Ω
Π	potential energy of solid; external potential (force) field
χ	anharmonicity coefficient
τ	time; durability
τ_c	cyclic life
τ_n	cycle period
$\tau_0 = 10^{-13}$ s	time of a complete revolution of dipole D on the orbital ring
μ_T	mechanodestructive sensitivity
Ω	Landau potential or macrowork for deformation resistance of solid
ω	resistance energy; angular velocity; potential
ω_1	Larmor frequency in Equation 1.6
ω_2	angular velocity of the positively charged nucleus around its own axis
Φ	increment in the phase of electromagnetic radiation
ϑ	relative radio-polarization coefficient of mechanical stress
ψ	bulk (volumetric) deformations

Contents

1

Structural Mechanics and Electrodynamics of Interatomic Bonds

1.1 Introduction

Modern physics explains the thermal, electrical, and magnetic properties[6,7] of various solid bodies used in everyday practice, but somehow fails to provide reasonable explanations for mechanical properties.[2-4] In our opinion, the edged notions used today in the physics of solids present a psychological barrier to further developments in the field, so it is necessary to revise and correct them in order to describe the behavior of solids in force, thermal, and other external fields.

The crystalline structure of a solid is the carrier of its physical–mechanical properties. This structure consists of molecules composed of atoms which, in turn, are the dynamic systems comprising positively charged nuclei and negative electrons. Naturally, the higher the degree of itemization, i.e., up to the AM level, the more detailed is the description of properties and the more reliable is their evaluation.

From the microscopic standpoint, a solid can be considered a gigantic molecular formation containing about 10^{25} interacting atoms per cubic centimeter of its volume. The laws of electrodynamics, classical, and quantum mechanics describing the interactions within this formation have been well established. As known, electromagnetic forces govern the interaction. Physical science has reached a very high level of understanding of the nature of such forces.[25] The available knowledge about the microlevel of solids is sufficient for providing insight into the physical nature of engineering properties of materials. However, many stereotypes, excessive simplifications, and delusions have accumulated in this area for years. They distort realistic ideas about the processes occurring in the discussed structure and thus restrict further advances in the science of strength of materials to explain the resistance of solids to external fields and aggressive environments.

1.2   The State of Electrons and Nuclei in Isolated Atoms

An atom consists of charged particles: a massive nucleus and the electrons orbiting it. The number of electrons is exactly the same as that of protons in the nucleus. Properties of a single atom and atomic complexes (solid, liquid, and gaseous materials) are entirely defined by the parameters of motion, the charge of the nucleus, and the state of the electrons rotating about this nucleus.

In Mendeleev's table, all known chemical elements are arranged in increasing order of their nucleus charge. Electrons in atoms can only be in a stationary state, which is characterized by a set of four quantum numbers: n, l, m, and s.[26] The principal quantum number n can be any positive integer $(1, 2, 3 \ldots)$. Orbital or azimuthal quantum number l designates the magnitude of the angular momentum M_l according to the equation

$$M_l = \frac{\hbar}{2\pi}[l(l+1)]^{1/2}, \tag{1.1}$$

where $\hbar$ is Planck's constant. Number l can be zero or any positive integer up to and including $(n - 1)$. A distinctive feature of an isolated atom is its central symmetry. The electron energy in such an atom depends only on the two quantum numbers: n and l.

Interactions of isolated atoms disturb the central symmetry, while the external factors affecting a solid lead to a change not only in the spatial configuration of electrons but also in the motion of the nuclei. As such, the degree of disturbance or ionization of atoms is determined by magnetic m and spin s quantum numbers. In a solid, numbers n and l change according to m and s. For example, if an atom enters an external magnetic field, the direction of which coincides with the x-axis, the projection of the momentum of electron M_x into this axis can take only values that are multiples of magnetic quantum number m

$$M_x = \frac{\hbar}{2\pi} m. \tag{1.2}$$

It is apparent that projection of momentum M_x is smaller than its absolute value M_l. Number m can assume only the following values: $0, \pm 1, \pm 2, \ldots \pm l$. Spin quantum number s determines the projection of spin (intrinsic electron momentum) M_s into a given direction

$$M_s = \frac{\hbar}{2\pi} s, \tag{1.3}$$

where s may assume values of $+1/2$ or $-1/2$.

The total electron momentum M_e consists of the orbital and spin momentums. Inner quantum number j is used to determine M_e as

$$M_e = \frac{\hbar}{2\pi}[j(j+1)]^{1/2}. \tag{1.4}$$

Number j is determined by l and s; for a given l, it may take one of the two possible values: $j = l \pm 1/2$, depending on the relative orientation of the orbital and spin momentums.

According to the Pauli exclusion principle, discovered by Swiss physicist Wolfgang Pauli in 1925, no two electrons can occupy the same quantum-mechanical state. Therefore, all the electrons associated with an atom differ from one to another in at least one of the four quantum numbers: n, l, m, or s. In an unexcited atom the electrons have the lowest energy state. While filling the next shell, the electrons first take free states with the lowest energy corresponding to the lowest values of n and l. Then the remaining electrons must take vacant states with higher energy levels.

Only two states have the identical values of n, l, and m, but differ in the spin projection. The first is for n, l, m and $(s = +1/2)$ and the second is for n, l, m, and $(s = -1/2)$. While making elementary calculations of the sum of terms of the arithmetic progression, accounting for the sign of the magnetic quantum number, one can see that the $2(2l + 1)$ states, which differ in values of projection of the spin and orbital momentums, correspond to identical n and l. However, the $2n^2$ states correspond to identical n but different l, m, and s. The principal and orbital quantum states of electrons, their quantity, and commonly accepted designations are given in Table 1.1.

The distribution of electrons in various states is denoted by the principal quantum number (as the corresponding number) and the orbital quantum number (as the corresponding letter). The number of electrons in a given state is indicated as power over the letter symbol. For example, the electron configuration of an unexcited atom of carbon is written as follows: $1s^2 2s^2 2p^2$. The sum of all the numbers over the letters is the number of electrons in an atom, which for carbon is equal to 6. Distribution of electrons in the states with different values of n and l according to the increase in the atomic number for all elements of Mendeleev's table can be found in Pollock[6] and Bogoroditsky et al.[26]

TABLE 1.1

Principal and Orbital Quantum States

Quantization Level	Principal n							Orbital l			
Numerical value	1	2	3	4	5	6	78	0	1	2	3
Letter symbols	K	L	M	N	O	P	Q	s	p	d	f
Quantity of states	2	8	18	32	50	72	98	2	6	10	14

Of great importance is a layered structure of the electron shell. In a state with $n = 1$, the electrons have a much stronger bond to the nucleus than in states with $n = 2$, $n = 3$, etc. In fact, quantum number n distributes the electron layers according to their rigidity: the larger the distance between a layer and the nucleus, the smaller the rigidity of the layer.

In excitation, the electron configuration of an atom may change to a substantial degree. As such, the electrons, on acquiring extra energy, must leave their former place to transfer to the state with a higher energy level. Such a transition is often accompanied by a change in the energy and also in the magnitude and direction of momentum M_e. For this reason, the electrons of the interacting atoms may move along the closed paths, which may be circular or some other symmetrical shape.

The order determined by the Hund rules[26] holds in layer-by-layer filling of the states in the electron shell of an atom. One of these rules postulates that electrons take the states in such an order that their total spin M_s and orbital M_l momentums have the highest possible values. According to this rule, while filling the s-layer, one electron takes a state with one orientation of the spin and the second takes a state with an opposite orientation of the spin. In the p-layer, the electrons first take three layers with one orientation of the spin, and then three layers with the other. A similar order is maintained in filling the d- and f-layers. Evidently, the electrons with parallel orientation of the spin differ in the value of magnetic quantum number m. Successively filling the layers leads to a periodical dependence of all the properties of elements on their atomic numbers.

The layer-by-layer configuration of electron shells of all elements of Mendeleev's table is given in References 6 and 26. Its analysis enables us to arrive at three conclusions of fundamental importance. The first is that, for the internal, completely filled layers, the total momentums are equal to zero. The process of filling the layers obeys the second Hund rule: in the first half of the states the orbital M_l and spin M_s momentums are set antiparallel and, in the second half, parallel, to each other. Summing up, they form the total momentum equal to zero.

The second conclusion is that momentums of atoms are determined by the overall momentum of the electrons found in the external unfilled shell of this atom. Evidently, the spin momentum of the nucleus should compensate for the momentums of these electrons. The third conclusion is that, in contrast to the positive charge concentrated in the nucleus, the total negative charge may change its spatial position following the transformation of the electron shells.

Figure 1.1 shows the electron configuration of atoms of lithium, beryllium, boron, and carbon in unexcited (a) and excited (b) states. The cells are arranged from left to right in the order of increase in quantum numbers n, l, and m. The value of projection of the spin at $s = +1/2$ is marked by an arrow pointed upward and that at $s = -1/2$ by an arrow pointed downward. All the elements in the $1s$ state have an electron pair; therefore, the latter has a passive behavior in formation of the interatomic bond. The lithium atom

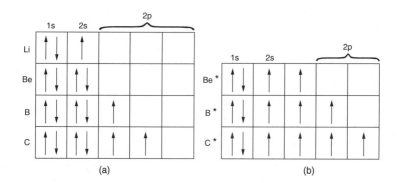

FIGURE 1.1
Electron configuration of atoms of lithium (Li), beryllium (Be), boron (B), and carbon (C) in nonexcited (a) and excited (b) states.

contains one unpaired electron in the $2s$ state, whereas boron and carbon in the $2p$ state have one and two unpaired electrons, respectively. Beryllium has no such electrons. Therefore, in the unexcited state, lithium and boron are univalent, carbon is bivalent, and beryllium has a zero valance.

In excitation, the chemical activity of atoms substantially changes (Figure 1.1b). So, beryllium becomes two-, boron three-, and carbon four-valent. Excitation of an atom consumes some energy. For example, for beryllium the transition of an electron from state $2s$ to state $2p$ requires about 2.7 eV of energy. The transition from state $1s$ to state $2p$ requires a very high energy to occur and thus $1s$ electrons in these elements are not excited. The described character of excitation is repeated in the third and subsequent periods. In the third period, an extra transition from the $3s$ and $3p$ states to the $3d$ state may occur, as it does not require too much energy.

While discussing the effect of valent electrons on the process of formation of interatomic bonds, one cannot neglect the paired character of existence of electrons. The electron pair can give rise to attractive (antiparallel spins) and repulsive (parallel spins) forces. So, the question to be answered is: What is the overall effect of interaction in coincidence of the binary and ternary electron pairs?

Consider an example of two electrons pairs 1 and 2 (Figure 1.2a) with opposite orientation of the spins (marked by open and filled circles), which have the same principal quantum number n (setting radius r of the electron orbit) but different orbital number l (setting the pairs in positions 1 and 2 differing in azimuths on the same orbit). Under these conditions the pairs are separated: the first rotates on a circular orbit with radius r in a downward direction, and the second rotates on the same orbit from left to right. As such, there is no interaction of the magnetic spin fields of both pairs (small circumferences, enclosing both pairs, do not intersect).

Placing a solid in force, thermal, magnetic, or other external energy field leads to disturbance of its atoms that results in drawing nuclei A and

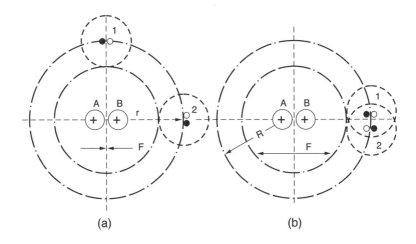

FIGURE 1.2
Schematic of symmetrical (a) and combined (b) electron configuration formed due to displacement of orbital planes.

B apart or closer and in displacements of the electron shells. As a result, orbital quantum numbers *l* of both pairs may become equal to each other (Figure 1.2b). Coincidence of their azimuths leads to superposition of the spin magnetic fields (circumferences enclosing pairs 1 and 2 intersect) and to the probability of interaction between these pairs and also between electrons of different pairs. The total spin of two electrons that are parts of different pairs has no definite value, which is why the resultant spin state is a superposition of the four states thus formed. So, the statistical weight of the state with parallel orientation of the spins is equal to three quarters; that with an opposite orientation is equal only to one quarter.[26]

The energy of interaction of the binary excited bond dissociates in the same ratio to the formation of compressive or tensile force **F**. Electrons with an excited configuration (Figure 1.2b) tend to take a free, higher orbit and more beneficial azimuthal location (at a simultaneous increase in the internal energy of a solid due to the external field). On the other hand, they may release the excessive energy in returning to the previous position (with the energy release to the environment).

Figure 1.3 shows a schematic of formation of molecular orbits between atoms of iron (Fe, nucleus charge is 26) and carbon (C, nucleus charge is 6). Located apart from each other at a distance of $r = \infty$ and having no interaction, the atoms of these elements (Figure 1.3a) have the following electron configuration:[26] Fe, $1s^2 2s^2 2p^6 3s^2 3p^6 3d^6 4s^2$ and C, $1s^2 2s^2 2p^2$ (see Figure 1.1). Electrons are located on the symmetrical circular orbits: two of each on the first orbit, and eight for iron and only four for carbon on the second orbit. The third and fourth shells in carbon are degenerated (marked by a dotted line), with 14 and 2 electrons, respectively, on them for iron. According to the Pauli

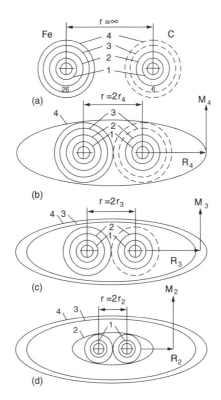

FIGURE 1.3
Schematic of formation of molecular orbits of electrons between iron (Fe) and carbon (C) atoms.

exclusion principle, electrons are arranged on the orbits so that the electric and magnetic charges are compensated for the mechanical momentum (Equation 1.4) is kept at minimum and formed only by valent electrons.

When these atoms are drawn to each other to a distance equal to the double radius of the fourth orbit (Figure 1.3b), the positive charges of nuclei Fe and C become indistinguishable for the two electrons located on this orbit. So, they must move around them on a common molecular elliptical orbit formed as a result of coalescence of the four allowable circular orbits of both atoms. (That such orbits do have an elliptical shape will be proved below.)

Binding the atoms by a common elliptical orbit leads to the formation of a weak interatomic bond. If $r < 2r_4$, its strength and rigidity grow with further rapprochement of the atoms. At the moment when the shells of the third orbits (occupied by 14 electrons in iron and by a free electron in carbon) come into contact, this level also splits (Figure 1.3c). After this moment, the third individual circular orbit of the considered iron atom vanishes, transforming into the common elliptical orbit 3, which again surrounds both atoms. Strength and rigidity of the bond reach their maximum after the formation of the third common molecular shell (Figure 1.3d). This occurs when the distance between nuclei of the atoms becomes equal to $r = 2r_2$.

The closer the atoms are to each other, the larger the number of the molecular orbits which surround them, the stronger and more rigid the bond, and, therefore, the greater total mechanical momentum of the bond, i.e., $M_4 < M_3 < M_2$. While approaching each other, the nuclei form a common central local magnetic field in which all the molecular electrons move. In such a field, the momentum is normal to radius vector **R**, which connects the center of the field with each of the electron orbits. The momentum forming due to the closed-loop motion of the nuclei compensates for the total momentum of all molecular electrons. Therefore, in mutual motion of electrons and nuclei, the law of constant momentum is fulfilled: M = const. This law plays a decisive role in formation of the mechanical properties of solids.

Figure 1.3 shows a plane projection of the spatial electron orbits; thus the value and direction of the overall momentum M are conditional. In fact, electrons are located in pairs in the planes having different spatial orientations and each particular orientation is set by orbital quantum numbers s, p, d, and f (see Table 1.1). Figure 1.4 shows a spatial position of azimuthal planes s, p, and d. Traces of the elliptical orbits, corresponding to different principal quantum (1, 2, 3, and 4) and orbital (s, p, and d) numbers, are marked by open circles (the orbit goes inside the plane in Figure 1.4) and filled circles (the orbit leaves the plane). If M_s, M_p, and M_d are the vectors designating the azimuthal components of the mechanical momentum and M_y is the total projection of the momentum to an arbitrary direction y, then

$$M_y = M_s + M_p + M_d. \tag{1.5}$$

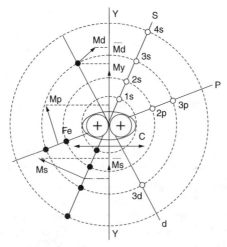

FIGURE 1.4

Spatial arrangement of azimuthal planes in iron–carbon compound (Fe–C).

So the interatomic bond can be represented in the form of a spatial system of electrons moving on the circular (nonintersected or individual) or closed noncircular (molecular or common) orbits around one nucleus or, simultaneously, two nuclei. According to the Pauli principle, in a state of dynamic equilibrium all the orbits preserve their shapes, i.e., they are rigid. Quantum mechanics[27,28] investigates in detail the problem of a rigid rotator, i.e., a material point rotating on a symmetrical trajectory (circumference or ellipse). The elliptical shape of molecular orbits leads to interrelated displacements of nuclei of elements Fe and C, which can be considered oscillatory only to the first approximation. In Figure 1.4 such movement is marked by a horizontal arrow. The material point oscillating due to a quasi-elastic force is called an oscillator. Thus, the interatomic bond is a system of the spatial set of elliptical rotators (electrons) and one paired oscillator (nuclei Fe and C).

A common opinion is that an individual atom is neutral as a whole. This is true from the electrostatic standpoint, according to which the positive charge concentrated in the nucleus is compensated for by the total negative charge of electrons moving on closed orbits around it; however, this is not entirely true from the electrodynamic standpoint. Only inert gases (helium, argon, etc.) should be considered absolutely neutral or passive. They have the compensated charges of an opposite sign, and the even number of electrons is distributed by principal n and azimuthal l quantum numbers on orbits so that their movements have no effect on the nucleus. The situation is different in chemically active elements: they have an odd number of electrons or their valent electrons form the asymmetrically filled external shells. Disturbance in the spatial symmetry of a dissipated and mobile negative charge due to the asymmetrical configuration of orbits of valent electrons leads to excitation of the massive nucleus. Mechanical mobility of the nucleus is the second cause of physical–chemical activity of all the elements known in nature (except for inert gases).

The system of valent electrons, which move on the closed orbit in the central symmetrical field of a massive nucleus, generates a weak uniform magnetic field of intensity of H_1. The internal, completely filled shells do not generate such a field. The presence of a weak magnetic field leads to a slow rotation of the overall mechanical and magnetic momentums of the electron system at the Larmor frequency[25]

$$\omega_1 = \frac{e}{2m_e c} H_1 \tag{1.6}$$

around the electric field of the nucleus, preserving their absolute value and the angle formed with this direction. Here, e and m_e are the charge and mass of an electron, and c is the velocity of light.

The Larmor procession of the magnetic field, ω_1, causes a spin rotation of the positively charged nucleus around its own axis at angular velocity

$$\omega_2 = \frac{ez}{2m_n c} \mathbf{H}_1, \tag{1.7}$$

where z is the valence of a chemical element and m_n is the mass of the nucleus.

A similar phenomenon is known from engineering: time variation of the magnetic field of the stator causes a circular rotation of the rotor of an electric motor. Because masses of an electron and nucleus differ by a factor of almost 2000, thus accounting for the z value, it can be considered that ω_2 is lower than ω_1 by one to two orders of magnitude.

Charge ez, rotating on the circumference, can be regarded as a circumferential electric current $i = ezv$. Expressing the linear velocity of the charge v in terms of nucleus radius r_n and its angular velocity ω_2, we can write $i = ez(\omega_2 r_n)^{1/2}$; substituting ω_2 from Equation 1.7, we obtain

$$i = (ez)^{3/2} \left(\frac{r_n \mathbf{H}_1}{2m_n c} \right)^{1/2}. \tag{1.8}$$

In turn, the circumferential electric current generates the magnetic field of the nucleus with the following intensity[29]

$$\mathbf{H}_2 = 0.2 \frac{i}{r_n}. \tag{1.9}$$

Substitution of Equation 1.8 into Equation 1.9 yields

$$\mathbf{H}_2 = 0.14 (ez)^{3/2} \left(\frac{\mathbf{H}_1}{m_n c r_n} \right)^{1/2}. \tag{1.10}$$

Existence of the magnetic fields around rotating objects is observed, for example, in astronomy: moving in space and turning around its axis, the Earth has its own magnetic field, whereas the moon, which does not have such a rotation, has no magnetic field.

Therefore, an individual atom can be assumed to be an electromagnetic cell (Figure 1.5) of a dual nature. On one hand, the negative charge of electrons (not shown in Figure 1.5) uniformly distributed in space may change its configuration under certain conditions, being grouped within the preset orbital planes and converting an atom into an electrical dipole. On the other hand, a weak magnetic field of valent electrons $\mathbf{H}_1$, oscillating at frequency ω_1, causes a spin rotation of the nucleus at an angular velocity ω_2 (for example, relative to the horizontal axis). The circumferential electric current generated by this rotation creates a magnetic field of the nucleus, $\mathbf{H}_2$ (horizontal magnetic force lines). This field transforms the atom into an elementary magnetic dipole with poles N-S.

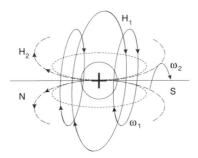

FIGURE 1.5
Electromagnetic nature of an isolated atom.

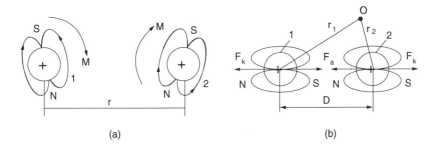

FIGURE 1.6
Schematic of interaction of approaching atoms.

1.3 Diagram of Formation and Energy of the Paired Bonds

Atoms exert no effect on one another if they are located at large distances from one another. In the case of a random approach of at least two of them (1 and 2) to distance r comparable with their sizes, they start interacting through their magnetic fields H_2 (Figure 1.6a). This interaction shows up as the formation of the torque: $M = M_1 \cdot M_2/r^3$, where M_1 and M_2 are the magnetic momentums of nuclei of the individual atoms. Momentum M turns the nuclei with the opposite magnetic fields toward each other, i.e., in the same way as that observed in experiments with magnets.[29]

Two pairs of forces act on atoms oriented in their own magnetic fields (position b in Figure 1.6): electric repulsive forces (Coulomb forces):

$$\mathbf{F}_k = e^2 \frac{z_1 z_2}{r^2} \tag{1.11}$$

and magnetic attractive forces (Ampere forces):

$$\mathbf{F}_a = 2\pi r_n \mu_0 \mathbf{H}_2, \tag{1.12}$$

where μ_0 is the magnetic permeability of vacuum. Substituting Equation 1.10 into Equation 1.12 yields

$$\mathbf{F}_a = 2\pi r_n \mu_0 \mathbf{H}_2 (e^2 z_1 z_2)^{3/2} \left(\frac{\mathbf{H}_1}{2m_n c} \right)^{1/2}. \tag{1.13}$$

The atoms start approaching each other under the effect of this system of forces.

At a distance

$$D = \frac{(m_n c e^2 z_1 z_2)^{1/4}}{2(\mu_0 \mathbf{H}_2)^{1/2} r_n^{3/4} \mathbf{H}_1^{1/4}}, \tag{1.14}$$

which is found from condition $\mathbf{F}_a = \mathbf{F}_k$, the rapprochement ceases and the individual electron shells of both atoms intersect to form molecular orbits (following the diagram shown in Figure 1.3). Electrons located on these orbits cover both nuclei to form the constant electromagnetic (EM) field with central symmetry.[25] Charges 1 and 2, located at a distance

$$\mathbf{D} = \mathbf{r}_1 - \mathbf{r}_2 \tag{1.15}$$

from each other and performing their own rotation, form the EM dipole, which can change its configuration depending on the transformation of a spatial shape of the negative charge.

For a closed system consisting of the EM field and particles located in it, the sum of the kinetic energy of the particles and energy of the field remains invariable. The equation of motion of charges e with mass m in the EM field at velocities $\dot{\mathbf{D}} = d\mathbf{D}/d\tau$, which are low as compared with that of light, c, is written as[25]

$$m\ddot{\mathbf{D}} = e\mathbf{E} = \frac{e}{c}[\dot{\mathbf{D}}\mathbf{H}] \tag{1.16}$$

where τ is the time and $\mathbf{E}$ and $\mathbf{H}$ are the intensities of the electric and magnetic components of the field.

The energy of a charged particle can be changed only by the electric field. The magnetic field does not yield in any work, but just causes a distortion of the path because it generates forces normal to the velocity of the particles (the second term in the right-hand part of Equation 1.16). For this reason nuclei 1 and 2 are localized in the central EM field. They start performing a closed movement about a common center of inertia (point 0 in Figure 1.6b) located at distances r_1 and r_2 from them in the plane normal to the magnetic component of the magnetic field, which is a superposition of fields H_1 and H_2.

Intensity of the electric field generated by all charges e_i located in a given elementary volume at distance r is determined as[25]

$$E = \frac{\sum_{i=1}^{n} e_i}{4\pi\varepsilon_0 r^2} = \frac{\alpha}{r^2},$$ (1.17)

where

$$\alpha = \frac{\sum_{i=1}^{n} e_i}{(4\pi\varepsilon_0)}$$

is the configuration factor formed by charges of nuclei 1 and 2 and valent electrons in a space with dielectric permittivity ε_0 as the vector sum of the electric fields generated by each separate charge. In other words, the charge of many particles is equal to the sum of charges of all particles and does not depend on their relative locations and motions.

This results in the formation of a dynamic system called "rotos," i.e., oscillating rotator, which is as stable as an individual atom. This system consists of two massive nuclei, 1 and 2, and three charges: two positive spot-like charges and one negative charge distributed around them on the electron orbits. Earlier, the notions of electric dipole and magnetic domain[6,7,26,30] were introduced in physics to explain the behavior of solids in electric and magnetic fields. Using the rotos to characterize the interatomic bond allows one to account for the dualism of the structure-forming particles (which have mass and, at the same time, electric charge), simplifies understanding of the processes occurring in their structure in various external fields, and, as shown further on, brings new ideas, concepts, approaches, and methods in the science of strength of materials. It can be easily shown that our considerations are well based on facts that became self-evident as early as the middle of the 20th century.

The mass of a rotos is concentrated in two massive positively charged nuclei. Because of the electric repulsive forces generated by the merging nuclei, the electrons, which have charges of the same sign and mass much smaller than that of the nuclei, shift to the points of the resulting electric field, where the forces affecting them are mutually compensated. So, the internal, filled electron layers of atoms (which come into contact when drawn together) and the external or valent layers participate in the formation of the rotos. As such, the former determine stability and rigidity of the interatomic bond and the latter determine its type and peculiarities.

The joint rotation of the nuclei around a common inertia center (point 0 in Figure 1.6b) excites the molecular magnetic field **H**. The collectivized electrons start moving along the equipotential lines of this field; as such, the paired arrangement of electrons and nuclei on the orbits is preferable. The pairs are formed of particles having spins oriented in opposite directions and overlapped wave functions.[6,26] In such a position the attractive forces act between them. Parallel orientation of the spins leads to a mutual repulsion of particles and the effect of repulsion leads to their location on different orbits. A similar phenomenon is observed in electrical engineering, where conductors having opposing directions of electric currents attract each other while those having the same direction repel each other.

In the noninertial reference system, which is rigidly connected with particles 1 and 2 and participates in all their movements, and whose origin coincides with the system inertia center 0, electrostatic forces are acting between the charges. In the inertia reference system (fixed with respect to the rotos), because the charges are in permanent motion, they generate the EM field. Therefore, interaction of the mobile structure-forming particles with their immediate surroundings must occur only through the EM field.

The discussed interrelated closed mechanical motion of the nuclei (similar to that in the solar system planets and their satellites) is the primary cause of binding atoms of a solid into a whole. In such a state the atoms are localized in nodes of the crystalline lattice, where they can perform only rotating and oscillating movements. The ion, covalent, metal, or other types of electric interaction[6,7,26,30] enhance or weaken the interatomic bonds thus arranged in a material. Atoms joined by a mechanical movement cannot separate on their own; to separate them, the application of a certain amount of external energy is needed.

The rotos is an elementary closed dynamic cell of solids. It allows us to explain behavior of solids in the force and thermal fields in the same way as the behavior of the domain in the magnetic fields[6,29,30] or as the electrostatic dipole in the electric fields.[6,26] In such a system the sum of all forces is equal to zero. The work ω due to force **F** at an infinitely small displacement

$$dr = vd\tau \tag{1.18}$$

of any particle is equal to

$$d\omega = \mathbf{F}dr = \mathbf{F}vd\tau. \tag{1.19}$$

Substitution of the left-hand side of the equation of motion (1.16) into Equation 1.19 yields

$$d\omega = mvdv = d\left(\frac{mv^2}{2}\right) \tag{1.20}$$

or

$$d\omega = eEV\,d\tau \qquad (1.21)$$

because the vector product is equal to zero: $v \times [vH] = 0$.

Comparison of Equations 1.20 and 1.21, accounting for Equation 1.18, makes it possible to write that

$$d\left(\frac{mv^2}{2}\right) - e\mathbf{E}d\mathbf{r} = 0, \qquad \text{i.e.,} \qquad d\left(\frac{mv^2}{2} - e\mathbf{E}\mathbf{r}\right) = 0.$$

This results in an expression for the overall internal energy of the rotos

$$u = \frac{mv^2}{2} - e\mathbf{E}\mathbf{r}. \qquad (1.22)$$

It has the following form in the polar system of coordinates[25]

$$u = \frac{m}{2}(\dot{r}^2 + \mathbf{r}^2\dot{\varphi}^2) - \xi, \qquad (1.23)$$

where $\dot{r} = dr/d\tau$ is the radial and $\dot{\varphi} = d\varphi/d\tau$ is the angular velocities of a given mass of charges of the rotos, m; $\xi = \alpha/r$ is the electrodynamic potential of interaction of the atoms following from Equation 1.17, which depends on the configuration of charges in the r direction.

Using the relationship between $\dot{\varphi}$ and $\dot{r}$[31] and determining the phonon radiation in the local region of space as a measure of the radial part of the kinetic energy of particles located in it as[32]

$$\dot{\varphi} = \dot{r}^2/\mathbf{r} \quad (\text{I}), \qquad t = m\dot{r}^2/2 \quad (\text{II}) \qquad (1.24)$$

and introducing the notation

$$ur = u - t, \qquad (1.25)$$

we can express Equation 1.23 in the following form

$$\xi = \frac{2}{m}t^2 + t - u \quad (\text{I}), \qquad u_r = \frac{M^2}{2mr^2} - \xi \quad (\text{II}), \qquad (1.26)$$

where $M^2/2m\mathbf{r}^2$ is the centrifugal part of energy and M is the mechanical momentum

$$M = mr^2\dot{\varphi}. \qquad (1.27)$$

The overall energy (Equation 1.23) depends only on the atomic parameters m, z, and e; therefore, it is constant for a given material, i.e., $u = \text{const}$. The interaction potential ξ (Equation 1.26.II) is characterized by the parabolic dependence on the local temperature t (1.24.II). In turn, the latter is determined by the rate of variations in linear sizes $\mathbf{D}$ of the EM dipole (Equation 1.15).

The obtained expressions are of fundamental importance for physics of deformation and fracture. Their analysis enables us to make the following cognitive and practical conclusions:

1. Structure-forming forces and processes at the AM level are of EM nature.

2. Relationship 1.22 represents the law of energy conservation of the rotos. As seen, its kinetic part (the first term) depends only on the velocity of particles, while its potential part depends only on their coordinates.

3. The dynamic bond is formed between atoms of the rotos. Deviation of the distance between them, $\mathbf{D}$, to either side from the equilibrium immediately causes radiation or absorption of phonons, t. This is a fundamental principle of the microworld, rather than a random occasion caused by fluctuations, as maintained by the authors of the kinetic concepts of strength.[10,33] Therefore, any interpretation of the deformation and fracture processes without consideration of the thermal processes is incorrect.

4. The useful work is done by the electric part of the EM field; the magnetic component does not contribute to this work. Because it is normal to the direction of the velocity of the structure-forming particles, the magnetic force distorts their paths, localizing them in a limited region of the space. All the particles move along flat, chaotically oriented closed paths. Only to a very rough approximation can this movement be regarded as the unidimensional unharmonic thermal oscillations considered in the Debye heat capacity theory.[6,7,23] This simplification limits our notions of the internal, thermal, and structure formation processes so that relationships between them and the accompanying phenomena cannot be established.

5. Movement of the particles occurs in the central electric field (Equation 1.17).

6. Equation 1.22 relates atomic electrodynamics and structural mechanics (Equation 1.23).

7. The overall internal energy of the rotos, u, is a constant value formed during solidification of a body. While this body exists, only its kinetic u_t and potential u_r parts vary.

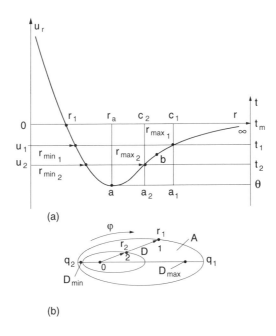

(a)

(b)

FIGURE 1.7
Relationship between potential, U_r, and kinetic, t, parts of the overall internal energy of electromagnetic dipole (a) and schematic of the paired motion of atoms making it in the potential well (b).

At t = const, the radial part of motion can be considered a uniform displacement in the potential field with the energy given by Equation 1.26.II. The graphical representation of this function is shown in Figure 1.7a.[25,30] As seen, when $r \to 0$, it tends to $+\infty$; when $r \to \infty$, it tends to zero on the side of negative values. The potential part of the overall internal energy of the dipole, u_r, is plotted on the left axis of ordinates, and the quantitative measure of the kinetic part, t, is plotted on the right axis. The graph shows flat section $r_1ab\infty$ of the three-dimensional potential well (Figure 1.8) in an arbitrary direction r, whose gently sloping branch $ab\infty$ shows the character of variations of t in depth of the well. The overall energy is determined by rectangle $r_a a\theta t_m$. The radial $\dot{r}$ and angular $\dot{\varphi}$ velocities change only its kinetic and potential components (lengths $a_1 r_{max_1}$ and $r_{max_1} c_1$ at the u_1 level, respectively).

Atoms approaching each other to form a bond occurs on solidification of materials. Because the only substantial difference between the atoms lies in the mass of the nucleus and the number of valent electrons, each excited atom supplies the outside electrons to form the bond (elliptical loops in Figure 1.3). In quantomechanical calculations,[20,30] instead of as given by Equation 1.26.II, the interactions normally set by the empirical law are in the form of a sum of two terms. One (negative) expresses the energy of attractive forces and the other (positive) expresses the energy of repulsive forces, which

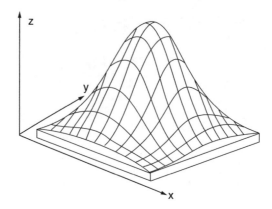

FIGURE 1.8
Spatial configuration of the internal potential field.

are assumed to monotonically increase with a decrease in the distance between the atoms, $\mathbf{r}$, from ∞ to 0:

$$u_r = \frac{A}{\mathbf{r}^n} - \frac{\alpha}{\mathbf{r}^m},\qquad(1.28)$$

where

$$A = \frac{\beta\hbar^2 z^{5/3}}{8m}\quad\text{(I)},\qquad \alpha = \beta z^2 e^2\quad\text{(II)}\qquad(1.29)$$

are the positive values expressed in terms of atomic parameters and the Madelung constant β, which characterizes the degree of interaction of the separated pair of atoms with their nearest neighbors. It is close to a unity in the order of magnitude;[6,20] α is the analogue of configuration factor (see Equation 1.17). The n and m values are assumed to be equal to $m = 6$ and $n = 7$ to 12. This is done to describe the behavior of atoms of specific materials as accurately as possible. Inequality $m < n$ means that the repulsive forces vary with r more rapidly than the attractive forces, leaving the latter behind at low values of r and lagging behind at high values of r.

The law of energy conservation (Equation 1.23) is valid not only for classical but also for quantomechanical approximation, with only one substantial amendment: levels of the energy (e.g., horizontal $r_{max_1} - r_{max_1}$ in Figure 1.7a) are quantized, i.e., the state of the rotos varies not continuously but assumes only the following discrete series of values

$$u_r = \frac{\hbar l(l+1)}{2m\mathbf{r}^2} - \xi,\qquad(1.30)$$

where l is the azimuthal quantum number for the nucleus that, by analogy with quantization of an electron (Equation 1.1), can assume integer values

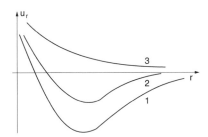

FIGURE 1.9
Dependence of shape of the potential well on the azimuthal number.

such as 0, 1, 2, 3.... From comparison of Equations 1.23 and 1.30 we find the formula for momentum (Equation 1.27) expressed in terms of the l number:

$$M = \hbar[l(l+1)]^{1/2}. \tag{1.31}$$

The plot of function $u_r = f(l)$ for different l is shown in Figure 1.9. As seen, if $l \cong 0$, $u_r \cong -\xi$(position 1). If the velocity of rotation of the nuclei increases insignificantly and the l number is not much greater than 1, u_r is not much different from position 1; it becomes just a bit distorted (position 2). If l is too high ($l \gg 1$), function $u_r = f(l)$ assumes the form of curve 3. In this case, u_r is always positive. The rotos cannot exist in such a state, which means that the nuclei, like electrons (Table 1.1 and Figure 1.4), arrange their movement in a small number of orbital planes s, p, and d.

A dynamic triad can be regarded as the simplest model of a two-atomic molecule; both electrons and nuclei radiate simultaneously. (The negative charge dissipated among the electron orbits is not shown in Figure 1.6b.) Electron radiation occupies a shortwave region of the EM spectrum and, under conventional states of macroworld, it does not show up.[6] The nuclear (dipole) or phonon radiation is in the thermal region.[6,25] So, it is not accidental that any change in the state of a solid caused by external factors is accompanied by thermal effects.[2,3,10] It can be seen from Equation 1.16 that the nucleus charges radiate only when they move with acceleration. They do not radiate phonons at uniform movement.

The dipole (Equations 1.14 and 1.15) radiates the EM waves whose electric **E** and magnetic **H** components can be determined by the following formulas:[25]

$$\mathbf{E} = \frac{1}{c^2 R}[[\ddot{\mathbf{D}}\mathbf{n}]\mathbf{n}] \quad \text{(I)}, \qquad \mathbf{H} = \frac{1}{c^2 R}[\ddot{\mathbf{D}}\mathbf{n}] \quad \text{(II)}, \tag{1.32}$$

where R is the distance from the center of field 0 (Figure 1.6b) to a point of measurement of the intensities of fields **E** and **H**; **n** is a unit vector in the R direction. According to Equation 1.26.I, the electric component of the field ξ (Equation 1.17) and thermal radiation of the rotos t are related by a parabolic

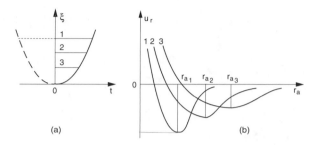

FIGURE 1.10
Dependence of the interaction potential (a) and shape of the potential well (b) on the thermal radiation of rotos.

dependence (Figure 1.10a). Therefore, an increase in its radiating power transforms the shape of the potential well in direction 1, 2, or 3 (Figure 1.10b).

1.4   Character of Movement of Bound Atoms

Extensive experiments conducted by the authors of the kinetic concept of strength revealed a close relationship between mechanical and thermal properties.[10,33] They showed how significant the role of the thermal movement of atoms is in the resistance of solids to deformation and fracture. In the practical range of temperatures, the electrons do not contribute to heat capacity.[30] The known models of solids (Drude-Lorentz, Fermi-Sommerfeld, etc.) have focused on their electron structure rather than on peculiarities of movement of the nuclei.[6,7,34]

The problem of motion of two particles interacting with each other according to the law given in Equation 1.26.II (Figure 1.6) is solved in Landau and Lifshits.[25] It can be substantially simplified if motion of the bound system is resolved into motion of the inertia center 0 and motion of atoms 1 and 2 with respect to this center. The potential energy of interaction depends only on the distance $\mathbf{D}$ between them (Equation 1.15). Combining the origin of coordinates with the inertia center yields $m_1\mathbf{r}_1 + m_2\mathbf{r}_2 = 0$ and thus

$$\mathbf{r}_1 = \frac{m_2}{m_1 + m_2}\mathbf{D}, \qquad \mathbf{r}_2 = \frac{m_1}{m_1 + m_2}\mathbf{D}. \tag{1.33}$$

By adding a designation of

$$m = \frac{m_1 m_2}{m_1 + m_2},$$

which is called the normalized mass, Equation 1.26.II can be considered as formally describing the motion of a single material particle having mass m

in the external field of $u_r = -\xi$, which is symmetric about a fixed origin of coordinates, i.e., the mass inertia center. Therefore, the problem of motion of the bound atoms is reduced to the solution of the problem of motion of one particle with normalized mass m in the central field, where its potential energy u_r depends only on distance $\mathbf{r}$ to a certain fixed point.

If this problem is solved, i.e., path $\mathbf{r}$ of the normalized particle is found, paths of the atoms comprising the rotos can be found from Equation 1.33. Landau et al.[31] show that atoms will move about a fixed inertia center of the system along geometrically similar paths, differing only in sizes, which are inversely proportional to their masses

$$\frac{\mathbf{r}_1}{\mathbf{r}_2} = \frac{m_2}{m_1}. \tag{1.34}$$

In motion, they will always be at the ends of some line $0\mathbf{r}_2\mathbf{r}_1$ that passes through the inertia center 0 (Figure 1.7b).

In motion in the central field, the system of particles retains its momentum M (Equations 1.27 and 1.31) relative to the common inertia center.[25] Because the vectors M and $\mathbf{r}$ are perpendicular, the constancy of M implies that radius vector $\mathbf{r}$ of a particle always remains in one plane, i.e., the plane normal to M. Therefore, the entire path of movement of atoms in the central field is in one plane. Orientation of planes in one direction causes anisotropy of mechanical properties, a well-known phenomenon of materials science and engineering.[2,3]

The law of conservation of momentum written in the form of Equation 1.27 allows a simple geometrical interpretation (Figure 1.11). Expression $1/2\mathbf{r} \cdot \mathbf{r}d\varphi$ is the area of the sector formed by two neighboring radius vectors, $\mathbf{r}$ and $\mathbf{r} + d\mathbf{r}$, and a segment of the arc of path dw. This makes it possible to write the momentum formula (Equation 1.27) in the following form

$$M = 2m\dot{w}, \tag{1.35}$$

where derivative $\dot{w}$ is called the sector velocity.[25]

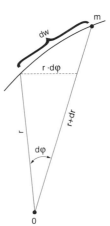

FIGURE 1.11
Geometrical interpretation of the constancy of momentum of a particle, m, with respect to the center of the local potential field, 0.

By dividing variables and integrating Equation 1.23, we obtain the formula for the path of a particle in the central field:[25]

$$\frac{r_a}{\mathbf{r}} = 1 + g\cos\varphi.$$ (1.36)

This is the equation of the conic section having its focus at the origin of coordinates, where

$$r_a = \frac{M^2}{m\alpha} \quad \text{(I)}, \qquad g = \left(1 + \frac{2uM^2}{m\alpha^2}\right)^{1/2} \quad \text{(II)}$$ (1.37)

are the parameter and eccentricity of the elliptical orbit. According to the known formulas of analytical geometry, minor a and major b semi-axes of an ellipse are equal to

$$a = \frac{r_a}{1 - g^2} = \frac{\alpha}{2|u|} \quad \text{(I)}, \qquad b = \frac{r_a}{(1 - g^2)^{1/2}} = \frac{M}{(2m|u|)^{1/2}} \quad \text{(II)}.$$ (1.38)

It should be noted that the major axis of the ellipse depends only on the energy, whereas the minor axis depends on energy and momentum. At $g = 0$, where the ellipse is transformed into a circle of radius $\mathbf{r} = r_a$, the potential energy reaches its minimum, equal to

$$u(r_a) = -\frac{\alpha^2 m}{2M^2}.$$ (1.39)

Figure 1.12a shows the formation of the conic section, and Figure 1.12b shows the main geometrical and energy parameters of the elliptical path. The largest and smallest distances to the field center (focus of the ellipse) are equal to

$$r_{min} = \frac{r_a}{1 + g} = a(1 - g) \quad \text{(I)}; \qquad r_{max} = \frac{r_a}{1 - g} = a(1 + g) \quad \text{(II)}.$$ (1.40)

Subtracting the expression for r_{min} from r_{max} yields

$$r_{max} - r_{min} = 2ag.$$ (1.41)

As can be seen from Figure 1.12b, this difference corresponds to the distance between the ellipse foci (points 0 and 0_2). Since the direction of radius vector $\mathbf{r}$ is not specified, the spatial location of the major semi-axis of the

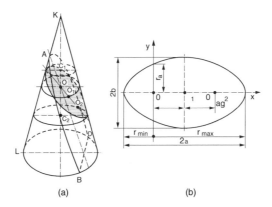

FIGURE 1.12
Schematic of formation of a conical section (a) and main geometrical and energy parameters of elliptical orbits (b).

ellipse is random and has a chaotic orientation in the majority of the interacting atoms. Coordinates of point b of the inflection of curve u_r (Figure 1.7a) and point r_1 of its intersection with the axis of abscissas r are obtained from the following conditions:

$$\frac{du_r}{dr} = 0 \quad \text{(I)} \quad \text{and} \quad u_r = 0 \quad \text{(II)}. \tag{1.42}$$

Hence, $r_1 = 1/2r_a$ and $r_b = 3/2r_a$. As the parameter of the elliptical path, r_a determines, at the same time, the characteristic size of the crystalline lattice.

The coordinate of inflection point b can also be obtained in a way other than presented above. It is located on the right branch of curve u_r, i.e., where its loop is set by Expression 1.40.II. Substituting here the values for a and g from Equations 1.37.II and 1.38.I and making simple transformations, we can obtain

$$r_{max} = \frac{\alpha}{2|u|}\left[1 + \left(1 + \frac{4|u|A}{\alpha}\right)^{1/2}\right] = \frac{\alpha}{|u|} + \frac{r_a}{2},$$

where parameters A and α are determined from Equation 1.29.

Subjecting this expression to double differentiation, equating the second derivative to zero, and then transforming and simplifying, we obtain the quadratic equation for the overall energy in the form

$$|u|^2 - \frac{2\alpha^2}{A}|u| - \frac{\alpha^4}{A^2} = 0.$$

Its solution is

$$u = 2|u(r_a)|, \tag{1.43}$$

where $u(r_a)$ is determined from Equation 1.39. Substitution of this equation into Equation 1.40.II yields

$$r_b = 1.5r_a. \tag{1.44}$$

Now we are ready to determine the eccentricity of an elliptical path at the inflection point b. In doing so, we can rewrite Equation 1.40.I for r_{min}, accounting for Equation 1.38.I as

$$r_{min} = \frac{\alpha}{2|u|}(1-g)$$

and substitute it into Equation 1.40.II, which yields

$$g_b = \frac{1}{3}. \tag{1.45}$$

By substituting Equation 1.44 into Equation 1.28, we can find the level of potential energy corresponding to inflection point

$$u(r_b) = -\frac{2\alpha^2}{9A}.$$

Rewriting and accounting for Equations 1.29 and 1.39, we obtain

$$u(r_b) = \frac{8}{9}u(r_a) \tag{1.46}$$

This proves a conclusion that the potential well (Equation 1.26.II) has a markedly asymmetric shape. It is characterized by the gently sloping right branch and the steep left branch. Its depth (the distance to the inflection point b) is equal to

$$\frac{1}{9}u(r_a),$$

while, behind this point, it gradually decreases to a value of

$$\frac{8}{9}u(r_a).$$

In addition, it is characterized by horizontal $u_r = 0$, at $\mathbf{r} \rightarrow \infty$, and vertical $u_r \rightarrow \infty$, at $\mathbf{r} \rightarrow 0$, asymptotes, minimum point a, and inflection point b.

From comparison of Equations 1.26.2 and 1.28, it follows that

$$M = \frac{1}{2} \beta^{1/2} \hbar z^{5/6}. \tag{1.47}$$

Substitution of Equations 1.29 and 1.47 into Equations 1.37.I and 1.39 yields

$$r_a = \frac{\hbar^2}{4mz^{1/3}e^2} \quad \text{(I)}; \qquad u(r_a) = -\frac{2\beta}{\hbar^2} mz^{7/3}e^4 \quad \text{(II)}. \tag{1.48}$$

It can be seen that the value of the mechanical momentum is almost linearly dependent on the valence of a chemical element, while the overall internal energy u of the interatomic bond is entirely determined by mass m, general charge e, and valence z of the constituent atoms, i.e., the position of a material in Mendeleev's periodic system. It is a constant value for a given material, equal to $u(r_a)$ (defined by length ar_a in Figure 1.7a). Within the entire range of variations in the radius vector from r_a to ∞, this length remains unchanged and formally displaces within rectangle $ar_a t_m \theta$ whereas its potential and kinetic components depend on radius vector $\mathbf{r}$. The variations of the former are defined by the ordinate of figure $ar_a t_m ba$ and those of the latter by $abt_m \theta$.

Integrating Equation 1.35 from zero to τ yields the time of revolution on the elliptical path, i.e., the motion period τ as $2mw = \tau M$, where w is the area of the orbit. Using Equation 1.38, we can find for ellipse $w = \pi ab$

$$\tau = 2\pi a^{3/2} \left(\frac{m}{\alpha} \right)^{1/2} = \pi \alpha \left(\frac{m}{2 |u|^3} \right)^{1/2} \tag{1.49}$$

and thus the frequency of revolutions is

$$\upsilon = \frac{1}{\tau} = \frac{1}{\pi \alpha} \left(\frac{2 |u|^3}{m} \right)^{1/2}. \tag{1.50}$$

Note that this frequency depends only on the rotos energy. At $M = \text{const}$, particles moving on the elliptical paths behave as flat oscillators. Chaotic spatial orientation of these oscillators, set by quantum number l (Equation 1.31), creates adherence allowing a solid to exist as a whole. In the electromagnetic spectrum (Equation 1.32), this motion corresponds to the thermal region. Therefore, it is logical that the temperature dependence of mechanical properties is observed at the macrolevel. In an extreme case, where $\varphi = 0$ or $\varphi = \pi$, the flat paths are degenerated into segments of straight lines; this case corresponds to unidimensional unharmonic oscillations.

On forming a bond and entering into contact on one of their individual orbits (e.g., on the third orbit in position *b* or on the second orbit in position *e*, Figure 1.3), ions m_1 and m_2 are drawn together by molecular orbits 4 and 3. During movement, the distance between them varies from D_{min} to D_{max} (Figure 1.7b). It follows from Figure 1.7a that the principle of variation in r_{min} is described by the left (steep) branch ar_1 of curve u_r, and, in r_{max}, by the right (gently sloping) branch $ab\infty$ of this curve. To the left from point r_a, in a region where $r_1 < r_{min} < r_a$, the electron shells are overlapped for a moment to give rise to powerful forces of repulsion between the nuclei. To the right (on the $ab\infty$ branch), at a considerable distance from vertical ar_a, the magnetic attractive forces return the nuclei to their initial positions. This is repeated an infinite number of times.

It can be seen from Equation 1.48.II that the smaller the number of electrons in the external layer, the weaker the interatomic bond. Naturally, an increase in the number of these electrons (at a simultaneous increase in the positive charge of an ion) leads to an increase in the total force that binds the ions and a decrease in sizes of the molecular orbits of the electrons. This results in an increase in the bond strength between the atoms.

While performing the bound motion on the elliptical paths about the fixed center 0, interacting atoms 1 and 2 may approach each other many times to a minimum distance D_{min} and then move away from each other to a maximum distance D_{max} (Figure 1.7b). At a preset energy level (u_1 and u_2 in Figure 1.7a), they cannot independently leave the localization zone because they are bonded to each other by the electrodynamic forces (Equation 1.16). Localization of motion and the forces that control this motion are primary causes leading to the formation of solids in rapprochement of the atoms.

Writing Equation 1.16 and accounting for Equation 1.24.II yields the temperature dependence of the interatomic interaction forces

$$\mathbf{F}t^{1/2} = \left(\frac{2}{m}\right)^{1/2} e\mathbf{E} + \frac{2e}{mc}[t^{1/2}\mathbf{H}], \tag{1.51}$$

where $\mathbf{F} = m\ddot{\mathbf{D}}$ is the Lorentz force (Equation 1.16) generated by the dipole (Figure 1.7b) at a rapid variation in its sizes inside the orbital ring from D_{min} to D_{max}.

Analysis of Equation 1.51 shows that it is meaningless to speak about resistance of a solid to external effects without correlation with its thermal conditions. Thus, the majority of the known strength theories[3,4] that do not consider this correlation are incorrect. The temperature has no effect on the energy of structure-forming particles (because the first term in Equation 1.51 remains unchanged), but changes the localizing ability of a magnetic component of the EM field. Until recently the correlation established by Equation 1.51 has been ambiguous, although it is natural for the macroworld and thus has been observed experimentally on many occasions.[2,3] The concentrated form of this correlation is found in the kinetic strength concept.[10]

Using the expression for total energy (Equation 1.23), we can show that at $u \geq 0$ the path of the motion becomes nonclosed (shape *BAD* in Figure 1.12a). When $u > 0$ and $g > 1$, the elliptical path is transformed into a hyperbola, whereas when $u = 0$ and $g = 1$, it transforms into a parabola. In both these cases the rotos disintegrates and the interatomic bond ceases to exist.

The law of energy conservation (Equation 1.23) is valid for any closed system of particles. Together with the laws of momentum conservation, $mv = $ const, and moment conservation, $M = $ const, they constitute the foundation of atomic mechanics. Because of its generality, this law can be applied to all phenomena and effects. In particular, it allows one to write the condition of destruction of the rotos into two atoms that form this rotos. If v_1 and v_2 denote the velocities at which atoms m_1 and m_2 move apart after decomposition of the bond, we can write the law of energy conservation for this process as

$$u = \frac{m_1 v_1^2}{2} + u_1 + \frac{m_2 v_2^2}{2} + u_2,$$

where u is the internal energy of the initial rotos and u_1 and u_2 are the internal energies of free atoms. Because the kinetic energy is always positive, it follows from the last relationship that $u > u_1 + u_2$. It is apparent that difference $u_w = u - (u_1 + u_2)$ is equal to the energy consumed for the formation of a local defect having the area $w = \pi ab$.

1.5 Localization Parameters and Rotos State Equation

Because u, m, α, and M are constants, differentiating Equation 1.23 and accounting for Equations 1.24 and 1.25, we can obtain

$$\mathbf{F} = \frac{C}{\mathbf{r}^3} - \frac{\alpha}{\mathbf{r}^2}, \tag{1.52}$$

where $C = M^2/m = $ const.

Condition $\alpha = $ const stabilizes the spatial configuration of charges of the triad (Figure 1.6b) in a certain position, and condition $M = $ const fixes the paths in one of the planes propagating through the center of a local EM field (Equation 1.26.II, point 0 in Figure 1.7). The latter condition does not prevent the other similar pair from arranging its movement about the same center (crystalline lattice node), but in a different plane. On solidification of a material, quantum numbers l determine its azimuth, forming multi-atomic molecules and governing the process of filling the three-dimensional volume (in analogy with Figure 1.4).

Atoms perform thermal motions along planar elliptical paths following the equipotentials of the potential well (Figure 1.7).[25,30] Steep ar_1 and gently sloped $ab\infty$ branches of the well define the boundaries of the region within which atoms move from the field center 0 (vectors $\mathbf{r}_1$ and $\mathbf{r}_2$), i.e., the set of points where radial velocity $\dot{r}$ vanishes. However, this does not result in their freeze-up because their angular velocity $\dot{\varphi}$ is not zero. Equality $\dot{r} = 0$ associates with points r_{min} and r_{max} of turning of the paths, at which radius vector r changes from increasing to decreasing and vice versa.

The potential limits that region of the space in which atoms 1 and 2 are localized. At the u_1 level, local temperature t_1, and direction r, the position of atom 1 is determined by vectors $\mathbf{r}_{min_1}$ and $\mathbf{r}_{max_1}$ (Figure 1.7a). At $t = \text{const}$ in the plane of motion A, the difference between them sets geometrical dimensions of dipole $\mathbf{D}$ (Figure 1.7b). In a full revolution of atoms along the elliptical paths, the value of $\mathbf{D}$ varies from D_{min} to D_{max}, characterizing linear thermal oscillations according to Debye.[6,23] In the same direction $0r_2r_1$ and at the same moment of time, but at temperature t_2, atom 2 has different sizes of the path, i.e., $\mathbf{r}_{min_2} - \mathbf{r}_{max_2}$.

An increase in local temperature t (t_1, t_2, etc. and up to melting point t_m) brings the energy level of pair 1 and 2 closer to the surface of the well, which is accompanied by corresponding variations in u_r (u_2, u_1, etc.). Therefore, the bond forces $\mathbf{F}$ (Equation 1.52). Curve $r_1ab\infty$ in Figure 1.7a determines the complete set of its allowed states. To meet condition $\mathbf{D} = \text{const}$ at $t = \text{const}$, the atoms must move within the Cartesian space between points $\mathbf{r}_{min}$ to $\mathbf{r}_{max}$ of turn of the paths in the r direction (as in any other one) with a positive or negative acceleration $\ddot{\mathbf{D}}$, thus changing the bond forces $\mathbf{F}$ following the law set by Equation 1.16 during each cycle.

Representation of Equation 1.52 in the form

$$\omega = f - \xi, \tag{1.53}$$

where $\omega = \mathbf{F}r$ is the energy of resistance of the dipole to deformation in the $\mathbf{r}$ direction and $f = C/r^2$ is the kinetic energy of the rotational part of the motion, makes it possible to derive the thermodynamic equation of state of the paired interatomic bond. This means that the central electric field (Equation 1.26.II) is homogeneous. It is only in such a field that the $\mathbf{F}r$ product determines the potential energy and that a moving particle is affected by force $\mathbf{F}$ having an electrodynamic nature (Equation 1.16).

It can be seen from Equations 1.26 and 1.53 that, for a given material (in this case u, M, and m are constant), the state of the bond is determined by three parameters, such as primitive $\mathbf{r}$, its first derivative $\dot{r}$ (the quantitative measure of which is t), and configuration factor α, which characterizes the spatial position of electric charges of the triad (Figure 1.6).

Bound by closed thermal motion, the atom pair exhibits resistance only within the potential well (Equation 1.26.II), shown by dashed curve 1 in Figure 1.13. At given parameters α and C, writing Equation 1.52 in the form

$$\mathbf{F}r^3 + \alpha r - C = 0 \tag{1.54}$$

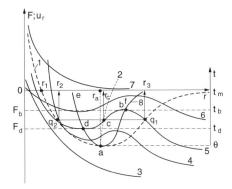

FIGURE 1.13
Temperature dependence of constitutional diagrams of the EM dipole in the potential well.

allows the characterization of transformation of the state diagrams of the pair (curves 3 to 7) over any range of variations in local temperatures (Equation 1.26.I). For example, between t_d and t_b the ranges of variations in **F** (between $\mathbf{F}_d$ and $\mathbf{F}_b$, respectively, on diagram 5 of Figure 1.13) are such that Equation 1.54 has three different roots: r_2, r_c, and r_3. Two of them become complex conjugates at the extreme points d and b.[31]

This is absolutely valid from the standpoint of atomic mechanics; as a function of an accelerated variation in sizes of dipole, $\ddot{\mathbf{D}}$ (horizontal section of the well, q_2cq_1 on diagram 5, Figure 1.13), the bond force **F** increases (in region q_2dc) during acceleration of atoms 1 and 2 in the direction shown by arrow φ (Figure 1.7b). It reaches its maximum at point d (somewhere midway between D_{min} and D_{max}) and then decreases to zero at point c (where **D** becomes equal to D_{max}). In the deceleration cycle (movement in a direction opposite to that shown by the arrow), the variation in **F** repeats with mirror symmetry (loop cbq_1 is located in the upper half-plane on diagram 5, Figure 1.13). Because parameters α and C depend only on the atomic characteristics, the family of diagrams 3 to 7 (Figure 1.13) sets many states in which the interatomic bonds for all solids, without any exception, may be found.

Some critical states exist among them. In diagram 5 (Figure 1.13), they are shown by points q_2, d, c, b, and q_1. At point a, where $g = 0$ and the circumferential motion is described by Equation 1.48, acceleration $\ddot{\mathbf{D}}$ is equal to zero. Radiation corresponding to the circular motion can be found by differentiating both parts of Equation 1.26.I:

$$d\xi = \left(\frac{4}{m}t + 1\right)dt. \tag{1.55}$$

At $d\xi = 0$, accounting for Equation 1.24, we have

$$v_{max} = |2r_a|, \tag{1.56}$$

where $v_{max} = 1/\dot{\varphi}_a$ is the maximum frequency of the phonon spectrum (Equation 1.50) corresponding to the characteristic Debye temperature θ,[6,7] and r_a, as in Equation 1.37, is the crystalline lattice parameter. On the other hand,

$$\frac{d\alpha}{\alpha} = \frac{d\mathbf{r}}{\mathbf{r}} \quad \text{(I)}, \qquad m\ddot{\mathbf{r}} = 0 \quad \text{(II)}. \tag{1.57}$$

By joining inflection point c of maximum b and minimum d in diagrams 4 to 6 (Figure 1.13) located inside potential well 1, we obtain curves 2 and 8. The latter has the only extreme at point a. As seen, at $t > \theta$, curve 2 divides the state of rotoses into two radically different phases: the center of the first is located along the line of minima ade, and that of the second along the line of maxima abf. This means that, in the region where $t > \theta$, a particular characteristic θ disintegrates into three subsets: Debye temperature of the first, θ_f (line ade), and, second, θ_s (line abf), and phase transition temperature T_c (curve 2). At a microlevel they are designated, accordingly, as v_f, v_s, and t_c. The bond force vanishes at t_c. In invariable thermal fields (where T = const), phase transitions are impossible. Therefore, in experiments T_c is regarded as the temperature at which brittle fracture occurs.[2,3]

For points inside and outside curve 8, the following is valid:

$$\frac{\partial \mathbf{F}}{\partial \mathbf{r}} > 0 \text{ inside curve 8} \quad \text{and} \quad \frac{\partial \mathbf{F}}{\partial \mathbf{r}} < 0 \text{ outside this curve.} \tag{1.58}$$

At high and low temperatures, the diagrams are transformed into monotonically decreasing isotherms 3 and 7. They characterize the state of a material where the bonds are of a nonthermal nature. At path turning points q_1 and q_2 (Figure 1.7b) where $\dot{\mathbf{r}} = 0$, the first derivative changes its sign and the bond force again becomes zero. This explains brittle failure of the bonds at cryogenic temperatures (point q_2) and plastic failure at high temperatures (point q_1), which are well known from practice.[3] Reaching the Debye temperature (Equation 1.56) is accompanied either by a phase transformation (Equation 1.58) that changes the spatial configuration of charges of the triad (Equation 1.57.I) or by loss in the interatomic bond (Equation 1.57.II) which, in turn, leads to the formation of vacancies. Figure 1.13 shows the character of variations in the state diagram (Equation 1.54) with t (Equation 1.26.I) being ignored. To be absolutely precise, each curve of 4 to 6 should have been located in its intrinsic potential well (Figure 1.10b).

Local temperature t is affected by two factors: internal (through accelerated variation in **D** from D_{min} to D_{max}, Figure 1.7b) and external background (synchronous increment in sizes of the elliptical orbits of atoms 1 and 2 with variation in t in depth of the well, Figure 1.7a). The first factor forms a two-branch diagram of the bond following the law set by Equation 1.54, while the second transforms its shape (diagrams 4 to 6 in Figure 1.13). Substituting equalities $\alpha/\mathbf{r} = \mathbf{E}\mathbf{r}$ and $dt = \mathbf{F}d\mathbf{r}$ (which follow from Equations 1.17 and 1.24.II) into Equation 1.54 yields the electrodynamic analogue of the equation of state of the rotos (Equation 1.53):

$$\omega = \frac{\mathbf{E}\left(\dfrac{d\mathbf{E}}{\mathbf{E}} + \dfrac{d\mathbf{r}}{\mathbf{r}}\right)}{\dfrac{4}{m}t + 1}, \tag{1.59}$$

where

$$\omega = \mathbf{F}\frac{d\mathbf{r}}{\mathbf{r}}$$

and the phonon dependence of the Lorentz resistance force, other than Equation 1.51,

$$\mathbf{F} = \frac{\mathbf{r}\left(\dfrac{\mathbf{E}}{\mathbf{r}} + \dfrac{d\mathbf{E}}{d\mathbf{r}}\right)}{\dfrac{4}{m}t + 1}. \tag{1.60}$$

Condition ω = const is always met at u = const. An increase in the heat background t causes displacement of the diagrams of state of the rotos to the surface of the potential well (change in positions in a direction of 4 to 5 to 6 in Figure 1.13). In this case the relative dimensions of the EM dipole, $\mathbf{D} = \mathbf{r}_3 - \mathbf{r}_1$, are increased and the resistance force $\mathbf{F}$ (Equation 1.60) is proportionally decreased to maintain condition ω = const. A decrease in t imposes a reverse character of changes on parameters $\mathbf{D}$ and $\mathbf{F}$. Maximum of Function 1.53 corresponds to the inflection point b of curve 1 with coordinate of $r_b = 1.5r_a$. Substituting this value into Equation 1.52 yields the maximum value of the deformation resistance energy:

$$\omega(\theta) = \frac{4}{27}\frac{\alpha}{r_a^2} \tag{1.61}$$

or, accounting for parameters α (Equation 1.29), M (Equation 1.47), and r_a (Equation 1.48.I), we can write

$$\omega(\theta) = \left(\frac{4}{3}\right)^3 \frac{\beta}{\hbar^4} m^2 z^9 e^6. \tag{1.62}$$

As seen, the resistance reaches its maximum when the bound atoms approach each other by a distance of $1.5r_a$. This distance directly depends on

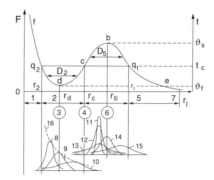

FIGURE 1.14
Stable and critical states of the EM dipole. Formation of structure of brittle and ductile bodies.

the mass of chemical elements, m. An increase in valence of elements leads to a drastic increase in the discussed resistance as seen from Equation 1.62, term z^9. A decrease in the r_a parameter causes an extra strengthening of the bond and an increase in its rigidity. In addition, the higher the quality of packing of structure of a solid, which is determined by the β parameter (Equation 1.29), the stronger the bond.

For our further consideration we select one of the state diagrams from a wide diversity of the probable ones (Figure 1.13)—one in Figure 1.7a to which temperature t_2 corresponds (Figure 1.14). Its analogue is diagram 5 in Figure 1.13. As noted earlier, it has three peculiar points d, c, and b, and consists of four essentially different regions: one-phase regions, i.e., shortwave fq_2 and long-wave q_1e, and two-phase regions, i.e., "cold" q_2dc and "hot" cbq_1. As follows from Equation 1.57, along with an increase in size of the dipole (displacement from the origin of coordinates to the periphery of the axis of abscissas), a change occurs in the spatial configuration of its charges (under certain conditions this leads to decomposition of the dipole), conditions of the rotational motion, and character of thermal radiation.

Substitution of the formula for determination of quantized orbits of electrons[20] $r_j = j^2\hbar^2 / 4\pi^2 me^2$ into Equation 1.40.I yields

$$r_a = \frac{1}{z^{1/3}} \left(\frac{\pi}{j}\right)^2 r_j,$$

where $j = 1, 2, 3 \ldots$ are the internal (Equation 1.4) quantum numbers determining the number of an orbit or, to be more exact, its distance from the nucleus. Because $(\pi/3)^2 \approx 1$ at $j = 3$, then

$$r_a \approx \frac{1}{z^{1/3}} r_3. \tag{1.63}$$

In the order of magnitude, r_a is equal to the radius of a nondisturbed atom on the third (counting from the nucleus) stable electron orbit. It determines minimum distance to which the neutral atoms of a given material should be drawn together (Equation 1.14) at temperature Θ to cause them to form a rotos. As such, the presence of valence z adds a periodical character to Equation 1.63, while a relatively free choice of the j numbers allows r_a to take any value near the third energy level of electrons.

It follows from Equations 1.42 and 1.44 that distances identified with the low-temperature boundary of the potential well (point r_1 in Figures 1.7a and 1.13) and with point b of maximum of the diagram of the dipole are, respectively,

$$r_1 = \frac{1}{2}r_a = \frac{1}{z^{1/3}}r_{1.5} \quad \text{(I)}, \qquad r_b = 1.5r_a = \frac{1}{z^{1/3}}r_{4.5} \quad \text{(II)}. \qquad (1.64)$$

Expressions 1.63 and 1.64 determine the zone in which atoms perform bound motion following the diagram in Figure 1.7b. They approach each other an infinite number of times to a distance commensurable with r_1 (corresponding to D_{min}) and then move away to the periphery of the potential well, where r is higher or lower than r_b. Every time they move through an electron cloud they change its configuration. Graphically, such a probability is shown in Figure 1.3 by a sequence of positions a, b, e, and d.

1.6 Electrodynamics of Interatomic Interaction

Calibration of the axis of abscissas (Figure 1.14) in relative units $r_{j'}$ equivalent to sizes of the stable electron shells (counting from the nucleus), makes it possible to estimate the size of the rotos and also to reveal the spatial configuration of charges in a two-atom molecule and explain objectively why its constitutional diagram has two extrema.

Figure 1.15 shows the variations in the configuration of charges with variations in the size of the dipole in accordance with Equation 1.57.I. Numbers of positions from 1 to 7 coincide with designations of the regions of the diagram in the lower part of the axis of abscissas in Figure 1.14. Positive charges of the nuclei are shown by solid circles and space charges of electrons are shown by open circles. At distances smaller than one-and-a-half the size of the first electron shell, r (1.5) (position 1), nuclei 1 and 2, together with a concentrated negative charge, are on a circular path ($r_1 = r_2 = r_0$). The space charge of electrons involves both nuclei (as a unit) into an accelerated circular motion ($+\ddot{\varphi}$), thus generating powerful Lorentz attractive forces $\mathbf{F} \to \infty$. By overcoming electrostatic repulsive forces, they stimulate drawing together of the nuclei ($\mathbf{D}_1 \to 0$), which is accompanied by weak bremsstrahlung radiation ($\ddot{\mathbf{D}}_1 > 0$).

The nuclei get into an acceleration well q_2dc between one-and-a-half r (1.5) and third r (3) electron shells. At this moment both nuclei turn out to be close to position D_{min} (Figure 1.7b). Here they are located on different elliptical orbits

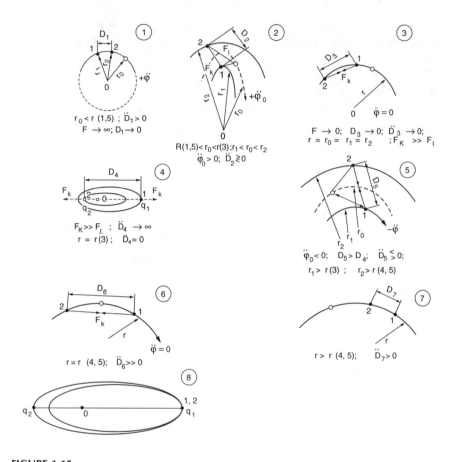

FIGURE 1.15
Schematic of configuration restructuring of the dipole. Electrodynamic nature of phase transitions and dislocations.

(position 2) between which a negative space charge ($r_1 < r_0 < r_2$) performs a planar motion. As shown earlier, it accelerates both nuclei ($\ddot\varphi_0 > \ddot\varphi_1 > \ddot\varphi_2$). The Lorenz attractive forces $\mathbf{F}_l$ compensate for the Coulomb repulsive forces $\mathbf{F}_k$ to form a stable two-atom structure with the dipole moment $\mathbf{D}_2$ in the plane of motion. Tendency of one of the atoms to increase (or decrease) orbital radius r leads to immediate bremsstrahlung ($\ddot{\mathbf{D}}_2 < 0$) or acceleration ($\ddot{\mathbf{D}}_2 > 0$) dipole radiation. By affecting $\mathbf{F}_l$, it restores the initial configuration of the charges.

At the bottom of the acceleration well (position 3), where $\mathbf{F} \to$ min and $D_{min} \to 0$, both nuclei, each rotating on its elliptical path, can be at the same point (q_2) for some moment of time (Figure 1.7b). Because $\mathbf{D}_3 \to 0$, the dipole does not radiate $\ddot{\mathbf{D}}_3 \to 0$ and therefore $\mathbf{F}_l \to 0$. The Coulomb repulsive forces ($\mathbf{F}_k \gg \mathbf{F}_l$) lead to its degeneration. Only absorption or radiation of a phonon can prevent a collision of the nuclei and, due to the formation of accelerated rotation $\ddot\varphi > 0$, draw the nuclei apart, as shown above, to the opposite branches of the acceleration well. This is the electrodynamic essence of phase transition, or, decomposition of the rotos at the characteristic temperature v_f in the low-temperature region of the thermal spectrum.

At inversion point c (position 4 in Figure 1.15), in the transition of the acceleration well $q_2 dc$ to the zone of braking barrier cbq_1 (Figure 1.14) under conditions of $\ddot{\mathbf{D}}_4 = 0$, the negative charge is regrouped and it can leave, for an instant, plane A of rotation of nuclei 1 and 2. In accordance with condition 1.34), the atoms of rotos A (Figure 1.7b) are always located at ends of the line that propagates through the center of inertia 0. At this moment they turn out to be simultaneously at points q_1 and q_2 of turning of the paths, where $\dot{r} = 0$ and thus $\mathbf{F}_l = 0$. The Coulomb repulsive forces $\mathbf{F}_k \gg \mathbf{F}_l$ initiate an immediate decomposition of the bond. This is the most unstable and dangerous state of the rotos, corresponding to the brittle fracture temperature t_c which slightly differs from the Debye temperature (Equation 1.55). The temperature dependence ξ is controlled by the law given in Equation 1.26.I, while its variation is set by curve 2 (Figure 1.13).

The overall energy of the interatomic bond (Equation 1.23) is a trinomial, the first term of which is purely kinetic, determining the radial part of motion of the atoms with radiation or absorption of a phonon; the third term is purely potential. The second term serves as a sort of kinetic-to-potential energy converter, or vice versa. In analogy with Equation 1.25, it is possible to distinguish the kinetic part of the overall energy as

$$t = -u(r_a) + \xi = t(\dot{\mathbf{r}}) + t(\dot{\varphi})\mathbf{r}^2, \tag{1.65}$$

where $t(\dot{\mathbf{r}}) = m\dot{r}^2/2$ is its radial part and $t(\dot{\varphi}) = m\dot{\varphi}^2/2$ is its rotational part. The following is true at the bottom of the potential well (point a in Figure 1.7b): $\mathbf{r} = \mathbf{r}_a$ and $\dot{\mathbf{r}} = 0$. Substitution of Equations 1.37.I and 1.39 shows that the potential part of the equality vanishes and the kinetic part takes the following form

$$0 = \frac{m\dot{\varphi}^2}{2} r_a^2. \tag{1.66}$$

Accounting for Equation 1.39, this yields

$$0 = \frac{\alpha}{2r_a} = \xi_a/2 = \mathbf{E}r_a/2. \tag{1.67}$$

It can be seen from Equations 1.37.I, 1.39, and 1.66 that motion on the circumference is characterized by the equality of the potential and kinetic energies. In this case the overall energy of the rotos becomes equal to zero, i.e.,

$$u = u(r_a) + 0 = -\varepsilon_a/2 + \varepsilon_a/2 = 0.$$

Atoms lose their bond and form a vacancy. At t_c the failure of the interatomic bond happens suddenly, in brittle manner. This is a low-temperature or brittle boundary of existence of the rotos.

At the bottom of the potential well, the kinetic energy contains just the rotational component $t(\dot{\varphi})$. It may be increased only due to the radial part (the first term in Equation 1.65). Comparison of Equations 1.66 and 1.67 shows that the following holds at the bottom of the well:

$$t(\dot{\varphi})_a = \frac{\alpha}{r_a^3}. \tag{1.68}$$

Because parameter α accounts for the configurational properties of the electric field,

$$\frac{\alpha}{r_a^3}$$

can be interpreted as the specific concentration of electric charges in the volume occupied by the rotos.

The region of deceleration, cbq_1 (Figure 1.14), is characterized by the opposite configuration of charges (position 5 in Figure 1.15), which is as stable as in the region of acceleration, q_2dc (Figure 1.14). Lagging behind nuclei 1 and 2 (Figure 1.15, position 5) during rotation, the negative charge creates deceleration due to the Coulomb attractive forces. Transfer of any of the nuclei to the other orbit (larger or smaller) is accompanied by phonon radiation or absorption, the corresponding reaction of the neighboring charge to the restoration of the dynamic equilibrium.

At point b of maximum in Figure 1.14 (coinciding with the inflection point at any shape of the potential well, Figure 10b), as in position 3 of Figure 1.15, all the charges are grouped at point q_1 of turning of the paths (position 6, Figure 1.15), located somewhere between the fourth and fifth electron shells. As such, the negative charge is found between nuclei 1 and 2. The Coulomb attractive forces $\mathbf{F}_k$, acting in a noninertia system of coordinates rotating at a constant angular velocity ($\ddot{\varphi} = 0$) relative to the center of inertia 0, cause a swift rapprochement ($\mathbf{D}_6 \to 0$) of both nuclei toward each other ($\ddot{\mathbf{D}}_6 \gg 0$). This rapprochement is accompanied by a thermal pulse and ends with a collision of nuclei 1 and 2 to form a vacancy and annihilation of the space electron charge. Therefore, it is logical that deformation and fracture of solids[2,3,35] are accompanied by thermal,[36] ultrasonic,[37] and electronic[38] effects.

Finally, in position 7 (Figure 1.15), a cycle of a configurational restructuring of the dipole is completed with return to position 1, but with the opposite arrangement of the charges: now massive nuclei 1 and 2 leave behind the electron charge in their rotational motion. This configuration may persist for an unlimited period of time, providing the thermal energy is continuously fed from the outside.

Thus, in the acceleration well q_2dc and in the deceleration zone cbq_1 (Figure 1.14), the separated charges on different orbits form a stable plane dynamic structure, which can preserve its size and shape due to an intrinsic EM radiation. This is the way the structure of solids forms; it corresponds to positions 2 and 5 in Figure 1.15. In positions 1 and 7 the circular paths are not that stable. They form one-phase systems, which can exist only if the thermal "feeding" is provided from the outside. Position 1 is identified with superbrittle overcooled solids and position 7 with liquids. Critical points d, c, and b in Figure 1.14 correspond to microscopic characteristic temperatures v_f, t_c, and v_s, where decomposition of solid structures or phase transformations may occur.

In the applied sciences,[2,3,35] a frequently debated question is the priority of resistance, deformation, and fracture, and the temperature dependence of these phenomena. Expressions 1.16, 1.24, and 1.52 give a clear and unambiguous answer to this question: geometrical sizes of the EM dipole **D** are the primitive parameters, the first derivative of which determines the local temperature t and the second derivative the resistance force **F**. Therefore, deformation and fracture resistance of solids are initially thermally dependent parameters with the following equivalence relationship between them:

$$D \to m\dot{\mathbf{D}}^2/2 \to m\ddot{\mathbf{D}}. \tag{1.69}$$

This allows Equation 1.58 to be written as

$$\frac{\partial \mathbf{F}}{\partial t} > 0 \quad \text{(I)}, \qquad \frac{\partial \mathbf{F}}{\partial t} < 0 \quad \text{(II)}. \tag{1.70}$$

As seen, atoms located closer to the center of inertia 0 in Figure 1.7b and on the left branches q_2d and cb in the constitutional diagram (Figure 1.14) exhibit resistance on heat absorption, while those located on the periphery exhibit resistance only on heat radiation. To distinguish them from each other, let us call the first *compressons* (C) and the second *dilatons* (D). Taken together, they form the *compresson–dilaton system* (CD), or rotos. These terms were first applied in studies[33,39] to define interatomic fluctuations of compression (compressons) and tension (dilatons). Phase transitions occur at extreme points d and b, changing the dynamics of motion of the bound atoms and their spatial configuration. When there is a shortage of supplied thermal energy, these points (Equation 1.57) are points of discontinuity of the function of state (1.52, 1.53, and 1.54).

The suggested interpretation of the atomic interaction fully accounts for a dual nature of the structurizing particles (simultaneously bearing mass and electric charges) and, as will be shown later, enrich materials science with new ideas, concepts, approaches, and methods. The preceding considerations are based on facts that became indisputable as early as the middle of the 20th century.

1.7   Thermal Radiation, Phase Transitions, and Formation of Vacancies

It is known from electrodynamics[25] that electric charges emit or absorb electromagnetic energy only when they move with positive or negative acceleration. Indeed, everywhere, except for the inflection point, differentiation of Equation 1.26.I yields

$$q = \frac{\alpha}{r^3} - 2t(\dot{\varphi}).$$ (1.71)

where

$$q = \frac{1}{r}\frac{dt}{dr}$$

is the heat flow generated by the rotos of geometrical size r when the specific concentration of the charge α/r^3 changes by $t(\dot{\varphi})$. Accounting for Equation 1.68 yields

$$q = -\frac{\alpha}{r^3}.$$ (1.72)

As seen, the rotos radiates phonons when the volumetric concentration of the electric charges changes. As follows from Equation 1.17, the wavelength of a phonon depends on the absolute sizes of the EM dipole

$$q = -\frac{E}{r}.$$ (1.73)

Under conditions of a dynamic equilibrium, $q = 0$, which is possible at $dt = 0$ or (accounting for Equation 1.24.II) at $D = \text{const}$. Such conditions may be provided only in the acceleration well $q_2 dc$ or in a region of deceleration cbq_1 (positions 2 and 5 in Figures 1.14 and 1.15). Substitution of the definition of parameter t (Equation 1.24.II) into Equation 1.73 yields

$$F_l = m\ddot{r} \approx E.$$ (1.74)

It follows from Equation 1.74 that radiation (or absorption) of phonons (Equation 1.72) is always accompanied by formation of the Lorentz forces F_l, which increase (or decrease) size D of the EM dipole, causing a corresponding displacement of horizontals D_2 and D_5 in Figure 1.14. This is physical essence of the equivalence relationship defined by Equation 1.69 and thermodynamic essence of deformation (whether due to applied force or thermal force).

As follows, in the zone of the minimum of the constitutional diagram q_2dc, the rotos can produce the resistance work only at acceleration of atoms with energy absorption; in the zone of maximum cbq_1, it does so at deceleration to radiate a phonon. Relationship 1.71 is of fundamental significance in physics of strength. It shows that any attempt to change the equilibrium arrangement of the atoms formed on solidification inevitably leads to the formation of a temperature gradient and, hence, to the radiation or absorption of phonon **q**. Therefore, variation in the stressed–deformed state of solids is always accompanied by thermal effects.[10,35,36]

The relationship between radial $\dot{r}$ and rotating $\dot{\varphi}$ velocities of a complex motion (Equation 1.24.I) allows expressing the kinetic energy (Equation 1.65) in terms of either radial

$$u(t) = \frac{m\dot{r}^2}{2} + \frac{2}{m}\left(\frac{m\dot{r}^2}{2}\right)^2 \tag{1.75}$$

or rotational

$$u(t) = \frac{m\dot{\varphi}}{2}\mathbf{r} + \frac{m\dot{\varphi}^2}{2}\mathbf{r}^2 \tag{1.76}$$

components.

Their differentiation by the corresponding variables, accounting for Equation 1.24.II, definition heat capacity given in Reif,[32] and force given by Equation 1.74, yields

$$\mathbf{F} = \frac{du(t)}{d\mathbf{r}} = \frac{m\dot{\varphi}}{2} + m\dot{\varphi}^2\mathbf{r} \tag{1.77}$$

$$C_v = \frac{du(t)}{dt} = 1 + \frac{m}{2}t. \tag{1.78}$$

Figure 1.16 shows temperature dependencies of kinetic energy (a) and heat capacity (b) of the rotos. Numbers 1 and 2 designate the curves for the first and second terms, respectively. In a low temperature field, the rotos appears to be more a rotator than an oscillator. The interacting atoms approach each other at such a short distance that, bound by several molecular orbits (Figure 1.3), they are capable of rotating relative to each other.

According to Equation 1.45, in the acceleration well the eccentricity of their orbits does not exceed one-third. Their nuclei become nondistinguishable by the external electrons and so they are located in azimuthal planes (Figure 1.4) determined by the corresponding quantum numbers, enclosing both atoms on all sides with a multilayer, almost spherical shell. Therefore, the total dependence follows a curvilinear path (solid line) to point b of the intersection

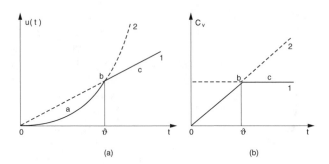

FIGURE 1.16
Phonon dependencies of the kinetic energy (a) and heat capacity (b) of rotos.

of curves 1 and 2. (In Figure 1.16a, the first curve shows direct proportionality and the second is a parabola.)

Passing point *b*, the rotos becomes more an oscillator than a rotator because its angular velocity decreases sharply while the radial velocity increases. This happens because atoms move apart with an increase in temperature, assuming prolate elliptical paths. As such, the rotos becomes less rigid (sequence of positions *a*, *b*, and *c* in Figure 1.3). This change in the kinetics of motion results in a change in the mechanism of formation of function $u(t)$: behind point *b*, it no longer follows direction 2 but follows straight line 1 (solid line). In other words, at point *b* the monotonic character of this function is disturbed and so it transforms into a complex dependence 0*abc*. The change of the curve corresponds to the microscopic level of the θ temperature (Equation 1.56). Similar explanations are also applicable to function $C_v = f(t)$ (Figure 1.16b).

One might expect a similar character of changes in Functions 1.73 and 1.74; however, this is not the case. They would have been such if $\dot{\varphi}$ remains invariable. But, in accordance with Equation 1.24.I, the angular velocity continuously decreases with an increase in radius vector **r** during variation in the shape of the elliptical path. As a result, for example, function $\mathbf{F} = f(\mathbf{r})$ would not follow the curve shown by broken line *abc* (Figure 1.16b) as its formal analogue heat capacity does. Instead, it follows a curve of a complex shape having extrema at points *d* and *b* (Figure 1.14). In this case, the coordinate r_b in the spatial system will correspond to temperature v_s in the energy domain. As will be shown later, these notions are confirmed by known physical theories (Debye, Dulong, Petit, etc.) and by the results of experiments.[6,7,34]

It can be seen from Figures 1.7b and 1.16 that, at point *b* of inflection of the curve of the potential energy and at the v_s temperature, all the energy parameters undergo substantial changes. Restructuring of the interatomic bond occurs at this point. The essence of this process is illustrated in Figure 1.15. Such a restructuring requires a supply of extra energy. As follows from Equation 1.71, to bring atoms to their new orbits and prevent their convergence and collision, it is necessary to give the rotos an extra angular momentum $t(\dot{\varphi})$ at positions 4 and 6. This can be achieved through absorption of a phonon (Equation 1.72) from the outside.

Disturbance of the monotonic character of Functions 1.74 and 1.75 (Figures 1.14 and 1.16) at the Debye temperatures θ, θ_f, and θ_s is in full agreement with the experimental facts. Indeed, a change in temperature of the majority of solids on heating, and their resistance to deformation, are not monotonic processes. In the former, the nonmonotony appears as a number of regions with a dramatic increase or decrease in temperature, whereas in the latter it shows up in the progressive saturation of a volume with dissipated local defects.[10,22,36]

The discussed thermal effects are indicators of phase transformations occurring in the structure of the solids.[30] Here, we will consider the mechanism and consequences of these transformations in detail because of their significant effect on the processes of deformation and fracture. Since they occur very rapidly and, while occurring, either absorb or release heat, the temperature at the indicated points drastically changes. Note that these abnormalities for each material are observed only at a very specific temperature—the Debye temperature. Only a small number of solids are thermally inert; to illustrate, Figure 1.17 shows thermograms of some minerals.[40] This suggests that the Debye temperature should be considered one of the most important structural characteristics of solids. For most of engineering materials this temperature is close to room temperature,[6,20] i.e.,

$$\theta = 290\text{--}300 \text{ K.} \tag{1.79}$$

Expressions 1.53 and 1.71 are different forms of differential of the overall energy of the rotos 1.23. As a result, the dynamic equilibrium of a bound pair of atoms is maintained by the interrelated system of forces: kinetic and potential. This allows arriving at a very important practical conclusion: *a kinetic counterforce can compensate for potential forces induced in the structure of*

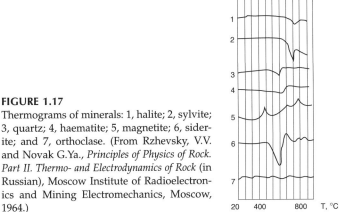

FIGURE 1.17
Thermograms of minerals: 1, halite; 2, sylvite; 3, quartz; 4, haematite; 5, magnetite; 6, siderite; and 7, orthoclase. (From Rzhevsky, V.V. and Novak G.Ya., *Principles of Physics of Rock. Part II. Thermo- and Electrodynamics of Rock* (in Russian), Moscow Institute of Radioelectronics and Mining Electromechanics, Moscow, 1964.)

a solid by external effects through generating the corresponding temperature gradi-ent. In this case the material structure will not fracture. Redistributing the internal energy between the kinetic and potential components preserves the initial order in the AM system.

In quantum mechanics,[27,28] the overall internal energy of the two-atom bond is expressed as

$$u = \frac{\hbar^2 l(l+1)}{2i} + \hbar v\left(n + \frac{1}{2}\right) - \xi, \tag{1.80}$$

where $i = mr^2$ is the moment of inertia of the two-atom bond and v is the classical frequency determined from Relationship 1.50. Unlike Equation 1.30, in addition to orbital quantum number l, this accounts also for principal quantum number n that defines the radial movement of the atoms. This means that the energy state of rotators and oscillators changes discretely. The former and the latter have essentially different intervals of discreteness:

$$q(l) = \frac{\hbar^2}{2i}(l+1) \quad \text{(I)}, \qquad q(n) = \hbar v \quad \text{(II)}. \tag{1.81}$$

For every chemical element, the oscillation energy quantum is larger than the rotation energy quantum, i.e.,

$$q(n) \gg q(l). \tag{1.82}$$

For example, for a hydrogen bond, this relationship is approximately equal to

$$\frac{\hbar v}{\hbar^2 / 2i} \approx 10^2 \div 10^3. \tag{1.83}$$

Because of this difference, the energy spectrum of the two-atom bond consists of a system of oscillatory (different values of number n) and rotational (different values of number l) levels; the distance between them, according to Equation 1.83 and depending on the type of chemical element, differs by two to three orders of magnitude (Figure 1.18). In this figure, bold lines correspond to the oscillation states with quantum numbers $n = 1, 1, 2, 3 \ldots$ and thin lines correspond to the rotation states with quantum numbers $l = 1, 2, 3, 4, 5\ldots$. This spectrum is called the band spectrum and is readily observed using a spectroscope.[20] Lines that fill out the bands between oscillation levels n are caused by a change in the rotational part of the kinetic energy (the second term in Equation 1.23 and the first term in Equation 1.80).

The rotos is a pure rotator near the bottom of the potential well, where the oscillation levels are not yet excited ($n = 0$). Here, in accordance with Equations 1.82 and 1.83, the increment in the kinetic energy can be caused only by the rotational levels (series of numbers l from 1 to 5). The distance

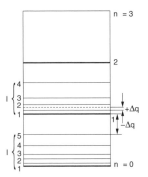

FIGURE 1.18
Energy spectrum of an interatomic bond.

between the energy levels of the rotator increases with number l, according to Equation 1.81.I, whereas that of the oscillator is constant and does not depend on number n (Equation 1.81.II). The next rotational level may be insufficient to excite the first oscillation level of the energy (coinciding with the Debye temperature). If this is the case, the rotos is found in an uncertain position: it is no longer a rotator and not yet an oscillator. The energy needed for transition can only be supplied from the outside.

Equal probability exists for the opposite variant, where the next rotational level (for example, the sixth, designated by a dotted line) turns out to be above the first oscillatory level, $n = 1$. This level is excited and the excess of the kinetic energy, $+\Delta q(l)$, is emitted into the environment. That is why the thermograms (Figure 1.17) show not only negative (e.g., curve 6), but also positive (position 5) spikes.

Thus, the laws of conservation of energy (Equation 1.23) and momentum (Equations 1.27 and 1.31) in the potential field with a central symmetry (Equation 1.26) allow explanation of the mechanics of motion of the interacting atoms so that the principles of stability (q_2dc and cbq_1) and critical points (d, c, b) can be established (Figure 1.14). Therefore, it is clear that the interatomic bond to the left and right from point c of inversion of the state diagram (Figure 1.14) is in fundamentally different energy states.

Phase transitions take place at inflection point b; here they are subjected to maximum forces of interaction, although the direction of these forces is not yet determined. An extra energy impact (thermal, force, or other) is required to move the bound pair of atoms to the q_2dc or cbq_1 zone. For example, to move from the first zone to the second, the rotos must acquire an additional portion of the kinetic energy, $+\Delta q$ (Figure 1.18), to transform from a rotator to an oscillator. However, no energy source (normally, neighboring rotos are such a source) may be available. As a result, the bond is disintegrated to form vacancies. Coming to the critical point b, the dilaton atom supplies the same amount of energy to the rotos. Because it is in the compresson state, it radiates this excess of energy. Thus, inflection point b (Figure 1.7a) is a type of one-sided valve that readily transmits atoms from the external orbit to the internal orbit. The reverse transition is possible only when extra heat energy is supplied from the outside.

The energy radiated or absorbed during phase transformations is in the infrared region of the electromagnetic spectrum, normally in a range of 100 to 4000 cm^{-1}.[28] Measuring the frequency of absorption or radiation EM energy by gas molecules, one can determine the following: natural frequency of the interatomic bond, intervals of oscillation $q(n)$ and rotation $q(l)$ (Equation 1.81) quantization, moment of inertia of the atoms, i, and other parameters needed for estimation of the character and value of the atomic interaction.[20]

1.8 Condition of Stability. Low- and High-Temperature Disintegration

An isolated atom is a stable electrodynamic formation. Motion of its electrons having mass m_e about the nucleus follows a circular orbit of radius r (Figure 1.2) so that the electric attractive force e^2/r^2 is balanced by the centrifugal force mv^2/r,[30] i.e., $e^2/R^2 = mv^2/R$, where v is the tangential velocity. Rewriting this relationship in the form $e^2/2r = mv^2/2$, we can see that the kinetic energy of an electron $t_e = mv^2/2$, taken with a minus sign, is equal to half of its potential energy,

$$\frac{1}{2}u_r = \frac{e^2}{2r}$$

with respect to the nucleus, i.e., $u_r = -2t_e$. Accounting for the last equality, its overall energy u can be represented in the form

$$u = t_e - 2t_e = -t_e. \tag{1.84}$$

Equation 1.84 states that, in the dynamic equilibrium of an atom, the overall energy of an electron is equal in value and opposite in sign to its kinetic energy.

The overall energy of a paired interaction of atoms is determined by Equation 1.23. If we transfer half of the first term from the right to the left: $u - mv^2/4 = mv^2/4 - \xi$, we see that $|u| = |\xi|$. Then, taking into account that both interacting atoms contribute to the kinetic energy of the rotos, we have

$$2|u| = 2\frac{mv^2}{2}$$

or, finally,

$$u = -\frac{mv^2}{2} = -t_r. \tag{1.85}$$

Comparing Equations 1.84 and 1.85, we come to a conclusion that this relationship is of a general nature so it is applicable to any dynamic system of charged particles that interact following the Coulomb law. If it is met, the system is stable; if not, it cannot exist in a bound form.

The only condition of stability of the bond is to maintain its overall energy at a certain level, i.e., u = const or, to be more exact, t = const and u_r = const. If any of these conditions is not met, the energy level shifts (horizontal line q_1 to q_2 in Figure 1.13) in depth of the potential well, changing the shape of the state diagram (positions 4 to 6). During such a shift, the energy state may be anywhere with an equal probability, including in the critical zones: at the bottom of the potential well, near its surface, or at critical points d, b, and c.

The two-atomic chain is stable everywhere, except for these points, where its behavior substantially changes. For example, as follows from comparing Equations 1.39 and 1.66, which determine the kinetic and potential energies of the rotos at the well bottom, the stability condition 1.82 is not met. As a result, at temperature t_c the bond is disintegrated into the constituent atoms.

Atoms perform a circular motion with different angular velocities at points r_1 and r_2. Writing Equation 1.64 for the circular motion in the form of

$$t = \frac{1}{2}mr^2\dot{\varphi}^2$$

yields that

$$t_1 = \frac{1}{2}mr_1^2\dot{\varphi}_1^2 \quad \text{and} \quad t_a = \frac{1}{2}mr_a^2\dot{\varphi}_a^2$$

at these points. Using the law of conservation of momentum (Equation 1.27) and accounting for the fact that $r_1 = 1/2r_a$, we have that

$$r_a^2\dot{\varphi}_a^2 = \frac{1}{4}r_1^2\dot{\varphi}_1^2 .$$

Hence,

$$\dot{\varphi}_1 = 2\dot{\varphi}_a \tag{1.86}$$

As seen, approaching the interacting atoms to a fixed center of inertia (point 0 in Figure 1.7a) causes an increase in the angular velocity of rotation. A similar relationship is also known in astronomy: the closer a planet is to the sun, the faster its rotation.[30] This means that displacement of the state of the rotos to the left from point c under the effect of any random fluctuation of the kinetic or potential parts of the overall energy leads to a change in sign of Lorentz forces from attraction to repulsion.

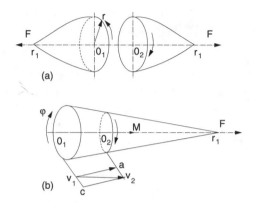

FIGURE 1.19
Diagram of disintegration of rotos on dramatic cooling.

At local temperatures t_c, v_f, and v_s the laws of conservation of energy (Equation 1.23) and momentum (Equation 1.27) are not valid at points d, c, and b. It can be assumed that, while rotating on decreasing circular paths and successively taking positions c_2, 0, c_1, every atom seems to "slide" along a generating line *KAL* of cone (Figure 1.12a). To be exact, this cone should be represented in the energy domain, rather than in the Cartesian space. In this case the generating line of the cone would have a curvilinear shape, corresponding to the left or right branch of curve u_r, while apex K is identified with point r_1 in Figure 1.7a.

Figure 1.19 presents a schematic illustrating the loss of the bond between atoms of the rotos. In transition to the circular paths with centers 0_1 and 0_2, they rotate with angular velocity φ. This leads to sliding of the atoms on the curvilinear paths under the effect of repulsive forces F. Sliding ends come to points r_1 and escape to infinity. Such a motion in the direction of momentum M leads to the formation of the velocity vector (Figure 1.19b). In accordance with Equation 1.86, the linear velocity v_2 in position 0_2 is always higher than that in position 0_1. Considering parallelogram v_1av_2c, we see that the velocity increment $\Delta v = v_2 - v_1$ is equivalent to the formation of two vectors: v_1a and v_1v_2. The first leads to sliding of the circular paths of particles along the generating line of the cone to its apex r_1, and the second to displacement of atoms to infinity in a direction normal to the rotation plane.

This is how the interatomic bond ceases to exist when its energy level enters one of critical states t_c, v_f, or v_s. At t_c and v_f, failure of the bond occurs suddenly and rigidly; this is the low-temperature boundary of existence of a solid. At low negative temperatures the interatomic bonds are susceptible to this kind of failure. Substituting α and r_a from Equations 1.29 and 1.48 into Equation 1.67 makes it possible to derive a formula for the θ temperature as

$$\theta = \frac{32\beta}{\hbar^6} m^3 z^3 e^8. \tag{1.87}$$

In a high-temperature region near the well surface, where $r \to \infty$ and $t \to t_m$, the potential energy $u_r \to 0$. Figure 1.7a shows that the motion is unlimited in this region. Therefore, deformation in this region occurs due to small forces and is accompanied by large changes in relative distances between the atoms. Indeed, if we express Equation 1.23 accounting for Equation 1.28 in the form

$$u = t(\dot{r}) + \frac{A}{r^2} - \frac{\alpha}{r} \tag{1.88}$$

and reduce it to a common denominator, we obtain a quadratic equation relative to r:

$$[u - t(\dot{r})]r^2 + \alpha r - A = 0.$$

Its solution is

$$r = \frac{-\alpha + (\alpha^2 + 4[u - t(\dot{r})]A)^{1/2}}{2[u - t(\dot{r})]}.$$

It can be seen that $t(\dot{r}) \to u$ as the energy level approaches the well surface. Thus, the denominator $[u - t(\dot{r})] \to 0$, and, as a result, $r \to \infty$. In this case, Equation 1.23 transforms into

$$u = \frac{m\dot{r}^2}{2} = t_m,$$

where $t_m = t(\dot{r})$ is the melting temperature. Assuming that $t_m \approx u(r_{a+})$ and accounting for Equation 1.48.II, we can write

$$t_m = \frac{2\beta}{\hbar^2} mz^{7/3} e^4. \tag{1.89}$$

Figure 1.20 shows the variations in a macroscopic value of t_m for most known chemical elements. The plot was developed using data tabulated in Chalmers.[41] In the plot, the numbers show locations of chemical elements in Mendeleev's table. For example, 3 is lithium; 42, molybdenum; 74, tungsten; etc. The melting temperature $T_m \cdot 10^3$ K of simple solids is shown on the axis of ordinates. The axis of abscissas contains relative atomic masses of chemical elements (upper series of numbers). In accordance with the periodic Mendeleev's law, it is broken into periods (designated by Roman numerals) and rows (Arabic numerals). Analysis of Figure 1.20 becomes simpler if one references Mendeleev's periodic table.

Function 1.89 has periodic variation over the entire known interval of the atomic masses. The number and spread of the periods are determined by

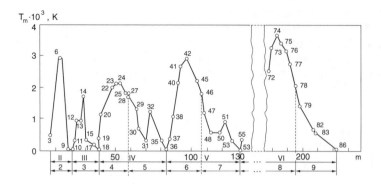

FIGURE 1.20
Dependence of the melting temperatures of chemical elements on their location in the periodic table.

the system of chemical elements. Inert gases—10, neon; 18, argon; 36, krypton; 54, xenon; and 86, radon—are located at the boundary. The span between numbers 55 and 72 is attributable to lanthanide elements located in this interval, which occupy places 58 to 71. Function 1.89 has a maximum inside each period. Small periods II and III comprise one maximum and large periods IV, V, and VI comprise two maxima. The major one corresponds to numbers 6, 14, 24, 42, and 74 of chemical elements and is located in even rows 4, 6, and 8, while the second (elements with numbers 34 and 52) is located in odd rows 5 and 7.

Extending this principle, one should expect formation of an additional maximum in the ninth row (between numbers 83 and 86). Melting temperature T_m is indeed linearly dependent on mass m and performs periodical oscillations inside each period as they increase. The latter can be explained by the almost quadratic dependence of T_m on the valence

$$\left(z^{2\frac{1}{3}} \right).$$

An increase in a common charge (e^4) causes formation of additional spikes in the odd rows.

At $t \to t_m$, the path of motion of the bound atoms can be easily distorted. If such a distortion occurs, it gradually changes its shape from an ellipse to a hyperbola (curve *BAD* in Figure 1.12a). Because this hyperbola is not a closed curve, particles assuming this path can leave the potential field. This is the mechanism of viscous flow and high-temperature disintegration of the interatomic bonds.

It is apparent that, in the system of chemical elements, the θ-temperature also has a periodic variation. At least, no known data contradict this postulate.

To summarize, in a solid every pair of atoms selected arbitrarily can exist in the bound form between the brittle (point 0) and ductile ($r \to \infty$) boundaries.

The character of the bond depends on the kind of path of a localized motion. Transition from low- to high-temperature boundary is accompanied by a change in the path from circumference to ellipse elongated in one direction. It follows from previous considerations that the atoms behave as pure rotators (in accordance with Equation 1.66, $\theta = f(\dot{\varphi})$) only at temperatures close to θ, and at high temperatures they behave as oscillators. (It follows from Equations 1.88 and 1.89 that $t_m = f(\dot{r})$.) Between these extrema, they have properties of both; their reaction to external effects greatly depends on which particular property dominates at a given moment. A logical question then arises: is it possible to determine the boundary that divides the bound atoms into rotators and oscillators?

1.9 Failure at the Debye Temperature

The potential well comprises all possible energy states of the rotos (Figure 1.13). In addition to the boundary θ and t_m, it comprises another special level $u(r_b)$. It passes through inflection point b of the curve of the potential energy u_r (Figure 1.7a) and corresponds to maximum (point b) of the state diagram (Figure 1.14). Equation 1.53 shows that the value and sign of the resistance energy depend on the state between kinetic f and potential ξ parts of the overall internal energy of the rotos. Parameters f and ξ are fully determined by chemical composition of a material (m, z, and e), geometric dimension of the rotos r, and kinetics of motion of the atoms which compose the rotos ($\dot{r}$ and $\dot{\varphi}$). If $f > \xi$, then $\omega > 0$; if $f < \xi$, then $\omega < 0$. Only force F can change the sign in the Fr product because r is always positive. Different inclination and shape of the convexity of the diagram to the left and right from the maximum point b indicate that the substantially different mechanisms of formation of the interatomic forces exist in these regions (Figure 1.15).

Now we are ready to answer a question about whether the state of the bound atoms at inflection point b meets the stability condition 1.85. Rewriting Equation 1.25 in the form of $u = t - \xi$ and substituting Equation 1.44, we see that

$$t - 2u = \frac{\alpha}{3r_a}.$$

Using Equations 1.39, 1.67, and 1.46, we find that

$$u_b = -\frac{3}{2} v_s. \tag{1.90}$$

Comparison of Equations 1.82 and 1.90 shows that the dynamic equilibrium of the bound atoms is violated at point b. Why do atoms lose the bond between them?

To find peculiarities of behavior of the rotos at the maximum of Function 1.53, let us write the total differential of its right- and left-hand sides by preliminarily substituting the expression for momentum (1.27) into the first term

$$\dot{F}r + F\dot{r} = -m\dot{\varphi}\ddot{\varphi}r^2 - m\dot{\varphi}^2 r\dot{r} - \frac{\alpha}{r^2}\dot{r}$$

or

$$Fr^2\left(\frac{\dot{F}}{F} + \frac{\dot{r}}{r}\right) = -\dot{r}\left(m\dot{\varphi}\ddot{\varphi}\frac{r^3}{\dot{r}} + m\dot{\varphi}^2 r^2 + \frac{\alpha}{r}\right).$$

The maximum of Function 1.53 emerges when the left- and right-hand sides become equal to zero, i.e.,

$$Fr^2\left(\frac{\dot{F}}{F} + \frac{\dot{r}}{r}\right) = 0 \ \ (\text{I}), \quad -\dot{r}\left(m\dot{\varphi}\ddot{\varphi}\frac{r^3}{\dot{r}} + m\dot{\varphi}^2 r^2 + \frac{\alpha}{r}\right) \ (\text{II}). \qquad (1.91)$$

By differentiating the law of conservation of momentum (1.27),

$$2mr\dot{r}\dot{\varphi} + mr^2\ddot{\varphi} = 0,$$

we find the condition for its fulfillment,

$$r^2 = 0 \ \ (\text{I}), \quad \frac{\dot{r}}{r} = -\frac{1}{2}\frac{\ddot{\varphi}}{\dot{\varphi}} \ (\text{II}). \qquad (1.92)$$

It can be seen that $M = \text{const}$ only at the moment when a relative increment in the radial velocity, $\dot{r}/r$ (whose quantitative measure is thermal radiation (Equation 1.24.II) is compensated for by a relative decrease in parameters of the rotational motion, $\ddot{\varphi}/\dot{\varphi}$. This means an increase in the Lorentz force (Equation 1.16). An alternative situation is the disintegration of the rotos and formation of a vacancy with area equal to r^2.

Relationship 1.92.II allows one to rewrite Condition 1.88.II in the form

$$\dot{r}\left(m\dot{\varphi}^2 r^2 - \frac{\alpha}{r}\right) = 0. \qquad (1.93)$$

Equalities 1.91.1 and 1.93 allow one to consider two possibilities. The first is that, at $\dot{r} = 0$, Equation 1.91.I transforms into $r^2\dot{F} = 0$. It cannot be that

$\dot{\mathbf{F}} = 0$ because $\mathbf{F} = m\ddot{r} = \text{const}$. The latter is possible only at $\dot{r} \neq 0$, which contradicts the initial premise. Therefore, we should assume $r^2 = 0$, leading to a conclusion that, if the rotos loses its thermal support at the point of maximum of function $\omega = f(r)$ (as $\dot{r} = 0$ and $t = m\dot{r}^2/2 = 0$), its atoms disintegrate to form a defect having a cross-sectional area of r_b^2 in the solid structure. On disintegration of the bond, the atoms rotate at a constant velocity. Indeed, it follows from Equation 1.92.II that

$$0 = -\frac{1}{2}\frac{\ddot{\varphi}}{\dot{\varphi}} \quad \text{at } \dot{r} = 0.$$

The primitive of the last equality has the form:

$$\text{const} = -\frac{1}{2}\ln\dot{\varphi}.$$

Hence, $\dot{\varphi} = \text{const}$.

The second possibility arises because

$$r^2 \neq 0 \quad \text{and} \quad \mathbf{F} \neq 0 \quad \text{at } \dot{r} \neq 0,$$

but

$$\frac{\dot{\mathbf{F}}}{\mathbf{F}} + \frac{\dot{r}}{r} = 0 \quad \text{and} \quad m\dot{\varphi}^2 r^2 - \frac{\alpha}{r} = 0.$$

The first equality yields

$$\frac{\dot{r}}{r} = -\frac{\dot{\mathbf{F}}}{\mathbf{F}}$$

and the second

$$m\dot{\varphi}^2 = \frac{\alpha}{r^3}.$$

Accounting for Equation 1.27 and the fact that $r_b = 1.5 r_a$, we can write

$$\frac{\dot{r}}{r} = -\frac{\dot{\mathbf{F}}}{\mathbf{F}} = -\frac{1}{2}\frac{\ddot{\varphi}}{\dot{\varphi}} \quad \text{(I)}, \qquad v_s(\dot{\varphi})_b = m\dot{\varphi}_b^2 = \frac{8}{27}\frac{\alpha}{r_a^3} \quad \text{(II)}. \tag{1.94}$$

Comparing Equations 1.94 and 1.68, we note that the rotational part of the kinetic energy at the bottom of the potential well is almost 1.7 times higher than that at the inflection point b. This means that, on deformation from r_a to r_b, the change of state of the rotos occurs only due to a change in the specific concentration of electric charges. At point b, however, an additional restructuring of the rotos takes place following diagram 6 (Figure 1.15) that changes the spatial configuration of the local potential field.

It follows from analysis of Relationships 1.63, 1.45, and 1.46 that, in the range of $r_a < r < r_b$ of states, the atoms that compose the rotos move on the elliptical paths whose dimensions do not exceed more than five quantized electron orbits. Therefore, the valent electrons perceive both nuclei of these atoms as a unit. In this case, the states of electrons are described by symmetrical wave functions,[6,7] which uniformly distribute these electrons in azimuthal planes (position a in Figure 1.2). Because they are densely enclosed by symmetrical electron layers, the atoms generate attractive forces that increase with an increase in the relative distance (region cb in the diagram of Figure 1.14). This is the compresson state of an atom.

At point b, the situation radically changes because restructuring of the rotos takes place. When $r > r_b$, the nuclei of both atoms become distinguishable for electrons. For some part of the rotation period they service only one atom and, then, only the other atom. Therefore, we have increasing probability of encountering electrons to locate only at one or the other nucleus at a given moment of time. Grouped near one or the other nucleus, they seem to be captured by them for some period of time. The wave functions of such electrons become asymmetrical (position b in Figure 1.2). The resultant configuration of the rotos produces an extra electrostatic repulsive force directed outward from its volume. It is this force that leads to a change in sign of the force in region bq_1e of the diagram in Figure 1.14. In such a state, the rotoses are susceptible to deformation enhanced by dilatons.

Rewriting Equation 1.25 and accounting for Equations 1.27 and 1.39, we find

$$-\frac{\alpha}{2r_a} - t(\dot{r}) = \frac{m\dot{\varphi}^2}{2}r^2 - \frac{\alpha}{r}.$$

Substituting in Equation 1.94.II and accounting for Equation 1.44 yields

$$t(\dot{r})_b = \frac{1}{6}\frac{\alpha}{r_a}.$$

At the inflection point b of the potential curve, the v temperature of the rotos, which consists of the radial $t(\dot{r})b$ and rotational $t(\dot{\varphi})b$ components is

$$v_s = t(\dot{\varphi})_b + t(\dot{r})_b = \frac{4}{27}\frac{\alpha}{r_a^3} + \frac{1}{6}\frac{\alpha}{r_a}.$$

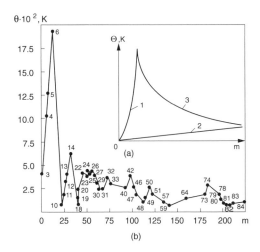

FIGURE 1.21
Theoretical (a) and actual (b) dependence of the Debye temperature Θ on the mass of interacting atoms, *m*, valence *z*, and common charge *e* of chemical elements.

Substitution of Equation 1.48.I makes it possible to obtain

$$v_s = 9.5 \frac{\alpha}{\hbar^6} m^3 z e^6 + \frac{2}{3} \frac{\alpha}{\hbar^2} m z^{1/3} e^6. \tag{1.95}$$

Figure 1.21a shows the dependence of the macroscopic θ_s temperature on the mass of the atoms. Note that the effect exerted by their valence z and charge e are not considered. In elements with comparatively small atomic masses, the dominant contribution to the temperature θ_s is made by the first, i.e., rotational, term (plot of a cubic parabola, Equation 1.87, is designated by number 1). These chemical elements are more rotators than oscillators. In more massive elements, the dominant is the second, i.e., radial, component; its plot is designated by number 2. Therefore, these become more oscillators than rotators. Over a wide range of variation of the atomic mass, function $\theta_s = f(m)$ has the form of curve 3. Comparison of Equations 1.87 and 1.90 shows the difference between the θ (Figure 1.13) and θ_s (Figure 1.14) temperatures and identity between the Modelung constant β and configurational parameter α (1.17).

Figure 1.21b shows the actual three-argument function $\theta_s = f(m, z, e)$ for most chemical elements. It was plotted using tabular data from Pollock[6] and Levich.[20] As before, the axis of abscissas shows relative atomic masses of chemical elements, and the axis of ordinates shows values of their Debye temperatures. By analogy with Figure 1.20, numbers in the plot designate consecutive numbers of elements in the periodic system. As shown, theoretical (Figure 1.21a) and actual data are in good agreement. Periodical oscillations about the mean value of function $\theta_s = f(m)$ are attributable to the effect of valence z and common charge e.

In our opinion, the Debye temperature θ and the Debye temperatures of the first θ_f and second θ_s kinds are so significant in the physics of solids that they should be ranked equal to other characteristic temperatures, i.e., temperatures of brittle fracture T_c and melting point T_m (Figures 1.13 and 1.14). In contrast to T_c and T_m, which change the aggregate state of a solid, the Debye temperatures change its phase or compresson–dilaton state. In the following chapters we will demonstrate that the Debye temperatures play a fundamental role in consideration of the behavior of solids in the external fields and aggressive environments, in general, and the processes of deformation and fracture, in particular. Unfortunately, the strength of materials, materials science, theory of reliability, and other applied sciences do not consider these temperatures significant. We shall show that this is incorrect.

As a part of the equation of state of the rotos (Equation 1.53) and that of a solid as a whole (see Chapter 2), the Debye temperature imparts a periodic character of variations to all other parameters of these aggregates. As early as 1911, Einstein was probably the first to pay attention to the relationship of the Debye temperature with mechanical properties of materials.[6] Later, although Lindeman found this correlation experimentally, no physical mechanism of this correlation was suggested. As a result, it has never before been used in studies on the strength of materials, remaining an uncalled-for factor of solid-state physics. It is only now, after comprehensive analysis of the equation of state of the rotos (Equation 1.53) as an elementary cell of a solid and deriving the equation of state of such a solid (Chapter 2), that its effect on behavior of structural materials in external fields (force, thermal, radiation, electromagnetic, etc.) and aggressive environments has become evident.

We can now formulate major peculiarities of the Debye temperature as follows:

1. From the kinematic point of view, condition

$$\frac{\dot{r}}{r} = -\frac{1}{2}\frac{\ddot{\varphi}}{\dot{\varphi}}$$

   transforms the rotos from a rotator to an oscillator.
2. From the electrodynamic point of view, a change in the concentration of charges in the volume enclosed by the rotos, α/r^3 indicates a change in its electron and ion (Figure 1.3) configurations.
3. In regard to the force and thermal fields, meeting condition $\dot{r}/r = -\dot{F}/F$ causes the interatomic bond to transform from the compresson to dilaton type (Equation 1.70).
4. The transformations can occur only under conditions of thermal interaction, i.e., where $\dot{r} \neq 0$.
5. In the absence of this interaction ($\dot{r} = 0$), the rotos disintegrates to form a lattice defect with an area of r^2.

1.10 Three Mechanisms of Disintegration of the Bonds. "Theoretical" Strength and Phenomenon of Brittle Fracture

Electrodynamic notions of the nature of interatomic bonds allow all phenomena and effects accompanying the process of deformation and fracture of materials to be explained from common positions. Moreover, they make it possible to formulate new approaches to resolving the problem of failure-free operation, reliability, and durability of structures, or activation of fracture when necessary (e.g., in metal cutting). Only these notions can answer basic questions of strength of materials and materials science because they clarify the relationship between strength and deformability, explain the fruitlessness of attempts to achieve theoretical strength, and reveal causes of variations in thermal conditions of materials under loading. They also explain the mechanism of formation of strength in compression and tension, the scale effect, phenomenon of brittle fracture, different behavior of metallic and nonmetallic materials in the force and thermal fields, etc.

As an example, consider the root cause of the phenomenon of brittle fracture, i.e., sudden failure of metal structures under a drastic decrease in the temperature of the environment. The crystalline structure of a solid should never be regarded as a static AM system because it undergoes continuous changes under the effect of external factors (force, thermal, etc.). Depending on the intensity of the external thermal field and conditions of heat transfer, any of the interatomic bonds may turn out to be in several critical situations: at the bottom of the potential well, i.e., where $t = \theta$ (Figure 1.7a), in a high-temperature zone (at least behind inflection point b of the potential curve), i.e., where $t \to t_m$, and at temperatures θ_f, t_c, and θ_s corresponding to the critical points of the state diagram of the rotos (Figure 1.14). The first two states take place at low and high temperatures of the external fields under conditions of free heat transfer, while the last three states take place under isothermal conditions. So, at the AM level, the possibility exists of realization of the θ, t_m, t_c, θ_f, and θ_s mechanisms of disintegration of the bonds. (Their electrodynamic essence is illustrated in corresponding positions in Figure 1.15.) For a wide range of structural materials, the mechanism should be considered characteristic and decisive.

The compresson–dilaton phase transitions occurring at the preceding temperatures can take place only with thermal support from the outside; with no support, the transition will not occur. Instead, the interatomic bond will fail, resulting in the formation of a submicroscopic defect. Because they are merged, such defects form a crack. Frequent and sudden temperature gradients lead to intensification of cracking processes and to a decrease in service life. Materials whose θ_s temperature is close to the service temperature are especially sensitive to phase cracking.

Durability of a structure made of such materials and used at these working temperatures is extremely low. Phase cracking of steels and alloys is similar, in its end result, to a well-known phenomenon of fracture of porous materials (e.g., concrete, ceramic, etc.) in seasonal variations of the ambient temperature about 0°C.[42] However, despite the very stringent requirement for low-temperature resistance imposed on such materials, no determination of the Debye temperature of metals and alloys is carried out. Moreover, this temperature is not even considered in specifying their service temperature. In our opinion, accounting for this factor is a major (though not yet utilized) reserve of increasing reliability and durability of engineering structures.

The situation becomes particularly dangerous if the Debye mechanism of fracture coincides with one of the temperature mechanisms. This takes place with either a drastic decrease or substantial increase in the ambient temperature. Both cases almost always involve catastrophic failures of the structure. Such failure appears as fracture for the first case while an excessive deformation of the structure causes its failure in the second case. At what ambient temperatures may we expect occurrence of the brittle fracture phenomenon?

We can write the law of conservation of energy at a paired interaction (Equation 1.23) for the θ and v_s Debye temperatures accounting for Equation 1.28 as

$$u_a = \theta + \frac{A}{r_a^2} - \frac{\alpha}{r_a}, \tag{1.96}$$

$$u_b = v_s + \frac{A}{r_b^2} - \frac{\alpha}{r_b}. \tag{1.97}$$

Equality 1.44 makes it possible to give Equation 1.97 the form:

$$v_s = u_b - \frac{4}{9}\frac{A}{r_a^2} + \frac{2}{3}\frac{\alpha}{r_a}. \tag{1.98}$$

Representing Equation 1.96 as

$$u_a = \theta + \frac{4}{9}\frac{A}{r_a^2} - \frac{2}{3}\frac{\alpha}{r_a} + \left(\frac{5}{9}\frac{A}{r_a^2} - \frac{1}{3}\frac{\alpha}{r_a}\right)$$

yields

$$\theta + \frac{5}{9}\frac{A}{r_a^2} - \frac{1}{3}\frac{\alpha}{r_a} = u_a - \frac{4}{9}\frac{A}{r_a^2} + \frac{2}{3}\frac{\alpha}{r_a}. \tag{1.99}$$

Comparing Equations 1.98 and 1.99 and taking into account equality $u_b = u_a$, which follows from the law of conservation of the overall internal energy, we have

$$v_s = \theta + \frac{5}{9}\frac{A}{r_a^2} - \frac{1}{3}\frac{\alpha}{r_a}. \tag{1.100}$$

Substituting

$$A = \frac{1}{2}r_a\alpha$$

obtained from Condition 1.42.I yields

$$v_s = \theta - \frac{1}{9}\frac{\alpha}{2r_a}.$$

Rewriting the last expression and accounting for Equations 1.39 and 1.67, we have

$$v_s = \theta + \frac{1}{9}\theta = 1.11\theta.$$

Hence,

$$\theta = 0.9v_s. \tag{1.101}$$

Therefore, brittle fracture begins developing where the ambient temperature becomes lower than θ_s by 10%. Accounting for Estimate 1.79 and the fact that $\theta \approx T_c$ (curve 2 in Figure 1.13), it can be considered that

$$T_c \approx 260 - 270 \text{ K}. \tag{1.102}$$

Temperatures θ_s and θ are determined by chemical composition of solids (Equations 1.87 and 1.95 and Figure 1.21). After solidification, they do not remain unchanged because the shape of the potential wells (Figure 1.10) and the state diagrams of the rotoses (Figures 1.13 and 1.14) depend on the temperature conditions in which solids are found. They always tend to preserve the energy balance between the internal and external thermal fields by absorbing or (most often) radiating phonons (Equations 1.71 to 1.73). Escape of phonons into the environment leads to a change in radial $\dot{r}$ and angular $\dot{\varphi}$ parameters of motion (Equation 1.65) of the atoms; therefore, it is accompanied by the transformation of the shape of the potential wells in direction 3, 2, 1 (Figure 1.10) and that of the state diagram in direction 6, 5, 4 (Figure 1.13). If the range between the θ and θ_s critical temperatures decreases, it inevitably provokes the simultaneous occurrence of the t_c and θ_s mechanisms of disintegration of the interatomic bonds (positions 4 and 6 in Figure 1.15) and substantial decrease in durability of metal structures.

The variation of durability is observed in practice. Figure 1.22 shows dependence of the rate of failures of steel structures on time $\tau(a)$ and temperature T_c (b) of brittle fracture.[43] As follows from Figures 1.13 and 1.14,

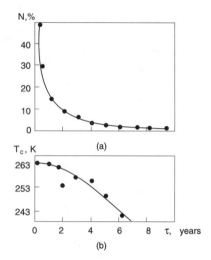

FIGURE 1.22
Dependence of the failure rate of steel structures on time τ (a) and temperature T_c (b) of service.
(From Silvestrov, A.V. and Shagimordinov, R.M., *Problemy Prochnosti* [*Problems of Strength*], 5,
88, 1972.)

it is the statistical mean of t_c of the phonon radiation; under nonisothermal
conditions, it varies following curve 2, with the beginning corresponding to
the θ temperature. The results of analysis of causes and conditions of fractures
of tanks, crane girders, parts of bridges, process columns, trestles, frames,
and coatings of industrial buildings (totaling more than 100 structures; most
of low-alloy steels) that took place from 1886 to 1971 in five countries were
used to construct this figure. The ratio of the number of failures over the
given period to their total number is given in $n\%$.

We came to the following conclusions:

- Fractures occurred, as a rule, under operational loads not in excess
  of the rated ones.
- Nominal stresses at the moment of fracture were lower than the
  yield stress.
- The overwhelming part of failures occurred during the first years
  of service (up to 80% of all failures occurred during the first 5 years).
- More than half of all fractures occurred at the first dramatic
  decrease in the service temperature.

With time, the rate of failures of statically loaded structures was markedly
decreased as well as the temperature at which they took place.

The mean temperature of the environment for all (more than 100) cases was
245 K, and the manner of variations in unit values of T_c (Figure 1.22b) follows
its microscopic analogue (curve 2 in Figure 1.13). Taking into account that

estimation of the θ temperature (Equation 1.79) was carried out using the results of testing of simple solids and the brittle fracture temperature T_c refers to engineering materials having multiple defects caused by their manufacturing, this agreement between the actual and calculated data is very fair. Even more important, it confirms the reliability of the notions developed in this book.

Explanation of the physical nature of the "cold shortness" phenomenon makes it possible to offer calculation, technological, structural, and combined methods for prevention of brittle fracture. Calculation methods provide for a necessary assessment of the θ_s temperature for all structural materials intended for use in regions with dramatic gradients of negative temperatures. Technological methods include manufacturing of materials with a developed dilaton phase; such structures have little sensitivity to dramatic change of temperatures. Structural methods include all arrangements intended for decrease in internal stresses, selection of materials with a maximum possible difference between the operational and Debye temperatures, freedom of deformation, minimization of areas of outside surfaces and their thermal insulation, periodical preventive (maintenance) heating (if possible), restoration of an overcooled structure applying the electric current of different frequency, ultrasonic oscillations, etc.

Combined methods are most effective. At the same time, they are more complicated and expensive: they require equipping critical structures with continuous-action diagnostic systems combined with the active methods of restoration of a damaged structure.[18,21]

1.11 Deformation, Coriolis Forces, and Inertial Effects

Atoms of a solid are always in thermal motion (Figure 1.6). When describing processes occurring with or inside solids, no consideration is normally given to the proper choice of a reference system, which often leads to wrong conclusions or incorrect estimates. The reference system associated with a free solid is inertial. A free solid is subjected to no external effects and is at rest or moves at a velocity having constant magnitude and direction. In the inertial system, the laws of motion of atoms have the simplest form (Equations 1.23 and 1.26). If a solid as a unit or its parts change their position in external fields (force, thermal, electric, radiation, etc.), this reference system becomes noninertial. In such systems it is necessary to take into account the inertia forces, in addition to others.

Until now consideration has been given to motion of atoms in a free solid. If this solid, however, is found in an external, e.g., force, field, such consideration is no longer adequate; its reaction to this field is deformation. Figure 1.23 shows typical cases of deformation: uniaxial tension a, pure bending b, and torsion c. Application of tensile load $\mathbf{F}$ causes atoms in position L_0 to move to position L. On loading, they pass a distance of $dx = L - L_0$ at a velocity

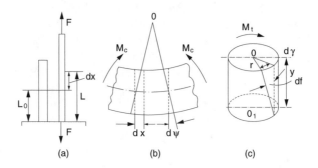

FIGURE 1.23
Different types of deformation.

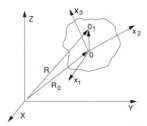

FIGURE 1.24
Diagram of motion of an elementary part of a
body during deformation.

of $\dot{x}$. Bending due to moment M_c may cause linear motion of atoms, dx, and their angular displacement $d\psi$. As such, they acquire linear $\dot{x}$ and angular $\dot{\psi}$ velocities. Torsion due to torque M_t causes angular displacement of atoms to move by $d\gamma$ angle. When $d\gamma$ is small, the displacement is equal to $rd\gamma = df$ and depends on the distance between planes 0 and 0_1. The motion velocity of the atoms in torsion is determined as

$$\dot{f} = r\frac{d\gamma}{d\tau} = \frac{df}{d\tau}.$$

As seen, deformation may involve linear, angular, and combined motion of the bound atoms.

Consider the most common case of deformation. In doing this, we conditionally disintegrate a solid (which is in the external force field of an arbitrary form) into a large number of small, but still macroscopic, parts. To describe the motion of the ith part of the solid, introduce two coordinate systems. One system is a fixed or inertial system, X, Y, Z, in which the solid and an observer are located, and the other system is moving system, x_1, x_2, and x_3. The latter is rigidly connected with the ith part of the solid and participates in all its movements during deformation. The origin of the mobile system, 0, is placed in the center of inertia of the ith part (Figure 1.24).

The position of the ith part of the deformed solid with respect to the observer (fixed system of coordinates) can be determined by assigning coordinates to the moving system. Let radius vector R_0 indicate position of the inertial center of the *i*th part of the solid prior deformation. Orientation of the *i*th part of the solid relative to the fixed coordinate system is determined by three independent angles which, together with three components of vector R_0, define six coordinates. Therefore, any small part of the deformed solid has six degrees of freedom.[25]

Consider an arbitrary, infinitely small displacement of the *i*th part of the deformed solid. It can be represented by the sum of two displacements. First is a translation of the inertial center 0 of the *i*th part from the initial to final position at a permanent orientation of axes of the mobile coordinate system. The second is rotation about the inertia center, as a result of which the *i*th part of the solid occupies its final state.

Designate the radius vector of an arbitrary atom of the *i*th part of the solid in the moving coordinate system as r, and the radius vector of the same particle in the fixed coordinate system as R. Then the infinitely small movement dR of particle 0_1 will consist of two parts: dR_0, together with the inertia center 0, and $[rd\psi]$ relative to the latter in rotation by the infinitely small angle $d\psi$; thus $dR = dR_0 + [rd\psi]$. Dividing this equality by time $d\tau$ during which the displacement takes place, and designating corresponding velocities as $\dot{R}$, $\dot{R}_0$, and $\dot{\psi}$, we obtain the following relationship between them:

$$\dot{R} = \dot{R}_0 + [r\dot{\psi}], \qquad (1.103)$$

where $\dot{R}_0$ is the velocity of the inertia center of the *i*th part of the deformed solid, $\dot{\psi}$ is the angular velocity of this part relative to the inertia center, and $\dot{R}$ is its velocity relative to the fixed coordinate system. As seen, for any kind of deformation, the displacement of each particle can be expressed in terms of translation $\dot{R}_0$ and rotation $\dot{\psi}$ components.

The important conclusion for the strength of materials follows from Equation 1.103: on deformation, the motion of the bound atoms of a solid occurs in a noninertial reference system. A distinctive feature of such systems is formation of the forces of inertia. If they are ignored, analysis of the AM system and the stressed–deformed state of a solid will be incomplete in general and not always correct.

It should be noted that, at a macrolevel, the movement of atoms during deformation is complex: bound into rotoses and rotating on the elliptical paths (Figure 1.7b), they simultaneously change their location in the bulk of the solid at velocity $\dot{R}$. As a result, the equation of motion of particles in an arbitrary noninertial reference system and expression for the energy have the following form[25] (instead of that defined by Equations 1.16 and 1.23):

$$m\ddot{r} = -\frac{\partial u_r}{\partial r} - m\ddot{r} + m[r\ddot{\varphi}] + 2m[\dot{r}\dot{\psi}] + m[\dot{\psi}[r\dot{\psi}]] \qquad (1.104)$$

$$u = \frac{m}{2}(\dot{r}^2 + r^2\dot{\psi}^2) - m\ddot{r}r - \xi \qquad (1.105)$$

As seen from Equation 1.104, the inertia forces caused by rotation of a local reference system consist of three parts. Component $m[r\ddot{\psi}]$ is due to nonuniformity of rotation (variable angular velocity), while two others (fourth and fifth terms) are due to rotation itself. Term $2m[\dot{r}\dot{\psi}]$ is known as the Coriolis force. Unlike all other forces considered earlier, this force depends on the velocity of a relative motion of particles, $\dot{r}$. Force $m[\dot{\psi}[\dot{r}\dot{\psi}]]$ is centrifugal. Note that the expression of energy (Equation 1.105) does not include terms due to nonuniformity of rotation, $m[r\ddot{\psi}]$, and the Coriolis force, $2m[\dot{r}\dot{\psi}]$. Although a number of explanations can be provided, the absence of these terms reduces the significance of Equations 1.105 and 1.23, which in turn limits understanding of some effects associated with deformation and fracture and sometimes leads to wrong considerations and conclusions.

Term $m[r\ddot{\psi}]$ characterizes acceleration and deceleration of particles on elliptical paths, which is accompanied by radiation or absorption of heat. Although this term is a part of Equation 1.105, it affects the energy in an indirect, rather than explicit, manner, i.e., through the velocity of variation in radius vector $\dot{r}$. In the case of variations in the stressed–deformed state, it is responsible for thermal effects. It is shown in Landau and Lifshits[25] that a relative movement of infinitely small parts of a solid on deformation applied at a velocity of $\dot{R}$ (Equation 1.103) causes an immediate change in the internal energy in accordance with relationship $1/t = v/R$, where $v = \dot{R} + [\dot{r}_i\dot{\psi}]$ is the average relative velocity of atoms which compose the ith part of the solid.

The Coriolis force $2m[\dot{r}\dot{\psi}]$ does not depend on the position of particles relative to the reference system. Therefore, in any local zone of the ith part of the solid, it affects the particles with a similar result. The direction of this force is normal to the axis of rotation and the direction of velocity $\dot{R}$, while its magnitude is equal to $F = 2m\dot{R}\dot{\psi}\sin\beta$, where β is the angle between $\dot{R}$ and $\dot{\psi}$. At the same time, it is also normal to radius vector r. Therefore, this force does not produce work, but deviates the direction of movement of particles without changing their velocities.

Because the acceleration of translation motion of the moving reference system, $\ddot{r}$ (in deformation it moves together with the ith part of the solid), is included in Equations 1.104 and 1.105, it may be thought of as equivalent to the presence an internal homogeneous force field. As such, the force acting in this field is equal to the product of mass of particles and acceleration $\ddot{r}$. Naturally, the direction of this force is opposite to that of $\ddot{r}$. In fact, this force "deforms" the intrinsic potential field (Figure 1.8) formed on solidification in a way that enhances the counteraction to that external factor which caused this acceleration, simultaneously stimulating orientation processes in the AM structure. The presence of this term in Equation 1.104 explains the known phenomenon of increase in the strength of materials under their dynamic loading, which is well known in strength of materials.[3,13]

A vast distance between micro- and macroworlds exists. A common opinion is that they are spaced at eight orders of magnitude. For this reason, displacements measured in experiments in millimeters or fractions of a millimeter

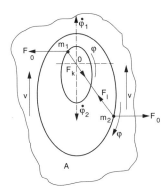

FIGURE 1.25
Diagram of formation of the Coriolis shear stresses.

look monstrous at the AM level. These displacements cause formation of powerful inertia forces in the moving AM system, and these forces have a fundamental effect on the deformation and fracture processes. The terms considered above are usually not the focus of attention of researchers because they are not included in Equations 1.104 and 1.105 in explicit form. However, it is precisely these terms that determine dislocational (term $2m[\dot{r}\dot{\psi}]$), dynamic $(-m\ddot{r})$, and thermal $(m[r\ddot{\psi}])$ effects.

It is known from experiments[44-47] that multiplication and migration of dislocations occur under conditions of plastic deformations, i.e., when the orientation processes are completed. Let the elliptical paths of motion of the bound pair of atoms m_1 and m_2 (Figure 1.25), which form the compresson bond, be oriented (e.g., in plane A) in the direction of deformation (designated by vector v). The compresson pair has an opposite direction of spin rotation of atoms m_1 and m_2 which compose this pair (indicated by arrows φ and located in the deceleration zone cbq_1, Figure 1.14).

Position of the stable dynamic equilibrium (position 5 in Figure 1.15) persists until the moment of formation of relative velocity v, when the plane moves in space during deformation. It causes formation of the Coriolis force $F_c = 2m[v\dot{\psi}]$ (designated by vector F_o in Figure 1.25), whose direction is determined by the left-hand rule. Force F_c violates dynamic equilibrium between Lorentz F_l and Coulomb F_k forces, causing rotoses to transform into the critical state (position 6 in Figure 1.15) and inducing dislocations and microdefects.

Because the Coriolis forces are normal to the deformation direction v, dislocations move, as a rule, in sections arranged normal to the external load, and fracture in tension often ends with shearing, rather than with tear.[44] Dislocations can be formed only in the deceleration zone cbq_1 (Figure 1.14). They are impossible in the acceleration well q_2dc, as here the nuclei of the bound atoms have the same direction of the spin rotation. Dislocation activity reaches its maximum at the yield point or near it, i.e., where the deformation velocity v reaches maximum.

The crystalline structure of a solid is not indifferent to the velocity of application of a mechanical load.[43-46] Increase in this velocity is equivalent to

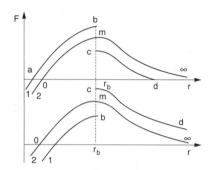

FIGURE 1.26
Formation of the "hardening–softening" effect under dynamic loading.

the formation of an extra internal potential field $(-m\ddot{r}r)$, whose value is determined by acceleration of deformation, $\ddot{r}$. Because it is opposite to the external force field, it enhances or weakens the effect of resistance of the solid. In this case the Le Chatelier-Braun principle[31] works for the potential processes.

Figure 1.26 shows the variations of the shape of the braking barrier in high-velocity tension (upper field) and compression (lower field). Number 2 designates function $\mathbf{F} = f(r_j)$ corresponding to stationary conditions; in Figure 1.14 it is designated cbq_1. Tension with acceleration leads to an equidistant displacement $m\ddot{r}r$ of the compresson branch *om* of curve 2 to position *ab*. As a result, the energy content of the compresson part of the solid volume increases by a value determined by the area of curvilinear figure *oabm*. This leads to an increase in the tensile strength. According to the energy conservation law, an increase in the compresson energy takes place at the expense of energy from the dilaton phase. Cooling of the latter leads to displacement of the dilaton branch of curve 2 from position $m\infty$ to position *cd*. As such, the energy conservation law requires that figure *oabm* be equal to figure ∞mcd. As a result, the smooth curve with maximum 2 is transformed into line *abcd* with a jump at maximum point *m*.

As seen, tension under dynamic conditions leads to redistribution of the internal energy between the dilaton and compresson phases: compresson saturation causes an increase in tensile strength and dilaton tension causes cooling, decrease in deformability, and embrittlement. This follows from the constancy of the resistance energy (Equation 1.53) $\omega = \mathbf{F}r = $ const. Potential jump *bmc* is responsible for another effect taking place under dynamic loading, namely, the interatomic bonds present here fail immediately, setting the direction of fracture. For this reason, the area of fractured surface under dynamic loading is always smaller than that in quasi-static loading.[3,24,45,48] Skipping similar considerations, we note that, in the case of a high-velocity compression, the processes occur in a reverse manner leading to embrittlement and decrease in strength.

1.12 Transition from Atomic Mechanics to Thermodynamics

The bound pair of atoms, which does not interact with other particles of the AM ensemble, is a two-atom gas molecule. As shown in quantum mechanics[27] and statistical physics,[11,23,32] the overall energy of such a system, u, consists of the translation of a molecule as a unit, u_1, rotational $u(\varphi)$, and oscillatory $u(r)$ motions of the constituent atoms. In addition, the collectivized electrons also contribute to this sum, $u(e)$. Therefore,

$$u = u_1 + u(r) + u(\varphi) + u(e). \tag{1.106}$$

There is no way for a free translation of the bound atoms to occur in a solid, and thus $u_1 = 0$. A certain temperature termed critical, starting from which the relative contribution of a particular term becomes significant, determines contributions of other terms. Critical electron temperatures are of an order of tens of thousands of degrees. Thus, within a range of the temperatures encountered in practice, $u(e) = 0$, so the fourth term in Equation 1.106 can also be ignored. As a result, Equation 1.106 becomes $u = u(r) + u(\varphi)$, which is absolutely identical to Equation 1.65. This means that behavior of the bound atoms is really described by Equation 1.23 and that the laws of motion in the central potential field (Equation 1.26) constitute the basis of atomic mechanics of a solid.

Writing Equation 1.23, accounting for definition 1.65, and defining the force as a differential of the potential energy by radius vector, $\mathbf{F} = du_r / d\mathbf{r}$ (Equation 1.74), we can write

$$\mathbf{F} = \frac{d(\mathbf{q} - u)}{d\mathbf{r}}. \tag{1.107}$$

By definition, the kinetic energy is always positive and so if $\mathbf{q} > |u|$, we have $\mathbf{F} > 0$. If $\mathbf{q} < |u|$, then $\mathbf{F} < 0$. Forces directed inward induce compressons while those directed outward induce dilatons. Considering difference $(\mathbf{q} - u)$ to be a generalized temperature and taking into consideration Equation 1.70, we can conclude that an increment in the compression force $\mathbf{F}_c$ occurs due to absorption of heat, i.e.,

$$\mathbf{F}_c = -\lambda \frac{d(\mathbf{q} - u)}{d\mathbf{r}},$$

whereas that of the dilaton force $\mathbf{F}_d$ is caused by radiation of heat, i.e.,

$$\mathbf{F}_d = \lambda \frac{d(\mathbf{q} - u)}{d\mathbf{r}},$$

where λ is the proportionality factor having physical meaning of thermal conductivity.

The differential form of the law of conservation of energy (Equation 1.107) enables transition from atomic mechanics to thermodynamics. Indeed, writing Equation 1.107 in the form of

$$du = d\mathbf{q} - \mathbf{F}d\mathbf{r} \qquad (1.108)$$

and replacing the infinitely small variation of radius vector $d\mathbf{r}$ by an increment of an elementary volume, dv, we obtain the first law of thermodynamics for the rotos:

$$du = d\mathbf{q} - \mathbf{p}dv, \qquad (1.109)$$

where $\mathbf{p}$ is the tensor of deformation resistance dv provided by the forces of atomic interaction $\mathbf{F}$.

It follows from Equation 1.109 that

$$\frac{du}{dv} = \frac{d\mathbf{q}}{dv} - \mathbf{p}.$$

As, by definition,

$$\mathbf{p} = \frac{d\mathbf{F}}{ds} \qquad (1.110)$$

where ds is the surface that encloses volume dv; then,

$$\frac{du}{dv} = \frac{d\mathbf{q}}{dv} - \frac{\mathbf{F}}{ds}.$$

Substituting $dv = d\mathbf{r}ds$, write

$$\frac{du}{dsd\mathbf{r}} = \frac{d\mathbf{q}}{dsd\mathbf{r}} - \frac{\mathbf{F}}{ds}$$

or

$$\frac{1}{ds}\left(\frac{du}{d\mathbf{r}} - \frac{d\mathbf{q}}{d\mathbf{r}} + \mathbf{F}\right) = 0.$$

Hence, we should assume that

$$\frac{1}{ds} = 0$$

or

$$\frac{du}{dr} - \frac{dq}{dr} + F = 0.$$

If $1/ds = 0$, then $(du/dr) - (dq/dr) + F \neq 0$.

It can be seen that meeting the first condition is equivalent to $ds \to \infty$ and thus the surface enclosing the material may change in an arbitrary manner. Writing the second condition in the form of

$$\frac{d(q - u)}{dr} = -F + \Delta F,$$

where ΔF is the noncompensated electrodynamic force generated by regrouping of charges (Figure 1.15), explaining free changing in shape of liquids. If $1/ds \neq 0$, then $(du/dr) - (dq/dr) + F = 0$. These relationships are valid for solids.

Let

$$\frac{1}{ds} = \text{const};$$

then $ds = \text{const}$. Providing that the first law of thermodynamics (Equation 1.109) is met, this signifies that an increment of the surface of a solid occurs in a discrete manner. In accordance with Equation 1.89.I, an interval of discreteness is equal to the area of a disintegrating rotos, i.e., $ds = r^2$. This is confirmed by investigations of the processes of deformation and fracture of solids.[10,33,36,39]

There is a vast variety of methods for packing flat EM dipoles (Figure 1.7b) to form a macroscopic body. Macroscopic bodies have mechanical, thermal, or other properties, depending on the shape and size of the dipoles and methods of their packing. Much evidence of this may be found in periodic,[1,5,33] special,[2,6,7,11,30,34] reference,[3,19] and encyclopedic[22] literature.

The short-range character of atomic interaction,[6,20] paired arrangement of motion of atoms in one plane A (Figure 1.25), and vectorial nature of the bond forces F allow a solid to be represented in the form of an infinitely small closed AM chain (Figure 1.27), which consists of N links (Figure 1.7b)

FIGURE 1.27
Infinite atom–molecule chain.

arranged very chaotically in volume V. Multiplying Equation 1.53 by N yields this widely known relationship in thermodynamics:[11]

$$\Omega = F - G, \qquad\qquad (1.111)$$

where $\Omega = -PV$ is the potential introduced into thermodynamics by Landau,[23] $G = \xi N$ is the Gibbs potential, and $F = fN$ is the free energy. In fact,

$$\Omega = \omega N = \beta \mathbf{D}^3 N \frac{\mathbf{F}}{\mathbf{D}^2} = PV,$$

where $\mathbf{P} = \mathbf{F}/\mathbf{D}^2$ is the internal pressure generated by each dipole from an orbital ring, $\mathbf{D}^2$ (Figure 1.7b). It keeps volume $V = \beta \mathbf{D}^3 N$ in dynamic equilibrium in structures whose packing density is characterized by Modelung constant β (Equation 1.29). As Equation 1.111 extends the equation of state of a unit rotos (Equation 1.53) to macroscopic volume V, it should be regarded *as the equation of state of a solid in thermodynamic potentials.*

That is, a solid can exist stably if it consists of interrelated rotoses randomly arranged in space. As such, the internal stresses of an opposite sign are compensated, maintaining sizes and shape of the solid unchanged for an unlimited period of time. Deformation is a result of violation of this equilibrium caused by external effects. It is only for this reason that the solids, as a mixture of compresson–dilaton pairs uniformly distributed in a volume (Figure 1.7b), have an equally good resistance to tensile and compressive loads.

It can be seen from Equation 1.107 that any attempt to increase $\mathbf{r}$ due to external forces is resisted only by the compresson components of the pair, and an attempt to decrease $\mathbf{r}$ is resisted by the dilaton components of the pair. Therefore, the nature of a solid is such that it involves the compresson–dilaton mechanism of resistance to any type of external loads. Compression is received by dilatons and tension by compressons. To involve compressons in the first case and dilatons in the second case, it is necessary to have appropriate thermal support. Existence of compressons and dilatons is not just an assumption; it is a fact following from fundamental physical laws, such as the law of conservation of energy (Equation 1.23) and the first law of thermodynamics (Equation 1.109). As will be shown in the next chapters, it is proved everywhere in practice.

2

Equation of State of a Solid and Its Manifestations at the Macroscopic Level

2.1 Introduction

This chapter describes the solution of the thermodynamic equation of state of solids, an equation that allows one to solve key problems in physics and engineering and to explain multiple and sometimes contradictory experimental facts from a common point of view that is well based physically. Using this equation, behavior of materials in various external fields (force, thermal, radiation, etc.) and aggressive environments can be predicted and efficient methods developed to control deformation and fracture. It is hoped that this equation will serve as an efficient vehicle for the advancement of knowledge in a number of fields and result in great practical applications.

2.2 Basic Thermodynamic Potentials

In Chapter 1 we considered properties of a paired interatomic bond. There is a countless set N of such bonds in volume V of a solid. Behavior of set N, whose order is estimated at 10^{25}, is described by the laws of statistical physics[23,32,49,50] and thermodynamics.[11,51,52] Transition from atomic mechanics (Chapter 1) to thermodynamics is carried out using Equations 1.09 and 1.112. Analysis of these equations shows no other way to understand the essence of processes occurring in the structure of actual solids but one that accounts for the basic definitions and relationships of thermodynamics. Let us consider some equations relevant to the present consideration.

The first law of thermodynamics (Equation 1.109) written for set N is

$$dU = dQ + d\Omega_p,$$

(2.1)

where

$$U = \sum_{i=1}^{N} u_i, \quad Q = \sum_{i=1}^{N} q_i$$

is the heat absorbed ($dQ > 0$) or released ($dQ < 0$) by the solid, $d\Omega_p = \mathbf{P}dV$ is the mechanical work done by the external forces (at $\mathbf{P} = const$) over the solid ($d\Omega > 0$) or is the mechanical work done by the solid over the external objects ($d\Omega < 0$). It follows from Equation 2.1 that a solid having the internal energy U, consisting of the potential Π and kinetic Q parts, can interact with the environment only through these parts of the internal energy. Their deviations to either side from the equilibrium values lead to a change in the energy state of the solid. When such a deviation occurs, terms of set N are transferred from one quantum state to another, following the diagram shown in Figure 1.15. These transitions, occurring close to the Debye temperature, underlie the compresson–dilaton transformations. Independently of the physical nature of a solid and its the initial state, it always interacts, even very slightly, with its environment through thermal radiation.

The probability δ that the interatomic bond will be in a certain energy state is determined by the Maxwell-Boltzmann (MB) distribution[49] as

$$d\delta = \Gamma \exp\left(-\frac{U}{kT}\right) d\mathbf{P} dr \tag{2.2}$$

where

$$T = \sum_{i=1}^{N} t_i / N$$

is absolute temperature, Γ is the parameter determined from the normality condition and accounts for the physical nature of a phenomenon, $d\mathbf{P}$ is the variation of tensor $\mathbf{P}$ in the r direction, and k is the Boltzmann constant. In analogy with Equation 2.1 and according to Equation 1.23, the overall energy can be represented as the sum of the kinetic Q and potential Π parts, which depend by definition on pressure $\mathbf{P}$ and relative distance $\mathbf{D}$ between the interacting atoms (Figure 1.7b). Substitution of Equation 1.23 into Equation 2.2, accounting for Equations 1.25 and 1.65, splits the Maxwell-Boltzmann distribution into two cofactors, kinetic and potential:

$$d\delta = \Gamma \exp\left(-\frac{\Pi}{kT}\right) dr \exp\left(-\frac{Q}{2mkT}\right) d\mathbf{P}. \tag{2.3}$$

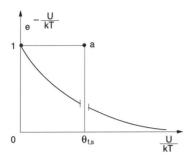

FIGURE 2.1
Boltzmann energy distribution of particles in a solid body.

Distribution by pulses (the first exponential cofactor after parameter Γ) can be considered independent of the character of interaction, as the energy of any interaction can always be introduced into the second cofactor. Because of this, this cofactor coincides with the Maxwell distribution for the noninteracting system of particles (an ideal gas)[49]

$$a\delta(\mathbf{P}) = \Gamma \exp\left(-\frac{Q}{2mkT}\right) d\mathbf{P} \tag{2.4}$$

This distribution is applicable for any AM system: gases, liquids, and solids.

The exponential function, which depends only on the relative location of particles in the potential field, is the Boltzmann distribution:

$$a\delta(r) = \exp\left(-\frac{\Pi}{kT}\right) dr \tag{2.5}$$

The plot of this function is shown in Figure 2.1. When the energy level (Figure 1.7a) tends to the bottom of the potential well, then $r \to r_a$, $T \to \theta$, and $\Pi \to U(r_a)$. In this case,

$$\frac{U(r_a)}{k\theta} \to \infty$$

and

$$\exp\left(-\frac{U(r_a)}{k\theta}\right) \to 0 \text{ (the extreme right end of the plot).}$$

When this level approaches the well surface,

$$r \to \infty \quad \text{and} \quad T \to T_m.$$

Accordingly,

$$\frac{\Pi_m}{kT_m} \to 0 \quad \text{and} \quad \exp\left(-\frac{\Pi_m}{kT_m}\right) \to 1.$$

The Debye temperature of the second kind, where

$$T = \theta_s \quad \text{and} \quad r_b = \frac{3}{2}r_a,$$

is characterized by the following processes:

$$\frac{\Pi_b}{k\theta_s} \to 0 \quad \text{and} \quad \exp\left(-\frac{\Pi_b}{k\theta_s}\right) \to 1. \tag{2.6}$$

Here the exponential function experiences a discontinuity (point *a* falls out of the plot), which means degeneration of the energy level. Therefore, in solids, within limits $(r_a - r_\infty)$, all energy states except one are available for the paired interaction. In this connection, the Boltzmann distribution for such solids should have the following form:

$$d\delta = \left[\exp\left(-\frac{\Pi}{kT}\right) - \exp\left(-\frac{\Pi_b}{k\theta_s}\right)\right] dr$$

or, accounting for Equation 2.6,

$$d\delta = \left[\exp\left(-\frac{\Pi}{kT}\right) - 1\right] dr. \tag{2.7}$$

Rewriting Equation 2.3 and accounting for Equations 2.4 and 2.7, we obtain

$$d\delta = \Gamma \exp\left(-\frac{Q}{2mkT}\right)\left[\exp\left(-\frac{\Pi}{kT}\right) - 1\right] d\mathbf{P} dr \tag{2.8}$$

Integration of the last expression makes it possible to determine the number of allowable states Z. This function is highly important for the thermodynamics of solids:

$$Z = Z_o I \tag{2.9}$$

where

$$Z_o = \int \exp\left(-\frac{Q}{2mkT}\right) d\mathbf{P} = (2\pi mkT)^{3/2}$$

is the integral of states of the system of noninteracting particles,[50] and

$$I = \int \left[\exp\left(-\frac{\Pi}{kT}\right) - 1 \right] dr \, , \tag{2.10}$$

which are called the configuration or interaction integral.

How does a solid react to the external, thermal, and force fields? To find the answer to this question, consider its behavior at quasi-static or reversible processes. In the course of such processes, the AM system passes through a series of successive equilibrium states, while the structure does not undergo any irreversible changes. As such, the variation of the energy is expressed in its most general form as

$$dU = d\Omega_p + dQ = -FdA + Td\left(\frac{U}{T} + \ln Z\right), \tag{2.11}$$

where F is the generalized force acting on the solid with variations in external parameter A. This parameter can be thought of as the intensity of electric, magnetic, force, and other fields (except for the thermal field).

If $dA = 0$, then

$$dQ = dU = Td\left(\frac{U}{T} + \ln Z\right). \tag{2.12}$$

The latter shows that, in quasi-static processes, which are realized inside the compresson–dilaton phase (positions 2 and 5 in Figure 1.15), the amount of heat received or released by the solid can be written in the following form:

$$dQ = Td\mathbf{s} \tag{2.13}$$

where $d\mathbf{s}$ is the change in certain function

$$\mathbf{s} = \frac{U}{T} + k\ln Z \, , \tag{2.14}$$

which is called the entropy of the AM system. By opening the differential of entropy in Equation 2.12, we find that

$$\frac{U}{T} = kT \frac{\partial \ln Z}{\partial T} \tag{2.15}$$

Therefore, when only the thermal fields affect a solid, the energy levels of the AM set remain unchanged (for example, U_1 in Figure 1.7a), i.e.,

$$\frac{U}{T} = \text{const}, \tag{2.16}$$

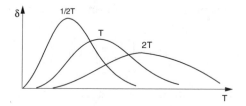

FIGURE 2.2
Temperature dependence of Boltzmann distribution.

and energy coming into the solid from the outside is consumed for changing the probability distribution, i.e., transformation of shape of the potential wells, following the diagram in Figure 1.10b.

Figure 2.2 shows how the shape of the Boltzmann distribution changes in this case (Equation 2.7) in different zones of the temperature interval. As the temperature increases from T to $2T$, the probability of the states with higher energy increases. Hence, on heating, the concentration of dilatons per unit volume increases and that of compressons decreases (positions D_2 and D_5 in Figure 1.14 approach the horizontal line q_2cq_1), leading to expansion of the solid. If the solid does not receive the energy but releases it (e.g., the temperature changes from T to $1/2T$), there is a reverse redistribution of the probabilities: the states with lower energy become more probable, and the compresson–dilaton (CD) proportion changes in an opposite direction. As a result, while cooling down, the solid decreases in size.

Variations in the CD proportion are accompanied by variations in strength and deformability. For example, oversaturation of the volume with dilatons during heating causes a decrease in the tensile strength and an increase in deformability. Alternatively, the compresson saturation during cooling increases the tensile strength but material becomes more brittle.

Equation 2.15 can be represented as

$$\frac{U}{T}\frac{\partial T}{T} = k\partial \ln Z,$$

which, after integration and accounting for Equation 2.16, becomes

$$\mathbf{s} = k\ln Z \tag{2.17}$$

That is, for the quasi-static processes, entropy becomes equal to the logarithm of the number of states, corresponding to the average energy of the entire AM system. As the number of the interacting particles per unit volume does not change, the area under each curve remains the same. It can be seen from Equation 2.15 that the ambient temperature determines the macro state of a solid and thus its internal energy depends on the ambient temperature.

A thermally insulated solid, whose potential well and state diagram shapes do not vary following the diagram shown in Figure 1.13, can exchange the potential energy $(dA \neq 0)$ with the environment through external fields that have natures other than the thermal one (force, radiation, electric, etc.). In this case Equation 2.11, accounting for Equation 2.17, transforms as

$$d\Omega_p = T d\mathbf{s} .\tag{2.18}$$

In such fields (at $T = const$), the solid does work only due to variations in the admissible energy levels. The initial Maxwell-Boltzmann (MB) distribution starts transforming due to variations in entropy. At $dT = 0$, one set of rotoses, while emitting heat and generating resistance forces, tends to the v_s-temperature (Figure 1.14) and fails following diagram 6 in Figure 1.15, whereas, on receiving heat from them, the others are deformed, bringing the sizes of dipole D_5 to the critical state of $q_2 c q_1$. In this case there is no way for the areas under the MB distributions to equal each other (Figure 2.2).

The MB character of distribution of the internal energy in the AM structure and the objective reality of existence of the compresson and dilaton types of the interatomic bonds result in the CD heterogeneity of structure at the atomic level. This can be traced at different scale levels. For example, spectroscopic investigations reveal it at the AM level.[10,33,39] At the macrolevel, it is identified by the indicators of hardness.[53,54] This heterogeneity is of fundamental significance for the formation of all physical–mechanical properties of materials; it was first reported in 1986[55] and then given detailed substantiation.[24]

The work done by a solid (or over a solid; Equation 2.1) is usually expressed in terms of changes in its size and shape as

$$d\Omega_p = -\mathbf{P} dV \tag{2.19}$$

It is evident that, for tension and compression,

$$dV > 0 \quad \text{(I)} \quad \text{and} \quad dV < 0 \quad \text{(II)},\tag{2.20}$$

respectively.

Substituting Equation 2.20 into Equation 2.1 and expressing the pressure of resistance to shape changing in the explicit form accounting for Equation 2.13, we can write

$$\mathbf{P} = T \frac{d\mathbf{s}}{dV} - \frac{dU}{dV}.\tag{2.21}$$

Condition 2.20 allows one to rewrite the last expression in the form

$$\mathbf{P} = \mathbf{P}_c - \mathbf{P}_d \tag{2.22}$$

where

$$P_c = T\left(\frac{d\mathbf{s}}{dV}\right)_u \quad \text{and} \quad P_d = -\left(\frac{dU}{dV}\right)_s$$

are the compressive (compresson) and tensile (dilaton) parts of the internal resistance that are generated by the corresponding phases (Equation 1.107). The first equality is valid for $U = \text{const}$, and the second for $s = \text{const}$. Using definition of the pressure (Equation 1.110), $\mathbf{P} = dF/d\mathbf{s}$, we obtain

$$dF_c = T\left(\frac{d\mathbf{s}}{dV}\right)_u dS \quad \text{and} \quad dF_d = -\left(\frac{dU}{dV}\right)_s dS.$$

In equilibrium state, the total resistance $\mathbf{P}$ is equal to zero. This is possible if $dF_c = -dF_d$. It can be seen that the internal compressive and tensile forces have the same magnitude and opposite directions. Leading to contraction of the external surface, the former prevents spontaneous tension of the solid; the latter prevents its spontaneous contraction. The application of the external force and thermal fields violates this equilibrium, causing deformation.

Chapter 1 showed that the state of interatomic bonds may change in a reversible manner (Figure 1.14) inside the acceleration well q_2dc or deceleration barrier cbq_1, or in an irreversible manner in which it passes in its evolution through the Debye temperature. In general, Relationship 2.1 takes the form

$$dU \le T d\mathbf{s} + d\Omega_p, \tag{2.23}$$

where the equality sign relates to the reversible processes and that of inequality to the irreversible ones. The variation in entropy can serve as a criterion of reversibility and irreversibility: it remains constant in the first case and increases in the second. An increase in entropy indicates a violation of the initial order in the AM system. In solids it shows up as deformation or disturbance of the initial structure. Adding the term ωdN to the left side of Inequality 2.23, which accounts for a destructive result of the irreversible processes, makes it possible to rewrite this equation as

$$dU + \omega dN = T d\mathbf{s} + d\Omega_p. \tag{2.24}$$

We can write Equation 2.24 for the reversible ($dU = Td\mathbf{s}_1 + d\Omega_1$) and irreversible ($dU + \omega dN = Td\mathbf{s}_2 + d\Omega_2$) processes occurring during the same period of time and then subtract the second equality ($-\omega dN = T(d\mathbf{s}_1 - d\mathbf{s}_2) + (d\Omega_1 - d\Omega_2)$) from the first. Because $d\mathbf{s}_2 > d\mathbf{s}_1$ by definition, $T(d\mathbf{s}_1 - d\mathbf{s}_2) < 0$. To preserve the sign in the left side, it is necessary that $d\Omega_1 - d\Omega_2 > 0$. Hence, $d\Omega_1 > d\Omega_2$ or $d\Omega_1 = d\Omega_2 + \omega dN$. As seen, the useful work done during a reversible process is always higher than that during an irreversible process. For the latter, its part equal to

$$d\Omega_3 = d\Omega_1 - d\Omega_2 = \omega dN \tag{2.25}$$

is spent nonproductively for the destructive processes. Hence, an important practical conclusion can be formulated: to prevent fracture of materials in the external energy fields, it is necessary to create conditions which would make it possible to avoid a change of the order established in the AM system of this material. It prevents occurrence of the destructive processes; however, when it is necessary to activate fracture, the opposite is true.

Major thermodynamic parameters of a solid are as follows: temperature T, internal energy U, its potential, Ω, and kinetic, Q, components, as well as entropy **s**. In principle, we can continue our consideration manipulating only these parameters. However, many results can be obtained more simply and would allow clearer interpretations if the thermodynamic potentials are used. We will limit our considerations to the most important of them.

The independent variables in the first law of thermodynamics (Equation 2.21) are entropy **s** and volume V, or, to be more exact, volume and temperature, because in accordance with Equation 2.13 **s** is a single-valued function of temperature. In practice, however, the technical state of solids is more often determined by the external mechanical stress σ (counteracted by internal stress **P**) and temperature T. Equation 2.21 can easily be transformed to accommodate these variables. By adding and subtracting the $Vd\mathbf{P}$ term to and from the right side of Equation 2.1, we obtain that

$$dQ = d(U + PV) - V d\mathbf{P}, \tag{2.26}$$

where the expression

$$H = U + PV \tag{2.27}$$

is called the thermal function, heat content, or enthalpy. Integrating Equation 2.26 at $d\mathbf{P} = 0$, we obtain

$$\Delta Q = (U_2 - PV_2) - (U_1 + PV_1) = H_2 - H_1, \tag{2.28}$$

where indices 1 and 2 designate two different states. Therefore, a change of the thermal function in the processes occurring at constant resistance **P** is equal to the amount of heat absorbed or released by a solid. It follows from the determination of enthalpy (Equation 2.27) that it can be represented as a sum of overall internal energy U and potential Ω (Equation 1.111), i.e.,

$$H = U + \Omega. \tag{2.29}$$

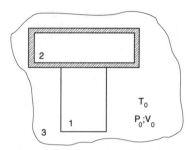

FIGURE 2.3
Physical nature of energy of resistance of materials to external effects.

This potential was introduced into thermodynamics for the first time by Landau and Lifshits.[23] Based on formal considerations, they used it to study the dependence of thermodynamic parameters on the number of particles in the AM system. In fact, it characterizes the ability of a solid to generate forces of internal resistance **P** (see Equations 1.16 and 1.53) rising when the external factors tend to change its size and shape V.

By differentiating Equation 2.27, $dH = dU + \mathbf{P}dV + Vd\mathbf{P}$, and substituting $dU = Td\mathbf{s} - \mathbf{P}dV$ from 2.21, we find the differential of thermal function as

$$dH = Td\mathbf{s} - Vd\mathbf{P}. \tag{2.30}$$

It follows from Equation 2.30 that enthalpy remains constant (H = const) only in stable structures ($d\mathbf{s} = 0$), which generate the constant force fields ($d\mathbf{P} = 0$).

Consider now the mechanical work (including its source) that a solid is able to do over the surrounding bodies and work that the surrounding bodies can do over it.[20] Let two solids, 1 and 2 (Figure 2.3), be in environment 3. Interactions, which can occur between solid 1 and environment 3, may be of different natures, such as thermal ($\vec{T}_o$), mechanical ($\mathbf{P}_o$), radiation, etc. Solid 1 behaves actively while solid 2, which is passive, has a reliable thermal insulation from the environment and can exchange only mechanical work with solid 1. We will call it the work object, and the work done over it the useful work.

Let solid 1 transform from one state into the other by doing the useful work ($-d\Omega$). During this transformation, it receives from or releases to environment 3 a certain amount of heat ($-dQ_o$). The volume of both solid 1 and environment 3 may be changed by respective values dV and dV_o. An increment in volume may be either positive or negative. In response, environment 3 will do work $d\Omega_o = \mathbf{P}_o dV_o + V_o d\mathbf{P}_o$ to solid 1. We can write the law of energy conservation for the considered closed system (active solid 1 plus work object 2 and environment 3) as

$$dU = -d\Omega + d\Omega_o - dQ_o, \tag{2.31}$$

because the volume of the closed system remains unchanged: $dV = dV_o$.

Assume that at any interaction with solid 1, an infinitely slow quasi-static process occurs in environment 3 at constant pressure $\mathbf{P}_o$ and temperature T_o. Therefore, $d\Omega_o = \mathbf{P}_o dV_o = -\mathbf{P}_o dV$. Substituting it into Equation 2.31 yields

$$dU = -d\Omega - \mathbf{P}_o dV - T_o \mathbf{s}_o. \tag{2.32}$$

Write the law of growth of entropy for the closed system (solid 1 plus environment 3) as follows: $d\mathbf{s} + d\mathbf{s}_o \geq 0$. Substituting the value of $d\mathbf{s}_o$ converts Equation 2.32 into inequality

$$(-d\Omega) \geq dU + \mathbf{P}_o dV - T_o d\mathbf{s}. \tag{2.33}$$

Assuming that temperature T_o and pressure $\mathbf{P}_o$ of the environment remain constant during the process, we can write

$$(-d\Omega) \geq d(U + \mathbf{P}_o V - T_o \mathbf{s}) = dG, \tag{2.34}$$

where the designation

$$G = U + \mathbf{P}_o V - T_o \mathbf{s} \tag{2.35}$$

is introduced. Solid 1 does the highest useful work over work object 2 when the reversible processes take place in the work object. This turns Equation 2.34 into an equality as

$$d\Omega_{max} = -d(U + \mathbf{P}_o V - T_o \mathbf{s}) = -dG. \tag{2.36}$$

As seen, the maximum useful work, which provides the maximum resistance of a solid to external effects, is equal to the decrease in the absolute value of parameter G. Also, it is fundamentally important that the value of G is determined not only by the independent variables directly related to the active solid 1 (namely: U, V, and $\mathbf{s}$) but also by those related to environment 3 (P_o, T_o). This consideration results in the following. First, this implies that the prediction of behavior of solids is incorrect without precisely accounting for the essential parameters of the environment. Second, it is possible to control the value of G (thus, resistance of solids to the external effects) by varying the active factors (P_o, T_o) in the purposeful manner.

Consider a particular case of practical application of solids. Assume that a structural material is in an external force field ($\mathbf{P}_o \neq \text{const}$) and operates at a constant temperature of the environment ($T = T_o = \text{const}$) under restrained conditions, i.e., where free deformation cannot occur ($V = \text{const}$). In this case, following from Equation 2.36, the material is capable of performing the following resistance work:

$$d\Omega = -d(U - T\mathbf{s}) = -d\mathbf{F},$$ (2.37)

where

$$\mathbf{F} = U - T\mathbf{s}$$ (2.38)

is a measure of the work done under the assigned conditions. It is called the free energy of the AM system. The other part of the internal energy equal to $T\mathbf{s}$ is called the bound energy. Therefore, under invariable thermal fields ($T = T_o = $ const), a solid resists the external effects ($\mathbf{P}_o \neq$ const) by spending the free (kinetic or thermal) part $\mathbf{F}$ of the overall internal energy U. Such processes are realized only if the solid can freely exchange heat with the environment. Otherwise (e.g., in adiabatic insulation), the resistance work is equal to the decrease in the overall internal energy. In fact, if in Equation 2.1 we assume that $dQ = 0$, then $d\Omega = -dU$. Therefore, reversible processes take place in the first case and irreversible processes occur in the second. (Entropy in the bound part of the energy increases due to a change in the state [Equation 1.53] or disintegration of the interatomic bonds.)

The result — that during the isothermal processes the resistance to external effects is formed due to a change, not in the overall internal energy of a solid, but only in its free or kinetic part — is explained by the fact that this change is immediately compensated for by heat exchange with the environment. Of course, the work of the material can be arranged so that the internal resistance forces are generated entirely due to absorbed heat or due to absorption of other kinds of the energy. As such, its structure serves as a converter of energy supplied from the outside to generate the resistance to external effects; by performing the reversible processes, it will never fracture. Equipped with a special system, these materials will adapt themselves to various service conditions, ensuring a 100% guarantee of failure-free operation of such engineering objects.

Writing Equation 2.38 in the form of

$$U = \mathbf{F} + T\mathbf{s}$$ (2.39)

and comparing the result with Equation 1.23, we can see that the overall internal energy U of a solid really consists of two parts: free or kinetic ($\mathbf{F} = Q$) and bound or potential

$$T\mathbf{s} = \Omega.$$ (2.40)

Differentiating Equation 2.39

$$dU = Td\mathbf{s} + \mathbf{s}dT + d\mathbf{F}$$

and substituting to the left part the differential of the overall energy from Equation 2.21, we can write

$$dF = -s\,dT - P\,dV. \tag{2.41}$$

Substitution of Equation 2.41 into Equation 2.37 results in an expression for the resistance work in stationary thermal fields:

$$(d\Omega)_T = s\,dT + P\,dV. \tag{2.42}$$

This function reaches its maximum at $(d\Omega)_T = 0$. Hence,

$$\frac{P}{s} = -\frac{dT}{dV}. \tag{2.43}$$

The last equality shows that a solid can exert resistance to the external factors **P** without irreversible restructuring and fracture of its structure (**s** = const) if temperature gradient

$$\left(-\frac{dT}{dV}\right)$$

is permanently maintained between this solid and the environment, i.e., due to replenishment of the AM structure with heat from the outside.

This behavior of solids can easily be explained on the basis of the concepts elaborated in Chapter 1. In accordance with Equation 2.18, the nonthermal (primarily mechanical) effects lead to a shift of the energy levels (e.g., D_2 or D_5) in the state diagram (Figure 1.14). They can return to the initial position provided that the compensating amount of heat is supplied simultaneously with the mechanical effects (e.g., in compression) to the solid or removed from it (in tension). If the induced loads are applied for some period of time, then, in accordance with Equation 2.43, the process of thermal support should be continued for that same time period. The free energy (its variation, to be more exact) is that particular part of the internal kinetic energy consumed for the immediate formation of response forces of resistance to the external load and then replenished by the inflow of heat from the environment.

That is, the resistance to the external loads in invariable thermal fields is realized due to the processes of heat exchange between the solid and its environment. The amount of heat supplied to the solid in this case is entirely spent on the formation of forces of resistance to the external mechanical effects. A part of this heat (designated as Δq in Figure 1.18) is consumed on the CD phase restructuring.

Returning to the resistance work (Equation 2.34), which a solid, freely interacting with the environment, can perform, note that, if temperature $(T = T_o)$

and internal stresses ($P = P_o$) are constant, Expression 2.35 is known in statistical physics as the Gibbs thermodynamic potential or the G-potential.[11,20,23] Its differential can be derived by substituting $P dV$, expressed in differential form

$$P dV = d(PV) - V dP, \quad \text{i.e.,} \quad dG = -s dT + V dP, \tag{2.44}$$

to the free energy formula (2.41). Equalities 2.27, 2.35, and 2.38 allow finding the relationships between the major thermodynamic potentials

$$\Omega = U - G - Ts \quad \text{(I)}; \quad \Omega = F - G \quad \text{(II)}; \quad G = H - Ts \quad \text{(III)}. \tag{2.45}$$

Consider the physical nature of the thermodynamic potentials. Written in a differential form,

$$d\Omega = -(P dV + V dP), \tag{2.46}$$

Ω (Equation 1.111) includes all possible types of resistance of solids to diverse external effects. The first term represents mechanical resistance. It is formed due to deformation dV and does not affect the ability of the AM structure to generate the resistance forces (P = const). The second term explains the physical nature of resistance of a solid to nonmechanical (radiation, chemical, electromagnetic, etc.) fields. These change the ability of the interatomic bonds to generate forces dP in the absence of deformation at V = const. Three other potentials, i.e., internal energy U, its free part F, and G-potential, contain these types of resistance; their differentials disclose the physical mechanism of their formation under diverse external conditions.

It follows from Equations 2.21 and 2.24 that U and F are consumed for generation of resistance to mechanical or force fields through deformation (they both contain term $P dV$) under special conditions: in thermally insulated solids U and under isothermal conditions F. It is of fundamental importance to note that:

- In the first case, the resistance is generated through increase in entropy ds, which is accompanied at T = const by destruction of the structure. (The mechanism of the destruction processes occurring under such conditions is described in Section 1.10.)
- In the second case, resistance is generated through a variation in the heat content, dT, continuously replenished from the external field, by preserving integrity of the structure at s = const.

Realization of a particular mechanism in actual bodies depends on the thermal–physical conditions of their local microregions.

The physical meaning of the G-potential is that its variation (Equation 2.44) represents that part of the internal energy consumed on resistance to the external effects of a nonmechanical nature. This resistance is provided by consumption of the kinetic energy dT, which is then compensated for by the heat exchange with the environment to an extent at which temperature T and the AM structure remain unchanged (s = const). Relationships 2.45.I and 2.45.II account for all possible variants of interaction of a solid with the environment. Because it relates structural parameters $\mathbf{P}$ and s to external T and V, it may be thought of as a different form of the equation of state (1.111), written in terms of the thermodynamic potentials.

2.3 Potentials of Systems with a Varying Number of Interatomic Bonds

Entropy s is the parameter that accounts for the state of the crystalline structure during interactions of a solid with external fields. Any interaction suggests the energy exchange and therefore the CD phase transitions, following the diagram shown in Figure 1.15. In turn, this may lead not only to a change in the intrinsic CD proportion and deformation but also to a failure of some interatomic bonds at the θ-temperature, when, in accordance with Equation 1.70, the probability of such an event is extremely high. Therefore, such an interaction very often leads to a decrease in the total number of interatomic bonds and to formation of submicroscopic defects in the structure.

Entropy is a generalized indicator of the order of the AM system formed during solidification and it characterizes the variations in the discussed order during reversible (force and thermal deformation) and irreversible (fracture) processes. To comprehensively describe the state of a solid, in addition to assigning thermodynamic potentials (Equation 2.45) characterizing the reversible processes, it is necessary to indicate the number of the bonds, N, participating in the interaction at a given moment.

Thus far we have assumed (except for Equations 2.24 and 2.25) that, during the interaction process, the number of the interatomic bonds does not change, i.e., N = const. In reality, this is far from the case: any change in the state of a solid almost always leads to some destruction of its initial structure. Therefore, N should be regarded as another independent variable that is a part of the additive thermodynamic potential, i.e.,

$$U = Nf\left(\frac{s}{N};\frac{V}{N}\right) \quad \text{(I)}; \qquad H = Nf\left(\frac{s}{N};\mathbf{P}\right) \quad \text{(II)}$$

$$F = Nf\left(\frac{V}{N};T\right) \quad \text{(III)}; \qquad G = Nf(\mathbf{P};T) \quad \text{(IV)}. \qquad (2.47)$$

The additivity means that any change in the amount of a material (hence, parameter N) leads to an equal change in these potentials. It follows from Equation 2.47 that a term proportional to dN, i.e.,

$$dU = T\,ds + \mathbf{P}\,dV + \omega\,dN \quad \text{(I)}; \quad dH = T\,ds + V\,d\mathbf{P} + \omega\,dN \quad \text{(II)};$$

$$dF = -s\,dT - \mathbf{P}\,dV + \omega\,dN \quad \text{(III)}; \quad dG = -s\,dT + V\,d\mathbf{P} + \omega\,dN \quad \text{(IV)}, \quad (2.48)$$

where ω designates a partial derivative,

$$\omega = \left(\frac{\partial U}{\partial N}\right)_{s;V},$$

and is actually determined by Expression 1.53, should be added to Equations 2.21, 2.30, 2.41, and 2.42. If a change in the potentials is considered regarding an appropriate pair of variables as constants, it can be seen from Equation 2.48 that

$$\left(dU\right)_{s,V} = \left(dF\right)_{T,V} = \left(dH\right)_{s,P} = \left(dG\right)_{T,P} = \omega\,dN. \tag{2.49}$$

It follows from this that

$$\omega = \left(\frac{\partial U}{\partial N}\right)_{s,V} = \left(\frac{\partial F}{\partial N}\right)_{T,V} = \left(\frac{\partial H}{\partial N}\right)_{s,P} = \left(\frac{\partial G}{\partial N}\right)_{T,P}. \tag{2.50}$$

As seen, ω can be calculated by differentiating any one of the thermodynamic functions with respect to the number of the paired bonds; as such, it can be expressed in terms of different variables. For example, by differentiating G written in the form of Equation 2.48.IV, we find that

$$\omega = \frac{\partial G}{\partial N} = f(\mathbf{P}, T).$$

Hence

$$G = \omega N \quad \text{or} \quad \omega = \frac{G}{N}. \tag{2.51}$$

Therefore, ω is the thermodynamic potential of a solid related to one interatomic bond. By expressing it in terms of functions $\mathbf{P}$ and T, we find that

$$d\omega = -s\,dt + v\,d\mathbf{p}$$

or, after integration, that

$$vp = \omega + st, \tag{2.52}$$

where s, v, and t are, respectively, the entropy, volume, and temperature of one rotos (Figure 1.7b).

Comparison of Equations 2.52 and 1.53 reveals the physical meaning of the ω-potential. Within the frames of classical atomic mechanics (Section 1.5), it serves as the converter of the kinetic part of the internal energy to the potential one, and vice versa. According to Equation 2.51, the G-potential determines the total number of such converters in the volume of a solid, V. For the same material, where $m = $ const and $M = $ const, the ω-potential depends only on the value of radius vector $\mathbf{r}$ (Equation 1.53). Solution of Equation 1.40 with respect to $\mathbf{r}$ at $\omega = $ const (e.g., at a level of U_1 from Figure 1.7a) yields

$$r_{c,d} = \{m\alpha \pm [(m\alpha)^2 - 2m\omega M^2]^{1/2}\}(2m\omega)^{-1}. \tag{2.53}$$

In the compresson and dilaton phases there are always rotoses with ω_c and ω_d that exert the same resistance. Because

$$\Omega = \sum_{i=1}^{N} \omega_i \tag{2.54}$$

and ω is a function of $\mathbf{r}$ alone, every value of the Ω-potential in the structure of a solid corresponds to volume V, within which function ω may have two values:

$$\omega_c = -\frac{M^2}{2mr_c^2} \quad \text{(I)}; \qquad \omega_d = \frac{M^2}{2mr_d^2} \quad \text{(II)}. \tag{2.55}$$

Accounting for this result, we write Equation 2.52 as

$$\omega_c - \omega_d = (st)_d - (st)_c. \tag{2.56}$$

In electrodynamics,[25] ω is known as the electric field potential, while in thermodynamics[11] it is called the ω-potential.

It can be seen from Equation 2.51 that the sense of the G-potential (Equation 2.45.III) can be thought of as the electrostatic potential of the space filled with an interacting system of charged particles α (Equation 1.17; electrons, nuclei, polarized atoms, asymmetric molecules, etc.).

To move a charge e from one position, ω_c, to the other, ω_d, e.g., in deformation of the EM dipole, D_5, in Figure 1.14, work

$$\omega_{c,d} = e(\omega_c - \omega_d) \tag{2.57}$$

should be done. Surfaces with equal potentials are called equipotential surfaces or equipotentials. The work of the field is equal to zero if the motion of the charge is along the equipotentials. Comparison of Equations 2.56 and 2.58 shows that work $\omega_{c,d}$ required for formation of the compresson and dilaton phases is performed during solidification of a body; that needed for separation is performed during its fracture. The work performed on equipotentials is equal to zero, so the force lines are always perpendicular to equipotentials. Because on a equipotential, $d\omega = 0$, then from Equation 2.52 we find that $\mathbf{s}dt = \mathbf{v}d\mathbf{p}$. This means that in the equilibrium state the constancy of entropy $\mathbf{s}$ and rotos volume $\mathbf{v}$ are maintained due to existence of a temperature gradient of internal stresses

$$\frac{\mathbf{s}}{\mathbf{v}} = \frac{d\mathbf{p}}{dt}. \tag{2.58}$$

This condition is valid in the potential well (Figures 1.7a and 1.8).

If a solid is subjected to elastic or reversible deformation, the number of interatomic bonds in it does not change ($N = $ const); however, its volume (geometrical dimensions and shape of the solid) does not remain unchanged ($\mathbf{v} \neq $ const). In contrast, when plastic or irreversible deformation takes place, the number of bonds changes. Consider a certain constant volume ($\mathbf{v} = $ const) inside an irreversibly deformed solid in order to trace the process of variation in the number of bonds contained in it ($N \neq $ const). In this case Equation 2.48.III reduces to

$$d\mathbf{F} = -\mathbf{s}dT + \omega dN. \tag{2.59}$$

This means that consumption of the free energy $d\mathbf{F}$ take place on account of a decrease in the intrinsic temperature dT of the solid and disintegration of the bonded structure dN. The G-potential (2.51) makes it possible to modify this expression so that the second independent variable is ω, rather than N. Substituting $\omega dN = = d(\omega N) - N d\omega$ for this, we obtain $d(\mathbf{F} - \omega N) = -\mathbf{s}\,dT - N d\omega$, or, accounting for Equation 2.46, we can write

$$d\Omega = -(\mathbf{s}dT + N d\omega) = -(PdV + V dP). \tag{2.60}$$

It can be seen from Equations 2.59 and 2.60 that the resistance of a solid to various external effects may be formed with destruction (Equation 2.59), where $\omega = $ const but $N \neq $ const, or without destruction (Equation 2.60), where

$\omega \neq$ const, $N =$ const, of the structure. In the second case, the CD transitions are impossible, whereas in the first the conditions created involve catastrophic consequences for integrity of the structure. The number of the bonds functioning at a given moment of time is determined through differentiating the Ω-function with respect to the ω-potential at constants T and V:

$$N = -\left(\frac{\partial \Omega}{\partial \omega}\right)_{T,V} = -V\left(\frac{\partial \mathbf{P}}{\partial \omega}\right)_{T,V},$$

from which a current density is calculated as

$$\rho = \frac{N}{V} = -\left(\frac{\partial \mathbf{P}}{\partial \omega}\right)_{T,V} \tag{2.61}$$

and the concentration of defects per unit volume of the solid is found as

$$\rho_\tau = \rho_o - \rho = \frac{N_o - N}{V}, \tag{2.62}$$

where $N_o - N$ is the number of the interatomic bonds that have failed by a given moment of time τ.

2.4 Thermodynamic Equation of State of a Solid

Expression of the interatomic bonds in terms of the EM dipole (Figure 1.7b) shows that Relationships 1.111 and 2.45.II, known in statistical physics,[21,23] are in fact the thermodynamic equation of state of a solid body. The speculations concerning Figure 2.3 logically lead to the same conclusion. Applying the notion of the "rotos" facilitates its interpretation. It is clear that the solid responds to external effects not with deformation V or resistance $\mathbf{P}$ taken separately, but simultaneously with both, i.e., with variations in the Ω-potential. In Troshenko,[3] we can find numerous proofs that these parameters indeed are in an inverse-proportional relationship.

The notions of free energy $\mathbf{F}$ and Gibbs potential G also acquire concrete physical content. The first is identified with a rotational part of the kinetic energy, related to the magnetic component of a local EM field (Equation 1.32.II) while its electrical component (Equation 1.32.I) forms the G-potential or, in other words, forms the spatial potential field (Figure 1.8) on the equipotentials of which the rotos atoms (ions) move along closed paths.

Equations 1.53 and 1.111 have absolutely identical forms and, so, correlate to micro- and macroscales. Although the latter is logically faultless, it is unsuitable for solving practical problems. Its interpretation and transformation from thermodynamic potentials to parameters measured in practice are impossible without utilizing major principles of fundamental and applied sciences: atomic and quantum mechanics,[27,28] statistical physics,[23] thermodynamics,[11] resistance of materials to deformation and fracture,[2,3] and theory of phase transitions.[30] Moreover, in doing this, it was found necessary to rule out some familiar concepts, specify others, and even involve new ones.[12,21,24]

Substituting F- and G-potentials expressed in terms of absolute temperature T of a solid into Equation 2.45.III and the logarithm of the function of state Z,[20,23] and accounting for their structure (Equation 2.47),

$$F = -kT \ln Z(N,V,T); \qquad G = -kT \ln Z(N,\mathbf{P},T) \qquad (2.63)$$

we can write

$$\Omega = -kT \ln Z(N,V,T,\mathbf{P}) \qquad (2.64)$$

or, using the definition of the Ω-potential (Equation 1.111) and the Boltzmann relationships (2.17), we obtain the equation of state of solids in the parametrical form

$$PV = \mathbf{s}(N,V,T,\mathbf{P})T \qquad (2.65)$$

The same follows from analysis of the main thermodynamic potentials (Equation 2.40).

Entropy of a solid $\mathbf{s}$ is the multifactorial and, therefore, is the most uncertain parameter in Equation 2.65. To reveal its physical nature and to confirm the validity of the equation of state (2.65), we derive it in a different way.

The function of state of a paired interatomic bond is determined by Equation 2.9. For a solid with the N quantity of such bonds, it can be written as[20]

$$Z = \frac{1}{N! \hbar^{3N}} Z_o^N J dv_1 dv_2 ... dv_N \qquad (2.66)$$

where Z_o and J are determined by Equation 2.10 and $dv_1 dv_2 ... dv_i$ are the products of the differentials of space coordinates for each of the interacting atomic pairs. Problem 2.66 is reduced to finding the integral of interaction, J (Equation 2.10). Integrating 2.66 with respect to each of $av_i = dx_i dy i dZ_i$ is performed for the entire volume of the solid, V. It follows from the physical nature of interaction of atoms (Figure 1.7) that potential energy U_r decreases so quickly with distance that it vanishes at a distance between the atoms of no more than several intrinsic radii r_a. (Normally Equation 1.64, this number is equal to three to four radii.)

The distance at which the interaction between the neighboring atoms becomes neglibly small is called interaction radius R.[23] If a sphere with radius R is plotted from the center of any microvolume dv_i, it will contain no more than a few atoms. Only these atoms can interact with each other because the rest are independent. By placing the focus of the sphere at the center of inertia, 0, of one of the interacting pairs (Figure 1.7b) and by decreasing its radius several times, we come to a paired interaction. By extending the process of separation of pairs to the entire volume of the solid v and connecting these pairs in series into links, we obtain an infinite atom–molecule chain (Figure 1.27). The pairs form the following series: 1–2, 2–3, 3–4 ... $(N - 3) \div (N - 2)$; $(N - 2) \div (N - 1)$; $(N - 1) \div N$; $N - 1$.

Let the odd pairs form the compresson sequence (Equation 1.70.I) and even pairs form the dilaton sequence (Equation 1.70.II). Alternation of the compresson and dilaton links arranged in one row compensates for the forces of the microvolumes directed inward and outward, which are inevitably formed in accordance with Equation 2.22. Such a chain has stable sizes and, contained in a preset volume and arranged in an extremely chaotic manner, is a chain-type compresson–dilaton model of a solid. Also it can be called the model with paired interaction. Despite the fact that it now allows for other (except paired) types of interaction, this model yields reliable and valid conclusions.

For such a model the potential energy Π can be represented as a sum in the form of

$$\Pi = u(r_{1,2}) + u(r_{2,3}) + \cdots + u(r_{(N-1),N}) = \sum u(r_{i,k}) \qquad (2.67)$$

where $k = i + 1$, and $u(r_{i,k})$ depends only on the coordinates of the neighboring atoms. This is determined by Equation 1.26.II and Figure 1.7a. The number of terms in this sum is equal to the number of the paired bonds formed in a given volume of the solid consisting of N atoms; it is evident that it is equal to $N - 1$. At $N \gg 1$, it can be considered equal to N. Then

$$\exp\left(-\frac{\Pi}{kT}\right) = \exp\left(-\frac{\sum u(r_{i,k})}{kT}\right) = \Pi \exp\left(-\frac{u(r_{i,k})}{kT}\right), \qquad (2.68)$$

where $u(r_{i,k})$ is a function of the relative distance between both atoms, i.e., D in Figure 1.7b, and the product is taken for all pairs, i.e.,

$$\exp\left(-\frac{\Pi}{kT}\right) = \exp\left(-\frac{u(r_{1,2})}{kT}\right)\exp\left(-\frac{u(r_{2,3})}{kT}\right)\exp\left(-\frac{u(r_{3,4})}{kT}\right)\cdots$$

It contains $N - 1$ cofactors.

We can express the subintegral expression in Equation 2.10 as

$$f_{i,k} = \exp\left(-\frac{u(r_{i,k})}{kT}\right) - 1. \tag{2.69}$$

This function tends to zero at $r \to \infty$ and is different from zero within the $(r - r_\infty)$ range, which limits the potential well (Figure 1.7a). Then, apparently,

$$\exp\left(-\frac{u(r_{i,k})}{kT}\right) = 1 + f_{i,k} \quad \text{and}$$

$$\exp\left(-\frac{\Pi}{kT}\right) = \Pi(1 + f_{i,k}) = (1 + f_{1,2})(1 + f_{2,3})(1 + f_{3,4})\cdots$$

$$= 1 + (f_{1,2} + f_{2,3} + f_{3,4} + \cdots) + (f_{1,2}f_{2,3} + f_{1,2}f_{3,4} + \cdots) + \cdots$$

The paired, ternary, and further products of function $f_{i,k}$, by definition and due to an assumption 2.67 of the significance of only the paired interaction, are always small and, moreover, lose their physical sense because of the effect of localization of atoms. Therefore, the following can be written with confidence:

$$\exp\left(-\frac{\Pi}{kT}\right) \approx 1 + (f_{1,2} + f_{2,3} + f_{3,4} + \cdots) = 1 + \sum f_{i,k}.$$

The number of terms in

$$\sum f_{i,k}$$

is equal to the N number of the pairs. Assuming for the first approximation that all the bonds are identical, we can also assume that all $f_{i,k}$ are also identical. Then

$$\exp\left(-\frac{\Pi}{kT}\right) = 1 + Nf(r_{i,k}). \tag{2.70}$$

Substitution of expression

$$\exp\left(-\frac{\Pi}{kT}\right)$$

from Equation 2.70 into Equation 2.66, accounting for Equation 2.10, yields

$$J = \int \left[\exp\left(-\frac{\Pi}{kT} \right) - 1 \right] dv_1 dv_2 \dots dv_N = N \int f_{i,k} dv_1 \dots dv_N$$

or, returning to Equation 2.69,

$$J = N \left[\int \exp\left(-\frac{u(r_{i,k})}{kT} \right) dv_1 dv_2 \dots dv_N - \int dv_1 dv_2 \dots dv_N \right]. \tag{2.71}$$

The first integral gives[20]

$$\int dv_1 \dots dv_{i-1} dv_{i+1} \dots dv_{k-1} dv_{k+1} \dots dv_N \int \exp\left(-\frac{u(r_{i,k})}{kT} \right) dv_i dv_k$$

$$= v^{N-2} \int \exp\left(-\frac{u(r_{i,k})}{kT} \right) dv_i dv_k$$

and the second integral is likely to be equal to v^N. Therefore,

$$J = N \left[v^{N-2} \int \exp\left(-\frac{u(r_{i,k})}{kT} \right) dv_i d_k - v^N \right] \tag{2.72}$$

Rewrite the last integral in the spherical coordinates, placing its origin in the center of one of the atoms, and add expression

$$\gamma = 4\pi \int \exp\left(-\frac{u(r)}{kT} \right) r^2 dr \tag{2.73}$$

where $r_{i,k} = r$ and 4π is the result of integration by angles. For J, we finally obtain that

$$J = Nv^N \left(\frac{\gamma}{V^2} - 1 \right). \tag{2.74}$$

Substituting this expression into Equation 2.66, we find that

$$Z = Z_0 N \left(\frac{\gamma}{V^2} - 1 \right). \tag{2.75}$$

All the preceding considerations concern the solids whose energy is made of the paired interaction of rotoses arbitrarily located in space. (Direction of axis r is not indicated in Figure 1.7a.) Therefore, while analyzing structural processes, it is necessary to take into consideration the linear dimensions of the EM dipoles (Figure 1.7b) and also to account for their relative orientation.

Using the known approaches of statistical physics and thermodynamics[11,20,23] and the function of state Z, we can find all the parameters that determine the behavior of solids under varying external conditions. The tensor of resistance **P** is found from the formula

$$\mathbf{P} = kT \frac{\partial \ln Z}{\partial V}. \tag{2.76}$$

Substituting Equation 2.75 into Equation 2.76, we obtain that

$$\mathbf{P} = kTN \frac{\partial \ln Z_o}{\partial V} + kTN \frac{\partial}{\partial V} \ln\left(\frac{\gamma}{V^2} - 1\right).$$

It can be seen from Figure 2.1 that $\gamma < 1$ over the entire range of variations in the exponent factor. Assuming that

$$\frac{\gamma}{V^2} \ll 1,$$

we expand logarithm of the second term in a power series of

$$\frac{\gamma}{V^2}.$$

Then, using only the first term of this expansion, we write

$$\mathbf{P} = \frac{kTN^2}{V} + kTN \frac{\partial}{\partial V}\left(\frac{\gamma}{V^2}\right)$$

because the logarithm of the function of state for a noninteracting system consisting of N particles is equal to $\ln Z_o = N/V$.[20]

Multiplication and division of the second term by V yields

$$PV = kTN^2 + kTNV \frac{\partial}{\partial V}\left(\frac{\gamma}{V^2}\right) \tag{2.77}$$

while displacing V under the sign of the differential converts (Equation 2.77) into

$$PV = kTN^2 + \frac{1}{2}kTN\frac{\partial \gamma}{\partial V}. \tag{2.78}$$

Differentiation of Equation 2.73 makes it possible to express the γ-function in terms of the Boltzmann distribution (Equation 2.5):

$$\partial \gamma = 4\pi r^2 d\delta, \tag{2.79}$$

where πr^2 is the area of orbital ring A (Figure 1.7b) defining the limits of existence of EM dipole D (Equation 1.15), and $d\delta$ determines that region in the energy spectrum of a molecule (Figure 1.18) to which it corresponds. In the equilibrium state, where $r^2 = \text{const}$, the position of the dipole in the state diagram (Figure 1.14) is stabilized in position D_2 or D_5. Representation of $V = \beta r^3 N$ (see explanations to Equation 1.111) allows the volume differential to be written in the form

$$\partial V = \beta(3r^2 N dr + r^3 dN). \tag{2.80}$$

Parameter V in Equation 2.78 (as well as in all its modifications) can be varied both at macroscopic ΔV and microscopic ∂V levels. In the first case the scale of a solid is varied as its shape remains unchanged. In the second case, the shape transformation occurs at a constant volume ($\Delta V = 0$). Deformation according to Equation 2.80 can occur in a reversible or an elastic manner without a loss in bonds (where $dN = 0$) or in an irreversible or a plastic manner, accompanied by the destruction processes (where $dN \neq 0$).

We can rewrite Equation 2.78 at $\Delta V = 0$ accounting for Equation 2.79 for conditions of elastic deformation, where Equation 2.80 is transformed into $\partial V = 3\beta r^2 N dr$

$$PV = kT\left(N^2 + \frac{2\pi}{3\beta}\frac{d\delta}{dr}\right). \tag{2.81}$$

From here, after substitution of Equation 2.5, we have

$$PV = kT\left[N^2 + \frac{2\pi}{3\beta}\exp\left(-\frac{u}{kT}\right)\right]. \tag{2.82}$$

Comparing Equation 2.82 with Equations 2.45.II, 2.63, and 2.65, we can write

$$F = kTN^2 \quad \text{(I)}; \quad G = \frac{2\pi}{3\beta} kT \exp\left(-\frac{u}{kT}\right) \quad \text{(II)}; \quad \ln Z(N,V,T) = N^2 \quad \text{(III)};$$

$$\ln Z(N,P,T) = \frac{2\pi}{3\beta} \exp\left(-\frac{U}{kT}\right) \quad \text{(IV)}; \quad s = k\left[N^2 + \frac{2\pi}{3\beta} \exp\left(-\frac{U}{kT}\right)\right] \quad \text{(V)}. \quad (2.83)$$

As seen from Equation 2.83.V, the atoms of a solid (if N implies their quantity in volume V) and their derivatives in the form of paired dynamic systems (Figure 1.7b) can exist in two fundamentally different states. The former, contained in a rotos, are subdivided by inequalities (1.70) into compressons and dilatons. The state of the latter, being parabolically varied (expression in brackets), is either in the acceleration well q_2dc or in the zone of a deceleration barrier cbq_1 (Figure 1.14) during solidification. In accordance with the rules of solving quadratic equations, the boundary between them is the apex of a parabola (point 0 in Figure 1.10a) having coordinates

$$\left[-\frac{\pi}{3\beta} \exp\left(-\frac{U}{kT}\right)\right] \quad \text{and} \quad \left\{\left[-\frac{\pi}{6\beta} \exp\left(-\frac{U}{kT}\right)\right]^2\right\},$$

which distributes the bonds along both branches of the parabola following the law

$$N_{1,2} = \pm\left[\frac{2\pi}{3\beta} \exp\left(-\frac{U}{kT}\right)\right]^{1/2}. \tag{2.84}$$

Thus, entropy of a solid in the nonequilibrium state (Equation 2.81) is determined by the quality of packing of the structure β, by the type (Equation 2.84) and number N of the interatomic bonds, their energy $d\delta$, and by linear dimensions and orientation of the EM dipoles in space dr.

Formally, Function 2.79 can be represented as the volume of a cone (Figure 1.12a) with base radius r and generating line $d\delta$ (Equation 2.5), where r varies over a range of r_1 to r_∞ (Figure 1.7a). Let us split it into two parts:

$$r_1 < r < r_a \quad \text{and} \quad r_a < r < r_\infty.$$

On solidification, the first comprises terms of set N, which have the minus sign (Equation 2.84), and the second comprises those with the plus sign.

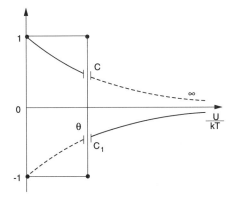

FIGURE 2.4
Diagram of distribution of Boltzmann probabilities for two-phase compresson–dilaton structures of solids.

The preceding considerations allow a conclusion that Boltzmann distribution of probabilities for a solid body is of a type other than that shown in Figure 2.1. In reality, it has a three-dimensional conical shape whose axial section is shown in Figure 2.4. In this section plane, the interpretation of the probability of finding the energy level in the compresson phase is determined by a curvilinear quadrangle $01C\Theta$ and that in the dilaton phase by triangle $C1\Theta\infty$. In the first case, the probability is positive and in the second it is negative; the forbidden boundaries are shown by dotted lines. In general, the figure delineated by contour $[01C\infty C_1(-1)0]$ is the axial section of the probability cone.

Let us compare the equation of state of a solid (Equation 2.77) with the van der Waals equation[11,23] written similarly:

$$PV = kTN\left(1 + \frac{bN}{V}\right) - \frac{aN^2}{V},$$

(2.85)

where a and b are constants. Despite the obvious differences, these equations clearly show the main peculiarity of the AM systems — CD phase heterogeneity, which shows up in the quadratic dependence of the $\mathbf{P}$ and V parameters of state on the number of the interacting particles N. For high densities of gases and liquids, it is necessary to disregard the constancy of values a and b and consider them as functions of temperature, in order to bring the solution of Equation 2.85 into quantitative agreement with the experimental data, whereas for solids this dependence is evident from Equation 2.65.

Nevertheless, an important advantage of the van der Waals equation is that it provides a correct description of behavior of gases and some liquids. As will be shown later, the equation of state of a solid (2.65) possesses the same advantage. For example, comprehensive experiments conducted by the authors of the kinetic concept of strength[10] showed that the ability of solids to resist variations in size and shape is closely related to their thermal properties.

Historically, however, the mechanical and thermal properties were studied separately.[2,3] In the Einstein and Debye theory of heat capacity,[11,20] the concepts of physical nature of heat, formulated as a result of investigations of gases and liquids, were almost automatically applied to solids. As a result, researchers overlooked the natural link between mechanical and thermal properties. Alternatively, Equation 2.65 allows explanation of thermal and many other physical–mechanical effects that accompany the behavior of solids in various external fields and aggressive environments.

2.5 Parameters of State, Relationship of Equivalence, and Entropy

In different scientific disciplines, different physical meanings are assigned to the term "state of a solid."[1-24] Following the definition by Frenkel,[56] we define it as "the shape of condensation of rotoses during the solidification process" and "the ability of a structure to preserve it afterwards under various external conditions." We can use this well-established term further on without fear of misunderstanding.

Because the equation of state (2.65) is of fundamental importance for solid-state physics, strength, reliability, and durability of structural materials,[21] we should consider the physical meaning of its parameters in detail. As will be shown later, entropy occupies a special place among these parameters (Equation 2.83.V).

Relationship 2.83.I allows Distribution 2.84 to be written as

$$N_{1,2} = \pm\left(\frac{F}{kT}\right)^{1/2}. \tag{2.86}$$

Therefore, the type of the bond depends exclusively on the kinetic part of the internal energy of a body. We can assume that the motion of atoms in solids is of a collective nature. This means that any change in the individual motion of an individual atom immediately affects the others. For example, an atom that has number n, quits the equilibrium state and moves over a certain distance dr in an arbitrary direction. As a result, it starts to become affected by its closest neighbors.

As the interaction between atoms is of a short-range character, a distur-bance will be transferred only to atoms with numbers $(n - 1)$ and $(n + 1)$. Atoms that follow, with numbers $(n - 2)$ and $(n + 2)$, will be affected very little because they are far from the considered excited atom and these inter-actions can be ignored. Displacement of the $(n + 1)$-th atom will lead to disturbance of the $(n + 2)$-th atom, while that of the $(n - 1)$-th atom will lead to displacement of the $(n - 2)$-th atom. In turn, these atoms will excite their neighbors. As a result, the entire AM set N (Figure 1.27) will start moving, which will result in waves propagating in different directions. Superposition of these waves will lead to formation of a standing wave.

Based on this idea and transforming our consideration from unidimen-sional to three dimensional, we can write the following equation for the logarithm of the function of state:[20]

$$\ln Z = -\frac{9N\theta}{8T} - 9N\left(\frac{T}{\theta}\right)\int_0^{\theta/T} y^2 \ln(1 - e^{-y}) dy , \qquad (2.87)$$

where

$$\theta = \frac{\hbar \nu_{max}}{k}$$

is the Debye temperature, $y = \hbar\nu/kT$ is the parameter of the integral of interaction, and ν is the oscillation frequency determined by Equation 1.50. The interaction integral in the kinetic representation (Equation 2.87) is iden-tical in the structure to the configuration integral (Equation 2.10). They differ only in the sign of the binomial in the subintegral expression. The θ-temperature (Equation 1.56) corresponds to the bottom of the potential well (Figure 1.7a). Following the path in Figure 2.4, we can break the integral in Equation 2.87 into two parts:

$$\int_0^{\theta/T} = \int_0^1 + \int_1^\infty \qquad (2.88)$$

As in the potential representation, this action corresponds to the separation of the integration limits into two intervals: $r_1 < r < r_a$ and $r_a < r < r_\infty$. Indeed, at point r_a (Figure 1.7a), $T = \theta$ and $\theta/T = 1$. The first integral is equivalent to a variation in the integration limits from 1 to ∞, since at $\Theta = $ const at the bottom of the potential well $T \to 0$, whereas $\theta/T \to \infty$. In contrast, near the surface of the well, $T \to \infty$ and $\theta/T \to 0$. This condition corresponds to the first interval. Certain reservations, however, should be considered in calculations using Equation 2.88.

Considering the second part of this equation, we see that at $T \to \infty$ the limit in the integral can be replaced by infinity, as the subintegral function remains very small regardless of how high its argument y may be. This yields

$$\int_0^{\theta/T} y^2 \ln(1 - e^{-y})dy = \int_0^{\infty} y^2 \ln(1 - e^{-y})dy.$$

The last integral was calculated in Levich.[20] Taking into account this solution and assuming that

$$\int_0^{\infty} y^2 \ln(1 - e^{-y})dy = \int_1^{\infty} y^2 \ln(1 - e^{-y})dy,$$

we find that

$$\int_1^{\infty} y^2 \ln(1 - e^{-y})dy = -\frac{\pi^4}{45}. \tag{2.89}$$

Considering the right-hand side of the integral (2.88), we notice that its limit at $T \to \infty$ is small. Therefore, if y is small in the subintegral function, this function can be expanded in a power series, i.e.,

$$\ln(1 - e^{-y}) = \ln y.$$

In this case,

$$\int_0^{\theta/T} y^2 \ln(1 - e^{-y})dy = \int_1^1 y^2 \ln y\, dy = \frac{1}{3}\left(\frac{\theta}{T}\right)^3 \ln\frac{\theta}{T} - \frac{1}{9}\left(\frac{\theta}{T}\right)^3. \tag{2.90}$$

Substitution of Equations 2.89 and 2.90 into Equation 2.87 yields

$$\ln Z = N\delta, \tag{2.91}$$

where

$$\delta = \left[-\frac{9}{8}\frac{\theta}{t_i} + 1 - 3\ln\frac{\theta}{t_i} + \frac{\pi^4}{5}\left(\frac{t_i}{\theta}\right)^3 \right] \tag{2.92}$$

is the MB distribution (Equation 2.2) expressed in terms of the kinetic parameters.

Comparison of Equation 2.83.III with Equations 2.91 and 2.92 shows that one of the terms, N, is distributed over the thermal interval through the relationship t_i/θ.

$$t_i = \Theta \quad \text{at} \quad \frac{t_i}{\theta} = 1,$$

so one of its parts has the Debye temperature and the other is distributed on both sides from it.

Function $\ln Z$ in a low-temperature region, where $t_i \ll \Theta$, can be represented as[20,23]

$$\ln Z_f = N_f \left[-\frac{9}{8}\frac{\theta}{t_i} + \frac{\pi^4}{5}\left(\frac{t_i}{\theta}\right)^3 \right] \text{(I)},$$

and in a high-temperature region (at $t_i \gg \theta$) as

$$\ln Z_s = N_s \left[-\frac{9}{8}\frac{\theta}{t_i} + 1 - 3\ln\frac{\theta}{t_i} \right] \text{(II)} \tag{2.93}$$

of the thermal spectrum.

An important note should be made concerning values of T and t_i, that make Equation 2.93 different from those in Levich[20] and Landau and Lifshits.[23] The absolute temperature of a solid T shows up in an experiment as the mean value of the phonon radiation, t_i, of the terms of set N (Equation 2.2).[32,49] As for the Debye temperature θ, by Definition 1.56[23] it is not a mean value. Moreover, it has a purely statistical nature and thus, in Equation 2.93, it should be related not to T, but to t_i (Equation 1.24.II). In any state of a solid, the entire set N does not have the Debye temperature, but rather some of its parts are determined by the Boltzmann statistics (Equation 2.92).[32,49] Equation 1.95 determines this temperature only at point a (Figure 1.13), where the CD phase transitions occur (Equation 1.57).

In intervals $T < \theta$ and $T > \theta$, it degenerates into the Debye temperature of the first θ_f (curve *ade*) and second θ_s (curve *abf*) kinds. The latter determines a brittle and plastic character of disintegration of the subsets in $(-N_1)$ and $(+N_2)$ formed in the acceleration well and in the deceleration zone. The brittle boundary of existence of the plastic subset N_2 is determined by the Debye temperature of the third kind or the temperature of separation of the C- and D-phases, t_c (Figure 1.14). Therefore, it seems impossible to measure the θ-temperature of a solid with an accuracy of more than 50 to 70%.[57]

The form of the δ-factor (Equation 2.92) indicates that $\ln Z$ is in fact a distribution of the terms of set N with respect to the characteristic temperature θ (positions 3 and 6 in Figure 1.14), depending on the radiating ability

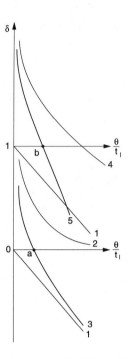

FIGURE 2.5
Determination of the Debye temperature for
brittle and ductile structures.

t_i (Equation 1.24.II) of each of them. Moreover, division of ln Z into two ranges (Equations 2.93.I and 2.93.II) is not accidental. It signifies that values of Θ in both of them are not the same (in Figure 1.14, designated as v_f and v_s). This is clearly proved by the existence of brittle and ductile classes of materials, which behave differently in the external fields.[2,3,26,48]

This is manifested most clearly in graphical interpretation of Function 2.92, shown in Figure 2.5. The θ/t_i ratios are plotted on the axis of abscissas, while the following functions are plotted on the axis of ordinates:

$$\delta_f = -\frac{9}{8}\frac{\theta}{t_i} + \frac{\pi^4}{5}\left(\frac{t_i}{\theta}\right)^3 \quad \text{(I);} \quad \delta_s = -\frac{9}{8}\frac{\theta}{t_i} + 1 - 3\ln\frac{\theta}{t_i} \quad \text{(II).} \quad (2.94)$$

Their following components are designated by numbers:

$$1: \left(-\frac{9}{8}\frac{\theta}{t_i}\right); \quad 2: \frac{\pi^4}{5}\left(\frac{t_i}{\theta}\right)^3; \quad 3: \delta_f; \quad 4: 3\ln\frac{\theta}{t_i}; \quad \text{and 5: } \delta_s.$$

It is seen that the plots of functions δ_f and δ_s are separated in the energy domain T by constant I, and that points a and b, where $t_i = \theta$, do not coincide on both axes. In Figure 1.14, they show the extreme states of the dipole and, for reasons that have not yet been formulated, they are located in opposite regions of the temperature spectrum. Therefore, parameters Θ in Expressions 2.94.I and 2.94.II are not identical. In analogy with Figure 1.14, by expressing it in terms of v_f for the low-temperature region and in terms of v_s for the high-temperature region, we can rewrite Equation 2.93 with respect to Equation 2.94 as

$$\ln Z_f = N\delta_f \quad \text{(I)}; \qquad \ln Z_s = N\delta_s \quad \text{(II)}. \tag{2.95}$$

Substitution of Equation 2.95 into Equation 2.65 allows the equations of state to be represented in the form of

$$PV = kTN\delta, \tag{2.96}$$

where δ is the MB factor, which may have the values of δ_f and δ_s (Equation 2.94). During solidification it distributes the atoms that form the dipole relative to temperatures v_f and v_s (Figure 1.14), depending on the radiating ability, t_i, of each, on the opposite branches of the acceleration well q_2ad or the deceleration barrier dbq_1. Comparison of Equations 2.91 and 2.64 makes it possible to express entropy in terms of measurable parameters:[26]

$$s_i = kN\delta. \tag{2.97}$$

In a general case, entropy (Equation 2.81) depends on the linear dimensions of the EM dipole (Equation 1.15). Consider the variation of this dipole under the effect of an instantaneous system of forces (Equation 1.52). In analogy with Equation 1.104, we can write the differential equations of motion of atoms, m, in an inertial system:

$$m\ddot{r} = \frac{C}{r^3} - \frac{\alpha}{r^2}.$$

Dividing both parts of the equation by m, adding designations of $a = C/m$ and $b = \alpha/m$, we finally obtain a nonlinear differential equation of the second order with constant coefficients:

$$r^3\ddot{r} + br - a = 0.$$

Unfortunately this equation does not have standard solutions. However, if it is multiplied by

$$2\frac{\dot{r}}{r^3},$$

the first term will be equal to a derivative of

$$\dot{r}^2$$

and the equation becomes equivalent to the following expression:

$$\dot{r}^2 + 2\int \frac{a - br}{r^3} dr = b.$$

Integrating the second term and simplifying, we obtain

$$\dot{r}^2 = \frac{a}{r^2} - \frac{2b}{r} + b.$$

If we select $b = b^2/a$, then

$$\dot{r}^2 = \left(\frac{a^{1/2}}{r} - \frac{b}{a^{1/2}} \right)^2.$$

Taking the square of both parts, we can obtain a nonlinear differential equation of the first order:

$$\dot{r}^2 = \frac{a^{1/2}}{r} - \frac{b}{a^{1/2}},$$

which is transformed into an equation with separate variables:

$$r\dot{r} = -\frac{b}{a^{1/2}} r - a^{1/2}.$$

Substitutions of

$$f(\tau) = r^2; \quad r = f^{1/2}; \quad \dot{f} = 2r\dot{r}; \quad r\dot{r} = \frac{1}{2}\dot{f}$$

yield

$$d\tau = \frac{df}{2a^{1/2}\left(1 - \frac{b}{a} f^{1/2} \right)}. \tag{2.98}$$

Integrating Equation 2.98 gives

$$\tau = \frac{1}{2a^{1/2}} \int \frac{df}{1 - \frac{b}{a} f^{1/2}} + b.$$

Substitution of

$$1 - \frac{b}{a} f^{1/2} = \cos\varphi$$

yields

$$f^{1/2} = \frac{a}{b}(1 - \cos\varphi). \tag{2.99}$$

Differentiating Equation 2.99 gives

$$\frac{b}{2a}\frac{1}{f^{1/2}} df = \sin\varphi d\varphi.$$

Substituting the expressions for $f^{1/2}$ and simplifying, we can write

$$af = \frac{2a^2}{b^2}(1 - \cos\varphi)\sin\varphi d\varphi. \tag{2.100}$$

By means of Equations 2.99 and 2.100, Equation 2.99 is expressed as

$$\tau = a^{3/2}b^{-2}\left(\int \mathrm{tg}\varphi d\varphi - \int \sin\varphi d\varphi\right) + b,$$

which, on integration, yields

$$\tau = a^{3/2}b^{-2}(\cos\varphi - \ln\cos\varphi) + b.$$

Returning to the substitution of Equation 2.99 yields

$$\tau = a^{3/2}b^{-2}\left[1 - \frac{b}{a}f^{1/2} - \ln\left(1 - \frac{b}{a}f^{1/2}\right)\right] + b$$

or, on substitution,

$$f(\tau) = r^2,$$

$$\tau = a^{3/2}b^{-2}\left[1 - \frac{b}{a}r - \ln\left(1 - \frac{b}{a}r\right)\right] + b.$$

Upon selecting the moment of the time reference such that $b = 0$ and accounting for the designations of C and a and Equation 1.37.I, we have

$$\tau = \frac{2\sqrt{2}m^{1/2}C^{3/2}}{\alpha^2}\left[1 - \frac{r}{r_a} - \ln\left(1 - \frac{r}{r_a}\right)\right].$$

The differential of this expression is

$$\frac{d\tau}{dr} = \frac{2\sqrt{2}m^{1/2}C^{3/2}}{\alpha^2}\left(\frac{1}{r_a - r} - \frac{1}{r_a}\right)^2.$$

From here, through simple transformations, we find the rate of variations in the dimensions of the EM dipole

$$\dot{r} = \frac{\alpha^2 r_a}{2\sqrt{2}m^{1/2}C^{3/2}}\left(\frac{r_a}{r} - 1\right)^2.$$

By means of Equations 1.24.II, 1.37.1, and 1.67, this is expressed as

$$t = \theta\left(\frac{r_a}{r} - 1\right)^2 \tag{2.101}$$

because, by definition,

$$r \ll r_a \quad \text{and} \quad \frac{r_a}{r} \gg 1, \quad \left(\frac{2_a}{r} - 1\right) \approx \frac{r_a}{r},$$

Equation 2.101 is transformed into

$$t = \theta\left(\frac{r_a}{r}\right)^2.$$

By multiplying and dividing the right-hand side by α, and noting that

$$\frac{\alpha}{r_a^2} = \frac{27}{4}F_b,$$

where F_b is the maximum resistance force which follows from Equation 1.52 at the θ_s-temperature (at point b in Figure 1.14), we have

$$t = x\frac{1}{F_b r^2},$$

where

$$x = \frac{4}{27}\theta\alpha$$

is the parameter that depends on the physical nature of a specific chemical element.

Then, by multiplying the numerator and the denominator of the right-hand side by r and expressing the rotos volume in terms of $\mathbf{v} = r^3$, we obtain

$$t = x\frac{r}{\omega}, \tag{2.102}$$

where, in analogy with Equation 1.53 and according to the definition of the Ω-potential (Equation 1.111), $\omega = F_b\mathbf{v}$ is the thermodynamic potential of one rotos. Hence,

$$\omega = x\frac{r}{t} \quad \text{(I)}; \quad F_b = x\frac{r}{\mathbf{v}t} \quad \text{(II)}. \tag{2.103}$$

Graphical representation of Equation 2.103 is given in Figure 2.6. Directly proportional dependence

$$r = \frac{1}{x}\omega t$$

was plotted at $\omega = $ const, and inverse dependence

$$F_b = \frac{xr}{\mathbf{v}}\frac{1}{t}$$

at $r = $ const and $V = $ const. These functions change in opposite ways: an increase in one inevitably causes a decrease in the other and vice versa. Their superposition curve (curve 1) has a minimum; to find it, we differentiate the left- and right-hand sides of Equation 2.103.I:

$$d\omega = \frac{tdr - rdt}{t^2} \tag{2.104}$$

and, recalling that, according to the law of conservation of energy and momentum, $\omega = $ const for a given solid,

$$\frac{dt}{t} = \frac{dr}{r}. \tag{2.105}$$

Similar to Equation 1.69, this equation is of fundamental importance in the physics of deformation and fracture; it indicates the equivalence of thermal and deformation processes. In this connection, it will be referred to as the *relationship of equivalence*. Moreover, it allows extension of the concept of entropy (Equation 2.14) to the deformation processes. From the differential form of entropy given by Equations 2.13 and 2.105, it follows that

$$ds(t) = ds(r) . \qquad (2.106)$$

The relative variation in temperature and linear dimensions of the rotos leads to the same result — an increment in entropy, i.e., to a violation of the initial order in the AM system. Most importantly, Equation 2.105 shows that deformation can have no nature other than thermal. Unfortunately, today not much attention is paid to this consideration.[2-4]

At extreme point B both functions (Equation 2.103) are equal (Figure 2.6). This means that the thermal energy supplied to the rotos is consumed only for resistance to deformation and for nothing else. This moment corresponds to the effective temperature t_o. Such a possibility was predicted earlier in the analysis of the physical nature of the G-potential (Equation 2.36). Relationship 2.105 can be rewritten as

$$t_o = r \frac{dt}{dr}. \qquad (2.107)$$

It is similar to the definition of heat flow (Equation 1.71) or

$$q = \lambda \frac{dt}{dr}. \qquad (2.108)$$

It turns out that, under any other conditions except $t = t_o$, thermal conductivity λ has an inversely proportional dependence on dimensions of the EM dipole: $\lambda = 1/r$, whereas at $t = t_o$ this dependence becomes directly proportional, i.e., $\lambda = r$. This indicates the possibility of obtaining the resistance work due to an external heat source (Equation 2.36).

As seen from Figure 2.6, the range of stability of the rotos in external fields is limited on both sides. A loss in stability in the zone of low temperatures is caused by a continuous increase in the resistance forces F_b and a catastrophic decrease in deformability r. This is explained by the fact that, as the temperature decreases, atoms constituting the rotos approach each other at such a distance that they share the surface shells as well as the deeper shells (Figure 1.3). In this case, a decrease in r inevitably leads to a substantial increase in λ in the escape of the kinetic part of the overall energy to the environment and, as a result, to a certain change in the shape of the potential wells in a direction of 3 to 2 to 1 (Figure 1.10b). This is accompanied by a

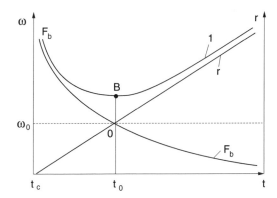

FIGURE 2.6
Temperature dependence of resistance energy of the paired interatomic bond.

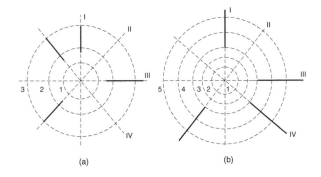

FIGURE 2.7
Condensation of rotoses into polyatomic molecules of brittle (a) and ductile (b) materials.

dramatic increase in F_b, which is the cause of embrittlement of all solids at low temperatures.[2–4,44]

Violation of stability on the right side (where $r \to \infty$ and $F_b \to 0$) occurs for another reason. An increase in dimensions of the EM dipole D leads to escape of the deep electron layers of atoms from "engagement" and to the possibility of their free movement in the outside shell (Figure 1.3). The rotos becomes plastic ($r \to \infty$) and its resistance catastrophically decreases ($F_b \to 0$). For this reason all bodies in a range of high temperatures can easily be deformed. These laws are valid also at submicroscopic depths; they penetrate through all of the scale levels and show up very clearly in the experiment.[2,3,44]

The rotos (Figure 1.7b) can be condensed into polyatomic molecules of brittle and ductile bodies at characteristic temperatures θ_f (Figure 2.7a) and θ_s (Figure 2.7b). Arabic numerals designate electron quantum numbers $j = 1 \ldots 5$ Equation 1.4, while Roman numerals stand for the nuclear ones $M = 1 \ldots$ IV (Equation 1.27). The former determine electron shells and the latter the azimuth of orbital planes, A. In accordance with Equations 1.63 and 1.64, solid bold lines on them show the location of an orbital ring within which the size of the EM dipole, D, varies from D_{min} to D_{max}.

From the formal standpoint, the formation of a solid as a result of series of elementary acts of addition of new and new flat dipoles to the initial system (Figure 2.7) can be represented as a space condensation of individual atoms into a cooperative structure, where the energy state and mobility of each link cannot be independent of the nearest neighbors. In terms of thermodynamics, this means that joining a large number of atoms into one super molecule, i.e., solid body, results in a decrease in entropy due to localization and fixation of the members of set N, which were independent before solidification. This adds new properties of the AM system that differentiate the solid state from other aggregate states. The form of condensation and subsequent processes associated with its conservation or variation cannot be explained from the standpoint of conventional concepts of the electrostatic character of interaction between atoms: ion, covalent, van der Waals, or other.[6,26,34]

The theories of elasticity and plasticity[58] and strength of materials[3] show that, for isotropic materials, the state of stress at a point is fully determined by the tensor of stresses, whose spherical component characterizes a change in volume and deviator — a distortion in shape of an element. The left-hand part of Equation 2.65 fills out these universally recognized but formal concepts with a clear physical content (Figure 2.7). Parameter $P = F/D^2$ is the Lorentz force (Equation 1.16) generated by each of the EM dipoles (Figure 1.7b) from the elementary plane D^2 (in fact, from a point),[24] which are located in space according to the law set by Equation 1.31 (the marked segments in orbital planes I to IV in Figure 2.7).

During solidification, the initial state of a body forms and is characterized by the initial entropy **s** and volume V. The external fields (force, thermal, radiation, magnetic, electrical, etc.), aggressive environments, and vacuum change them and thus change all parameters of Equation 2.65, including V. This volume is responsible for the scale effect in the case of macroscopic variations in volume ΔV[24] and change in the size and shape of the body during deformation dV under conditions where $\Delta V = 0$. Parameters **P** and V are interrelated and form the Landau thermodynamic potential Ω (Equation 1.111).[23] Owing to its variation, the body resists external effects. Numerous experiments on deformation of various classes of materials prove an inversely proportional relationship between these two parameters.[2,3,44,45]

The notion of $V = \beta r^3 N$ makes it possible to write, at $T = $ const, the differential of the Ω-potential in the following form

$$d\Omega = FrNd\beta + \beta NFdr + \beta rNdF + \beta rFdN . \qquad (2.109)$$

This equation's deviation from the equilibrium value of this potential is composed of orientation of rotoses as a whole (the first term) in a direction of the external effect (Figure 2.8a), followed by differentiation of their sizes

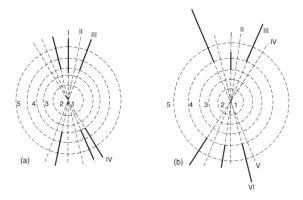

FIGURE 2.8
Schematic of orientation (a) and destruction (b) processes occurring in molecules under the effect of different external fields.

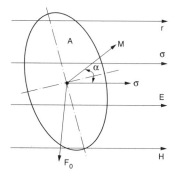

FIGURE 2.9
Schematic of the effect of the external force, σ, electric, $\mathbf{E}$, and magnetic, $\mathbf{H}$, fields on electromagnetic dipoles.

(the second term) and capability of generating the resistance forces (the third term; see Figure 2.8b), which eventually ends with disintegration of some of them (the last term).

Indeed, in electric $\mathbf{E}$, magnetic $\mathbf{H}$, force σ, and other fields, mechanical momentum M of a rotos (Figure 2.9) is affected by strong Coriolis forces F_o (Equation 1.104) which, by decreasing angle α, tend to orient orbital plane A in the direction of the field (or opposite to it) and transforming the isotropic configuration of a polyatomic molecule (Figure 2.7) into the anisotropic configuration (Figure 2.8a). In the true diagram of deformation of ductile materials (Figure 2.10), which has the identical form in tension, bending, and torsion[3,59] (Figure 1.23), the orientation mechanism (Figure 2.8a) forms elastic stage $0a$, while the degradation mechanism (Figure 2.8b) forms plastic stage ab, where ε are the longitudinal strains.

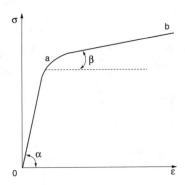

FIGURE 2.10
True stress–strain diagram for tension, bending, and torsion.

All parameters in Equation 2.96, except θ, are universal. Expressed in terms of mass m, valence z, and common charge of atoms e, (Equation 1.95), it determines properties of a particular structure. By relating the equation of state (2.96) to Mendeleev's table, the θ-parameter transforms this equation into the periodic law, which covers the entire diversity of solids.[24]

Differentiating Equation 2.94 gives

$$d\delta_f = \left[\frac{9}{8}\frac{1}{y} - 3y^2\right]\left(\frac{dt}{t} - \frac{d\theta}{\theta}\right) \qquad \text{(I)};$$

$$d\delta_s = \left[\frac{3\pi}{5}y^3\right]\left(\frac{dt}{t} - \frac{d\theta}{\theta}\right) \qquad \text{(II)}, \qquad (2.110)$$

where

$$y = \frac{t}{\theta}.$$

The definition of entropy (Equation 2.13) and the relationship of equivalence (Equation 2.106) allow one to write the expression in parentheses in the form

$$\left(\frac{dt}{t} - \frac{d\theta}{\theta}\right) = ds(t,r) - ds(\theta), \qquad (2.111)$$

where $ds(t,r)$ is the change in the configuration part of entropy of the inter-atomic bond in thermal t or force r deformation, and $ds(\theta)$ is the same caused by deviation of the Θ-temperature. Hence, varying parameters θ, t, and r can control entropy.

Figure 1.10b shows the variation in shape of a local potential field with the θ-temperature, i.e., in fact, $ds(\theta)$. Rotos parameter r_a was selected as an indicator of the position of this curve (Equation 1.26.II) on the axis of abscissas. For example, let the initial state of the interatomic bond be determined by a function of interaction with parameter a_1. Differential $ds(\theta)$ shifts it to position a_2 or a_3. At θ = const and, therefore, at r_a = const, the curve is fixed in the energy field, and its shape (positions 1, 2, 3, etc.) becomes dependent only on the value of the kinetic part of the overall internal energy (Equation 1.26.I).

In the range of $t > \theta$, for plastic structures the Θ-temperature is degenerated into the temperature of brittle fracture t_c, whose variation is set by curve 2 (Figure 1.13). As a result, the range of stability of a rotos $t_c - \theta_s$ (determined by the distance between curve 2 and the right branch of curve 8) becomes dependent on the heat content (Equation 2.27). Heat escaping into the environment initiates a shift of the state diagrams in a direction of 6 to 5 to 4 and thus increases the probability of failure of metal structures in the force fields (Figure 1.22).

The failure-free operation depends on how far the actual temperature of rotoses t is from the t_c and Debye θ_s temperatures. This distance is determined by Relationship 2.111. Also, it assigns the sign of the binomial in brackets, which radically changes the character of the interaction. A change in sign occurs at

$$\frac{9}{8}\frac{1}{y} - 3y^2 = 0 .$$

Hence, $y = 0.72$, and the temperature at which brittle fracture begins is equal to $t_c = 0.72\theta_s$. Therefore, the degradation changes in a structure occur within a temperature range of $0.72\theta_s$ to θ_s. The probability of disintegration of the interatomic bonds is the highest in this range. The estimation conducted earlier (Equation 1.79) and that proved by the experiment (Equation 1.102) show that, when the actual temperature decreases by 10% with respect to the Debye temperature θ_s, brittle fracture occurs in the majority of structural materials (Figure 1.22).

These consequences can be avoided by creating materials whose $\theta(m, z, e)$ and thus the location and shape of the potential wells (Figure 1.10) are such that, in the operational range of variations of the external thermal fields, which cause transformation of the shape following the law given by Equation 1.26.I, the actual temperature of rotoses t_i is always between θ_s and $0.72\theta_s$. This is a traditional approach in materials science,[60,61] where structures are normally upgraded by varying the qualitative (m, z, e) and quantitative indicators of alloying elements. Because such methods of controlling the entropy are implemented only at the manufacturing stage, they should be classed as passive or materials science methods.[12]

Equation 2.97 determines the microscopic level of entropy s_1, whereas in Equation 2.65 all the parameters (including s) are of the macroscopic level.

This formula expresses the degree of perfection of structure of a given body with respect to the environment at microscopic s_1, mesoscopic s_2, and macroscopic s_3 levels, i.e.,

$$s = s_1 + s_2 + s_3 \tag{2.112}$$

where

- s_1 characterizes the type of structural organization of the AM system in the Cartesian V (Figure 1.7b) space and energy T (Figures 1.13 to 1.15) domain.
- s_2 describes the degree of imperfection of a supermolecular complex: size and homogeneity of the grain composition, structure of the boundary regions, microdefects, etc.
- s_3 is composed of different types of macrodefects: cracks, porosity, dents, scratches, tool marks, drastic transitions from one geometrical shape to another, etc.

While the macroscopic part of entropy s_3 can be comparatively easily formed by structural or technological methods, the methods of controlling its microscopic, s_1, and mesoscopic, s_2, parts are not that evident. Under the effect of external factors, their parameters N and δ change in magnitude and direction (Figures 1.15 and 2.8) to form configurational s_c (affected by the orientation processes), phase s_f (changes under the effect of phase transitions), and destruction s_d (formed due to the flow of failures of rotoses) components of entropy, i.e.,

$$s_{1,2} = s_c + s_f + s_d \tag{2.113}$$

As follows from our considerations, entropy of a solid body is a multistage (Equation 2.112) and multifactorial (Equations 2.83 and 2.97) value.

The close analogue to Equation 2.65 is well known in physics[11,49] as the Clapeyron-Mendeleev equation of state for an ideal gas,

$$PV = RT, \tag{2.114}$$

with the only difference that the universal gas constant $R = kN$ is converted here into entropy (Equation 2.97) $s = kN\delta$. Apparently, nature is not so wasteful as to create a new set of laws for each aggregate state of a material (gaseous, liquid, and solid).

In the first case, N implies the quantity of paired interatomic bonds contained in volume V of a solid, while in the second, it is the quantity of noninteracting atoms of gas in the same volume. It can be seen that these equations are identical in form, differing only in parameter δ (Equation 2.92). The latter is responsible for interactions of atoms. This analogy is natural, highlights the internal unity of the aggregate states of a material, and reveals their fundamental difference. The internal unity of the aggregate states is that these states are of a material nature and include the same atoms and molecules. They differ only in the level of heat content.

The quantitative difference of solids grows into a new quality: formation of interatomic bonds. Unlike the unbound state (Equation 2.114), where the Ω-potential depends only on temperature T, the bound state (Equation 2.65) is determined by a character of distribution (Equation 2.83.V) of the free part (Equation 2.83.I) of the overall internal energy among the members of set N. This suggests that parameter V should be considered differently. If it determines the three-dimensional space limited by the external walls for gases and liquids, for solids it determines the capability of preserving their geometrical dimensions and shape (the order of the AM system acquired during solidification).

Summarizing the results obtained thus far, we note the following. Equation 2.65 has the same fundamental importance for solids as Equation 2.114 does for gases.[31,32] The state of gases is described by parameters **P**, V, and T, for solids; however, another one, though multifactorial (i.e., **s**), supplements them. Also, two of them become vectorial (**P** and **s**). As a result, the processes that occur in solid structures under the effect of external fields are characterized by extreme complexity and diversity.[2–4,19] Equation 2.65 describes the correlation between them adequately for all solids, without any exception, whatever conditions they are in[12,24] and allows solutions of many practical problems by not-yet-traditional methods. Internal stresses σ generated by a crystalline structure, as well as pressure **P** in a general physical sense,[31,32] initially have tensor nature (Figure 2.8).

Clear evidence of dependence of resistance of a body on external effects on the volume is the scale effect. It has a physical nature[24] rather than a statistical nature, as is currently thought.[3]

The effect of temperature on physical–mechanical properties is proved by numerous experiments[2,3] and forms the basis of the kinetic concept of strength.[10] Structural imperfections and various defects find their expression in entropy **s**.

2.6 Brittle and Ductile Structures

Regarding reaction to electric fields, solids are subdivided into conductors, semiconductors, and dielectrics,[6,26] whereas in magnetic fields they behave as diamagnetics, paramagnetics, and ferromagnetics.[7,30] When it comes to the reaction of solids to mechanical and thermal fields, there is no distinct classification of their behavior. The shape of the state diagram with two extrema (Figure 1.14) and the possibility of splitting the bonds into two different types (Equation 2.84) provide convincing evidence that they do not react to force and thermal fields as homogeneous and single-phase media.[24]

Therefore, solid-state physics and modern theories of strength do not allow explanation of many effects accompanying thermal and force deformation and fracture.[2,3]

During solidification of a solid, parameter δ (Equation 2.92) distributes paired interatomic bonds of set N in the energy (right-hand side of Equation 2.65) domain and Cartesian (left-hand side) space with respect to extrema (points a and b in the state diagram in Figure 1.14) following the law given in Equation 2.94. Positions 8 to 15 in the lower field of Figure 1.14 show various locations of the MB factor in the energy domain. Positions 8 to 10 refer to the clearly defined brittle materials (cast irons, natural and synthetic stones, ceramics, etc.) and those with numbers 11 to 15 refer to ductile materials (most metals and alloys). One might ask a logical question: what laws govern the process of distribution of the bonds?

Consider thermal interaction between a body and its environment in the absence of other fields. In practice, this is a very common case because artificial or natural materials very rarely have a directional thermal control. Assume that the left-hand part of the basic thermodynamic relationship (2.12), written for a rotos, becomes[49]

$$dq = \lambda dt \tag{2.115}$$

and the right-hand part is the differential of the sum of potential $u(r)$ and kinetic $u(t)$ parts of the overall energy

$$\frac{\lambda}{dv} - C_{p,v} = F, \tag{2.116}$$

where

$$C_{p,v} = \left[\frac{\partial u(t)}{\partial t}\right]_v = \left[\frac{\partial u(t)}{\partial t}\right]_p$$

is the heat capacity and

$$F = \left[\frac{\partial u(r)}{\partial v}\right]_t = \left[\frac{\partial u(r)}{\partial v}\right]_v$$

is the generalized force. Using the known relationships between thermodynamic parameters,[2] we can write

$$\frac{\lambda}{dv} - C_{p,v} = -\left(\frac{\partial \omega}{\partial v}\right)_t = t\left(\frac{\partial p}{\partial t}\right)_v, \tag{2.117}$$

where ω is the thermodynamic potential of an individual rotos (Equation 1.53) and p is the pressure it creates. Invariable parameters at which the

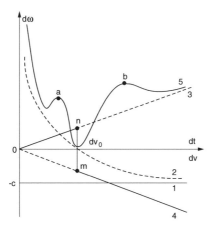

FIGURE 2.11
Differential character of variations in the potential of rotos with variations in the phonon radiation and volume.

partial derivatives in the brackets of Equation 2.117 are calculated are shown as subscripts.

Figure 2.11 shows the differential character of variation in potential $d\omega$ with variation in local temperature dt and volume dv. Number 1 designates the negative level set by heat capacity $C_{p,v}$. Curve 2 shows the inverse-proportional dependence of thermal conductivity λ on volumetric deformations dv. It intersects the axis of abscissas at point dv_o, where the expression given in brackets in the left part of Equation 2.117 changes sign to the opposite sign. Straight lines 3 and 4, outgoing from the origin of coordinates, are the two parts of the plot of function

$$d\omega = ydt \qquad (2.118)$$

where the expression in the left part of equality 2.117), i.e.,

$$y = \frac{\lambda}{dv} - C_{p,v}$$

serves as parameter y.

At point dV_o parameter y changes sign from plus to minus. Thus, the direct proportionality 2.116 interrupts here. In the upper half-plane it propagates upward from the origin of coordinates to point dv_o (region 0 to n). Its continuation from point m is located in the lower half-plane (solid line directed downward from point m). The product of two negative dependencies in the lower half-plane gives a positive value of the total function (left part of expression 2.117). It persists in the positive half-plane within the entire range of variation in arguments dv and dt.

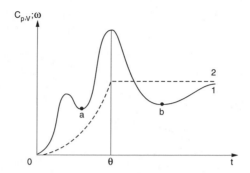

FIGURE 2.12
Phonon dependence of the thermodynamic potential, 1, and heat capacity, 2, of rotos.

Number 5 designates the plot of this function. It shows the variation in differential $d\omega$. The temperature dependence 1 of primitive ω is shown in Figure 2.12. For comparison, the temperature dependence of heat capacity $C_{p,v}$ (curve 2) is shown in the same field. Its original is shown in Figure 1.16b.

Our analysis of dependencies

$$d\omega = f(dv, dt) \quad \text{and} \quad \omega = f(t)$$

results in the following conclusions of high generalization. First, the fundamental feature of these dependencies is that they both have two local extremes at points a and b. The presence of these extremes creates some ambiguity in solidification: interatomic bonds of a specific material have two options, i.e., localization near one or other minimum. This ambiguity is the result of a two-phase (compresson–dilaton) structure of rotoses. Second, thermal interaction (presence of term dQ in Equation 2.1) causes a substantial change in kinetic characteristics of a solid — an interaction that always takes place and is an integral peculiarity of the surrounding world.

This is so natural that we sometimes pay no attention to it. It accompanies the body from the moment of solidification to destruction of its structure. For this reason the behavior of a material at any moment of time is determined by two parameters, such as heat capacity $C_{p,v}$ and thermal conductivity λ. Now we are ready to answer the question of what these parameters depend on and how they change.

All parameters of a solid are determined by the state function Z (Equation 2.66) of the set consisting of N rotoses. Any characteristic of a rotos, by definition (see Section 1.5), can be split into two components: rotational (index φ) and oscillatory (index r). To the first approximation, these components can be considered independent of each other so that the state function can be thought of as

$$Z = Z_r Z_\varphi. \tag{2.119}$$

The energy of a quantum harmonic oscillator is[50]

$$u_r = \hbar v \left(n + \frac{1}{2} \right)$$

and its state function is

$$Z_r = \exp\left(-\frac{\hbar v}{kt_i} \right) \sum_{n=0}^{\infty} \exp\left(-\frac{\hbar v n}{kt_i} \right) \tag{2.120}$$

Summing up an infinitely diminishing geometrical progression yields

$$Z_r = \exp\left(-\frac{\hbar v}{kt_i} \right) \left[1 - \exp\left(-\frac{\hbar v}{kt_i} \right) \right]. \tag{2.121}$$

Thus, the state function and all thermodynamic parameters of a solid are determined by the following relationship:

$$-\frac{\hbar v}{kt_i}. \tag{2.122}$$

Accounting for the definition of the Debye temperature (2.87), it takes the following form:

$$\frac{\theta}{t_i},$$

and it can be expressed as

$$Z_r = \exp\left(-\frac{\theta}{t_i} \right) \left[1 - \exp\left(-\frac{\theta}{t_i} \right) \right]^{-1}.$$

The mean oscillatory energy of N rotoses can be found using the formula

$$U_r = N u_r = N k T^2 \frac{\partial}{\partial T} \ln Z_r = \frac{N \hbar v}{2} \operatorname{Cth}\left(\frac{\hbar v}{2kT} \right). \tag{2.123}$$

Thus, the oscillatory part of heat capacity is

$$C_r = \frac{\partial U_r}{\partial T} = \frac{Nk}{4} \left(\frac{\hbar v}{kT} \right)^2 \frac{1}{\operatorname{Sh}\left(\dfrac{\hbar v}{kT} \right)}. \tag{2.124}$$

The mean values of the oscillatory part of the energy and heat capacity are the complex functions of temperature t and intrinsic frequency of rotoses, v. We can specify the limits of these functions on both sides of the Debye temperature. At $t \gg \theta$, expanding the exponential function into a series and limiting it to the first terms of the expansion yields

$$U_r = ktN \quad \text{(I)}; \quad C_r = kN \quad \text{(II)} \tag{2.125}$$

If $t \ll \theta$, $\exp\dfrac{\theta}{t} \gg 1$, so

$$U_r = \frac{1}{2}kN\theta + kN\theta\exp\left(-\frac{\theta}{t}\right) \quad \text{(I)}; \quad C_r = kN\left(\frac{\theta}{t}\right)^2 \exp\left(-\frac{\theta}{t}\right) \quad \text{(II)}. \tag{2.126}$$

As the temperature decreases, the oscillation energy tends in the limit to a constant

$$U(r_a) = \frac{1}{2}N\hbar v_a. \tag{2.127}$$

This limit characterizes the zero oscillation energy and corresponds to the moment of transfer of the elliptical path into the circular one at a temperature of brittle fracture, Θ, at the bottom of the potential well (Figure 1.7b). Equation 2.127 represents the statistical mean of individual values (Equation 1.39).

The dependence of the mean energy of an oscillator on the t/Θ ratio according to Equation 2.156 is shown in Figure 2.13. Designations of levels r_a and r_b correspond to temperatures Θ and θ_s (Figure 1.14). Although it determines critical temperatures, this dependence nevertheless fails to give a comprehensive idea of the location of the rotos on the temperature scale because the energy of the oscillator is determined by two independent variables: temperature t and frequency v. Therefore, setting the temperature (or level in the potential well) does not yet characterize its entire energy. It is

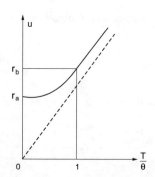

FIGURE 2.13
Dependence of the mean energy of rotos on the relative temperature.

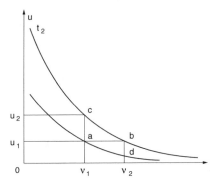

FIGURE 2.14
Dependence of the energy of rotos on the frequency of oscillations of its atoms at fixed temperatures t_1 and t_2.

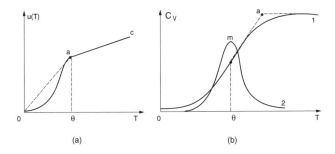

FIGURE 2.15
Temperature dependencies of kinetic energy (a) and heat capacity (b) of solids.

necessary to indicate additionally its natural frequency, i.e., we must show on which side of the Θ-temperature it is situated. Considered at the same temperature, two rotoses with different natural frequencies have different energy. Therefore, for performing the same useful work, they should have the same ω-potential (Figure 1.7a).

Figure 2.14 shows dependence of the energy of an oscillator on the frequency at a fixed temperature. As seen, the same energy level (e.g., U_1) is provided by the bonds having different frequency of oscillation of atoms, v_1 and v_2, or, accordingly, temperature t_1 or t_2 (points a and b). At the same frequency (e.g., v_1) but different temperature, it is possible to obtain different energy U_1 or U_2 (points a and c). This proves that interatomic bonds in a solid may be in two different quantum states, which differ in temperature or in frequency. The difference lies in the Debye temperature.

Figure 2.15 shows the known dependence of the oscillatory heat capacity of the AM system (curve 1) on the temperature.[49] Describing thermal properties of the system as a whole, this function, in fact, is the integral function of distribution of rotoses by energy. (Their individual characteristics are shown in Figure 1.16.)

Number 2 designates the corresponding differential distribution curve; it has a two-branch shape with maximum at point m. Its apex coincides with the point of inflection of curve 1 and is identified with the Debye temperature. The main contribution to the oscillatory heat capacity is made by the bonds whose intrinsic temperature lies on the right branch of curve 2 (with respect to point m). More rigid and "cold" bonds lie on the left branch; as the temperature decreases, their oscillations are gradually "frozen." That is, splitting the bonds of a solid into two phases, i.e., cold or compresson and hot or dilaton phases, can be shown in the direct and most natural manner if they are identified with the quantum oscillators.

Levich[20] presented proofs to the fact that heat capacity of the AM system, which can be in two quantum states, does have the form of a function with maximum. The function of state of such a system can be expressed as

$$Z = \sum_{i=1}^{n} \exp\left(-\frac{U_i}{kt}\right)\delta(U_i),$$

where U_i are the quantum energy levels and $\delta(U_i)$ is the number of possible states of a bond whose energy is equal to U_i (Equation 2.92). For simplicity, we limit our considerations to the two energy levels (e.g., U_1 and U_2, Figure 2.14) and, introducing designations of $\delta_f = \delta(U_1)$ and $\delta_s = \delta(U_2)$, write

$$Z = \delta_f \exp\left(-\frac{U_1}{kt}\right) + \delta_s \exp\left(-\frac{U_2}{kt}\right) = \delta_f \exp\left(-\frac{U_1}{kt}\right)\left(1 + \frac{\delta_s}{\delta_f}^{-\frac{U_2-U_1}{kt}}\right). \qquad (2.128)$$

Assuming that U_1 and U_2 belong to the compresson and dilaton phases, i.e., are located on the opposite sides of the Θ-temperature, we have $U_2 - U_1 = k\theta$. Then Equation 2.128 can be expressed as

$$Z = \delta_f \exp\left(-\frac{U_1}{kt}\right)\left[1 + \frac{\delta_s}{\delta_f}\exp\left(-\frac{\theta}{t}\right)\right]. \qquad (2.129)$$

As seen from Equations 2.128 and 2.129, in the left extreme zone of the cold branch of curve 2 (Figure 2.15b) where $\theta \gg t$, the second term can be neglected, so that

$$Z = \delta_f \exp\left(-\frac{U_1}{kT}\right).$$

This means that, at a given temperature, the probability that the system gets into the phase space with energy U_2 is very small. If N bonds are found in a unit volume, then the function of their states becomes

$$Z = \frac{1}{N!}\left[\delta_f \exp\left(-\frac{\delta_f}{kt}\right)\right]^N \qquad (2.130)$$

from where the energy and heat capacity are determined as

$$U = kT^2 \frac{\partial}{\partial t}\ln Z = U_1 N; \quad C_v = \left(\frac{\partial U}{\partial t}\right)_v = 0$$

As follows, the existence of the "hot" phase at low temperatures has no effect on thermodynamic properties of the "cold" phase. In fact, they both may behave as independent systems with no energy exchange. As follows from Equation 2.92, it is this situation that occurs between the compresson and dilaton phases in solidification of brittle and ductile materials. They are separated in temperature range, as shown in Figure 1.14.

When inequality $\theta \gg t$ is not fulfilled, it is necessary to keep both terms in the function of state 2.129:

$$Z = \delta_f \exp\left(-\frac{U_1}{kt}\right)\left[1 + \frac{\delta_s}{\delta_f}\exp\left(-\frac{\Delta U}{kT}\right)\right].$$

Equation 2.130 then becomes

$$Z = \frac{1}{N!}\left[\delta_f \exp\left(-\frac{U_1}{kt}\right)\right]^N\left[1 + \frac{\delta_s}{\delta_f}\exp\left(-\frac{\Delta U}{kT}\right)\right]^N.$$

From there, following the known rules, we find the energy and heat capacity

$$U = kT^2 \frac{\partial}{\partial T}\ln Z = NU_1 + \frac{N\delta_s \Delta U \exp\left(-\dfrac{\Delta U}{kt}\right)}{\delta_s\left[1 + \dfrac{\delta_s}{\delta_f}\exp\left(-\dfrac{\Delta U}{kt}\right)\right]} \qquad (2.131)$$

$$C_v = \left(\frac{\partial U}{\partial T}\right)_v = kN\frac{\delta_s}{\delta_f}\left(\frac{\theta}{t}\right)^2 \exp\left(-\frac{\theta}{t}\right)\left[1 + \frac{\delta_s}{\delta_f}\exp\left(-\frac{\theta}{t}\right)\right]^{-1}. \qquad (2.132)$$

The variation in heat capacity in accordance with Equation 2.132 is shown in Figure 2.16. Numbers 1 and 2 designate plots with different ratios of the

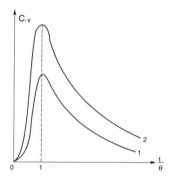

FIGURE 2.16
Variations in heat capacity of the two-phase structure.

energy levels $U_2/U_1 = 1$ and $U_2/U_1 = 2$. As the temperature increases, the heat capacity grows and eventually reaches maximum, where the t/θ ratio becomes equal to one, and then again vanishes with subsequent increase in this ratio. Such behavior is a characteristic feature of two-phase AM systems that are able to make interphase transitions. The cause of vanishing of C_v in the high-temperature region becomes clear from Equation 2.131, which, written for very high temperatures, becomes

$$U \approx NU_1 + \frac{\delta_s}{\delta_f} N \frac{\Delta U}{1 + \frac{\delta_s}{\delta_f}} = \text{const.}$$

As seen, the energy becomes independent of the temperature. Physically, this means that thermal excitation reaches saturation at $t \gg \theta$ because heat capacity of an individual rotos within a range of $t > \theta$ does not depend on the temperature (Figure 1.16b). The presence of maximum at $t/\theta = 1$ is attributable to an increase in the required heat to ensure the process of transition of part of the AM system from one phase into the other.

It is known in quantum mechanics[27] and statistical physics[50] that the function of state Z and thermodynamic parameters derived from it for a rotational component have a manner of variations similar to oscillatory components. Therefore, by performing multiplication in Equation 2.119, one should not expect any fundamental changes in the essence of the previous consideration on heat capacity. It just enhances them, especially in the compresson region, where $t < \theta$.

Now we are ready to elaborate a bit more on the primary reason of splitting of set N into the cold N_f and hot N_s subsets formed in the acceleration well q_2dc and in the zone of the deceleration barrier cbq_1 (Figure 1.14). It is likely that the cause is in the fact that Equations 1.30 and 1.80 do not account for the spin quantum number of nuclei, s (see Section 1.2). Like electrons (Figure 1.1),

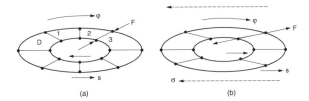

FIGURE 2.17
Rotoses with parallel and antiparallel spins of atoms in free (a) and excited (b) states.

united in pairs, they may have parallel and antiparallel natural rotation (comparable in astronomy to the 24-hour rotation of the Earth about its axis).

Figure 2.17 shows a schematic of the orbital rotation of the rotos nuclei (in a direction of arrow φ) with the antiparallel (Figure 2.17a) and parallel (Figure 2.17b) spins (designated by arrows s). Positive charges rotating on the elliptical paths create closed electric currents. Parallel currents generate magnetic repulsive forces (designated by arrows F), and antiparallel currents create attractive forces (Equation 1.13). As a result, atoms of rotoses shown in Figure 2.17a are always compressed (compressons) and those shown in Figure 2.17b are always subjected to the repulsive forces (dilatons).

In the equilibrium state, both nuclei rotate so that, in any of positions 1, 2, 3, etc., the distance between them always remains unchanged and equal to D (Equation 1.15). This allows representing the EM dipoles as segments D_2 and D_5 in the state diagram of the rotos (Figure 1.14). In polyatomic molecules (Figure 2.7), the orbital quantum numbers l distribute them over different azimuthal planes I, II, etc. creating, in the local space, a certain order characterized by entropy. The external fields (shown by arrow φ in Figure 2.17b) orient them in the direction of the field to deform isotropic molecules as shown in Figure 2.8, thus changing entropy s.

The MB factor (Equation 2.92) distributes rotoses N in the energy, T, domain and in the Cartesian, V, space. In fact, the evaluation of the probabilities allows the δ-factor (Equation 2.84) to be written as

$$\delta = n(x,y,z)/N = \frac{3\pi}{3\beta}\exp\left(-\frac{U}{kT}\right), \tag{2.133}$$

where $n(x, y, z)$ is the number of the interatomic bonds that have the same energy but are located at different points of the three-dimensional space. Hence,

$$n = \frac{2\pi}{3\beta}\exp\left(-\frac{U}{kT}\right)N.$$

The presence of the Boltzmann factor in the equation of state (2.65) leads to a fundamental change in the conventional concepts of the energy level of a material and its structure affected by the external fields and aggressive

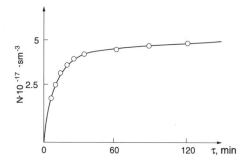

FIGURE 2.18
Time dependence of the concentration of free radicals in loaded oriented Capron ($\sigma = 71$ MPa, $T = 20°$C).

environments. In fact, compresson–dilaton structure should be considered because it inevitably shows up at micro- and macroscopic scale levels.

Results of investigation into the kinetics of accumulated damages in the molecular structure of polymers using the method of electron paramagnetic resonance are discussed in Regel and Slutsker.[62] According to this method, breaks of chemical bonds in a polymer are revealed by the formation of free radicals with nonpaired electrons at the broken ends of the chains. Figure 2.18 shows the typical dependencies of the concentration of the broken bonds in oriented Capron N during the time when a specimen is kept under a load τ. This work contains evidence that this dependence is exponential.

Mass spectrometry of polymers also evidences that this type of dependencies in mechanical–chemical processes is universal.[63] This method allows evaluation of the rate of accumulation of broken molecules by the kinetics of formation of volatile products from a solid polymer. As proven, the rate of their formation exponentially increases depending on the value of tensile stresses (Figure 2.19). As noted, volatile products start forming from the moment of load application. This is natural, because, even in the free state, some bonds have the Θ-temperature. They are the first to initiate the CD transitions from the moment of application of the load. The destruction of these bonds leads to the formation of volatile products.

Subsequent stages of the mechanical destruction process are characterized by continuous coarsening of the scale of fracture: from centers of nucleation of submicroscopic defects (tens and hundreds of angstroms) to micro- (few and tens of microns) and macro- (tenths of and few millimeters) cracks. As proven by experiments,[62–64] a change in scales does not violate the exponential law that governs occurrence of the thermodynamic process of disintegration of a structure. Figure 2.20 shows the kinetics of formation of submicrocracks in oriented Capron. The results were obtained by measuring the scattering of x-rays at small angles, which allows detection of defects from 10 to 10^3–10^4 angstroms in size and estimation of their concentration, which can reach to 10^{12}–10^{17} per cubic centimeter. It was found that the concentration of

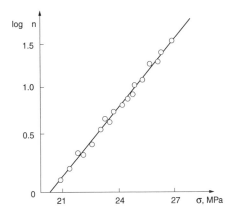

FIGURE 2.19
Dependence of logarithm of the yield of volatile products with respect to tensile stresses in a polystyrene specimen at $T = 20°C$.

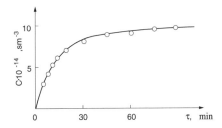

FIGURE 2.20
Time dependence of concentration C of submicroscopic cracks in oriented Capron at $\sigma = 23$ MPa and, $T = 20°C$.

submicrocracks at the moment of saturation increased to such an extent that the distance between them became comparable with their sizes. This leads to interaction of microdefects and their accelerated growth toward each other.

Results of investigations of the kinetics of cracking of glassy polymers are given in Regel et al.[10] and Regel and Slutsker.[62] Typical diagrams of growth of length of a microcrack, l, with time τ at a constant test temperature and variable stresses are shown in Figure 2.21. As seen, the CD transitions that underlie the destruction processes can take place at any level of external stresses. Their physical mechanism remains constant and only the intensity of the process varies.

For macroconsiderations, we write the differential of both parts of the equation of state (2.65) as

$$V d\mathbf{P} + P dV = T d\mathbf{s} + s dT . \tag{2.134}$$

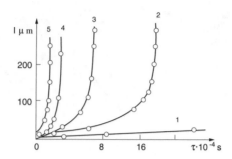

FIGURE 2.21
Time dependence of length of an avalanche crack in a polymer at $T = 20°C$ and different stresses σ, MPa: 1, 5.8; 2, 7.1; 3, 7.4; 4, 7.6; 5, 7.8.

This shows that any change in state of the AM structure always occurs in two successive stages:

$$V d\mathbf{P} = \mathbf{s} dT \quad (I); \quad \mathbf{P} dV = T d\mathbf{s} \quad (II). \quad (2.135)$$

Expressing absolute values of the macroscopic parameters and their differentials in terms of microscopic variables in Equations 1.24.II, 1.111, 2.97, and 2.109 makes it possible to reveal their mechanism:

$$-2\beta N dt + \beta r N dF = k N \delta F dr \quad (I);$$

$$3\beta N dt + \beta r N dF = kT(N d\delta + \delta dN) \quad (II). \quad (2.136)$$

The first stage is associated with ordering the location of the EM dipoles in space dr at constant r, β, and N (Figure 2.8a). This leads to an increase in the resistance forces dF and is accompanied by absorption of heat ($-dt$). The second stage is characterized by redistribution of the internal energy within the MB-factor, $d\delta$ (change in positions in a direction of $1/2T$ to T to $2T$ in Figure 2.2) or a decrease in the number or type of the AM bonds, dN. Such processes are inevitably accompanied by differentiation of sizes of the EM dipoles (Figure 2.8b) dN and by intensive heat release ($+dt$).

Numerous experiments[2,3,19] prove that the state of materials (at least in the force fields) changes following the law set by Equation 2.136. This can be easily proved if we rewrite Equation 2.134 as

$$dP = \frac{d(\mathbf{s}T)}{V} - P \frac{dV}{V}$$

and substitute dV expressed in terms of the cross-sectional area of the specimen S and its length L:

$$dV = d(SL) = V(dS/S + dL/L) = V(\varepsilon + \eta), \quad (2.137)$$

where ε are the longitudinal and η-transverse strains

$$-d\mathbf{P} = \mathbf{P}(\varepsilon + \eta) - \frac{d(sT)}{V}. \qquad (2.138)$$

Standard tests of typical specimens (V = const) in uniaxial tension $\sigma = -d\mathbf{P}$ in constant thermal fields $dT = 0$ under conditions where transverse strains are not measured ($\eta = 0$) and measured entropy is disregarded ($ds = 0$) transform Equation 2.138 into the universally known Hooke's law:

$$\sigma = P\varepsilon. \qquad (2.139)$$

In this case parameter P acquires the meaning of the Young modulus E_1.[3,59] In the true deformation diagram (Figure 2.10) it is characterized by angle α. The region of plastic deformation, ab, and its modulus E_2 (angle β) determine the second term of Equation 2.138.

Arranged chaotically in the bulk of a body (Figure 2.7), the compresson and dilaton bonds, after application of the external force field, become oriented with respect to the field (Figure 2.8a). Both brittle (curve 1 in Figure 2.22) and ductile (curve 2) materials pass through the elastic stage of deformation. Under conditions of existence of tensile pressures (Figure 2.17b) the shape of brittle solids does not change because of a chaotic location of rotoses in volume V. This system is immediately disintegrated after completion of the orientation processes (Figure 2.8a).

As proven by experience,[3] the deformation diagrams of all brittle materials consist only of the orientation region ($0b_1$ in Figure 2.22), which is natural. Displacement of cold atoms from position q_2d to cd (Figure 1.14) occurs only due to absorption of heat. Compressons from region cd, with heat capacity by an order of magnitude higher than 15, supply this heat. As a result,

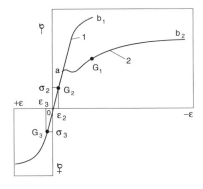

FIGURE 2.22
Flow diagram of deformation of brittle (1) and ductile (2) materials in tension (upper field) and compression (lower field).

decreasing their resistance, they tend to the v_f-temperature where they lose the mutual bond. Brittle structures have no other alternative.

Ductile materials behave differently (positions 11 to 15 in Figure 1.14). They comprise the cold compressons (region cb) and hot dilatons (q_1b). The latter, while giving heat to atoms and moving to the v_s-temperature (point b), exert, jointly with the atoms, growing resistance to the external loads. Under conditions of thermal support, compressons are transformed, without any consequences to structure, into the D-phase, which leads to development of the deformation process (long segments in Figure 2.8b). At the lack of heat (shown by short segments), they disintegrate to form a microdefect. The inelastic stage of deformation (ab in Figure 2.10 and ab_2 in Figure 2.22) is attributable particularly to the capability of ductile structures to involve the D-phase into the resistance and to make the CD-phase transitions.

Thermal fields have a substantial effect on behavior of solid bodies in force fields, causing a change in shape of the MB factor from position 12 to position 15 during heating, and in an opposite direction during cooling. Therefore, under certain conditions they can behave as brittle materials and under other conditions as ductile materials. This is proved by engineering experience.[2,3,44]

2.7 Temperature Dependence of Mechanical Properties

Prediction of the temperature dependence of mechanical properties of solids is of high scientific and practical importance.[2,3] It can be seen from the equation of state (2.96) that they are entirely determined by the Ω-potential (Equation 1.111) which, in turn, is a function of kinetic parameters T and δ. This makes the temperature dependence of the Ω-potential multifactorial. First, the mean temperature of a solid T indicates the location of its MB distribution in the energy domain. Second, it determines its shape (see change of positions from $1/2\,T$ to $2T$ in Figure 2.2, as well as shape variation in a direction of 8 to 10 in Figure 1.14. In the equation of state (2.96), linear term T is responsible for displacement of the apex of distribution and parameter δ_f or δ_s (Equation 2.94), which is responsible for the shape variation. Third, of fundamental importance is the fact where the MB distribution is located with respect to the Θ-temperature.

The differential of Equation 2.134 can be written as

$$\mathbf{P}dV + Vd\mathbf{P} = kTN\delta\left(\frac{dN}{N} + \frac{dT}{T} + \frac{d\delta}{\delta}\right). \tag{2.140}$$

Deviation of the dT temperature to one side or the other from the equilibrium value is equivalent to displacements of the center of the MB factor within positions 8 to 10 or 11 to 15 (Figure 1.14). It is accompanied by qualitative and quantitative variations in the interatomic bonds dN during the process of CD phase transitions at the Θ temperature[12] and its spread $d\delta$.[53] In the macroworld, this shows up as thermal deformation dV and variation in resistance of a solid, dP.

If a solid whose temperature differs from the Debye temperature by not more than 12% is in the environment, then

$$\frac{9}{8}\frac{\theta}{t} = 1.$$

It can be seen from Figure 1.21 that the majority of natural solids at room temperature are under such conditions. Then Equation 2.94 takes the form

$$\delta_f = \frac{\pi^4}{5}\left(\frac{t_i}{\theta}\right)^3 - 1 \quad \text{(I)}; \qquad \delta_s = -\ln\left(\frac{\theta}{t_i}\right)^3 \quad \text{(II)}. \qquad (2.141)$$

It follows that, first, the spread δ of the MB distribution of brittle and ductile materials is governed by the same law—the cubic parabola (Figure 2.23)—and, second, that the logarithm of the number is much smaller than the number itself, which indicates its substantial difference in both structures ($\delta_f \gg \delta_s$). Moreover, a change of the variable from

$$\frac{t_i}{\theta} \quad \text{to} \quad \frac{\theta}{t_i}$$

means that the first is made of the compresson (Figure 2.17a) type of the bonds and the second from the dilaton type (Figure 2.17b).

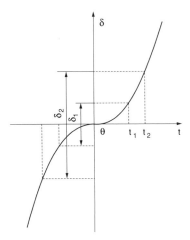

FIGURE 2.23
Temperature dependence of spread of the MB distribution.

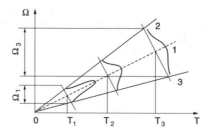

FIGURE 2.24
Temperature dependence of the Ω-potential.

To relate our considerations to the temperature interval, the origin of coordinates is conditionally combined with the θ-temperature. At temperature t_1, let its spread be equal to δ_1. Then an increase in temperature t_2 will lead to an increase in the spread to δ_2. In reality, the apex of the MB factor can be located at any point of the temperature range. Figure 2.24 shows the effect of both temperature parameters (T and δ) on the Ω-potential. Number 1 designates a direct proportionality, while straight lines 2 and 3 bound the spread of the MB distribution at a preset temperature (in accordance with Figure 2.23). It is apparent that the higher the temperature of a solid, the higher the Ω-potential and the larger its spread.

For example, at temperature T_1 the MB distribution has an acute apex and on the axis of abscissas determines a small spread of Ω_1, while at $T_3 > T_1$ the shape of the MB factor becomes smeared. This causes a corresponding increase in the spread of the Ω-potential to a value of Ω_3. The Ω-potential is a two-parameter function (Equation 1.111). In accordance with the relationship of equivalence (2.105), an increase in the spread of parameter T in the high-temperature range inevitably leads to an increase in deformability V and the equivalent decrease in resistance P (Figure 2.6). Positions 8 to 10 and 12 to 15 in Figure 1.14 show particularly this character of variation in the MB distribution.

Figure 2.24 illustrates only one aspect of such a multiaspect phenomenon as reaction of solids to temperature variations. The other, no less important, aspect is associated with an unavoidable escape of the right end of the MB distribution during heating outside v_s and during cooling outside the t_c temperature (dashed regions under curve 13 in Figure 1.14), with a possibility of transformation of the D-rotoses (Figure 2.17a) into the overheated dilatons (in the first case) or compressons (in the second case). Differences in their structure are illustrated by positions 2 and 7 in Figure 1.15. This leads to an increasing change of signs of internal stresses from compresson to dilaton and to loss in the resistance to external loads. Existence of residual tensile and compressive stresses is a clear proof of this fact.[3,44]

Let us assume that a solid has the CD structure, one part of whose rotoses, N_c, is in the compresson and the other, N_d, in dilaton state. From the known function of state (Equation 2.91) and using basic thermodynamic relationships,[20,23] we obtain the generalized Debye and Dulong-Petit formula for an overall internal energy of the solid:

$$U = kT^2 \frac{\partial}{\partial T} \ln Z = \frac{9}{8} kN\theta + 3kNT + \frac{3\pi^4}{5} kNT \left(\frac{t_i}{\theta}\right)^3 \qquad (2.142)$$

and its heat capacity:

$$C_v = \left(\frac{\partial U}{\partial T}\right)_v = 3kN + 240kN \left(\frac{t}{\theta}\right)^3.$$

As pressure and volume are not included in the explicit form in the expression of the internal energy, heat capacity at constant volume C_v and pressure C_p coincide, i.e.,

$$C_v = C_p = 3kN + 240kN \left(\frac{t}{\theta}\right)^3. \qquad (2.143)$$

In general, this became clear from the explanations to Equation 2.116. As proved by practice, both heat capacities are equal for solids.[20] Remaining within the frames of restrictions specified in derivation of Equation 2.141, one can consider that

$$240kN \left(\frac{t}{\theta}\right)^3 \gg 3kN.$$

Then

$$C_v = 240kN \left(\frac{t}{\theta}\right)^3 \approx 245kN \left(\frac{t}{\theta}\right)^3 = T_c kN \left(\frac{t}{\theta}\right)^3, \qquad (2.144)$$

where, according to Equation 1.102 and Figure 1.22, $T_c = 245$ K is the critical temperature of cold shortness. It can be seen that only those rotoses whose temperature is above T_c contribute to heat capacity of the solid. Giving Equation 2.142 the form of

$$U = 3kN \left[\frac{3}{8}\theta + T\left\{1 + \frac{\pi^4}{5}\left(\frac{t}{\theta}\right)^3\right\}\right],$$

we note that the same restrictions make it possible to write

$$U = 60kNT \left(\frac{t}{\theta}\right)^3. \qquad (2.145)$$

By substituting Expressions 2.91 and 2.142 for $\ln Z$ and U into Equation 2.14, we find entropy of the solid:

$$s = 4kN\left[1 + \frac{\pi^4}{5}\left(\frac{t}{\theta}\right)^3 - \frac{3}{4}\ln\frac{\theta}{t}\right].$$

Assuming, as above, that T differs from θ by no more than 12%, we can consider that $T \approx \Theta$ and

$$\ln\frac{\Theta}{T} \approx 0.$$

As such,

$$\frac{\pi^4}{5}\left(\frac{t}{\theta}\right)^3 > 1.$$

Therefore,

$$s = 80kN\left(\frac{t}{\theta}\right)^3. \tag{2.146}$$

Comparison with Equation 2.144 allows us to write

$$s = \frac{1}{3}C_v. \tag{2.147}$$

Accounting for Expressions 2.144, 2.145, and 2.146, we can express Equation 2.65 as

$$\Omega = \frac{1}{3}U(T) \text{ (I)}; \quad \Omega = \frac{1}{12}C_v T \text{ (II)}. \tag{2.148}$$

The authenticity of these equations is proved by the classic Debye theory,[50] in accordance with which the thermal energy of a solid equals

$$U(T) = \frac{3\pi^4}{5}kNT\left(\frac{t}{\theta}\right)^3 \approx 60kNT\left(\frac{t}{\theta}\right)^3. \tag{2.149}$$

Comparison of Equations 2.96 and 2.149, accounting for Simplification 1.41, proves validity of Relationship 2.148. As the Debye theory is in full agreement with the experiment, the same should be expected for Equation

2.148. Note that Equation 2.128.II is of highest importance because it comprises thermal–physical characteristics of a material, which are measured in practice as independent variables.

If the right-hand side of the equation of state (2.148.I) is written using Equation 2.142, then the expression accounting for the CD character of structure of solids becomes

$$PV = \frac{\pi^4}{5} kTN_c \left(\frac{t}{\theta}\right)^3 + kTN_d + \frac{3}{8} k\theta N_\theta, \qquad (2.150)$$

where N_θ is the number of rotoses that, at a given moment, have the temperature of phase transition, θ. Differentiating this expression, assuming that parameters P, N, and θ are constant, we have

$$PdV = \frac{4\pi^4}{5} kN_c \left(\frac{t}{\theta}\right)^3 dT + kN_d dT . \qquad (2.151)$$

Considering that the elementary volumes of the C and D rotoses hardly differ from each other, i.e., $\mathbf{v} = \mathbf{v}_d = \mathbf{v}_c$, and the C and D parts of the volume of a solid are proportional to their number, i.e., $V_d = \mathbf{v}_d N_d$; $V_c = \mathbf{v}_c N_c$, we can rewrite Equation 2.151 as

$$P = aT^3 V_c \frac{dT}{dV} + V_d \frac{dT}{dV},$$

where

$$a = \frac{4\pi^4}{5} \frac{1}{\theta^3} .$$

Using definitions of thermal conductivity (Equation 2.115) and that of coefficient of thermal expansion,[20,23]

$$\alpha = -\frac{1}{V}\left(\frac{dV}{dT}\right)_p,$$

we have

$$P = -aT^3 \frac{1}{\alpha_c} + V_d \frac{Q}{\lambda} . \qquad (2.152)$$

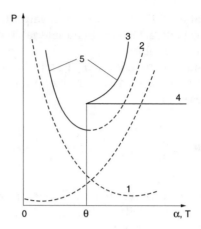

FIGURE 2.25
Temperature dependence of tensor of resistance **P** of the CD structures.

The structure of Function 2.152 is shown in Figure 2.25. Numerical designations in this figure correspond to the following components:

$$(1) \qquad P_1 = f\left(\frac{1}{\alpha_c}\right)$$

$$(2) \qquad P_2 = f(aT^3)$$

$$(3) \qquad P_3 = aT^3 \frac{1}{\alpha_c}$$

$$(4) \qquad P_4 = V_d \frac{Q}{\lambda}$$

$$(5) \qquad P = -aT^3 \frac{1}{\alpha_c} + V_d \frac{Q}{\lambda}$$

As seen, in the C region (0 to θ range), resistance P continuously decreases (plot 5). At the Debye temperature function Equation 2.152 has discontinuity. Its jump at this point is equal to

$$P_4 = V_d \frac{Q}{\lambda}.$$

It is required to ensure continuity of the process of CD transitions. In the D region function 2.152 changes sign that causes further decrease in the resistance. Plot 3 shows the variation in the absolute value of P_3.

There is a very tight link between thermal and mechanical properties; at the AM level, it appears as the relationship of equivalence (2.105). This link persists over huge scale-time distances that separate this level from the macroworld, and transforms into the equation of state (2.65). Let us elaborate on the mechanism of this link and prove its existence by experimental data.

If, in the first law of thermodynamics (Equation 2.1), we express a change in the heat flow dQ in terms of thermal conductivity (Equation 2.155), we obtain $dU = \lambda dT - PdV$. By taking derivatives $dTdV$ in the right-hand part out of the brackets and expressing, in analogy with Equation 1.25, the overall energy U in terms of the sum of the potential $U(r)$ and kinetic $U(T)$ parts, we can write

$$\frac{dU(T)+dU(r)}{dTdV} = \frac{\lambda}{dV} - \frac{P}{dT} .$$

Accounting for the definitions of heat capacity 2.142 and internal resistance (2.116), we can express this as

$$\lambda - C_v = 2P\frac{dV}{dT} .$$

If the right-hand part is divided and multiplied by V, then

$$\Omega = \frac{1}{2\alpha}(\lambda - C_v), \tag{2.153}$$

where α, as in Equation 2.152, is the thermal expansion coefficient.

It can be seen that the solid can perform the work of resistance to deformation, Ω, in two directions—direct or negative (where $\lambda < C_v$)—due to consumption of internal reserves of the kinetic energy, and reverse or positive (where $\lambda > C_v$) when heat is received from the outside. In both cases its value is proportional to rigidity of the solid inverse to the thermal expansion coefficient

$$\Delta = 1/\alpha \tag{2.154}$$

and which, at any moment of time, depends on the relationship between λ and C_v.

If we are interested in thermal conductivity, we should first be concerned with the causes of formation of thermal resistance. The energy exchange between the bonds is realized because the atoms that constitute them perform thermal motion following the law set by Equation 1.23. It is this law that establishes the mode of thermal equilibrium between particles oscillating at different frequencies (Figure 2.14). Also, this law is "responsible" for heat resistance of the solid. In mathematical description, however, the process of interaction of the set of oscillating atoms is usually replaced by the system of moving imaginary particles, i.e., phonons (Equation 1.71).

An expression of thermal conductivity within the frames of the CD notions can be obtained if we relate thermal radiation with the phonon gas. The elementary theory of thermal conductivity of conventional gases can be applied to this gas. As shown in the kinetic theory, the coefficient of thermal conductivity of a gas can be expressed as[30]

$$\lambda = \frac{1}{3}lvC_v,\qquad(2.155)$$

where l is the length of the mean free path of phonons and v is the velocity of sound in a given material.

The length of the mean free path of phonons, l, can be represented as a value inverse to the coefficient of scattering of sound waves:

$$l = \frac{1}{\zeta}.\qquad(2.156)$$

This coefficient is determined by the quantity of phonons per unit volume of a solid, n, and the effective area of their scattering, f, i.e.,

$$\zeta = nf.\qquad(2.157)$$

The total quantity of photons consists of the compresson n_c and dilaton n_d parts, i.e., $n = n_c + n_d$, while the scattering area is equal to the area of the elliptical orbit of atoms constituting the rotos and thus found from Equation 1.49.

To the first approximation, the concentration of phonons in volume can be considered to be proportional to the temperature of this volume. If we multiply this temperature by the mean energy of one phonon, we obtain the following expression for the energy of the compresson and dilaton phases:

$$n_c\hbar v_c = \frac{3\pi^4}{5}kNT\left(\frac{t}{\theta}\right)^3;\qquad n_d\hbar v_d = 3kNT.\qquad(2.158)$$

Assuming that, near the Debye temperature, $v_c = v_d = v_\theta$, and expressing n_c and n_d in terms of the above relationships, we find their sum:

$$n = \frac{3N}{\hbar v_\theta}kT\left[1 + \frac{\pi^4}{5}\left(\frac{t_i}{\theta}\right)^3\right].$$

Using the definition of the θ-temperature (Equation 2.87) and considering, as before, that

$$\frac{\pi^4}{5}\left(\frac{t_i}{\theta}\right)^3 > 1,$$

we can write

$$n = \frac{3\pi^4}{5}N\left(\frac{t_i}{\theta}\right)^3\frac{T}{\theta}.$$

Multiplication and division of this expression by the Boltzmann constant k, accounting for Equation 2.144, yields

$$n = \frac{1}{4k}\frac{T}{\theta}C_v. \qquad (2.159)$$

Substituting Equations 2.156, 2.157, and 2.158 into Equation 2.155 yields

$$\lambda = B\frac{\theta}{T}, \qquad (2.160)$$

where

$$B = \frac{4}{3}\frac{kv}{f}$$

is a constant for a given material.

It can be seen that λ is determined by the θ-temperature and, according to Equation 1.95 and Figure 1.21, is individual for each material. The ratio between its maximum and minimum values can reach a factor of $2 \cdot 10^4$.[64] Thermal conductivity of dielectrics is much lower than that of metals and has an inverse-proportional dependence on the temperature. In metals, this dependence is less pronounced than in dielectrics. As an example, Figure 2.26 shows such a dependence for marble in a range of high temperatures,[40] where λ is measured in J/m s K.

Transition to the zone of low temperatures is accompanied by its substantial increase. According to the data given in Galyas and Poluyanovsky,[15] thermal conductivity reaches maximum in a range of 5 to 30 K, after which it falls to zero. This indicates that, unlike metals (Figure 1.21), brittle rock materials have the Debye temperature close to absolute zero. This effect results from existence of the acceleration well and the deceleration barrier in the state diagram of the rotos (Figure 1.14).

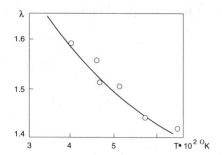

FIGURE 2.26
Character of variations in thermal conductivity of marble in the high-temperature range.

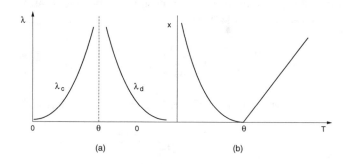

FIGURE 2.27
Variations in thermal conductivity (a) and thermal expansion (b) in the compresson and dilaton phases.

Substituting n_c and n_d from Equation 2.158 into Equation 2.155, we find the variations in thermal conductivity of the compresson and dilaton phases:

$$\lambda_c = \frac{5}{9\pi^4} \frac{vC_v\theta^4}{fN} \frac{1}{T^4} \quad \text{(I)}; \qquad \lambda_d = \frac{vC_v\theta}{9fN} \frac{1}{T} \quad \text{(II)}. \qquad (2.161)$$

The reciprocal of thermal conductivity is called thermal resistance

$$\Lambda = \frac{1}{\lambda}.$$

In accordance with Equation 2.161, this parameter varies as shown in Figure 2.27. In a solid body, however, the interatomic bonds can be isolated by thermal resistance from the surrounding thermal field, and exist in the acceleration well q_2dc and the zone of the deceleration barrier cbq_1 (Figure 1.14), separated over the entire energy range by the phase transition temperature θ (curve 2 in Figure 1.13). At this temperature, functions

$$\lambda = f(T) \quad \text{and} \quad \Lambda = f(T)$$

have singular points, caused by the CD phase differences. If, in the C phase, the crystalline structure hampers escape of heat to the outside, in the D phase, it hampers its inflow from the outside. The thermally active rotoses are grouped near the θ-temperature. They immediately react to any change in the concentration of phonons by absorbing or releasing them and, as a result, these rotoses find themselves in the dilaton or compresson phase. The thermally inert rotoses are located at the ends of the well and inertly react to a change in the temperature conditions.

Substituting Expression 2.148.II for the Ω-potential in the left-hand part of Equation 2.153, multiplying both parts of the equality by ρ, and adding the designation for thermal diffusivity,

$$a = \frac{1}{\rho}\left(\frac{1}{6}\alpha T + 1\right),$$

we obtain the known relationship between thermal–physical parameters of a solid:[30]

$$a = \frac{\lambda}{C_v \rho}, \tag{2.162}$$

which determines thermal–insulation properties of solids.

Generalized data on the variations in a for the majority of the known classes of structural materials can be found in Troshenko[3] and Ashby.[65] All are grouped along the line of $C_v\rho = 3 \cdot 10^6$ J/m^3 · K. Depending on the class of material, parameters a and λ vary within the limits of almost five orders of magnitude, creating different conditions for formation of thermal fields in materials and, therefore, providing them with a different degree of protection from external effects.

Coefficients of volume and linear expansion characterize the ability of materials to change their geometrical sizes with temperature. Thermal expansion coefficient α of metals is related to its heat capacity C_v through the Gruneisen rule[30]

$$\alpha = bC_v, \tag{2.163}$$

where b is a proportionality coefficient particular for each metal. It can be obtained from Equation 2.153 by taking

$$\alpha = \frac{1}{2\Omega}\left(\frac{\lambda}{C_v} - 1\right)C_v$$

out of the brackets in the right-hand part and assuming that

$$b = \frac{1}{2\Omega}\left(\frac{\lambda}{C_v} - 1\right).$$

As follows from Equation 2.148, solids perform the work of resistance Ω by spending their own kinetic energy. Its variation is set by the temperature dependence of heat capacity (Figure 2.15b). As the temperature in interval 0a increases, both terms in the right-hand part of Equation 2.148.II increase. This leads to an increase in Ω and, hence, improvement in the resistance and deformability of materials. At temperatures above the Debye temperature (especially after point *a*), the growth of heat capacity stops, while further increase in Ω occurs only due to an increase in temperature. As follows from the equivalence relationship (2.105), this may lead only to catastrophic deformation.

Equation 2.148 converts Equation 2.96 into a symmetrical form: its left-hand part is a parametric expression of the potential portion of energy of a solid, while its right-hand part is the kinetic portion. This allows the equation of state to be written in the simplest form:

$$\Omega = \Omega(T), \qquad (2.164)$$

where

$$\Omega(T) = \frac{1}{12} C_v T.$$

In analogy with Equation 2.134, we can write differentials of its parts as

$$PdV + VdP = \frac{1}{12}(C_v dT + TdC_v). \qquad (2.165)$$

Equivalence relationship 2.105 allows a conclusion that the first terms of each side of Equation 2.165 characterize the elastic part of the resistance work (region 0a in the deformation diagram in Figure 2.10), and the second ones the plastic part (region ab). Such a representation allows us to make a very important practical conclusion: maintaining the values of thermal–physical parameters C_v and T at certain levels can control deformation and resistance of materials.

Modern concepts of the mechanism of deformation and fracture disregard the fact that thermal processes play very significant roles in this mechanism.[2-4] Therefore, it is not a surprise that thermal–physical properties of structural materials are not given due consideration. Of the entire set of solid bodies, the only exception is likely to be rocks.

Because of the necessity to develop thermomechanical methods for their production, the preceding properties of different types of rock were investigated in detail.[15,40] As an example, Figure 2.28 shows experimental dependencies of heat capacity on the temperature for basalt 1, diabase 2, and rock salt 3. The axis of ordinates for the first two is the left vertical line calibrated to read in J/g· K, while that for the third is the right vertical line calibrated to read in kcal/mol.

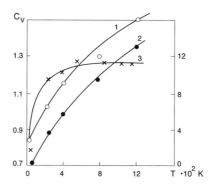

FIGURE 2.28
Experimental temperature dependencies of heat capacity for marble (1), diabase (2), and rock salt (3).

It can be seen from comparison of Figures 2.15b and 2.28 that the theoretical and experimental dependencies within the investigated temperature range almost coincide in shape. Comparison of Figures 2.10, 2.22, and 2.28 shows that Equation 2.165 gives the correct reflection of the two-stage mechanism of deformation and fracture based on the thermomechanical processes. However, even for rocks, a very limited number of investigations were carried out, so it is impossible to consider generalized results. The common tendency follows.[40]

The most significant increase in heat capacity for all minerals and rocks takes place in a range of 200 to 400°C. As such, it increases by a factor of 5 from 0.4 to 2 kJ/g · K. This means that particularly these temperatures develop in the structure of these materials during the process of deformation and fracture. When a rock is cooled from room temperature to 90°C, specific heat decreases for all studied bodies by 23 to 59%. It is apparent that in this case one might expect a corresponding increase in the strength. A rather close correlation of an inverse-proportional character exists between heat capacity and specific mass of minerals, which is predicted by Equation 2.148.

We can rewrite the equation of state (2.148.II) accounting for the relationship of Equivalence 2.105 in the form

$$P = \frac{1}{12} C_v \frac{dT}{dx}, \tag{2.166}$$

where dx is change in geometrical sizes of a deformed body in the direction of action of external forces (Figure 1.23). Considering the definition of thermal flow (Equation 2.108), it holds that

$$P = q, \tag{2.167}$$

where

$$q = \frac{1}{12} C_v \frac{dT}{dx}$$

is the thermal flow formed due to the internal reserves of the kinetic energy C_v. It follows that, in its physical nature, mechanical resistance **P** (Equation 1.111) of a body is its *thermal–physical* characteristic, rather than purely *mechanical*.

If the temperature gradient (which is equivalent to internal stresses) is kept constant, i.e.,

$$j = \frac{1}{12} \frac{dT}{dx} = \text{const}$$

(this condition is always met under quasi-static loading conditions or in constant force fields), then

$$\mathbf{P} = jC_v. \tag{2.168}$$

This shows that the variations in the resistance of a material are controlled by its heat capacity. Comparison of Figures 2.10 and 2.15b, which illustrate the true deformation diagram and temperature dependence of heat capacity, proves that this statement is correct.

Approximation of curvilinear region 0a of the heat capacity plot (Figure 2.15b) by a straight line makes it possible to write

$$C_v = iT, \tag{2.169}$$

where i is the angular coefficient of this region. Multiplying the left- and right-hand parts of Equation 1.168 by T and assuming that $i = j$ (confidence in its validity gives the relationship of equivalence 2.105, accounting for Equation 2.169), we have

$$\mathbf{P} = \frac{1}{T} C_v^2. \tag{2.170}$$

Equation 2.170 is *the law of formation of the resistance of materials to deformation and fracture* in quasi-static force and thermal fields.

Figure 2.29 presents a nomographic chart of the two-parametric dependences (2.170). Number 1 designates plot given by Equation 2.169 and number 2 function $\sigma = f(T)$. This nomogram clarifies the mechanism of formation of the temperature dependence of strength and the role played in it by heat capacity.

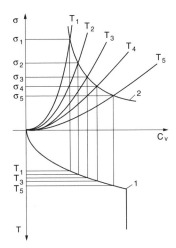

FIGURE 2.29
Thermal–physical mechanism of formation of strength of solids.

The discussed law is of common nature and thus is universally justified in practice for a wide diversity of materials: refractory metals and their alloys (niobium, titanium, molybdenum, tantalum, tungsten, etc.),[66,67] many steels and alloys, including multicomponent ones,[68–71] nonferrous metals and alloys,[69] nonmetallic materials,[72] polymers,[73] composite materials,[73,74] single crystals,[75] rock,[39,40] and building materials.[76] It manifests itself at the following types of the stress–strain state: tension, bending, and shear,[3] as well as compression.[15,76] Not only static strength but also the dynamic,[3] specific,[66] and sustained[70] strength, as well as creep,[73] yield, fatigue,[44] durability,[2,3] and heat resistance,[68] and even hardness,[71] have a similar type of temperature dependence.

To illustrate, Figure 2.30 shows the typical temperature dependence of elasticity modulus E for alloy EI 652.[68] The data presented in Khimushin[68] on the temperature dependencies of E for several tens of heat-resistant alloys repeat the contour of plot 2 in Figure 2.29. Figure 2.31 shows the dependence of the yield stress, σ in the range of low temperatures for (0.1%C, 2%Mg, 1%V, 1%B) steel at four strain rates $\dot{\varepsilon}$, s^{-1} (curves 1, $8 \cdot 10^{-4}$; 2, $8 \cdot 10^{-3}$; 3, $8 \cdot 10^{-2}$; and 4, $8 \cdot 10^{-1}$)[3] and steel St E47 (curve 5).[77] Despite the fact that curves 2 and 5 concern different steels, they almost coincide in a temperature range from 100 to 300 K.

Figure 2.32 shows curves of shear deformation for steel 1018 under quasi-static ($\dot{\varepsilon} = 5 \cdot 10^{-4}$ s^{-1}, lower row) and dynamic ($\dot{\varepsilon} = 10^{3}$ s^{-1}, upper row) loading conditions in a low-temperature range.[3] As seen, with change in temperature to both sides from 0°C, the curves are shifted equidistantly relative to each other. By connecting characteristic points in these curves (e.g., yield serration), we obtain a hyperbolic dependence (dashed line) described by Equation 2.170.

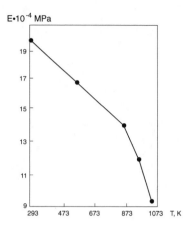

FIGURE 2.30
Temperature dependence of elasticity modulus for alloy EI 652.

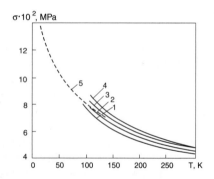

FIGURE 2.31
Dependence of low-temperature yield stress at four strain rates for steel 10G2FB (1 to 4) and steel St E47 (5).

The deformation diagrams for the most diverse materials in high-temperature range have a character identical to that previously described. This can be readily seen from the appropriate section of Troshenko.[3] To illustrate, Figure 2.32 shows temperature dependencies of long-time strength (on a base of 1000 hours) for carbon, 1, and stainless steels, 2, and short-time strength for carbon steel, 3.

If investigation of a certain mechanical characteristic is conducted over a wide temperature range, the hyperbolic dependence 2 (Figure 2.29) has, as a rule, a local maximum. For example, for specific strength of a molybdenum alloy (0.5% Mo, 0.1% Ti, Zr) it is located near 1200 K,[3] for short-time strength of many steels and alloys it lies within a range of 673 to 873 K,[68] and for fiberglass at 20°C.[72] It has been noted[3] that the presence of such a maximum is a characteristic feature of many high-melting-point alloys.

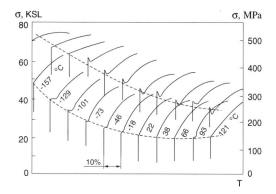

FIGURE 2.32
Curves of shear deformation for steel 1018 under quasi-static (lower line) and dynamic (upper line) loading in the low-temperature range.

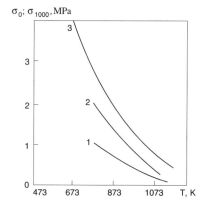

FIGURE 2.33
Temperature dependence of short-time (3) and long-time strength on a base of 1000 hours for carbon (1) and stainless steel (2).

As an example, Figure 2.34 shows data for hot-pressed materials based on titanium carbide, 1 and 2, silicon carbide, 3, and carbon, 4. For these materials, the maximum of strength is observed at 2050 to 2300 K. Analyzing experimental data on the effect of temperature on static and cyclic strength, Ekobori[44] concluded that fatigue resistance, as well as static tensile strength, has a maximum within a temperature range of 230 to 350°C (Figure 2.35). In this figure, 1 designates the strength in static tension; 2, the yield strength; 3, the fatigue limit in bending at a loading rate of 2000 min^{-1} ($n = 5 \cdot 10^5$); and 4, the same at 10 min^{-1} ($n = 5 \cdot 10^5$) for steel with 0.17%C.

Formation of maximum on curve 2 (Figure 2.29) is attributable to the temperature transformation of the MB factor (Figure 2.2) and the complex shape of the state diagram of a rotos with two extrema (Figure 1.14). Indeed, with

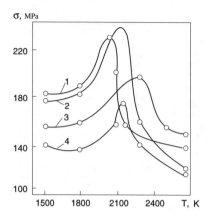

FIGURE 2.34
Temperature dependence of strength of refractory compounds based on titanium carbide (1 and 2), silicon carbide (3), and carbon (4).

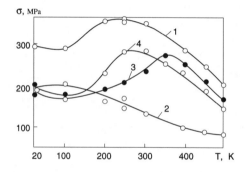

FIGURE 2.35
Effect of temperature on static and cyclic strength of carbon steel.

a variation in temperature, the MB factor takes positions 3 to 10 or 11 to 15, tracing curve 16 in the energy domain in the first case while the state of each rotos is determined by diagram *fdcbe*. As such, the location of its apex formed on solidification with respect to θ_f and θ_s temperatures is very important. For brittle bodies, it is in the dilaton region q_2dc, and for ductile bodies in the compresson region cbq_1 (Figure 1.14). For this reason the temperature dependence of strength of the former comprises a minimum and that of the latter a maximum. As dependence $\sigma = f(T)$ is a superposition of diagram and a curve of 16 type in a corresponding region, these extrema are more pronounced for some materials than for others.[3,44]

Figure 2.36 shows the mechanism of formation of temperature dependence of the Ω-potential of ductile materials. Let the MB distribution have shape 1 at temperature T_1. Its apex in the state diagram 2 corresponds to point *a*, which locates in Figure 1.14 in the zone of the deceleration barrier cbq_1. As

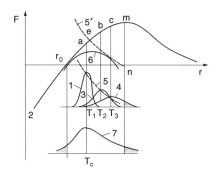

FIGURE 2.36
Mechanism of formation of temperature dependence of tensile strength.

temperature increases, this distribution changes in a direction of 1 to 2 to 4 (according to the diagram in Figure 2.2) and the locus of its apex follows path *abk* on force curve 2. By connecting maxima of distributions 1 to 2 to 4, we obtain curve 5 (dotted line), which is the locus of the MB distribution maxima when temperature increases. Curve 5′ was obtained by equidistantly displacing curve 5 to the zone of force curve 2. Superposition of the compresson part of curve 2 and curve 5 forms a function having a maximum at the point of their intersection, *e*. The temperature at which the maximum resistance is observed is called effective or most favorable and designated as T_n.

Substituting Equation 2.142 into Equation 2.148 and differentiating with respect to T at $N = $ const, we obtain

$$d\Omega = kN + 80kN\left(\frac{t}{\theta}\right)^3. \tag{2.171}$$

In accordance with the law of energy conservation (Equation 1.23) and that of momentum (Equation 1.27), at any change of the state, the Ω-potential is a constant value, i.e., $d\Omega = 0$. From Equation 1.171, we then find that

$$T_n = 0.24\theta. \tag{2.172}$$

This shows that the effective temperature T_n is equal to a quarter of the Debye temperature θ. Substituting Equation 2.172 into Equation 2.142 and after simplification, we obtain

$$\mathbf{P}_n = -0.5k\rho\theta. \tag{2.173}$$

Equation 2.173 presents the maximum resistance that a material can develop at the effective temperature. The minus sign corresponds to tensile resistance.

Dependencies $\sigma = f(T)$ and $E = f(T)$, modified to account for these results, are shown as a single curve in Figure 2.37. In the low-temperature, *ab*, and

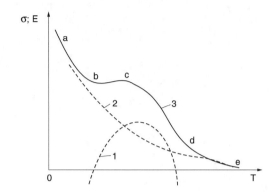

FIGURE 2.37
Temperature dependence of elasticity modulus and strength of ductile (compresson) materials.

high-temperature, *de,* zones this curve coincides with hyperbola 2 in Figure 2.29. Curve 1 corresponds to line 6 in Figure 2.36. This shape of the temperature dependencies is highlighted in Troshenko,[3] Ekobori,[44] and Mack Lin.[45] Point *c* corresponds to the resistance maximum observed at temperature T_n.

Modern design and calculation methods tend to ensure the required strength of parts and structures by varying the mass of a material.[51] In fact, these methods utilize only the potential part of the internal energy. By maintaining the service temperature at the T_n level, it is possible to provide an increase in strength due to the effective consumption of its kinetic part. So far, this is a not-yet-claimed reserve of materials. With an appropriate design, it can be directed to increase reliability and durability of engineering objects.

It may be noticed (Figure 2.36) that, in accordance with Figure 2.6, the temperature transformation of the MB distribution forms the opposite character of variation in the strength (curve 5′) and deformation properties (determined from the spread) of materials (its mechanism is illustrated in Figures 2.8b and 2.24). Typical temperature dependencies of mechanical properties of chrome–nickel austenitic steels are shown in Figure 2.38a and that for niobium in Figure 2.38b; these cover the low- and high-temperature regions.[3] It is seen that they support theoretical results: strains ε vary opposite to strength σ.

During heating the MB distribution can be transformed so much that its hot end (dashed zone of curve 4 in Figure 2.36) is found in plastic zone 7 (Figures 1.14 and 1.15). In this zone, bonds cannot withstand tensile load and thus are deformed. For this reason the tensile resistance decreases and ductility increases. The material fails by ductile fracture at the moment when, in accordance with Equation 2.158, the apex of distribution 4 in Figure 2.36 coincides with critical vertical line *mn.* As such, the deformation diagrams change as shown in Figure 2.39, where the initial regions in tension of titanium high alloy B95T (a) and duralumin D16ARN (b) are depicted at different temperatures.[3] The deformation curves in compression of rock materials, e.g., in compression of concretes,[78] exhibit similar characteristics.

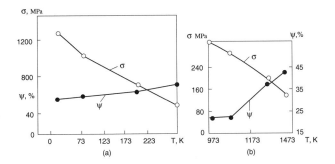

FIGURE 2.38
Dependence of strength and deformation properties of chrome–nickel austenitic steel Kh23N18 (a) and niobium (b) in low- and high-temperature ranges.

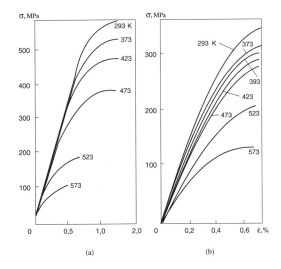

FIGURE 2.39
Stress–strain diagrams in tension of alloys V95T (a) and D16ATN (b) at different temperatures.

Curve 7 in Figure 2.36 demonstrates the variation in the MB distribution under the effect of two or several destructive factors (e.g., static or quasi-static tension in a variable thermal field). They tend to shift the right-hand end of the distribution to the plastic zone. Quenching forms an extra left-side asymmetry. As a result, the left "tail" of the curve may be found in the zone of brittle fracture (behind critical point c [Figure 1.14] where the temperature is equal to T_c). A situation in which the destructive processes simultaneously affect both branches of the MB distribution leads to a catastrophic brittle fracture of a material (Figure 1.22).[43]

Unlike ductile materials, the structure of brittle materials is composed of substantially different rotoses (Figure 2.17) having different thermal–physical

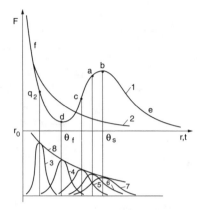

FIGURE 2.40
Diagram of formation of the temperature dependence of mechanical properties of brittle (dilaton) materials.

characteristics (Figures 2.27 and 2.28). Moreover, parameter δ_f, in contrast to δ_s (1.141), does not distribute them in the zone of the deceleration barrier but in the acceleration well (Figure 1.14). In this connection, the temperature dependence of their Ω-potential is of a different character. The state diagram of one of the rotoses that are part of the dilaton set N_d is shown in Figure 2.40 by curve 1. In accordance with Equation 2.96, the absolute temperature of a solid T fixes the MB factor δ_f (curves 3 to 7) at certain points of the diagram (q_2, d, c, a, etc.). At constant temperature, constitutional diagram 1 is degenerated into isothermal line 2 (Figure 1.13).

On heating, the MB factor successively takes positions 3 to 7, describing curve 8 in the energy domain. Unlike ductile structures, in order to move to the plastic zone (region be of the diagram) the dilaton rotoses must pass a much longer path, q_2dcab, in the thermal range, passing through its extreme points (d and b). Therefore, temperature dependencies of the Ω-potential of such materials must comprise both maxima and minima.

The effect of temperature on strength and deformation properties of concrete based on portland cement is discussed in Ryzhova.[78] Figure 2.41 shows the variations in the short-time strength in compression (curve 1) and tension in bending (curve 2) expressed in terms of relative units $n\%$, while Figure 2.42 shows the deformability of the compression zone in bending of reinforced concrete elements for the temperature range from 20 to 250°C.

In accordance with Figure 2.6, to maintain the Ω-potential at a constant level during the process of progressive decrease in the tensile strength (curve 2 in Figure 2.41), it is necessary to increase deformability. (Parameter V is responsible for it.) The opposite variations of strains (Figure 2.42) prove the theoretical prediction that follows from the equation of state written for both the microscopic (Figure 2.6) and macroscopic (Equation 2.96) scale. Such variations are due to the opposite direction of variation of both parameters that determine the Ω-potential. This versatile prediction is valid for metallic

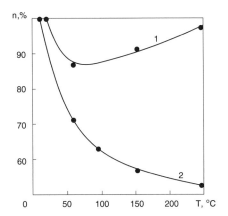

FIGURE 2.41

Effect of temperature on strength of concrete in compression (1) and tension (2).

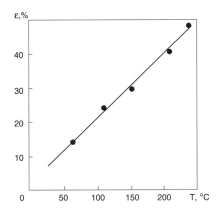

FIGURE 2.42

Variations in mean strains of end fiber of the compressed zone of concrete in bent reinforced concrete elements during heating.

(Figure 2.38) and nonmetallic (Figures 2.4 and 2.42) materials. Figure 2.43 shows temperature dependencies of the compression strength of concrete over a wide temperature range for several types of binders.[76] Inorganic glasses also have a similar variation of strength with temperature.[79]

It can be seen from analysis of Figures 2.4, 2.29, and 2.37 that, at low temperatures (below 250°C), the tensile strength in bending of concrete (curve 2 in Figure 2.41), similar to the majority of metals, follows the hyperbolic dependence (curve 2 in Figure 2.37). In compression (curve 1), it tends to maximum, which lies above 250°C. This is fully verified by the data shown in Figure 2.43. It is evident from this figure that concretes based on a high-alumina cement have maximum compression strength near 400°C (curve 3), whereas that of Portland and alumina cements lies above 1200°C.

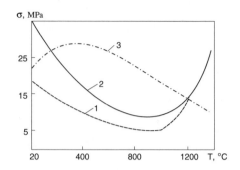

FIGURE 2.43
Variations in strength of concrete on heating of Portland (1), alumina (2), and high-alumina (3) cements.

Each material has its own peculiar MB distribution. In the structure of any material one can always find a certain amount of compresson and dilaton bonds. Their proportion determines the compression and tensile strength. They are not the same for most materials, which proves the validity of the CD concepts. In literature, the compresson (or compressive) and dilaton (or tensile) stresses are called residual, which is not quite correct. As shown by experience,[3] in practice there is no way to avoid them.

At present, emphasis in engineering research is placed on the processes of deformation and fracture because these are regarded as responsible for reliability and durability of engineering objects. However, this is not entirely so. Underestimation of the role of phase transformations, which radically change the resistance and deformation of materials in external fields, is one of the major mistakes in traditional concepts of the nature of strength. This may lead to a sudden and "unpredictable" or "accidental" failure of structures — engineering practice can offer a large number of examples of these.[2,3] The known values of compression and tension strength can be used as indicators of the concentration of compressons and dilatons per unit volume of a material under given test conditions. This concentration can be varied using directed external effects, thus controlling deformation and fracture processes by activating them, where necessary (e.g., in production and processing of mineral resources), or avoiding them (e.g., in operation of critical parts and structures under extremely severe conditions).

2.8 Periodic Law of Variations in State

In the equation of state (2.96), individual peculiarities of the structure of a body are determined by the Θ-temperature (Equation 1.95). Substitution of expressions of thermal conductivity (Equation 2.160) and heat capacity (Equation 2.144) into Equation 2.153 yields the θ-temperature dependence

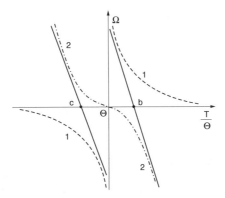

FIGURE 2.44
Dependence of the Ω-potential on the relative temperature.

of the Ω-potential:

$$\Omega = B_1 \frac{\theta}{T} - B_2 \left(\frac{t_i}{\theta}\right)^3 , \tag{2.174}$$

where $B_1 = 2B\Delta$ and $B_2 = 120kN\Delta$. This dependence is shown in Figure 2.44, where number 1 designates function

$$y_1 = B_1 \frac{\theta}{T}$$

and 2 designates cubic parabola

$$y_2 = B_2 \left(\frac{t_i}{\theta}\right)^3 .$$

Solid lines show how the Ω-potential changes in the dilaton (first and fourth quarters) and compresson (second and third quarters) phases.

At points c and b in Figure 2.44, the Ω-potential is equal to zero, whereas at the Debye temperature it has discontinuity. The loss of continuity of parameters **P** and V occurs because of an inadequate temperature sensitivity of the C and D rotoses (Figure 2.27). Macroscopic examinations show that an abrupt change in all thermal–physical characteristics occurs here (Figure 1.17). Local mechanical overstresses occur at $T = \theta$ that are attributable to the competing character of the compresson (compression) and dilaton (tension) pressures (Figure 2.17).

Assume that one of the neighboring C rotoses transforms into the D-state, whereas the other has no such possibility. At the moment of the CD transition, the uncertainty leads to disintegration of the bond between them and, therefore, to formation of a local defect. This is the cause of loss of continuity of the Ω-function at the Debye temperature. As such, both parameters of the Ω-function, **P** and *V*, experience discontinuity. Because of this, often and sudden (sometimes even isolated) fluctuations in the ambient temperature near the Debye point cause catastrophic consequences for a structure (Figure 1.22).

Point *c* coincides with the temperature of brittle failure v_f (Figure 1.14), while point *b* coincides with that of plastic flow v_s (corresponding to the inflection point of curve 1 in Figure 1.13). At v_f a body can generate resistance forces but cannot be deformed. At v_s, it loses the first capability, although it can be readily deformed (Figure 2.6). This situation is especially dangerous when the MB distribution is between T_c and θ_s (perhaps only for metals and their alloys) and increases its spread under the effect of any external factor. As such, the process of disintegration of the bonds extends to both ends and the body catastrophically fractures (dashed regions at position 13 in Figure 1.14). In strength of materials,[2,3] this phenomenon is called cold shortness.

Formula 2.174 is in fact an analogue of the equation of state in thermodynamic potentials (1.111). Its first term (F-potential) establishes the direct dependence and the second (G-potential) the inversely proportional dependence of the Ω-potential on the θ-temperature. In accordance with Equation 1.95), the θ-temperature for every chemical element is determined by the atomic constants: mass *m*, valence *z*, and common charge of nucleus *e*, i.e., its place in the periodic system of elements. Now we can show that this equation correctly describes the state of all solid bodies, without any exception, and therefore is general. In fact, as a form of a periodic law of variation in state, it is supported by multiple evidence in practice.

In elastic deformation the **P** parameter is equivalent to the elasticity modulus (Equation 2.139). If, in the range of medium and heavy elements, the θ-temperature decreases with an increase in the atomic mass (see Figure 1.24), the elasticity modulus *E* and deformation resistance σ should increase, retaining periodical oscillations of their values with variations in *z* and *e*. Figure 2.45 shows variations in the elasticity modulus *E* and strength σ in

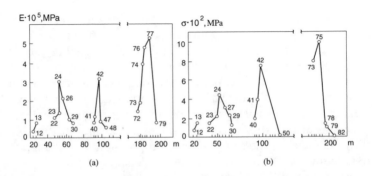

FIGURE 2.45
Variations in elasticity modulus (a) and tensile strength (b) in simple solids.

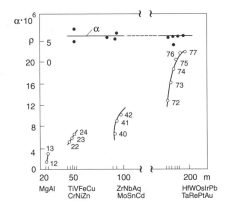

FIGURE 2.46
Dependence of density ρ and thermal expansion coefficient α (horizontal line) on the atomic mass of chemicals.

tension for simple solids, depending on their atomic mass m. It was plotted using tabular data given in Troshenko.[3]

As in Figure 1.21, numbers near points stand for the consecutive numbers of elements in the periodic system. Equation 2.174 provides the correct description of the variations in state of solids in the force fields for the entire periodic system. This fact, in turn, supports the versatility and validity of the equation of state (2.96).

By rewriting Equation 2.96 in the form

$$P = \Omega_N(T)\rho, \tag{2.175}$$

where $\Omega(T)_N = kT\delta$ is the potential expressed in terms of kinetic parameters, and $\rho = N/V$ is the density of a solid, we obtain the directly proportional relationship between fracture resistance P and density ρ of a solid.

Figure 2.46 shows the densities (g/cm^3) for some chemical elements located in different periods of Mendeleev's table.[3] They are designated by circles. The horizontal line connecting solid points shows, for comparison, the variation of the coefficient of thermal expansion, α (deg^{-1}), for these elements, which does not depend on θ (2.152). It is seen that the larger the mass of an element in the table as a whole and in each of its periods, the higher the density. This is natural because an increase in the number of protons is accompanied by an increase in the total number of electrons in an atom. Therefore, the number of molecular orbits (Figure 1.3) that unite atoms into a rotos, as well as the probability of participation of internal electron layers in formation of a bond, increases.

The revealed principles are valid not only for simple solids, but also for engineering materials. The proof of this statement can be found in Ashby,[65] where unique experimental data on physical–mechanical properties of different classes of structural materials are analyzed. In this work, the results of tests of materials of cardinally different types (metals and alloys, polymers, ceramics, wood, etc. — nine classes in total), arranged on the coordinates, form a linear correlation of the type of Equation 2.175. The natural spread of indicators between groups and inside the groups is attributable to the varying character of the $\Omega\ (T)$ parameter.

The generalized equation of state (2.174) opens avenues for the purposeful control of resistance of materials to external fields: force, thermal, electromagnetic, radiation, etc.[12,21] Mendeleev's periodic law can formally be regarded as a combination of 108 elements considered separately. The equation of state allows formulation of periodic systems consisting of 2, 3, etc. combinations of these elements, as well as prediction of their physical–mechanical properties.

Intuitively, a similar approach is used in materials science,[59–61] where the variation of properties of steels over wide ranges is achieved by varying combinations of alloying elements and their contents. For example, iron–carbon compounds provide a diverse spectrum of physical–mechanical properties. In the periodic system, their initial components are located near the maximum of the state function (see positions 6 and 26 in Figures 1.21b and 2.45). It is impossible to obtain a similar effect with elements located in the zone of its minimum (e.g., with nitrogen 7, phosphorus 15, sulfur 16, etc.).

2.9   Phase and Aggregate States of Materials

The bond system of atoms may be in different phase (compresson and dilaton) and aggregate (solid, liquid, and gaseous) states. Under which conditions may such states occur? Let the interatomic bond be in the C state and locate energy U_1 (Figure 1.7a) at distance t_1 from the bottom of the potential well. Assume that its energy changes by $dU = U - U_1$. We can expand the energy as a series in powers of difference $(r - r_1)$. For this case, it is not necessary to have the exact expression of function $U = f(r)$. It is enough to know its variation with an increase in the interatomic distance (plot $r_1ab\infty$ in Figure 1.7a), assuring its continuation near level U_1.

Limiting to terms of the fourth infinitesimal order, we can write

$$U = U_1 + \left(\frac{dU}{dr}\right)_1 (r - r_1) + \frac{1}{2}\left(\frac{d^2U}{dr^2}\right)_1 (r - r_1)^2 + \frac{1}{3}\left(\frac{d^3U}{dr^3}\right)_1 (r - r_1)^3 + \cdots. \qquad (2.176)$$

We can express a change in linear sizes of a rotos in terms of $dD = r - r_1$, surface in terms of $ds = (r - r_1)^2$, and volume in $dv = (r - r_1)^3$ and account for the structure of the equation of state (1.53). Using the definition for the resistance force (1.107) $F = dU(r)/dr$ and parameter of the quasi-elastic force or the rigidity coefficient of the bond (2.154) $\Delta = d^2U(r)/dr^2$, as well as

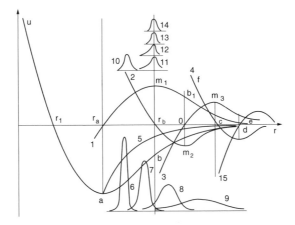

FIGURE 2.47
Variations of the potential energy of a compresson interatomic bond and its four derivatives.

parameter of ductility or anharmonicity coefficient $\chi = d^3U(r)/dr^3$ up to the fourth derivative of $\psi = d^4U(r)/dr^4$, we can write

$$d\omega = FdD + \frac{1}{2}\Delta dS + \frac{1}{3}\chi dV + \frac{1}{4}\psi dDdV + \cdots, \tag{2.177}$$

where $d\omega = U - U_1$.

It follows from comparison of Equations 2.177 and 1.53 that the last expression is differential of the equation of state of the interatomic bond. It shows that deviation of the internal energy from the equilibrium value causes changes in linear sizes of the rotos, and also in its surface and volume. As such, the surface and volume variations are controlled by the higher order derivatives. This allows one, using the graphical structure of function $U = f(r)$ (Figure 1.7a), to trace the change of phase and aggregate states with the interatomic distance r.

Figure 2.47 shows the variations in the potential energy of the C rotos (Figure 2.17a; curve U) and its four derivatives designated by corresponding numbers 1 to 4. Curve 5 determines the location of the instantaneous center of inertia of atoms, which constitute the rotos, at each of the energy levels (positions 4 to 6 in Figure 1.13), and each point of this curve corresponds to the center of length of the EM dipole D_5 in Figures 1.14 and 1.15 (position 5). The process of expansion of a body is characterized by continuous increase in r with simultaneous variation in the space configuration of the rotos charges (Figure 1.15). In mathematical terms, this means that during this process dD has a positive sign. In region $(r_a - r_b)$ curve 2 is in the upper half-plane, and curves 3 and 4 are located outside this region. Therefore, here χ and ψ are equal to zero, and Equation 2.177 transforms into

$$d\omega = FdD + \frac{1}{2}\Delta dS. \tag{2.178}$$

After differentiation of the ω-potential (Equation 1.53) and simplification, we obtain the equation that can be thought of as an analogue of the known Hooke's law (Equation 2.139) for an interatomic bond:

$$dF = \frac{1}{2}\Delta dD. \qquad (2.179)$$

The atoms of the dipole pair, whose state is in this region of the energy spectrum, perform the work of resistance $d\omega$ exclusively due to orientation effects (following the diagram in Figure 2.8a), i.e., in reversible or elastic manner.

While passing through point m_1, parameter Δ changes its sign to the opposite sign (region $r_b - m_2$ of curve 2 is located in the lower half-plane). In region $m_1 - b_1$ of the descending branch of force curve 1, the deformation process is affected by parameter χ. Up to point 0 in the axis of abscissas (its position is set by point m_2 of the minimum of curve 2 or point b_1 of inflection of force characteristic 1) it is negative. Therefore, an increment in the internal forces here is determined by the following expression:

$$dF = -\frac{1}{2}\Delta dD - \frac{1}{3}\chi dS. \qquad (2.180)$$

The presence of the parameter of anharmonicity χ in Equation 2.179 radically changes the character of the bond. Losing the capability of internal compression and acquiring the trend to spontaneous tension, it transforms from compresson to dilaton (Figure 2.17). The CD transitions occur at the Θ-temperature at critical point m_1 of force curve 1, and may have catastrophic consequences for a crystalline structure, depending on whether they are accompanied by thermal support from the outside or not.

The discussed thermal support is realized in its direct sense on heating of the body. In this case the MB distribution, by retaining all the energy levels, changes in shape. Transformed in a direction of 6, 7, 8, 9 (lower field in Figure 2.47), it limits the same area. This means that all the bonds successfully pass critical point m_1 by transforming from compressons to dilatons. The presence of the anharmonic term in Equation 2.180 indicates that the body surface increases, i.e., the body expands during heating. Tension in the force fields is also an expansion, but one-sided, and, as a rule, without heat supply from the outside; it also occurs following the law set by Equation 2.180. The only difference is that deformation of one part of rotoses (long lines in Figure 2.8b) takes place due to degradation (short lines) and degeneration of the other part. As such, an increment in the surface of a body occurs due to the process of cracking and change in the shape and also in the area of the MB factor.

The expansion due to an external force proceeds to point 0 along the axis of abscissas, while the thermal one continues further. This point corresponds to two critical situations: points of inflection b_1 of force function 1 and point m_2 of maximum of parameter Δ in the negative region. At point 0, parameter

χ (curve 3), while passing through zero, changes its sign: it becomes positive instead of negative. Since this moment the expansion process is affected by the fourth derivative ψ, while the force interaction is described by the following formula:

$$dF = -\frac{1}{2}\Delta dD - \frac{1}{3}\chi dS + \frac{1}{4}\psi dV. \tag{2.181}$$

The last term has the plus sign because shoulder c_f of curve 4 is located in the positive zone.

The jump-like increment in the volume at point 0, together with the formation of spherical molecules (following the diagram in Figure 2.7) out of symmetrical rotoses whose atoms rotate along circular paths (position 7 in Figures 1.14 and 1.15), implies the transition of a material from the solid to liquid state. Elementary forces dF are formed within the MB distribution (position 9) and generate that internal pressure, which is the primary cause of existence of a liquid. As follows from Equation 2.181 and Figure 2.47, this pressure is positive, i.e., directed into an object. It is this pressure that can preserve the liquid state for an unlimited period of time. The direction of the pressure can be easily determined by comparing the values of ordinates of functions Δ, χ, and ψ in the negative and positive regions in the $0 - c$ range. The negative direction is set by region $m_2 - d$ of curve 2, while the positive one is set by the sum of ordinates in regions $0m_3$ (curve 3) and fc (curve 4). The compresson nature of the internal pressure is responsible for the surface phenomena in a liquid.

The volume effects taking place on melting and solidification, for which the third term of Equation 2.181 is responsible, may get both positive and negative direction. They are estimated by values from 1% for antimony to 6% for aluminum. For copper, the change in volume during solidification equals 4.1% and for iron 2.2%.[41] Explanation of these differences can be found based on CD concepts.

Under conditions of one-axis or two-axes tension and moderate temperatures at which fracture of solids usually occurs, the deformation processes do not go that far. Even at the maximum possible service temperatures they occur near the θ_f and θ_s temperatures (points m_1 and b_1 on force curve 1, Figure 2.47). At different stages of loading, their course is determined first by Equation 2.179 and then by Equation 2.180. Unlike thermal expansion, where addition of the anharmonic term to Equation 2.180 leads to an enlargement of the outside surface because of the volumetric expansion, in this case its increment occurs because of failure of part of the compresson bonds at the moment of passing through critical point m_1 and formation of the fracture surface. The variation of the MB distribution taking place in this case is shown in the upper field of Figure 2.47 and designated by numbers 10 to 14.

Due to failure of the interatomic bonds, the distribution changes in shape and in filling out of the energy levels. (The area under the probability curve continuously decreases.)

The shift of the hot end of the distribution to the region of critical states is characterized by formation of an asymmetric curve (position 11), which is aggravated by growth of external stresses (position 12). Due to the use of the internal kinetic reserve, the number of dilatons in volume gradually increases. This leads to an increase in the area under the right dilaton branch of the distribution from position 11 to position 13. Equality of the dilaton and compresson concentrations (position 13) corresponds to reaching the critical state by a stressed–strained body. Even an insignificant excess in the dilaton concentration over the compresson concentration (position 14) leads to a catastrophic completion of the fracture process under the effect of an excessive dilaton pressure. The role of the external load (e.g., in tension) is reduced only to formation of the excessive dilaton concentration in the bulk of a material. The fracture process under conditions of thermal deficiency develops due to the intrinsic internal dilaton pressure.

If the first four derivatives (by radius vector r) of the potential energy (Equation 1.26.II) are set equal to zero, where they acquire extreme values (points r_b, 0, c, and d on the axis of abscissas, Figure 2.47), we have the sequence of critical points with coordinates expressed in terms of parameter of the crystalline lattice, r_a (Equation 1.37.I) (r_a; r_b = 1.5 r_a; r_o = 2 r_a; r_c = 2.5 r_a) and the corresponding energy levels (U_a; U_b = 0.88 U_a; U_o = 0.75 U_a; U_o = 0.64 U_a).

A very important characteristic of the interatomic bond is its initial energy U_a determined by the value of the work required for splitting one mole of a solid body into individual atoms at temperature T_c, and for drawing them apart to an infinite distance. These relationships make it possible to get an idea about the variations of the binding energy with an increase in the interatomic distance. Redistribution of the internal energy between the kinetic and potential components and the accompanying variations of physical–mechanical properties of a material, depending on the interatomic distance, are given in Table 2.1. The following designations are used in this table: C, compresson phase; CI and CII, compresson state of the first and second kind; D, dilaton phase; G and L, gaseous and liquid states; m, melting; e, evaporation; $P > 0$, pressure directed inside of the volume; and $P < 0$, volume expansion.

It is seen that the overall internal energy of a solid U all the time remains constant. However, an increase in the interatomic distance requires a continuous replenishment of the thermal energy from the outside, which is necessary for reduction of the potential interaction, especially at the phase transition points where it jumps by δU_a, δU_c, δU_d, etc. Naturally, this part of the energy returns to the environment with a decrease in r. Physical–mechanical properties of solids depend not so much on the value of the overall internal energy as on the ratio of its potential and kinetic parts. An increase in the interatomic distance by factor of 1.5, 2.0, and 2.5, as compared with the solidification parameter r_a, occurs due to an increase in the kinetic energy by 12, 25, and 36%, respectively (see line 2, Figure 2.47), and a corresponding decrease in its potential part.

TABLE 2.1

Redistribution of Internal Energy in Phase and Aggregate Transformations of a Rotos

No. 1	Physical-Mechanical Parameters 2	$r < r_a$ 3	r_a 4	$r_a < r < r_b$ 5	$r_b = 1.5 r_a$ 6	$r_b < r < r_o$ 7	$r_o = 2 r_a$ 8	$r_o < r < r_c$ 9	$r_c = 2.5 r_a$ 10	$r > r_c$ 11
					Critical Values of Interatomic Distance					
1	Levels of internal energy	$U > U_a$	δU_a	$U = U_a$	δU_a	$U = U_a$	δU_m	$U = U_a$	δU_i	$U_i = U_e$
2	Values of potential part of internal energy	$U_r < U_a$	$U(r_a)$	$U_{r_a} < U_r < U_{r_b}$	$0.88\,U_{r_a}$	$U_{r_b} < U_r < U_{r_a}$	$0.75\,U_{r_a}$	$U_{r_o} < U_r < U_{r_c}$	$0.64\,U_{r_a}$	$U > U_c$
3	Signs of derivatives of potential energy with respect to interatomic distance	$F < 0$, $\Delta > 0$	$F = 0$	$F > 0$, $\Delta > 0$	$F = max$, $\Delta = 0$	$F > 0$, $\Delta < 0$, $\chi < 0$	$F > 0$, $\Delta = min$, $\chi = 0$	$F > 0$, $\Delta < 0$, $\chi > 0$, $\psi > 0$	$F > 0$, $\Delta \to 0$, $\chi = min$, $\psi = 0$	–
4	Values of kinetic part of energy	–	$U(\Theta)$	$U_{r_c} < U_T < U_{T_b}$	$0.12\,U_{T_c}$	$U_{\theta_s} < U_T < U_\theta$	$0.25\,U_\theta$	$U_{r_o} < U_T < U_{T_c}$	$0.36\,U_\theta$	$U_T > 0.36\,U_\theta$
5	Characteristic temperatures	Critical values of interatomic distance $T \le \theta_f$	Θ	$T_c < T < \theta_s$	θ_s	$\theta_s < T < T_m$	T_m	$T_m < T < T_c$	T_e	$T > T_e$
6	Phase transition heat	$T = \theta_f$	L_o	–	L_{cd}	–	L_m	–	L_e	–
7	Phase and aggregate state	D	–	C I	–	D	–	C II	–	G
8	Energy range of phase state	$\infty - U_a$	–	$0.88\,U_{r_a}$ for U_r, and $0.12\,U_a$ for U_T	–	$0.13\,U_{r_a}$ for U_r, and U_T	–	$0.11\,U_{r_a}$ for U_r, and U_T	–	–
9	Direction of internal pressure (determined by sign of parameters Δ, χ and ψ)	$P < 0$	–	$P > 0$	–	$P < 0$	–	$P > 0$	–	–
10	Temperatures at which load compressive tensile	– $\theta_f - \theta$	–	$T_c < T < \theta_s$ Freely deformed	–	Freely deformed $T \leftarrow 0_s$	–	Do not perceive Do not perceive	–	–
11	What stresses are induced at cooling heating	σ_c σ_t	–	$\sigma_c \to 0$ Tension	–	Compressive $\sigma_t \to 0$	–	–	–	–
12	What size parameters are changed in deformation	l, S, V	–	Linear l	–	Surface S	–	Volumetric V	–	–

The state of any material is characterized by four critical points: r_a, r_b, r_o, and r_c. The last two are well studied and find wide engineering applications. They determine aggregate transformations of a material from the solid to liquid state, r_o, and from the liquid to gaseous state, r_c. Unlike a solid body, liquid cannot resist variations in its size and shape, but retains connectivity. When it transforms into the gaseous state, it loses this property: atoms and molecules of a gas exist separately, performing chaotic motions in a given volume. Interaction between them is possible only at collisions. If the temperature is increased by another order of magnitude, atoms and molecules will lose connectivity.

Gas is ionized, becoming plasma in which there are no longer atoms and molecules, but only individual particles of what they consisted of previously: nuclei, ions, and electrons. Therefore, solid, liquid, and gas differ only in the level of the kinetic energy of their atoms, which is determined by the corresponding characteristic temperatures: melting T_m, evaporation T_e, and ionization T_i (see row 5 of Table 2.1).

The first critical point r_a corresponds to the temperature of brittle disintegration of the bond, T_c. It can be identified with the moment of formation of a solid from the gaseous state. Indeed, if the temperature of gas exceeds T_c, its atoms and molecules can have no interaction other than dynamic. For the potential interaction to occur between them, their kinetic energy should be decreased to T_c, where the circular motion becomes dominant, which favors formation of a rotos. Formation of the potential interaction leads to transformation of the material from the gaseous state, whose characteristic feature is domination of the translation motion, into the solid state, where the main type of the motion is a joint motion and oscillation of atoms.

As far as the second critical point determined by the Debye temperature θ is concerned, the role it plays in structural processes is not yet fully understood. In statistical and quantum physics[27,28,49-51] and thermodynamics,[11,52] it is often used for descriptions of properties of a solid. However, its physical nature has not yet been revealed; processes that it characterizes or states that it shares are unknown. Within the CD concepts of structure of a solid, its physical implication becomes as evident and clear as the temperature of melting T_m or evaporation T_e. It characterizes transition of a crystalline structure from the compresson to the dilaton phase, or vice versa. Such transitions do not change the aggregate state of a material or its chemical composition but lead to a change in sign of internal stresses. As a rule, these phases are indistinguishable by instrumentation methods; therefore, they escape the attention of researchers. Only a change in phase composition of metals and alloys under certain external effects (deformation, heating, cooling, etc.) provides convincing evidence in favor of their existence.

Derivatives of $U(r)$ change sign in intervals indicating change in the direction of action of the resistance forces between the critical temperatures: between r_a and r_b they are directed into the volume (compresson state of the I kind) and in an interval of $r_b - r_o$ they cause a volumetric expansion (dilaton state). A liquid (zone $r_o - r_c$) may also comprise an internal compressive pressure (compresson state of the II kind). Qualitative changes in the structure are not as evident here as the quantitative changes because the equation for the overall bonding energy (1.80) does not, unfortunately, contain the term comprising the spin quantum number s (Equation 1.3). It is this term that should account for the CD phase transformations occurring following the diagram in Figure 2.17.

Existence of the compresson and dilaton states was predicted by Gerfeld and Hipper-Meyer in 1934.[80] However, this fact was not given much consideration. While developing the melting theory, these authors established that the sum of elastic and thermal pressures in a solid on changing its volume has a minimum. It corresponds to vanishing of the modulus of volumetric compression. Further increase in the volume of the body makes it negative. The authors concluded that the body favors compression instead of resisting it. Therefore, they classified the state of the body corresponding to the transition point as mechanically unstable. In the CD terminology, it corresponds to critical point m_1, and a change in sign of internal stresses implies transition of a material from the compresson to the dilaton state.

Varner and Boley[81] analyzed the opposite process, i.e., the process of formation of thermal stresses on solidification of a metal ingot having a square cross section. They found that, immediately after solidification, the metal undergoes deformation under the effect of tensile stresses induced in the plane of the cooling front. Plastic deformations persist on further cooling. Then tensile stresses affecting the elementary volume of a material decrease, reduce to zero, and transform into compressive stresses, which eventually reach sufficiently high values to cause flow in compression. Evaluating these variations of the internal stresses as surprising, the authors nevertheless failed to give any explanation within the frames of traditional concepts. However, this picture becomes natural with the CD approach, as it describes the actual change of the dilaton pressure into the compresson pressure during the process of solidification and cooling of metal, i.e., transition of the D-phase structure from zone $(b_1 - m_1)$ to the C-phase structure, located in zone $(m_1 - r_a)$ (Figure 2.47).

Chalmers[41] presented examples of formation of defects in castings whose shape prevents compression of metal during cooling. In the casting practice, this phenomenon is called "hot break." Cracking of welds caused by compression of deposited metal in the transition of the dilaton to the compresson phase during solidification and subsequent cooling falls within the same category of defects.[1]

In the literature we can also find direct proof of existence of the C and D states. For example, the time dependence of the strength of glass is explained by its two-phase structure, i.e., the presence of elastic and plastic zones.[82] Different behavior of different phases leads to redistribution of stresses and, therefore, formation of fractures in zones with an increased elasticity. In terms of the CD concepts, this means that fracture of glass develops due to dominant failure of the C bonds.

The binding forces between atoms depend on r_a and $U(r_a)$. These parameters determine the thermomechanical stability of a chemical compound; mechanical and thermal stabilities are just its different manifestations.

The former dominate at low temperatures near the Debye temperature (point m_1), and the latter at high temperatures (within a range of 0 to c). At the beginning it hampers mechanical fracture of a solid, generating resistance forces, and then it determines temperatures and heat needed for melting and evaporation. Therefore, thermal stability should be regarded as a continuation of the mechanical stability. It is natural that they have an identical character of variations: strength, temperature, and heat of melting of crystals increase with a decrease in lattice constant r_a and an increase in its energy $U(r_a)$. Their variation in the reverse direction weakens interatomic bonds and is accompanied by a decrease in the melting temperature.[83]

Although the process of melting solids has been considered from different standpoints,[84] the nature of the liquid state is far from being fully understood and thus no common opinion exists concerning this problem.[85] However, while developing the theory of nonequilibrium states and thermodynamics of fluctuations, Frenkel[86] predicted that different aggregate states, in principle, could be directly correlated along a common line. In fact, Table 2.1 gives such a line, based on a successive loss of order of the AM system from a far order to a near approximation. It accompanies transition of a material from the solid to liquid state. The latter is characterized by fracture of the lattice while the compresson bonds of the II kind are preserved. Mention of the existence of such bonds can be found in Born's work.[87]

The processes of transition from one aggregate state to the other (for example, from solid to liquid state, and vice versa) are well studied and find wide application in engineering practice. Although many of them are controllable, the possibility of active control of physical–mechanical properties of solids has not yet been perceived in full and no effective control methods have been developed so far. In our opinion, this cannot be achieved in principle without understanding internal CD nature. Without this understanding, it is impossible to explain from unified positions the diversity of experimental facts that has been accumulated so far in science of resistance of materials to deformation and fracture.[2,3,19,22] Only revealing the physical nature of CD transitions and realization of the role played by the Debye temperature in them will allow development of methods for activation or prevention of fracture in order to decrease energy consumption of the technological processes (in the first case) or increase reliability and durability of engineering objects (in the second case).

2.10 Mechanical Hysteresis: Causes of Formation and Practical Consequences

Let us return now to the state diagram of a rotos (Figure 1.14), its variation in the thermal field (positions 3 to 7 in Figure 1.13), and the diagram of configuration restructuring of its charges (Figure 1.15). The physical meaning of different regions of the diagram becomes clear. In regions fq_2 and q_1e the dependence of internal stresses on the local volume is conventional for liquids and gases: F_i grows if V_i decreases, and vice versa. Here a material behaves as a single-phase AM system: in the first case, as the overcooled D-structures, and in the second case, as liquids or gases. Their electrodynamics is determined by positions 1 and 7 in Figure 1.15. The states located in regions q_2d and bq_1, as well as in region db of the diagram, are characterized by Inequalities 1.58 and 1.70. This is how the dilaton and compresson parts of the volume behave (Figure 2.17).

In force and thermal fields the MB factor of ductile and brittle structures, while changing in shape and moving along the axis of abscissas, can take one of the possible positions shown in the lower field of Figure 1.14. Terms of set N can change the state smoothly along the two-branch curve q_2dcbq_1 and also along a straight-line region q_2cq_1, missing the intermediate states. As such, a change in the local volume occurs at constant stresses due to the destruction of one part of the interatomic bonds and CD transitions in the other part.

In fact, the work of resistance of a body to external effects is determined by the differential in Equation 2.134. CD transitions occur with disintegration of the bonds at the inelastic stage of deformation (region ab in Figure 2.10) under isothermal conditions ($dT = 0$). These processes are determined by the second term in Equation 2.135. The position of the straight-line region q_2cq_1 in the system of coordinates is determined by the Maxwell rule,[31] according to which areas q_2dc and cbq_1 in each local volume should be identical in order to preserve the shape of a body. This rule establishes equality of thermodynamic potentials of the compresson and dilaton phases (Equation 2.161) in a geometrical shape.

If during deformation the state of the local part of a body is found at point q_1 or q_2, the compresson in the first case and the dilaton in the second case have two options: they may transform into a new state following path q_2dcbq_1 (in compression it happens in the reverse order), by filling on their way all intermediate values of the volume from r_2 to r_1, or pass from point q_2 to q_1, taking the shortest way — along straight line q_2cq_1 through point c, leaving interval $r_2 - r_1$ open. The first option is taken by the overheated compressons, i.e., those that can receive an extra portion of the thermal energy to pass the Debye temperature successfully and come to the dilaton phase. Compressons without such a possibility lose the bond with their nearest environment and form a local defect with size equal to $r_2 - r_1$.

In compression, on the other hand, the dilatons, which cannot by any reason transfer their excessive heat to their nearest neighbors, follow path q_1cq_2. Graphical integration of Equation 2.134 allows vivid interpretation of the processes occurring in the bulk of a material during its deformation and fracture. For this purpose it is more convenient to arrange the diagram in Figure 1.14 as shown in Figure 2.48. Integral

$$\int (Vd\mathbf{P} + PdV)$$

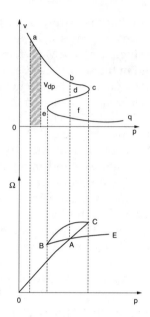

FIGURE 2.48
Graphical integration of the constitutional
diagram.

is shown by the area located between the ordinate, the axis of abscissas, and
diagram *abceq*. In integration along the *abc* branch, the area continuously
increases, corresponding to region *OAC* in the lower field of the figure. As
such, region *AC* corresponds to the metastable state of the dilaton phase.
Integration along *cde* yields a decrease in the area; however, it always remains
larger than in the previous case of integration. In the lower field, this is
shown by the CB branch, which describes the processes of the CD transitions.
Then, the area again grows from point *B* (region *BAE*). In this case, region
BA is equivalent to the metastable state of the compresson phase.

Note that the Ω-potential of the dilaton phase (branch *AC*) is larger than
that of the compresson (region *AB*); this leads to pronounced macroscopic
consequences. For instance, the compression strength of brittle materials is
always higher than tensile strength. At point *A*, where the dilaton *AC* and
compresson *AB* phases intersect, their potentials are leveled. This position
corresponds to critical point m_1 in Figure 2.47.

Compresson and dilaton pressures are equal, providing constancy of the
shape of a body in the equilibrium state. In the isothermal compression, the
state of a body, varying in a direction of *abc*, is found at point *b* with potential
Ω_o. After rounding cape *c* of the diagram, it again returns to the same pressure
(point *d*), but this time with potential $\Omega_1 = \Omega_o +$ area *bcd*. Passing point *e*,
this state for the third time is characterized by the same pressure (point *f*).
However, the potential at this point is equal to $\Omega_2 = \Omega_o +$ area *bcd* − area *def*.
As $\Omega_2 = \Omega_o$, area *bcd* is equal to area *def*.

The existence of metastable states in a two-phase structure leads to formation of the mechanical hysteresis and the associated phenomena. The body
is found in a tensile force field. Its state starts changing along the downward
arrow: from point *q* in a direction of point c_1 (Figure 2.49). In region qc_1, the

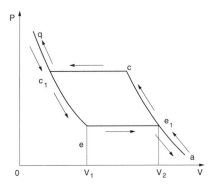

FIGURE 2.49
Diagram of formation of the mechanical hysteresis loop.

entire CD set participates in the resistance. Here the useful work is performed due to orientation processes (Figure 2.8a), i.e., exclusively due to variation in the first term of Equation 2.134. They are accompanied by an intensive redistribution of the kinetic part of the internal energy between local micro-volumes of the body. At point c_1, the compresson flow is divided into two. The part absorbing phonons Δq (Figure 1.18) successfully passes the Debye temperature θ (Figure 1.14) and follows the path $c_1 c$. The other part, which does not have such a possibility or, even worse, spends its energy for defor-mation of its neighbors (following the diagram in Figure 2.8b), transforms into the metastable state and enters region $c_1 e$. Further development of the process leads to overcooling of the metastable compressons. As a result, they are found at point e and, forced to follow their neighbors of the first part, join them at point e_1. This changes their state in a jump-like manner while following path ee_1.

Starting from this moment, the second term of Equation 2.136.II, compris-ing parameter dN, starts affecting the process of deformation and fracture. Disintegration of the interatomic bonds results in formation of local submi-crodefects $\Delta v = v_2 - v_1$. Distributed in the volume in a random manner in accordance with the MB factor (Equation 2.3), they increase porosity of a material at this stage.[62] After point e_1 only the dilatons participate in the resistance to deformation, as the compressons either have transformed into the *D*-phase (on the $c_1 c$ path) or ceased existing (in region $e_1 e$). Processes that occur in compression are identical (movement along the upward arrow on the $ae_1 cc_1 q$ path). In this case splitting the dilaton flow into the deformation (in the direction of $e_1 c$) and destruction ($e_1 e$) parts occurs at point e_1.

The previous discussion suggests that phase transformations in the direct and reverse directions form a hysteresis loop. In thermal deficiency, it deter-mines an irreversible or destructive character of the deformation processes within an interval of $(v_1 - v_2)$. The area of loop $c_1 ee_1 c$ is equal to the amount of heat released (in compression) or absorbed (in tension). Its formation is related exclusively to the kinetics of the processes occurring in the two-phase

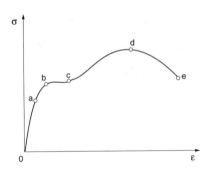

FIGURE 2.50
Typical diagram of tension of a ductile material.

CD system. Understanding the physical nature of mechanical hysteresis allows identification of the measures intended for correction of the stressed–deformed state. To eliminate fracture of the structure, it is necessary to transform the destructive part of the CD flow into the deformational part. Supplying to or removing from the system the amount of heat equivalent to the CD transitions can accomplish this. Substantiation of the methods for controlling the stressed–deformed state can be found in Komarovsky.[12,24]

Let us consider how the mechanical hysteresis manifests itself in practice. Technical tests of materials are normally limited to the measurement of longitudinal strains ε, while the determination of the heat balance in such tests is rare and limited to research purposes. Figure 2.50 shows a typical tension diagram for ductile materials used to find the major mechanical properties of materials. Voluminous literature is dedicated to its external or phenomenological description, which addresses practical problems.[2,3,44,48] Our objective, however, is to address the questions of principal importance such as why diagram $\sigma = f(\varepsilon)$ for ductile materials takes this particular form but not any other, whether it is possible to purposefully change it or not, and, if so, then how this can be accomplished.

Region $0a$ in Figure 2.50 corresponds to the elastic (reversible or orientation) stage of deformation. It is described by Hooke's law (Equations 2.139 and 2.179). Under the effect of the external force field, the chaotically arranged AM set (whose mechanical analogue is an infinite interatomic chain shown in Figure 1.27) formed on solidification is oriented in the direction of the field (according to the diagram in Figure 2.8a), adjusting the CD proportion to new varying conditions. At point a, the orientation process is completed and the state of the body is at critical point c_1 (Figure 2.49).

From this moment, the deformation can occur only due to variations in the distance between the interacting atoms and, therefore, due to an increase in size of the elliptical paths of their natural thermal motion (see Figure 1.7b). Internal forces start concentrating on compressons, whereas, when heat is supplied from the outside, dilatons cannot perform useful work in tension. They freely deform (illustrated by a descending branch bq_1e of the diagram in Figure 1.14). Both dilatons and compressons become strong absorbers of heat required for continuation of the deformation process; however, the specific demand for heat of the former is much higher than that of the latter (Figure 2.27).

This means that heat is absorbed first by compressons, which take it even from the neighboring dilatons. Giving up this heat, the latter also contribute to the resistance, while moving along branch eq_1b toward compressons to the Debye temperature (point b in Figure 1.14). The heat exchange process divides compressons into two types: those which, while absorbing heat and deforming, can bear the load (they follow the c_1c path in Figure 2.49), and those which, while giving up their heat to active neighbors, fall into the metastable state (in the direction of c_1e). Similar processes, but with loss of heat, occur also in the D-phase. The process of formation of the destructive and deformational flow takes place in region ab of diagram $\sigma = f(\varepsilon)$ (Figure 2.50) and ends at point b. Irreversible strains appear at this moment, caused by the beginning of irreversible transformations in the structure, i.e., disintegration of the metastable dislocation phase.

Standard, as well as special, tests of materials are almost never accompanied by estimation of the heat balance and continuous verification of the degree of fracture of the structure. This is not because of underestimation of the significance of such measurements but rather due to costs of the tests, including time, labor, and testing machines. As a result, the absence or underestimation of this information does not allow obtaining the complete hysteresis loop (Figure 2.49). In diagram $\sigma = f(\varepsilon)$ (Figure 2.50), it degenerates into a horizontal region bc that characterizes yield of the material.

At point c the metastable part of the compresson phase is fully destroyed. As a result, the volume of the material is saturated with a certain amount of dispersed microscopic defects.[10] Now the structure is made of two parts: active compresson (capable of deforming and inducing internal stresses) and cold dilaton. Plastic deformation, emergence and movement of dislocations, and hardening caused by cooling of dilatons characterize region cd. Point d corresponds to the maximum load that the specimen can withstand before fracture. In region de the so-called "neck" may form when the majority of rotoses oriented in the field clear out the opposite direction (Figure 2.8b). At point e the specimen is divided into parts. At this moment the MB distribution is arranged symmetrically about the Θ-temperature, as shown in position 13 in Figure 2.47.

Consider the causes of formation of the neck. Its emergence is attributed to the CD phase transitions in the oriented set of rotoses (Figure 2.8). Brittle materials (curve 1 in Figure 2.22) cannot have such transitions and thus the neck is not formed in their testing. After passing point c (Figure 2.50), all rotoses become field oriented. This can cause a transverse necking of the specimen (following the diagram in Figure 2.8).

Figure 2.51 shows the cross section of a cylindrical specimen made of a ductile material subjected to axial tension. In accordance with Equation 2.105, in region cd of the deformation curve of Figure 2.50 such stressed–deformed

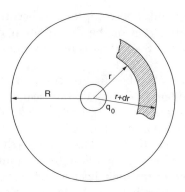

FIGURE 2.51
Mechanism of formation of the "neck" in tension of ductile material specimens.

state generates the temperature gradient in the axial direction (according to the diagram in Figure 2.8b). In region *de* a radial flow of phonons, directed from the periphery to the center, appears:

$$q_r = -n\frac{dq}{dr},$$
(2.182)

where *n* is the coefficient of diffusion of the flow along radius *R*. The minus sign indicates a decrease of the flow along the radius with an increase in *r*. As the formation of new phonons takes place in any conical coaxial layer limited by the surfaces with radii *r* and *r* + *dr*, the densities of flows through surfaces S_r and S_{r+dr} are different. The total number of phonons flowing through the cylindrical surface S_r of a unit height per unit time is equal to $2\pi r1qr$. In the coaxial layer, new phonons are formed between *r* and *r* + *dr*. Their quantity is found from relationship $2\pi r dr1q$, where *q* is determined by Equation 2.108.

The balance of radiation in the radial direction can be written as

$$2\pi r q_r + 2\pi r dr q = 2\pi (r + dr) q_{r+dr}.$$
(2.183)

The value of the function in the vicinity of a point can be determined as[80]

$$q_{r+dr} = q_r + \frac{dq_r}{dr} dr.$$
(2.184)

From Equation 2.182, we find

$$\frac{dq_r}{dr} = -n\frac{d^2q}{dr^2}.$$
(2.185)

Substituting Equations 2.182 and 2.165 into Equation 2.184, we obtain

$$q_{r+dr} = -n\left(\frac{dq}{dr} + \frac{d^2q}{dr^2}dr\right). \tag{2.186}$$

Substituting Equation 2.186 into Equation 2.183, simplifying, and omitting values of the second order of infinitesimal, we find that

$$rq - q_r - r\frac{dq_r}{dr} = 0. \tag{2.187}$$

Accounting for Equations 2.181 and 2.185, Equation 2.187 can be transformed as

$$\frac{d^2qr}{dr^2} + \frac{1}{r}\frac{dq_r}{dr} + \frac{q_o}{n}q_r = 0,$$

where q_o is the concentration of phonons in any unit volume along the axis of the specimen and q_r is the same at any distance from it.

This equation has the following solution:

$$q_r = q_o B\left(\sqrt{\frac{q}{n}}r\right), \tag{2.188}$$

where $B(x)$ is the Bessel function with argument

$$x = \left(\frac{q}{n}\right)^{1/2}r.$$

Its graphical form is shown in Figure 2.52. As seen, $B(x) = 0$, if $x = 2.4$, and $B(x) = 1$, if $x = 0$. Prior to the formation of the neck $q_r \approx 0$,

$$\left(\frac{q}{n}\right)^{1/2} = \frac{2.4}{R_o}, \tag{2.189}$$

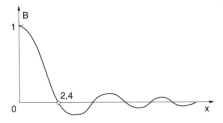

FIGURE 2.52
Graphical representation of the Bessel function.

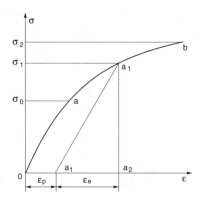

FIGURE 2.53
Manifestation of mechanical hysteresis through strain hardening.

where R_o is the radius of the specimen between points c and d of the diagram (Figure 2.50). After substitution of Equation 2.189 into Equation 2.188, we have that

$$\frac{q_r}{q_o} = B\left(2.4\frac{R}{R_o}\right). \tag{2.190}$$

It can be seen that geometrical dimensions of specimen R turn out to be related to the radial heat flow q_r through the Bessel function B. To maintain the deformation process in region de, phonons must migrate from the periphery of the specimen to its center. This leads to the trend of relationship q_r/q_o from 0 to 1. As follows from Figure 2.52, with such a variation of function B its argument R/R_o tends from 2.4 to 0. This means that coaxial cooling is accompanied by the transverse compression of the specimen.

Mechanical hysteresis shows up, in particular, as the effect of strain hardening. In the elastic region $0a$ of the deformation diagram (Figure 2.53), the orientation processes take place following the diagram shown in Figure 2.8a. After removal of the external force field having intensity, which does not exceed the σ_o level, the body returns to the initial state (point 0), and the molecules again take a symmetrical shape (Figure 2.7). In this case the modulus of state **E** (Equation 2.139) varies only due to an increment in terms dr in Equation 2.136.II. The CD proportion transforms reversibly from the initial position (position 10 in Figure 2.47) in the direction of 10 to 11. If the curvilinear region ab is reached, not only the dr-terms but also the subsequent terms of Equation 2.136 are involved into the work.

After load removal, the associated irreversible configuration changes in the rotoses (Figure 2.8b) fix the CD distribution in the state resulting from the effect of external loading σ_1. Emergence of plastic deformation ε_p, in addition to elastic deformation ε_e, makes its return to the initial state, determined by the origin of coordinates, impossible. After removal of the load, irreversible plastic deformations now fix the changed CD distribution (positions 11 to 12 in Figure 2.47) rather than the previous one (position 10). It then determines the subsequent behavior of the body. This explains the fact that in reinitiation of loading, the stressed–deformed state is formed not in the direction of $0a$, but along path 0_1a_1 (Figure 2.53). Hardening due to the orientation effect can occur only in region ab until yield stress σ_2 is reached. After that, the progressive contribution to the deformation process is made by the flow of failures of the interatomic bonds (term dN in Equation 2.136).

2.11 Compresson–Dilaton Nature of Dislocations

The overall internal energy (Equation 1.80) and, therefore, the ω-potential of the rotos (Equation 1.53) are the multifactorial parameters depending on quantum numbers n, l, and s (Equations 1.1 to 1.3) and frequency v (Figure 2.14). That is why the ideal arrangement of the C and D rotoses (positions 3 and 6 in Figure 1.15), where the electric charges of the pair perform a close movement in one plane and the electrodynamic forces are balanced, is an exception rather than a rule. In reality, deviation of the s, n, l, and v variables from the equilibrium values changes the direction of action of the binding forces (Figure 2.17), geometrical dimensions, and shape of the orbital paths (Figure 1.7b) accordingly. This generates various kinds of dislocations[34,88,89] in an ideal crystalline structure.[44,47,90]

Consider a small (compared to the dimensions of a solid) volume inside a solid. This volume can be thought of as the AM system immersed in a medium with constant temperature and pressure. According to Equation 2.49, the state of this part, as well as of the entire body (Equation 2.45), is determined by assigning it two thermodynamic values. Analysis of thermodynamic potentials does not suggest that, at any selected pair of independent variables (e.g., P and V, s and T, or others), only the homogeneous state of the system can characterize thermal and mechanical equilibrium of the body. Such equilibrium is established most readily when the AM structure is a two-phase formation.

Mechanical and thermal equilibrium of the considered volume and of the entire body is determined by the equation of state (Equation 2.45 or 2.65). The phase equilibrium of the compresson and dilaton components is characterized by

$$\omega_c\left(P,t_c\right) = \omega_d\left(P,t_d\right), \tag{2.191}$$

where $P = P_c = |-P_d|$. It is written accounting for Equation 2.47.I and follows from the obvious speculations. As ω is the thermodynamic potential of an atomic bond, according to Equation 2.35, it can be written as

$$\omega = h - u(r), \tag{2.192}$$

where, according to Equation 2.27, $h = u + pv$ is the heat content of one rotos and

$$u(r) = T\mathbf{s} \tag{2.193}$$

is the bound or potential part of its internal energy. Differentiation of Equation 2.192 at a constant heat content $h = $ const with respect to the selected direction dr or elementary volume dv yields

$$F = \left[\frac{\partial u(r)}{\partial r} \right]_t = -\left[\frac{\partial \omega}{\partial v} \right]_t. \tag{2.194}$$

The graphical form of this equation with no accounting for thermal interaction is shown in Figure 1.14, which represents the state diagram of the rotos.

Substituting Equation 2.193 into Equation 2.194 and accounting for[7]

$$\left(\frac{\partial \mathbf{s}}{\partial v} \right)_T = \left(\frac{\partial \mathbf{P}}{\partial T} \right)_v,$$

we have

$$F = \left[\frac{\partial u(r)}{\partial r} \right]_T = -\left[\frac{\partial \omega}{\partial v} \right]_T. \tag{2.195}$$

$\mathbf{P}$ and T of compressons and dilatons, which are in equilibrium, can be expressed as functions of each other. This means that compressons and dilatons can form a rotos not at any $\mathbf{P}$ and T. Given parameters of one of them determine the parameters of the other. Such a CD pair at levels D_3 and D_5 occupies the opposite branches of the acceleration well or deceleration barrier (Figure 1.14).

In the P–V coordinate system, the points at which the formation of the bound dilaton–compresson pairs is possible fall in a certain region limited by curve *cbkad* (Figure 2.54). This curve is an exact copy of curve 8 located inside the potential well in Figure 1.13. The field on the left of branch *cbk* is a homogeneous compresson state while the dilaton phase is concentrated to the right of *kad*. Line *kme* identifies the locus of points corresponding to the critical state in which the entire body or its individual bonds can be at the Debye temperature. Figure 2.55 shows graphically Equation 2.195 at $T = $ const (e.g., at level *bma*) in integral, 1, and differential, 2, representations. In fact, it repeats the temperature dependence of heat capacity (Figure 2.15b). This is another proof of the kinetic nature of mechanical properties (Figure 2.29).

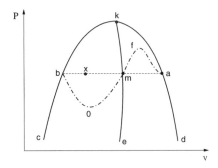

FIGURE 2.54
Representation of the compresson–dilaton zone in isothermal space.

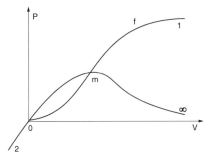

FIGURE 2.55
Integral (1) and differential (2) representation of internal stresses.

It was shown in Section 2.10 that transition from the compresson zone to the dilaton, and vice versa, can occur in a reversible manner along path *bomfa*, or with fracture of the bond along straight line *bma*. Its failure can take place at any of the points of line *kme*, except for point *k* (Figure 2.54). Here the C and D phases couple, and through this point the safe transition from one phase to the other can occur. This allows an important practical conclusion: by controlling one of the fields (force, thermal, or any other) in response to a random or expected variation in the other, we can eliminate or activate, if necessary, the process of fracture of a material.

The bell-shaped curve *cbkad* limits the state space in which, at P = const and T = const, both phases (compresson and dilaton) level each other. For example, at any point x the specific volume of the C and D phases is determined by abscissas of points a and b, lying on a horizontal straight line that passes through this point. In accordance with the proportionality rule, which characterizes the balance of the amount of a material, it can be easily concluded that the specific volumes of the C and D phases in this case are inversely proportional to the ax and bx lengths.

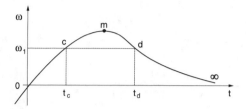

FIGURE 2.56
Temperature dependence of the ω-potential.

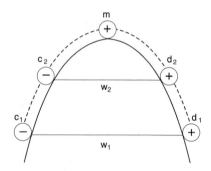

FIGURE 2.57
Diagram of formation of vacancies in synchronous motion of atoms of the CD pair at the Debye temperature.

The relationship of Equivalence 2.105 allows one to represent the temperature dependence of the ω-potential (Figure 2.56) using the known function $P = f(V)$ (curve 2 in Figure 2.55). Points c and d of intersection of the branches at a given potential (e.g., ω_1) determine the compresson, t_c, and dilaton, t_a, temperatures, at which both phases can form a stable coupling. As such, the condition of thermodynamic equilibrium (2.191) should be met, by all means, because it stabilizes the structure and maintains the preset size and shape of a body. Equilibrium is preserved when the variation of the external pressure is within a certain range determined by critical point m; however, at each level it is characterized by different values of t_c and t_d.

Under the given P and T, the ω-potential always tends to minimum. This means that, at any of the points of branches $0m$ and $m\infty$ (e.g., c and d), the trend of sliding the energy state in a direction to the axis of abscissas exists. Only the mutual bond withstands this trend, preserving the integrity of the structure. Figure 2.57 shows a mechanical analogy of the CD nature of the bond. The C and D atoms (shown in the form of negative and positive circles) can be kept from sliding down on both sides from apex m of the convex curve only by a rigid fiber (in Figure 2.57, shown by a dashed line). The length of the fiber can vary, moving the energy state of the CD bond from one position (e.g., $c_1 d_1$) to the other ($c_2 d_2$). When a compresson and a dilaton simultaneously come to critical point m, where the temperature is equal to the Debye temperature, they annihilate to form a local microdefect.

It can be seen from Figure 2.56 that $t_d > t_c$. This means that transition from the C to the D phase requires thermal support, whereas the reverse transition is accompanied by heat radiation. According to the condition of equilibrium of the phases (2.191), any transition takes place at P = const and ω = const. Application of the first to Equation 2.52 yields

$$s = -\frac{\partial \omega}{\partial t}. \tag{2.196}$$

The second transforms Equation 2.192 into the following relationship:

$$h_d - h_c = (ts)_d - (ts)_c.$$

As the difference in thermal functions h is equivalent to a variation in the amount of heat, q, it holds that

$$q = (ts)_d - (ts)_c. \tag{2.197}$$

Because $\omega = f(t)$ is convex upward (Figure 2.56), the derivative of Equation 2.196 in compresson region ocm is always positive, and that in the dilaton region $md\infty$ is negative. This yields

$$s_d = \left(\frac{\partial \omega}{\partial t}\right)_d \quad \text{and} \quad s_c = -\left(\frac{\partial \omega}{\partial t}\right)_c. \tag{2.198}$$

A continuous energy exchange occurs between the D and C phases. The first radiates heat, whereas the second absorbs it, providing the equilibrium state. In such a state, $q = 0$ and $(ts)_d = (ts)_c$. Equality of the bond energy in the C and D states implies that structure of a body during solidification can be arranged from the CD pairs (Figure 1.7b), which mutually compensate for their own attractive and repulsive forces (positions 3 and 6 in Figure 1.15). As a result, under stable external conditions the body preserves its shape and size.

It follows from Equation 2.198 that the equilibrium of a body is disturbed by varying temperature t or entropy s, or both simultaneously. Let some amount of heat q be supplied to a body. Then Equation 2.197 is transformed into inequality

$$-(ts)_c + q < (ts)_d + q. \tag{2.199}$$

This means that if excessive heat is supplied, a part of the compressons transfers to the dilaton phase. As a result, the MB distribution is transformed in the direction from $1/2T$ to $2T$ (Figure 2.2) and thus the internal dilaton tension will no longer be compensated for by the compresson compression so that the body expands on heating. The expansion continues until Equality 2.197 is restored at a new level.

Naturally, upon cooling the process reverses. That is, the external thermal fields lead to potential consequences: expansion or compression of the body. It might be expected, in turn, that potential interaction of the body with its surroundings (tension, torsion, compression, etc.) should be accompanied by thermal phenomena. This has been proven experimentally.[3,4,35,36]

For example, to ensure tension, it is necessary that the same inequality (2.199) be met. To ensure the deformation process, it is necessary to supply the same amount of heat q. At initial stages of loading, where the demand for heat is low (region $0abc$ of the $\sigma = f(\varepsilon)$ diagram in Figure 2.50), it is readily met by heat absorbing from the outside. In region cde the deformation process is supported from the internal reserves. Dilatons start giving their heat to the compresson neighbor in the bond (e.g., in Figure 2.57 in a direction of $d_1 \rightarrow c_1$). As a result, while moving synchronically to critical point m corresponding to the θ_s-temperature, both (dilaton, by giving the heat, and compresson, by receiving it) bear the external load. If at the previous stages it was distributed only on compressons, the forced involvement of dilatons into this process causes short-time hardening of a specimen in region cd. The progressing escape of the CD pairs to critical point m causes an increase in the number of the failed bonds (according to the diagram in Figure 2.8b) and, hence, the fracture of the structure. Point d (Figure 2.50) characterizes an exhaustion of the CD reserves and a decrease in cross section of the specimen to such an extent (Figure 2.51) that it can no longer withstand any increase in the external load.

As deformation and fracture in tension occur due to redistribution of the kinetic part of the internal energy, the specimen is cooled. Formation of other types of the stressed–deformed state (compression, bending, torsion, etc.) can be considered in an identical manner, and it can be shown that compression, for instance, is accompanied by release of excessive heat.

A characteristic feature of solids is localization of atoms in an arbitrary space region where they are caught by the solidification process and which they cannot leave afterwards, unlike in gases or liquids. For this reason they are able to exchange energy only with their nearest neighbors. Solidification is a random process. In some space regions solidification starts at potential ω_1, in other regions at ω_2, in still others at ω_3, etc. (Figure 2.57). As a result, the crystalline structure of solids almost always has grain structure (Figure 2.58).

It is clear that, at $\omega = $ const, the crystalline structure inside a grain has a perfect arrangement. Consider how the mutual bond is arranged along the grain boundaries. By analogy with Equation 2.191, to ensure the internal equilibrium at any boundary (e.g., 1 and 2), it is required that

$$\omega_1 = \omega_2. \tag{2.200}$$

First, let us clarify stability conditions of a two-phase system at the boundary. By analogy with Equation 2.192, we can write variation of the ω-potential as

$$\Delta\omega = \Delta u - t_1 d\mathbf{s}_2 + P_1 dv_2, \tag{2.201}$$

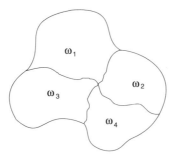

FIGURE 2.58
Grain structure of actual solids.

where Δu is the sudden change or jump of the internal energy at the grain boundary; t_1 and P_1 are the temperature and pressure of one of the grains; s_2 and v_2 are the entropy and volume of the adjacent grain. Expanding Δu in a power series in powers of ds_2 and dv_2 and keeping the terms of the second order yields[23]

$$\Delta u = \frac{\partial u}{\partial s_2}ds_2 + \frac{\partial u}{\partial v_2}dv_2 + \frac{1}{2}\left(\frac{\partial^2 u}{\partial s_2^2}ds_2^2 + 2\frac{\partial^2 u}{\partial s_2 \partial v_2}ds_2 dv_2 + \frac{\partial^2 u}{\partial v_2^2}dv_2^2\right). \quad (2.202)$$

Because, by definition,[20]

$$\left(\frac{\partial u}{\partial s_2}\right)_v = t_2 \quad \text{and} \quad \left(\frac{\partial u}{\partial v_2}\right)_s = -P_2$$

are the temperature and pressure established in the CD set of the second grain, then

$$\Delta u = t_2 ds_2 - P_2 dv_2 + A, \quad (2.203)$$

where

$$A = \frac{1}{2}\left(\frac{\partial^2 u}{\partial s_s^2}ds_2^2 + 2\frac{\partial^2 u}{\partial s_2 \partial v_2}ds_2 dv_2 + \frac{\partial^2 u}{\partial v_2^2}dv_2^2\right).$$

Substituting Equation 2.203 into Equation 2.201, we have that

$$\Delta \omega = (t_2 - t_1)ds_2 + (P_1 - P_2)dV_2 + A.$$

As seen,

$$\Delta \omega = d\omega + d^2\omega, \quad (2.204)$$

where

$$d\omega = (t_2 - t_1)ds_2 + (P_1 - P_2)dV_2, \quad \text{and} \quad d^2\omega = A.$$

It is apparent that the state of stable equilibrium at the grain boundary can be well determined by two relationships:

$$d\omega = 0 \quad \text{(I)}; \quad d^2\omega < 0 \quad \text{(II)}. \tag{2.205}$$

The first equation sets the required condition of equilibrium while the second provides its stability.

Requirement $d\omega = 0$ leads to equality

$$\left(\frac{ds_2}{dv_2}\right)_t = -\frac{P_1 - P_2}{t_2 - t_1}.$$

It is known from the relationship between the main thermodynamic values[20] that

$$\left(\frac{\partial s}{\partial v}\right)_T = \left(\frac{\partial P}{\partial T}\right)_V.$$

Therefore,

$$\left(\frac{\partial P_2}{\partial t_2}\right)_V = \frac{P_1 - P_2}{t_1 - t_2}. \tag{2.206}$$

Condition $d^2\omega < 0$ or $A < 0$ is met at any ds_2 and dv_2, if

$$\frac{\partial^2 u}{\partial s_2^2} < 0 \quad \text{and} \quad \frac{\partial^2 u}{\partial s_2^2}\frac{\partial^2 u}{\partial v_2^2} - \left(\frac{\partial^2 u}{\partial s_2 v_2}\right)^2 < 0. \tag{2.207}$$

The first of these inequalities can be expressed as

$$\frac{\partial^2 u}{\partial s_2^2} = \left(\frac{\partial t_2}{\partial s_2}\right)_v = \frac{t_2}{C_2} < 0,$$

where C_2 is the heat capacity of the second grain. Because t_2 is always positive, then

$$C_2 < 0, \tag{2.208}$$

The second of the inequalities (2.207) written in the form of the Jacobian is

$$\frac{\partial(t_2; P_2)}{\partial(s_2; v_2)} > 0,$$

which, in variables t_2 and v_2, becomes

$$\frac{t_2\left(\dfrac{\partial P_2}{\partial v_2}\right)_T}{C_2} > 0.$$

Because $C_2 < 0$ from Equation 2.208, it follows that

$$\left(\frac{\partial P_2}{\partial v_2}\right)_T < 0. \tag{2.209}$$

Condition 2.200 provides thermal, and Condition 2.209 mechanical, stability of structure of a material at the grain boundary.

Assume a leak of some amount of heat, $C_2 < 0$, through the boundary because of a random fluctuation on the side of the second grain. A decrease in temperature t_2 in this zone is accompanied by a decrease in the local pressure. This means that

$$\left(\frac{\partial P_2}{\partial t_2}\right)_v < 0.$$

Then, according to Equation 2.206, the following inequality can be written:

$$\frac{P_1 - P_2}{t_1 - t_2} < 0.$$

It is valid only when $P_1 - P_2 > 0$ and $t_1 - t_2 < 0$, and at $P_1 - P_2 < 0$ and $t_1 - t_2 > 0$. The first pair of the inequality cannot take place, so $P_1 < P_2$ should take place at $t_1 > t_2$. That is, the first grain should react in the dilaton manner to a random loss of heat and subsequent decrease in the internal pressure in the second grain: it should decrease pressure with an increase in temperature.

The described behavior, which takes place at the boundary, can only provide the type of CD bond shown in Figure 2.59. Its characteristic feature is that every compresson and every dilaton are bound with at least two rotoses

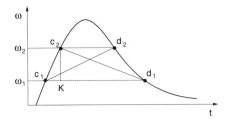

FIGURE 2.59
Jumps of potentials in the formation of dislocations.

of the opposite type, rather than with just one (as in formation of a regular structure). For example, compresson c_1 in grain with ω_1 binds simultaneously with a dilaton from its own grain, d_1, and with that from the neighboring one, $d_2(\omega_2 > \omega_1)$. The interaction of compresson C_2 and dilatons d_1 and d_2 is also formed according to the same diagram. Only with the discussed type of bond is the CD equilibrium condition (2.191) met inside each of the grains and also at their boundaries. Designating $\Delta\omega = \omega_2 - \omega_1 = KC_2$, we can write it using potentials of both grains:

$$\omega_{c_1} = \omega_{d_1} = \omega_{d_2} - \Delta\omega; \qquad \omega_{c_2} = \omega_{d_2} = \omega_{d_1} + \Delta\omega.$$

Summing up both expressions yields

$$\omega_{c_1} + \omega_{c_2} = \omega_{d_1} + \omega_{d_2} \quad \text{or} \quad \omega_c = \omega_d.$$

The CD bonds in which the continuity of the ω potential (with step KC_2 in plot $\omega = f(t)$) is violated form dislocations that cause a localized distortion of the ideal crystalline lattice (Figure 2.60). Symbol $\perp$ designates the jump of the ω-potential and corresponds to segment KC_2 of the ordinate in Figure 2.59.

Voluminous literature is dedicated to dislocations,[44,47,90] so it is not our intent to consider this issue in detail. We tackle it here because it is directly related to the CD nature of solids, being one of the characteristic manifestations of this nature. It follows from Figure 2.59 that dislocations with equal probability may exist in compresson and dilaton phases. The former can be regarded as negative, with the latter as positive. Recalling considerations related to Figure 2.57, we can readily show that dislocations of opposite signs, on reaching the θ_s temperature, destroy each other.

To illustrate the latter statement, Figure 2.60a shows the variation of the crystalline structure due to the formation of vacancies (both shown by symbol $\perp$). Formation of a compresson vacancy is explained as a dilaton that

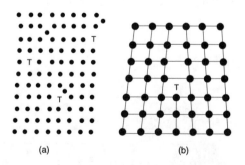

(a)   (b)

FIGURE 2.60
Dislocation diagrams.

has lost its compresson neighbor, which compensated before for its positive pressure. To restore equilibrium, the dilaton must redistribute its energy among remaining nearby compressons. This results in the formation of a step of the ω-potential (Figure 2.59) and, therefore, dilaton dislocation (Figure 2.60b).

Dislocations in structure of solid bodies are common. Their amount and mobility are determined by external conditions; therefore, the force and thermal fields lead to substantial activation of the processes of multiplication, interaction, and movement of the dislocations.[44,47]

3

Interaction with External Fields

3.1 Introduction

The thermodynamic equation of state (2.65) establishes the relationship between the ability of a crystalline structure to generate resistance forces and the temperature distributed inside the AM set in accordance with the Maxwell-Boltzmann (MB) factor formed on solidification. Resistance forces are generated by polarized atoms moving along closed paths and interacting with each other (Figure 1.7b) that maintain a certain shape and dimensions of a body within the preset volume and counteract the external factors that try to change them. This chapter presents experimental verifications of the equation of state showing that it is applicable to all solids, without exception, under various external conditions. This proof is based on the author's experimental results and on data of other investigators.[91–169]

The initial state of materials in parts and structures forms during solidification and then is modified by technological methods. External fields and aggressive environments affect it during operation through one or several parameters accounted for by the terms of Equation 2.96. When applied, the equation of state is transformed into the interaction formula (Section 3.2).

Structure-forming particles possess electric charges in addition to their masses. While moving, they form local electromagnetic (EM) fields through which bodies interact with external fields of a similar nature (electric, magnetic, thermal, or other) (Section 3.3). A structure formed on solidification consists of polyatomic molecules of spherical shape. The external fields, which cause orbital planes to orient to a certain direction, give them an ellipsoid shape (Section 3.4). Deformation, as well as resistance, is of a thermal-active nature. In constant thermal fields, an increase in size of some interatomic bonds may take place only due to heat removal from the others. As such, the former maintain the deformation process while the latter degrade. In this sense, the deformation and the destruction processes are always interrelated (Section 3.5).

The volume of a material has an ambiguous effect on its stressed–deformed state, which shows up as the scale effect (Section 3.6). An increase in the mass (volume) of a material decreases the level of internal stresses; however, it also decreases resistance to crack propagation. Unfortunately, traditional design methods are based on the assumption that resistance of a material increases proportionally to its volume. In our opinion, such methods comprise a danger not yet fully realized, resulting in premature fractures of seemingly safe engineering objects (events wrongly considered accidental). The scale effect is a principal obstacle to improvement of available design methods. It can be overcome, but 100% guaranteed reliability and durability can be achieved only through development and application of energy methods to control stressed–deformed state.[12]

It is known that solid bodies are heterogeneous at all scale levels. Heterogeneity is embedded in their structure at the stage of solidification due to the MB-factor, which is responsible for distribution of energy inside the AM set. Section 3.7 presents the results of investigation of the extent to which the MB factor affects resistance of materials to deformation and fracture. Also, it substantiates the feasibility of developing materials with desirable properties by varying their shapes. Physical–mechanical properties depend on the density of a body; Section 3.8 shows that this dependence directly follows from the equation of state. The correlation between the mechanisms of an increase in density during solidification and its decrease during destruction processes has been established.

Mechanical stresses change the state of a body. An analysis of such a change in quasi-static tensile and compression is given in Section 3.9, and change under complex stressed state conditions is considered in Section 3.10. The effect of quasi-static strengthening in compression, which can be applied for deceleration of fracture is described.

The results of this chapter allow us to conclude that the equation of state provides the correct description of behavior of materials in force and thermal fields.

3.2   Equation of Interaction

Fractures and premature failures of engineering objects occur in practice. This is attributable[2,13] to insufficient accuracy of design methods, incomplete knowledge of properties of materials, ungrounded criteria of fracture and selection of safety factors, imperfection of materials and technological processes, deviations from normal service conditions, inopportune and low quality of maintenance and repair, etc. Although true to a certain extent, this cannot be a prime cause for discussed failures. The most common cause of failure of parts and structures is a change in the internal state of materials with service (operating) time caused by the physical–chemical processes occurring in materials and activated by external factors.

The result of these processes is called by different terms: *aging* (thermal, strain, etc.), *ebrittlement* (radiation, thermal, etc.), or *damage* (force-, moisture-, chemical, hydrogen-induced, etc.).[2,3,44] In our opinion, the internal mechanism of these processes is determined by peculiarities of the compresson–dilatons (CD) layout of a structure, and their rate by external conditions: loading mode, working temperature, effect of environment, etc. Unfortunately, estimation of durability and strength of parts and structures is often carried out using oversimplified methods that account only for particular characteristics of static strength.

As seen from the equation of state written in the form of Equation 2.148.I, the Ω-potential, which determines strength and deformation properties of solids, depends on the kinetic part of internal energy, $U(T)$. This energy can change under the effect of various external fields: mechanical, thermal, chemical, radiation, electric, magnetic, gravitational, electromagnetic, and other. The thermal energy occupies a special place in this list because of its capability of concentrating in local volumes and transforming into other types, high rates of transfer from one object to the others, the absence of a preferable direction of propagation, etc.

All other kinds of energy can transform into thermal. Principles of such transformations are based on the laws of thermodynamics and statistical physics.[11,23,51] As follows from the equation of state (2.148.I), all physical–chemical processes associated with improvement or deterioration of service properties of materials, or with failures of parts and structures, are thermally activated; practice proves this.[2,3]

In modern technical physics, the processes of deformation and fracture and the associated structural changes are insufficiently studied[6,7,10,30,36,44–46] and opinions about their mechanisms differ. Understanding physical–chemical processes that can lead to failures creates prerequisites for scientifically grounded selection of the most efficient design-technological methods for improving reliability of parts and structures. Such understanding also requires development of new procedures for design, calculation, and estimation of their durability, substantiation of up-to-date methods for control of strength and deformability of materials, and elimination of failures under emergency situations.[12,18]

During solidification, the material forms a specific order of the AM system (determined by entropy s, Equation 2.112) and thus stores certain internal energy U (whose value is found from Equations 2.142 and 2.145), which it spends during the following period for resistance to external effects. Revealing the nature of physical processes occurring in the material from the moment of manufacture of parts and structures up to the time they fully perform their service purpose is the principal part of a general technical problem of decreasing material consumption and increasing reliability and durability.

The generalized functional diagram shown in Figure 3.1 describes the life of a material under varying external conditions. As seen, its structure is formed in two stages. The design stage (part 1 of the diagram) includes selection of quantitative and qualitative characteristics of initial elements (x_1, x_2, ... x_n) and prediction of properties of the finished product. The latter determine the design value of the Ω_1-potential.

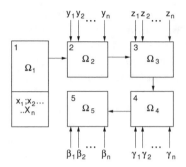

FIGURE 3.1
Diagram of development and functioning of materials under variable external conditions.

The actual AM structure is formed during the manufacturing process (part 2) under the effect factors (y_1, y_2, ... y_n) that characterize conditions of solidification. For example, for metals and their alloys, these factors are material of a mold or ingot, rate of casting, conditions of cooling of ingots, etc.[41] After solidification, the state of the material is determined by the initial Ω_2-potential. The structure is formed by atom–molecule and supermolecular complexes randomly oriented in space and interacting with each other—rotoses (Figure 1.7b) and grains (Figure 2.58).

During the period that follows (parts 3, 4, and 5 in Figure 3.1), structure of the material experiences continuous (controllable or noncontrollable) quantitative and qualitative changes caused by a variety of factors. Processing (part 3) is accomplished by mechanical, thermal, thermomechanical, or other effects (z_1, z_2, ... z_n) in order to improve structure and provide desirable shapes and properties. As a result, the thermodynamic potential changes from Ω_2 to Ω_3. At the stage of storage, transportation, and assembly (part 4), parts and structures are subjected to random mechanical, climatic, and other effects (γ_1, γ_2, ... γ_n). They contribute to a change in the state of the material by transforming the Ω-potential from Ω_3 to Ω_4. During service, the material is affected by a variety of external fields that are constant and alternative in nature: electric, thermal, magnetic, radiation, electromagnetic, chemical, climatic, etc. (β_1, β_2, ... β_n). The service level of the thermodynamic potential Ω_5 describes its state.

Quantitative indices of the process of variations of the state can be obtained by solving either the complete differential equation (2.109) or the first law of thermodynamics (2.1), which is part of it and valid at T = const and **P** = const. The direction of the discussed variations is predetermined by the second law of thermodynamics, according to which ordered structural systems tend to disintegrate, i.e., lose their order embedded during manufacturing under the effect of external factors. This trend is aggravated under the simultaneous effect of several external factors. Until now it has been impossible to take into account their combined impact because of insufficient knowledge of destruction processes; therefore, they seem irregular and random and their consequences appear unexpected and unpredictable.

During service time, the technical state of a material undergoes continuous changes and thus the Ω-potential assumes a series of values Ω_i, where $i = 1$, 2, 3, ... n. This series forms a four-dimensional-phase space of the states, Ω, determined (in accordance with the MB distribution, Equation 2.2) in three Cartesian one-energy coordinates. This space is limited on one side by initial state Ω_2 (the time of manufacturing) and on the other by failure state Ω_f (the moment of exhaustion of functional fitness).

Let us set a certain region Ω in the space of the energy states. Each point in this region, Ω_i, corresponds to a certain definite structure of a material, and any real structure corresponds to a certain definite point in this region. By applying the known rule of analytical geometry[91] to this space, we find distance L between two neighboring Ω_i and Ω_j:

$$L(\Omega_i, \Omega_j) = \left[\sum_{i=1}^{n} (\Omega_i - \Omega_j)^2 \right]^{1/2}. \tag{3.1}$$

Proximity of states Ω_i and Ω_j can be estimated from this formula. Let us distinguish some critical subset Φ in region Ω. This subset is defined as follows: if a material enters into any state belonging to Φ, it cannot perform its functions. Introduction of the notion of proximity (Equation 3.1) for a material allows one to predict the probability of its entering into critical region Φ knowing the current state of this material at a given moment of time, or, possibly, even by a trend of development of $\Omega(\tau)$. For practical applications, behavior of function $\Omega(\tau)$ is of key interest.

The state of a material varies under the effect of external factors. The space of states can be considered to be their reflection, and each of its points is an image of a point of the space of external effects. Let $\Omega(x_i; y_i; z_i; \gamma_i; \beta_i)$ be the state of a material at a certain moment of time τ. At a moment of $\tau + d\tau$, it will be determined by vector $\Omega + d\Omega = (x_i + d_{xi} ... \beta_i + d\beta_i)$. Then

$$\frac{d\Omega}{d\tau} = f\left(\frac{dx_i}{d\tau}; ... \frac{d\beta_i}{d\tau} \right)$$

characterizes the rate of variation of the state, and

$$\Omega(\tau) - \Omega_2 = \int_0^\tau \Omega d\tau$$

indicates its state by time τ if, at the initial moment of $\tau = 0$, its state is Ω_2.

Functioning of a part can be described as a process of conversion of the information on the external effects into the state parameters of the material of which this part is made. Structure of this material serves as the materials matrix of the space of states and can be regarded as a converter of the vector of the external effects, $A = f(y_i, z_i...\beta_i)$ into the state parameters

$$\Omega = f(A; A_x), \tag{3.2}$$

where A_x characterizes structural changes under the effect of configuration factors $(x_1, x_2,... x_n)$.

Expression 3.2 is the generalized equation of state. For materials working in different external fields, Equation 2.164 can be written as

$$\Omega = \Omega(T) \pm \sum_{i=1}^{n} A_i, \tag{3.3}$$

where A_i are the parameters of the force σ, thermal T, magnetic H, electric E, or any other field. Hereafter Equation 3.3 is referred to as the equation of interaction.

For an isolated body affected by none of the preceding external fields, U = const, in accordance with the law of conservation of energy and, therefore, $dU = 0$. As $U = U(r) + U(T)$, then

$$dU(r) = -dU(T). \tag{3.4}$$

Equation 3.4 can be written as

$$\frac{\partial^2 U(T)}{\partial T \partial V} dV\,dT = -\frac{\partial^2 U(r)}{\partial T \partial V} dV\,dT.$$

Because

$$\frac{dU(r)}{\partial T} = C_V \quad \text{and} \quad \frac{\partial U(r)}{\partial V} = P_V,$$

then $C_V dT = -P_T dV$ by multiplying both parts of the last equality by V and by noting that

$$\alpha = \frac{1}{V}\left(\frac{\partial V}{\partial T}\right)_P$$

is the thermal expansion coefficient (Equation 2.152) and $\Omega_T = P_T V$ is the thermodynamic potential, we obtain

$$C_V = -\Omega_T \alpha. \tag{3.5}$$

This equation describes the mechanism of mutual transformation of the kinetic and potential parts of the internal energy at any kind of interaction. As seen, these transformations take place in opposite directions. Heating leads to deformation α and deformation causes a change in heat capacity C_V and, hence, generation or absorption of heat. The coefficient of proportionality between the direct (heating) and reverse (deformation) processes is the Ω-potential. The correlation between the processes of thermal expansion and heat generation under mechanical loading was noted earlier.[35]

Relationships 3.1 to 3.5 are very general and macroscopic; they contain no reference to the physical nature of atoms and molecules and so are applicable to all solid bodies. Of course, the extent of understanding the problems of strength and durability of materials, as well as the possibility of controlling their properties, will dramatically grow if the macroscopic approach is combined with the knowledge of microscopic properties of interatomic bonds considered in Chapter 1. This allows promising new ways to be identified in solving the problem of deformation and fracture of solids.[24]

Let us transform the equation of interaction (3.3) into one suitable for practical purposes. All physical–mechanical properties of solids are initiated at the AM level and thus they are interrelated.[24] Mobile charges, joined in a rotos (Figure 1.7a), are surrounded by the EM field (Equation 1.16 and Figure 1.8) through which they can interact with their neighbors and also with external fields. Interaction always and everywhere shows up in the thermal range of the EM spectrum (right-hand side of the equation of state [2.65]). In the theory of dielectrics,[26] the static model of a rotos is called the dipole. An electric component of the EM field can perform the work of resistance (Equation 2.134); the latter forms the Ω-potential of an individual rotos (Equation 1.53) and a solid as a whole, Ω (Equation 2.109).

Orientation processes develop in dielectrics placed in electric fields following the diagram shown in Figure 2.9. From the mechanical standpoint, the impact of the field on a material is equivalent to elastic deformation in the force fields (Figure 2.8.).[24] Its quantitative measure is the polarization vector:[26]

$$p = KN\mathbf{E}, \tag{3.6}$$

where K is the polarization coefficient of a given material in the field having intensity E. On the other hand, p is determined[40] as a value proportional to dielectric permittivity ε_p

$$p = (\varepsilon_p - 1)\mathbf{E}. \tag{3.7}$$

Comparison of Equations 3.6 and 3.7 shows that the ability of dielectric materials to polarize in the electric fields is estimated by parameter

$$\varepsilon_p = KN + 1. \tag{3.8}$$

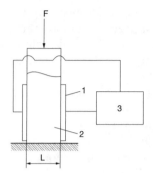

FIGURE 3.2
Diagram of polarization of dielectrics
in a force field.

Information on the value and variation of ε_p with frequency of the polarization current, temperature, and other factors for different classes of materials
can be found in References 6, 26, 40, and many others.

Let us calculate the work done by an electric field over dielectric 1 located
between plates of capacitor 2 charged using electric system 3 (Figure 3.2).
Assume that, at a certain moment of time, the difference of potentials at the
capacitor plates has become equal to φ. The work on imparting the electric
charge de to capacitor is expressed by the following formula known from
electrostatics:[31]

$$A = \varphi de. \tag{3.9}$$

Designating the area of the capacitor plates as S and the distance between
them as L, and expressing the charge in terms of density f: $de = Sdf$, and the
electric potential in terms of the field intensity, $\varphi = EL$, we can write

$$A = VEdf \tag{3.10}$$

Density of the charge f is expressed in terms of induction D, $f = D/4\pi$,[25] so
it holds that

$$D = E + 4\pi p. \tag{3.11}$$

By substituting Equation 3.11 into 3.10, we obtain

$$A = d(E^2/8\pi)V + dpEV. \tag{3.12}$$

The first term of Equation 3.12 does not depend on the presence of a
dielectric and thus it represents the work performed in electric system 3 and
consumed for excitation of polarization. The second term is the work spent
on anisotropy of a material, following the diagram shown in Figure 2.8. This
term is of particular interest for our further consideration. The work done

by the electric field over a dielectric is considered as having the "plus" sign, and that over external bodies has the "minus" sign. If a process involving a simultaneous change in parameters V and P occurs in a structure (e.g., in deformation of a polarized dielectric by compressive force F, following the diagram shown in Figure 3.2), then, according to Equation 2.134, the work done is equal to

$$A_1 = \mathbf{P}dV - V\mathbf{E}dp \qquad (3.13)$$

The first law of thermodynamics (Equation 2.1) and the basic thermodynamic equilibrium (Equation 2.13) allow us to reexpress Equation 3.12 as

$$V\mathbf{E}dp = dU + \mathbf{P}dV - T d\mathbf{s} \qquad (3.14)$$

Accounting for the equation of state of a solid body expressed in terms of thermodynamic potentials (Equation 1.111) and their definition,[11] we can note that the electric field changes all the parameters that characterize the state of a material and thus affects its resistance to deformation and fracture.

Let us assume that some elementary volume of dielectric 1, located in the force and electric fields (Figure 3.2), has received polarization p and, deformed in the quasi-static manner by compressive load F, moves in the transverse direction to distance dL. Movement of polarized microvolumes along the lines of force of the electric field (Figure 1.8) results in performance of the following work:

$$A_2 = -p\frac{\partial \mathbf{E}}{\partial L} dL = -pd\mathbf{E} \qquad (3.15)$$

It follows from Equations 3.13 and 3.15 that, in addition to polarization work, the electric field performs work against the external force fields; the total amount of work done is equal to their sum:

$$A = \mathbf{P}dV - Vd(\mathbf{E}p) \qquad (3.16)$$

Using the same approach as in Equation 3.14, we have that $Vd(\mathbf{E}p) = dU + \mathbf{P}dV - Td\mathbf{s}$. Substitution of the differential of free energy $\mathbf{F}$ (Equation 2.38) yields

$$d\mathbf{F} = dU - d(T\mathbf{s}) - Vd(\mathbf{E}p).$$

Hence, after integration at $V =$ const, we find that $U - T\mathbf{s} = \mathbf{F} - V\mathbf{E}p$. Determination of the Gibbs potential G (Equation 2.35) makes it possible to obtain an analogue of the formula of interaction (Equation 3.3) expressed in terms of thermodynamic potentials[24] and describing behavior of dielectric materials in force and electric fields

$$(\mathbf{P} \pm p\mathbf{E})V = \mathbf{F} - G. \tag{3.17}$$

Tensor parameter $\mathbf{P}$ has become independent of the intensity of the external electric field. We can control the Ω-potential and, therefore, resistance to deformation and fracture by varying its value and direction.

Similar to the force effects (Equation 2.148), polarization also changes thermal conditions of a material. Let us now calculate the thermal effect caused by isothermal polarization, ignoring a change in the specific volume and assuming validity of Equation 3.8.[52] Based on these assumptions, Equation 3.14 assumes the form of $T ds = dU - V\mathbf{E}dp$. Solving it with respect to s and regarding $\mathbf{E}$ and T as independent variables under conditions of existence of differential ds yield

$$\left(\frac{\partial U}{\partial \mathbf{E}}\right)_E = V\mathbf{E}\left(\frac{\partial \mathbf{P}}{\partial \mathbf{E}}\right)_T + T\left(\frac{\partial \mathbf{P}}{\partial T}\right)_E.$$

Hence, under these assumptions,

$$dQ = VT\left(\frac{\partial \mathbf{P}}{\partial \mathbf{E}}\right)_E d\mathbf{E}.$$

After substituting Equation 3.7 and integrating with respect to $\mathbf{E}$, we find that

$$Q = \frac{\mathbf{E}^2}{8\pi} VT \frac{\partial \varepsilon_p}{\partial T}.$$

For orientation polarization (Figure 2.8),[26,40]

$$\varepsilon_p - 1 = \frac{\text{const}}{T}.$$

In this case,

$$Q = -\frac{1}{2} p\mathbf{E}V \tag{3.18}$$

As seen, at

$$\frac{d\varepsilon_p}{dT} < 0,$$

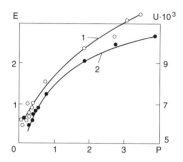

FIGURE 3.3
Effect of the electric field on a force component of the Ω-potential.

the dielectric generates heat. From comparison of Equations 3.17 and 3.18, up to half of the energy of the electric field converts into heat, whereas its second half is spent changing the Ω-potential.

To understand the influence of the electric field on the stressed–deformed state of a material, consider the following experimental facts. Internal pressures, which might exceed mechanical strength, are formed in dielectrics at a high voltage of the electric fields.[92] Thus, mechanical stresses of about 5 MPa were formed in polyethylene samples at E = 7 MV/cm, whereas their compression strength was 13 MPa. In lithium fluoride crystals, stresses were equal to 3.2 MPa at E = 1 MV/cm, whereas their yield strength was 5 MPa. Figure 3.3 shows the relationship between pressure P (MPa) formed in crystals of alkaline–halide compounds (curve 1), electric field intensity E (MV/cm), and crystalline lattice energy U (J/mol) (curve 2). As seen, mechanical pressure P increases, with an increase in E changing proportionally the energy of the lattice, U; this fully agrees with the theory (Equation 3.17).

A material can be affected simultaneously not only by one but also by i external fields. By defining their nature through generalized parameters a_i and A_i and the direction of their action by the plus or minus sign, we can rewrite Equation 3.17 as

$$\left(P \pm \sum_{i=1}^{n} a_i A_i\right) V = sT. \tag{3.19}$$

If parameter A_i designates the intensity of electric **E**, magnetic H, force σ, and thermal T (not to be confused with the mean value of internal phonon radiation T), fields, then a_i acquires the meaning of polarizability p, magnetic susceptibility μ, and rigidity Δ. The latter is expressed in terms of ratio of volume V, deformed in the presence of a field to initial volume V_0, i.e.,

$$\Delta = V / V_0. \tag{3.20}$$

All these serve as quantitative measures of resistance of a structure to a change in the shape of the MB factor and its location in Cartesian and energy spaces under the effect of the preceding fields (positions 8 to 15 in Figure 1.14).

3.3 Analogy between Polarization, Magnetization, Force, and Thermal Deformation

A distinctive feature of any bound AM system is the presence of a large number of charged particles (Figure 1.15), such as nuclei, electrons, polarized ions, and asymmetric molecules, their movement along the closed paths (Figure 1.7b), and the associated EM fields (Equation 1.32). Different physical nature, value and degree of mobility, character of movement, and mass of charge carriers induce internal electromagnetic, thermal, mechanical, electric, and magnetic fields. These particles can interact with each other through these fields and, together, all can interact with external fields of similar natures.

Thermal and electromagnetic fields have the same physical nature: thermal radiation occupies the long-wave range of the electromagnetic spectrum. Electric and magnetic fields are particular cases of the electromagnetic field. A mechanical field can exist between any (not only charged) particles or their clusters. For this reason, it is the most common type of the field, as it can show up in mechanical (Equation 2.153) and also electric (Equation 3.18), magnetic, or any other interaction (Equation 2.65).

Joined by binding forces into a crystalline structure, atoms may have symmetric or asymmetric electron shells. In the first case, the centers of gravity of the electric and magnetic charges of opposite signs coincide. Such microvolumes are neutral with respect to electric and magnetic fields. Separated in space, they initiate constant electric and magnetic moments. The former are formed where there are no free charge carriers, i.e., in dielectrics, and the latter are formed in metals.

For example, ion compounds have a constant dipole moment. Schematically, an ion molecule can be represented as consisting of positive and negative ions apart from each other at distance r and bound with each other by the electric attractive forces (Figure 3.4). Such a moment, similar to the mechanical one (Equation 1.27), is a vector value. It is directed from the negative charge to the positive and its magnitude equals

$$M(E) = zer, \tag{3.21}$$

where z is the valence and e is the ion charge. Moments of the two-atomic bonds are approximately proportional to the difference in electronegativity of their atom components. Consisting of several atoms, the dipole moment of complex compounds depends on their structure; symmetric molecules do not have a constant moment.[26]

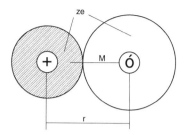

FIGURE 3.4
Diagram of formation of a constant dipole moment.

Moments can be formed also in neutral microvolumes under the effect of external fields: electric, magnetic, or mechanical. Opposite electric or magnetic charges (Figure 1.6), while shifting in such fields, violate the initial symmetry of the interatomic bond, and it acquires an "induced" moment For example, induced dipole moment M_i is proportional to the intensity of the external electric field $\mathbf{E}$

$$M_i = p\mathbf{E}, \tag{3.22}$$

where p is the polarization vector of a given material (Equation 3.6). Dielectrics (like all other materials) differ in symmetry of structure of the interatomic bonds and degree of order of their arrangement in a volume, s. Therefore, their polarizability depends on the direction of the external field.

Substituting Equation 3.21 into 3.22 and accounting for the intensity of the electric field of a point charge defined by Equation 1.17, we have that

$$p = 4\pi\varepsilon_0 zr^3. \tag{3.23}$$

This shows that polarizability has the dimension of volume and is proportional to the valence, i.e., it is determined by the location of a chemical element in the periodic system. It is also related to movement of all electrons in atoms. However, valent electrons make the largest contribution to polarizability. For this reason, Equation 3.23 is valid for elements in the same group of Mendeleev's table. Within one period, polarizability repeats the character of variation of the valence. To compare, consider its values ($p \cdot 10^{-24}$ cm³) for ions of the following elements: carbon, 0.012; aluminum, 0.067; titanium, 0.272; copper, 1.81; and lead, 4.32.

Polarizability of the dipole bond can be represented as a sum of three terms,

$$p = \mathbf{p}_e + \mathbf{p}_i + \mathbf{p}_0, \tag{3.24}$$

where $\mathbf{p}_e$ and $\mathbf{p}_i$ are the polarizabilities associated with displacement of the electron and ion charges relative to the equilibrium position, and $\mathbf{p}_0$ is the

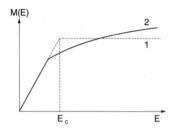

FIGURE 3.5
Dependence of the mean value of the dipole moment on the intensity of the external electric field.

orientation polarizability of constant dipoles. In accordance with the Langevin theory,[26] the last component is

$$\mathbf{P}_0 = \frac{M_0^2}{3kT}. \tag{3.25}$$

The mean value of projection of the dipole moment $M(E)$ into the direction of the electric field having intensity E is found from Equation 3.22. Figure 3.5 shows this dependence.[26]

Except for very strong fields ($E > E_c$), the mean moment grows rapidly and almost linearly with an increase in the field intensity E. At $E < E_c$, where constant dipole moments are oriented along the direction of the field (Figure 2.8a), contribution of the first two terms to polarization (Equation 3.22) is not great. At $E > E_c$, Function 3.29 tends to saturation (horizontal dotted line 1 in Figure 3.5). A change in the inclination angle of line 2 near the E_c value indicates a change in the polarization mechanism: instead of orientation (the first term in Equation 2.136), it becomes electron–ion, and Equation 3.23 takes the following form: $p = \mathbf{p}_e + \mathbf{p}_i + \text{const}$. In this range of the intensities, polarization processes are accompanied by CD phase transitions, disintegration of interatomic bonds, and destruction of dielectrics.[92] The second term in Equation 2.136 is responsible for these processes.

All solids possess certain magnetic properties that are more pronounced in metals because they contain collectivized electrons. Elementary carriers of magnetic properties are electrons or, to be more exact, their spin and orbital magnetic moments. A consistent analogy exists between electric and magnetic properties. For example, in some metals the internal forces of interaction between electrons may lead to parallel spontaneous (at the absence of the external magnetic field) orientation of the spin magnetic moments or to their antiparallel, but ordered, arrangement. The first are known as ferromagnetics and the second are called antiferromagnetics; both are formal analogues of dielectrics with constant dipole moments.

With respect to the external magnetic field, the rest of the materials can be subdivided into diamagnetics and paramagnetics. In fact, the latter are analogues of dielectric with the induced dipole moment. Extending the analogy to Equation 3.22 allows a relationship between magnetic moment $M(H)$ induced by the external magnetic field and intensity of this field H to be written as

$$M(H) = \mu H, \tag{3.26}$$

where μ is the magnetic susceptibility. Diamagnetics have a negative value of μ while μ is positive for paramagnetics. According to the Langevin theory,[26]

$$\mu = \frac{NM^2(H)}{3kT}, \tag{3.27}$$

where N is the number of interatomic bonds and $M(H)$ is the magnetic moment of one of them.

Paired distribution of electrons at energy levels is a rule (see Figure 1.2); however, it has some exceptions that are observed in ferromagnetics and can be explained from the energy standpoint.[30] The potential component of the internal energy depends on the mean distance between charges (Equation 1.26) and, according to wave mechanics, the orientation of projections of the spin magnetic moments of electrons exerts a strong impact on it. It describes behavior of electrons in a statistical manner. Advantage or disadvantage of their certain state is determined by the minimum of Function 1.53. Blokhintsev[27] has shown that this condition is met at a similar direction of projections of spins of all z electrons. In this case, the overall energy decreases due to formation of repulsive electric forces between electrons.

As in dielectrics, where the presence of constant dipole moments does not show up under normal conditions, ferromagnetic bodies also seem nonmagnetized in the absence of the external magnetic field. In reality, they are made up of domains, i.e., local magnetized regions, that have initial chaotic orientation. The domain structure of ferromagnetic bodies conforms to the existence of magnetic forces, which hamper one-sided orientation of the spin moments of electrons.

The external magnetic field changes the state of a ferromagnetic body, i.e., it becomes magnetized. Transformation of chaotically oriented domains into the ordered state with respect to the direction of the external field takes place. Figure 3.6 shows a typical form of a magnetization plot (Equation 3.26).[30] Region OA corresponds to the processes of displacement of domains, and AB corresponds to their rotation. The processes of displacement of the domains closer to the direction of the external field continues to the right from point B. However, their mechanism changes because very strong fields lead to turning of magnetic axes of electrons.

The mechanical moment of momentum (Equation 1.27) is induced by charges, which have mass m and move along closed paths (Figure 1.12). Polarized

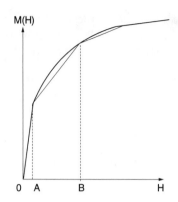

FIGURE 3.6
Typical plot of magnetization of ferromagnetic materials.

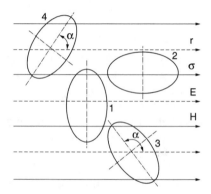

FIGURE 3.7
Possible orientations of rotoses of polyatomic molecules in different external fields.

ions moving along such paths (Figure 1.6) are the carriers of the mechanical moment (1.31).

Finally, consider the physical effect of a thermal field on a body. The intensity of this field is estimated by temperature T and the result of the effect on the body is estimated by a change in the kinetic energy $U(T)$. We can rewrite Equation 1.24.2 in the following form: $dt = Fdr$. If radius vector r coincides with the direction of action of force F (shown in Figure 3.7), then increment dt expresses the work of resistance in terms of a change in the shape or size of the elliptical path of a rotos, i.e., $d\omega = Fdr$. If their directions do not coincide (Figure 2.9), increment dt generates moment $M(t)$, which tends to turn the rotos along the direction of the thermal field, i.e., $dM(t) = Fdr = dt$. Using the definition of heat capacity (Equation 2.116), we can write that $dU(t) = C_v dM(t)$ or, after differentiation with respect to the entire set N,

$$U(r) = C_v M(T) = C_v T.$$

As seen, the variation of the mechanical moment excited by a thermal field is determined by the temperature dependence of function $U(r)$. Figure 1.16a shows the typical shape of such dependence. The micromechanism of orientation (region *oab* of plot 1) changes into that of deformation (region *bc*) at the Debye temperature *v*.

Equation 2.65 can be expressed as

$$PV = U(T). \tag{3.28}$$

Differentiation and accounting for obvious equality $dV \approx dT$ (which follows from the relationship of equivalence, Equation 2.105), as well as for Transformation 2.137, yield

$$\frac{d\mathbf{P}}{\eta + \varepsilon} + \mathbf{P} = C_v. \tag{3.29}$$

The typical temperature dependence of heat capacity, confirmed experimentally,[15] can be found in any course of thermodynamics[51] or statistical physics[31,49,50] (Figure 2.15). It consists of the ascending Debye region (determined by the first term of Equation 3.29) and the horizontal region (where $\mathbf{P}$ = const), obeying the Dulong-Petit law.[32] The plot of Function 3.28 has an absolute analogy with Figure 2.15a.

Electric, magnetic, and mechanical moments can be mutually compensated or noncompensated in formation of a solid. For example, segnetoelectrics[26] are characterized by noncompensated electric moments and ferromagnetics[29] by noncompensated magnetic moments. Noncompensated mechanical moments show up in anisotropy of mechanical properties.[3] Ordering of chaotically oriented electric and magnetic moments occurs in the external electric and magnetic fields. Force fields lead to deformation of solids. Deformation, regardless of its origin (force or thermal), is the process of ordering of mechanical moments.

We can conclude, therefore, that polarization, magnetization, and deformation at the AM level are phenomena of the same order, and thus must occur following the same laws—Formula 2.134 and comparison of Figures 2.10, 2.15, 3.5, and 3.6 leave no doubt about it. Voluminous literature is dedicated to substantiation of the AM nature of the first and second phenomena.[6,7,26] As far as investigation of deformation and fracture of solids is concerned, these correlations have never been considered.[2,3,19,26]

The proposed analogy allows us to suggest technologies that allow activation[16] or elimination[18,24] of fracture of solids. This creates required conditions for solving typical practical problems using nontraditional methods that prevent accidents and catastrophes.[14,17]

3.4   Orientation Nature of Elastic Stage of Deformation

It is apparent that solids, interacting with external fields and environments following the law given by Equation 3.19, should retain their integrity. As a result of a rather simple chain of reasoning, we can very quickly conclude that the smallest element capable of doing this is the EM dipole (Figure 1.7b). Impact on a solid by any field is directional. If it were not for moment M (Equation 1.27) distributing dipoles in the bulk of a body during solidification in a chaotic manner (Figure 3.7), the external field (changing geometrical sizes of the dipoles in a sequence shown in Figure 1.15) would immediately lead to macroscopic deformation, disintegration of bonds, saturation of the volume with defects, and increase in the s_d component of entropy (2.113). Because the dipoles are initially arranged in the volume in an extremely chaotic manner (2.3), the applied external field tends to orient them to its direction.

During the solidification process, rotoses (Figure 1.7a), like molecules (Figure 3.7), have a random arrangement in the volume. As a result, their mechanical moments acquire a chaotic orientation. Therefore, at the moment of application of an external force field with intensity σ on such a system, rotoses have different orientations. Part is oriented at angle α to the direction of action of the external field (Figure 2.9), whereas the other part is located in planes coinciding with vector σ (Figure 3.7). As such, the type of macro-stressed state (tension, compression, bending, torsion, or their combination) does not have a substantial effect on behavior of the rotoses; at the microlevel, any of them is transformed into an increase in the intensity (compression) of a local potential field (Figure 1.8) or its decrease (tension). In the second case, the large axis of the ellipse may be located normally to the lines of force of the field (position 1), coincide with their direction 2, or be at angle α to this direction (positions 3 and 4, Figure 3.7).

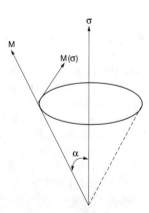

FIGURE 3.8
Precession of the mechanic moment of a rotos
in the external force field.

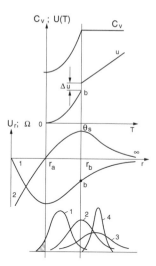

FIGURE 3.9
Graphical representation of parameters of the equation of state: kinetic C_v and $U(T)$ and potential U_r and Ω, and variants of MB distribution δ inside the AM set N (positions 1 to 4).

Some time after termination of loading, the exchange processes in the interatomic bonds form a new MB distribution, differing from the initial one in energy; a change will occur in positions inside sets 8 to 10 or 11 to 15 in Figure 1.14. For a given body, the physical nature of M and the manner of action of σ are the same for all the rotoses, so Distribution 2.2 will be formed such that the maximum possible number of the moments are oriented near $\alpha = 0°$.

If the vector of the external force field intensity forms angle α with the moment (Figure 2.9), a pulse of the Coriolis force (Equation 1.104) $F_0 = ma_0$, where $a_0 = 2M\sigma \sin \alpha$, i.e., Coriolis acceleration (Figure 3.8), will form in the direction normal to the plane in which vectors σ and M lie. Direction of this pulse depends direction of rotation of atoms in a rotos formed in plane A, with respect to the vector of the field intensity.

The moment of the Coriolis force $M(\sigma)$, normal to the direction of field σ and moment M, starts affecting the natural moment of the rotos M in the force field; this leads to precession of the moment of the rotos. Because of this precession, the total moment of the rotos (designated as M_y in Figure 1.4) tends to take a position directed against the external force field. This action is similar to the diamagnetic effect, which takes place in any body located in a magnetic field.[26] It is counteracted by a direct reaction of the rotos to the external force field aimed at the formation of the preferred orientation of the moment in a direction of the field.

As a result of the existence of two opposite trends, an isolated rotos can orient neither along nor against the field. In the volume of a body, each member of set N has random orientation. In a free state, the energy of thermal motion is distributed between them in accordance with the MB factor (Equation 2.2). After application of the external field to this body, the local potential field (Figure 1.8) deforms, and rotoses, having different orientation of the moment with respect to its direction, according to Equation 2.17, will now have different energy. This is determined from the formula

$$U = -[M\sigma] = -M\sigma\cos\alpha$$

and characterizes a change in the initial MB distribution of the internal energy (a change in positions in a direction of 11 to 15 in Figure 1.14).

The Ω-potential of the rotos (Equation 1.53) becomes apparent in two ways. If force F lies in the orbital plane A (Figures 1.7a and 2.9), it has the meaning of work performed by the EM dipole $\mathbf{D}$ with variation in radius vector $\mathbf{r}$. In all other cases, force $\mathbf{F}$ generates orientation moment $M(\sigma)$ (Figure 2.9), which turns the orbital plane along the direction of field σ. In accordance with the universal Le Chatelier principle, [31] moment $M(\sigma)$ will enhance the orientation processes occurring in the structure, which tend to counteract the external field. The result of the external effect can be written as

$$U = -[M \times M(\sigma)] = -M \cdot M(\sigma)\cos\alpha. \tag{3.30}$$

A body reacts to the external force field so that the change in its internal energy is minimal.

The orientation theory of paramagnetism of metals and polarization of dielectrics was developed by Langevin and its description can be found in any course of statistical physics,[49,50] theory of dielectrics,[26] or metals.[30] So, we must apply it to the deformation processes occurring in force fields. According to quantum mechanics,[27] the mechanical moment of structurization particles, M, may have a definite orientation in the external force field. Its projection into the direction of the field, $M(\sigma)$, can have only the following values:

$$M(\sigma) = M\frac{j}{l},$$

where l and j are the orbital and internal quantum numbers, respectively, and M is determined from Equation 1.31. The energy of resistance of a rotos to the external field is equal to

$$u = -\left(M\sigma\frac{j}{l}\right).$$

In accordance with Table 1.1, number l may have one of four values: 0, 1, 2, or 3, and number j may have one of eight values:

$$\pm\frac{1}{2}; \quad 1\pm\frac{1}{2}; \quad 2\pm\frac{1}{2}; \quad 3\pm\frac{1}{2}.$$

Therefore, the number of combinations of 2 of 12 determines the number of the energy states, in which a rotos may be, i.e.,

$$C_{12}^2.$$

The probability of finding it in a state with mechanical moment M, in the case of thermal equilibrium, is set by

$$\omega(M) = \text{const} \cdot \exp\left(\frac{M\sigma}{kT}\frac{j}{l}\right).$$

The mean projection of the moment to the direction of the field is expressed in terms of sum of the states as[26]

$$M(\sigma) = \frac{\sum\limits_{j=-l}^{j=+l} M\frac{j}{l}\exp\left(\frac{M\sigma j}{kTl}\right)}{\sum\limits_{j=-l}^{j=+l}\exp\left(\frac{M\sigma j}{kTl}\right)} = kT\frac{\partial}{\partial\sigma}\ln\left[\sum\limits_{j=-l}^{j=+l}\exp\left(\frac{M\sigma j}{kTl}\right)\right].$$

Calculation of the sum of terms of a finite geometrical progression (in square brackets) yields

$$M(\sigma) = kT\frac{\partial}{\partial\sigma}\ln\left[\frac{\exp(j+1)a - \exp(-ja)}{\exp a - 1}\right], \qquad (3.31)$$

where

$$a = \frac{M\sigma}{kTj}.$$

For nonintense force fields σ, i.e., when

$$a = \frac{M\sigma}{kTj} \ll 1,$$

we can differentiate Equation 3.31, then expand the result into a series. Limiting our consideration to the first term of the series, we write

$$M(\sigma) = \frac{M^2\sigma}{kT}\frac{(j+1)}{j}.$$

Because

$$j = C_{12}^2 \gg 1,$$

the latter expression is transformed into an analogue of Langevin's formula, which is known from the polarization theory:[26]

$$M(\sigma) = \frac{M^2 \sigma}{3kT}. \tag{3.32}$$

This dependence is shown in Figure 2.10, which provides the correct description of elastic deformation at $T = $ const on an assumption that $E = M(\sigma)$ and $\varepsilon = 3kT/M^2$.

Figure 2.10 shows that the projection of the mean moment, which determines resistance of a body according to Langevin, increases rapidly and almost linearly with the intensity of the external field. When the external force field is strong, i.e., when $\sigma > \sigma_f$, it tends to saturation (line *ac*), i.e., to such a position where all paired interatomic bonds are oriented along the field direction (Figure 2.8a). However, experience shows[2,3] that the true diagrams of deformation in tension, compression, torsion, and other types of the stressed state have the form of broken line *oab*, consisting of two almost linear segments *oa* and *ab* and a small transition zone near point *a*.

Langevin's theory is approximate. It considers solids as AM systems consisting of a set of independent rigid elements called dipoles or domains in electric and magnetic fields. With respect to force fields, they serve as rotoses (definition of the latter is given in Chapter 1). Absolute coincidence of the theory with an experiment for region *oa* (described by Hooke's law) indicates that deformation processes in this case are generated by turning of the rotoses in the direction on the external force field and transformation of polyatomic molecules from the spheroidal (Figure 2.7) to ellipsoidal shape (Figure 2.8).

An experimental confirmation to the orientation mechanism of elastic deformation can be found in the study by Berry,[93] who reports that deformation of a rubber-like polymer occurs due to stretching and orientation of chain molecules in the direction of the field. If the AM system is fixed for some time in such a state (e.g., by rapid cooling), the material becomes anisotropic. This orientation mechanism is additionally confirmed by the fact, widely known to every experimenter, that, at any kind of stressed state, the fracture surface always has a peculiar, definite orientation with respect to the direction of action of the external force that caused this fracture. In tension, for example, it is normal to the effective force and, in compression, it is parallel to this force.

At considerable levels of stresses σ, where Condition 3.32 is no longer met and Langevin's formula is not capable of explaining a change in the deformation mechanism (at point *a* in Figure 2.10), processes occur in region *ab* of the diagram and cause fracture of materials at point *b*. The force orientation is a reversible phenomenon. After relief of loading, the initial state of a submicrostructure is restored. After orientation saturation, however, the irreversible degradation processes start playing decisive roles. At point a, the total moments M_y (see Figure 1.4) of the overwhelming majority of rotoses turn out to be arranged in the direction of the field (coinciding with the y-axis) or, to be more exact, against the field.

It follows from the quantum–mechanical rule of formation of azimuthal planes s, p, and d that its components M_s, M_p, and M_a will not coincide with this direction. Further increment in the total projection of moment M_y may take place only due to rotation of its azimuthal components in the direction normal to the field, i.e., because of restructuring of the rotoses. These processes occur in region ab of the deformation diagram (Figure 2.10) and are accompanied by accumulation of damaged structure, which eventually ends with fracture.

The structure of many bodies always consists of chaotically oriented compresson and dilaton rotoses. In region oa their moments participate equally in the orientation process. Starting from point a, the behavior of both substantially differs. Whereas compressons have the possibility to become dilatons in region ab of the diagram due to orientation along the field of the azimuthal planes, in principle, dilatons do not have such a possibility. Consider now the physical nature of the processes occurring in region ab of the deformation diagram (Figure 2.10).

The most sensitive to the external field are the rotoses whose orbits are oriented normally to its direction (position 1 in Figure 3.7). Because of the existence of the force field gradient, a force acting on the rotating particles in a direction of a small axis of the ellipse tries to change the shape of the orbits from elliptical into round. In accordance with Equality 1.38, the large axis of the ellipse is not indifferent to such a transformation of the path, as its value is determined by the overall energy of a two-atomic system at a given time moment.

It follows from the previously mentioned equalities that the intensity of variation of the small axis of the ellipse b is higher than that of the large axis a (value $|u|$ in the formula for b has a power of $\frac{1}{2}$). Therefore, when both axes are forced to change, there is a moment when $a = b$ and the ellipse is transformed into a circle. At this moment when

$$u = \left| \frac{mb^2}{2M^2} \right| = u(r_a),$$

the process of deformation of the orbit is accomplished with decomposition of the rotos into constitutive atoms. That is, rotoses whose orbits are oriented normally to the tensile force field are the first to fail. At the moment when the elliptical orbit is degenerated into a circle, an atom loses the bond to its closest neighbor in the rotos. This results in the formation of vacancy or a point microdefect.

This is the behavior of "cold" compressons. In tension, the rotoses with a normal orientation to the direction of the external field (type 1 in Figure 3.7)

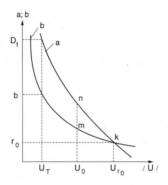

FIGURE 3.10
Dependence of sizes of large, *a*, and small, *b*, axes of elliptical paths of atoms on the value of the overall energy of a rotos.

become cold compressons. In the case of a conditional representation of the MB distribution in the energy domain (positions 1, 2, 3, and 4 in Figure 3.9), they will be situated at their left ends (dashed region in position 1). Descending to the bottom of the well, they continuously generate excessive heat consumed by the active part of compressons (those located near the distribution apex) and by dilatons. The first take the load at the initial stages of loading, and the second join them only at the final stages.

What happens to interatomic bonds with other orientation of orbital planes? To find an answer to this question, consider variations of the large, *a*, and small, *b*, axes of the ellipse with variation of the overall internal energy $|u|$. Graphical representation of Dependencies 1.38 is shown in Figure 3.10. The magnitude of the overall energy of particles decreases with an increase in the kinetic part; curves *a* and *b* approach each other at an infinitely close distance (but do not intersect). The path of movement of atoms becomes an ellipse deformed in one direction (at level u_T). In this case, the ends of vectors r_{min} and r_{max} "slide" up the branches of the curve of the potential energy (Figure 1.7a) and the energy level seems to emerge from the potential well. This is how compressons behave in tension — absorbing the thermal energy emitted by degrading dilatons (position 1 in Figure 3.7).

Let T = const and the internal energy be u_0 for a given rotos at a given time moment. With an increase in the intensity of a tensile force field, all the rotoses that have been oriented by this time will undergo configuration changes, although the scales of these changes depend on the degree of their proximity to the direction of the field. For example, a transversely oriented rotos (position 1 in Figure 3.7) requires for restructuring energy that is numerically equal to the surface area of a curvilinear rectangle (Figure 3.10)

$$u_0 m k u_{r_0}$$

because the small axis of the ellipse is the first to undergo changes.

A much higher energy is required for transformation of a rotos oriented as shown in position 2 (Figure 3.7). As such, deformation follows a path of $n - k$ (Figure 3.10). In the tensile force field, both axes of such a rotos increase simultaneously. However, these changes have different intensities: for the large axis they follow a path of k to n and for the small axis a path of k to m. For this reason its shape becomes elongated and the amplitude of oscillations of atoms that form the rotos increases in a field direction equal to an increase in the local temperature. This rotos becomes a local consumer of thermal radiation.

Rotos 1 has lower sensitivity to tensile stress; to induce the process of deformation of its neighbor 2, it can give a part of its kinetic energy. Thus, rotos 1 provides rotos 2 with the energy required for phase transformations to occur (following the diagram in Figure 1.15). This is the internal relationship between heat exchange and deformation processes. These cannot be separated in their nature and time — the only reason, at the macroscopic level, the deformation process is always accompanied by thermal effects.[35,36] At the AM level both processes occur simultaneously and have the same physical nature, i.e., mutual exchange of the kinetic energy between neighbors under conditions of its shortage. The opposite character of changes in configuration of the rotoses leads to elongation of a sample in the longitudinal direction and to narrowing in a transverse direction. In ductile materials these processes often lead to formation of the neck (Figure 2.51).

The crystalline structure's reaction to compressive loads is different. In this case, rotos 2 supplies the kinetic energy to rotos 1, and material, being compressed, increases its transverse sizes. In fact, steepness of the curves in the n to k direction is lower than to the left from point m (Figure 3.10). Therefore, movement of point n to k is preferable to movement of m to the left. A change in the shape occurs along the upper curve in a direction of n to k, up to disintegration of the bond; this process consumes energy whose value is equal to the area of curvilinear trapezoid

$$u_0 n k u_{r_0},$$

but larger than the area of figure

$$u_0 m k u_{r_0}$$

(which characterizes the energy of fracture of the rotos in tension) by the area of curvilinear triangle *mnk*.

Therefore, at the initial stages of loading, a longitudinal increase in density of a material sometimes leads to a local increase in temperature of the specimen. A short-time phonon saturation of the volume in compression takes place if the structure has a much greater number of rotoses with orientation of type 2 than of rotoses of type 1 (Figure 3.7). The latter

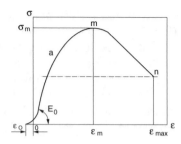

FIGURE 3.11
Diagram of compression of concrete.

have no time to absorb released kinetic energy, leading to an excess of heat, which sometimes causes a short-time elongation of the samples, rather than their shortening.

For example, Freisine and Berg[48] described the diagram of compression of concrete, paying attention to the initial region of the development of deformations (Figure 3.11). They stated that it has a negative effect on the behavior of a structure under actual service conditions. In laboratory tests, this region usually escapes researchers' attention because it is attributed to the alignment of specimens; thus these measurements, as a rule, are not taken into consideration.

The flow of phonons adopted by all the bonds causes a volumetric expansion of a body including that in the direction of action of the external force, leading to its short-time elongation at initial region ε_0. Using inserted sensors of stresses and temperature, Sammal[94] was able to provide experimental proof of this phenomenon. The effect of reverse deformation at the initial stages of loading is universal and observed not only in concretes during compression, but also in metals during tension.[95] Preliminary tentative loading of critical structures to a low level of stresses, ensuring free deformation, is a reliable method for eliminating undesirable initial effects.

Compresson–dilaton phase transitions in region ab of the true diagram (Figure 2.10) end with fracture of the body at point b. This is preceded by a change in the resistance mechanism. After point a, the main contribution to it is made by the second term of Equation 2.136. In this case, the linear character of Function 2.139 does not change. What has changed is the inclination angle of the second region of the diagram. Instead of α, it has become β. It is apparent that fracture starts when P in Equation 2.139 becomes equal to zero.

A graphical representation of this situation is shown in position 2 of Figure 3.9, where the apex of the MB distribution coincides with the θ_s-temperature. At this moment, by overcoming the compresson compression, the internal tensile pressure breaks the sample. The primary effective factors of fracture are not the external mechanical forces but the internal dilaton and compresson pressures, which always exist in the bulk of a body. In particular, violation of the dynamic equilibrium between them in a local microvolume generates the elementary fracture acts that, by coalescing, divide the body into parts. The external force provides conditions for progressive development of this process over the entire working volume of the body up to its fracture in one of the preceding sections.

3.5 Plastic Deformation and Destruction Processes

The first reaction of each member of the bound AM system to the external field is orientation: while adjusting to new conditions, all rotoses of poly-atomic molecules (Figure 2.7) start taking the most beneficial energy state to be oriented along the field (Figure 2.8a). This reaction is described by the first term of differential equalities (2.135). Contribution of the second term to resistance at the first stage is negligible. It characterizes variations of natural sizes of the rotoses (Figure 2.8b; movement to the right from axis r_j in Figure 1.14). This is accompanied by CD phase transitions and inevitable disintegration of the bonds under conditions of thermal insufficiency (particularly at T = const); therefore, it starts playing a decisive role only after completion of the orientation processes.

In practice, the orientation effect shows up in the formation of anisotropy.[3] In other words, the structure of a body is capable of changing in the direction of decreasing the s_c component of entropy (2.113). However, this does not contradict the second law of thermodynamics because, with a decrease in the configuration part of entropy of s_c, its phase component s_f grows. Moreover, any change in entropy of the initial system always involves a consumption of the energy from an external source. Generation of any type of energy is inevitably accompanied by an increase in entropy of a large macrosystem.[56] The ability of the entropy of solids to cause changes in their shapes puts them in a special position in comparison with other types of the aggregate state of a material.

The true diagrams of polarization, magnetization, force, and thermal deformation coincide in shape (Figures 2.10, 3.5, and 3.6). This means that the preceding processes obey the same law (Equation 3.19). As to their reaction to the electric fields, solids are subdivided into conductors, semiconductors, and dielectrics,[26] and in the magnetic fields they behave as dia-, para-, and ferromagnetics.[30] Plenty of evidence is available that their reaction to the force and thermal fields is other than that of homogeneous media.[24] That is why modern theories of strength fail to explain many effects that show up in the force and thermal fields.[2,3,44]

In the case of orientation saturation (point *a* in Figure 2.10), the brittle bodies fracture and the stressed–deformed state of the ductile bodies continues to develop following the law given in Equation 2.135.II. At the inelastic stage of deformation, entropy changes due to displacement of the MB factor in the direction of 11 to 15 (Figure 1.14; the first term in brackets) with a simultaneous increase in its span (the second term). This inevitably transforms part of the rotoses into critical states T_c and θ_s.

Equating both parts of Equation 1.52 to zero and accounting for the expression for the moment (Equation 1.27), we obtain that

$$F_c = 0; \quad T_c \geq 0; \quad \alpha_c = T(\dot{\varphi})r_c, \tag{3.33}$$

where

$$T(\dot{\varphi}) = mr^2\dot{\varphi}^2/2$$

is the rotational part of the kinetic energy at T_c. The θ_s-state is characterized by the following relationships:

$$\frac{dr}{r} = -\frac{dF}{F} \quad (I); \quad \alpha_{\theta_s} = 2T(\dot{\varphi})r_{\theta_s} \quad (II); \quad \frac{d\alpha}{\alpha} - \frac{d\dot{\varphi}}{\dot{\varphi}} = 2\frac{dr}{r} \quad (III). \tag{3.34}$$

At the T_c temperature (point c in Figure 1.14), the rotos, on transforming into a pure rotator, loses all its capability of generating resistance forces (Equation 3.33) and disintegrates. At the CD phase transition (Equation 3.34), a change occurs in the direction of variation of resistance force F: from an increase in the compresson phase to a decrease in the dilaton state (3.34.I).

To change configuration of the local electric field $d\alpha/\alpha$ and rotation acceleration $(-d\dot{\varphi}/\dot{\varphi})$ it is necessary to have a pulse of thermal energy, which should be two orders of magnitude higher than that consumed for a regular change of the state inside the compresson phase (Equation 3.34.III). The transformation of a rotator into an oscillator really requires such consumption of energy.[20] At T = const, not all the rotoses can receive the energy required for the CD transition. Therefore, some are replaced by vacancies,[36] accompanied by release of a kinetic energy impulse:

$$q_T = k\theta_s. \tag{3.35}$$

The thermodynamic nature of fracture finds experimental proof in numerous studies dedicated to the kinetic concept of strength.[10] To ensure dynamic equilibrium (Equation 2.65), the structure of any material should be made up of compressons (internal atoms of a rotos [Figure 2.17] for plastic materials and external atoms for brittle materials) and dilatons (external atoms for ductile materials and internal ones for brittle materials). This adds fundamental differences to the behavior of these bodies in external fields; otherwise a body cannot retain preset sizes and shape even under equilibrium conditions. In fact, positions 8 to 15 in Figure 1.14 show different variants of the CD ratio in the same set N. Its shape and position in the Cartesian, V, and energy, T, spaces depend on state parameters (Equation 2.96) and external fields (Equation 3.19); they entirely determine the physical–mechanical properties of materials.

The critical states of a rotos determine characteristic temperatures θ_f, T_c, θ_s, and T_m. The last corresponds to a point of the q_1-transition of the state from two-phase (position 5 in Figure 1.15) to one-phase or ductile (position 7). If, at T_m the AM bond loses rigidity (and a body becomes incapable of maintaining its shape on its own), and at T_c it cannot generate the resistance forces (here $F = 0$), the θ_f and θ_s temperatures cause it to transform from one phase into the other.

Deformation, as well as resistance, is initially a thermoreactive process (Equation 1.70). The system of notions developed in this book allows an elementary deformation act to be identified with an increase in size of the EM dipole (Figure 2.8b) — after completion of the orientation processes, of course (Figure 2.8a). In this case, dipoles of brittle materials move up to horizontal line q_2cq_1, and those of ductile materials move down to it (Figure 1.14). At $T = $ const, the deformation process can take place only due to dilaton radiation.

Because they are heat traps and absorb phonons, compressons serve as a sort of deformation brake. As a result, the rotos, which supplies heat to the deforming neighbors (positions III, IV, V, and VI in Figure 2.8b), tends to reach critical temperatures θ_f and θ_s to decrease the size of the dipole (positions I and II). This results in size differentiation of the dipoles (compare Figures 2.8a and 2.8b).

The θ_f-temperature is in the low-temperature region of the EM spectrum. Therefore, brittle structures have a low storage of kinetic energy and, hence, deformability. Moreover, regardless of how the energy state of the rotos in the acceleration well develops, it always involves a decrease in the resistance forces. Such structures are in a critical situation (positions 3 and 4 in Figure 1.15) and have no alternative but brittle fracture. They immediately decompose on completion of the orientation processes (at point a in Figure 2.10). When tested to tension, bending, and torsion, brittle materials have exactly the shape of these deformation diagrams.[3,59]

Rotoses of ductile materials behave differently in force fields (positions 11 to 15 in Figure 1.14). When dilatons, which radiate phonons, and compressons, which absorb them, approach the θ_s-temperature (position 6 in Figure 1.15) they show an increasingly high resistance to external factors and thus form region ab of the deformation diagram (Figure 2.10). Under conditions of excessive heat, both atoms pass through the θ_s-temperature with no consequences to the integrity of a structure, entering a new phase (position 5 in Figure 1.15) and developing the deformation process. This is the only cause of a substantial difference in the angles of inclination of elastic, α, and plastic, β, regions of the diagram. At a shortage of heat, they lose connectivity to form a submicrodefect.

Differentiation and activation of dipoles (Figure 2.8b) in a tensile force field at $T = $ const cause a change in the shape of the MB distribution (for ductile materials in a direction of 11 to 12 to 13 in Figure 1.14). By providing heat to the entire set N, the thermal field moves its major part from the compresson (C) to dilaton (D) phase (sequence of positions 8 to 9 to 10 for brittle materials and 12 to 14 to 15 for ductile materials).[18,24] Cooling changes the positions in the reverse direction. Tension of ductile materials (change in positions in a direction of 11 to 12 to 13) moves the hot end closer to the θ_s-temperature and the cold end closer to the T_c-temperature.

Compression, on the other hand, by increasing steepness of both fronts (reverse change in positions 13 to 12 to 11), moves its end out of dangerous zones. Consumption of the kinetic part of the overall internal energy for the CD transitions inevitably leads to the formation of dislocations.[47] Under conditions of insufficient heat, they multiply to form local microdefects in a crystalline structure. The MB parameter δ distributes them chaotically over the volume.[10] Periodic change in compression–tension of states increases the probability of movement of the end regions of the MB factors out of critical temperatures T_c and θ_s, thus activating the process of accumulation of damage (dashed regions in position 13) and initiating fatigue of materials.[24]

Therefore, three fundamentally different fracture mechanisms take place at the AM level, such as brittle at T_c (following diagram 4 in Figure 1.15), Debye at θ_f and θ_s (following diagrams 3 and 6), and ductile at T_m (position 7). Because the θ_s-temperatures of most metals and alloys are close to room temperature,[6,10] Debye fracture is the most characteristic and decisive fracture mechanism for most metals under normal working conditions.

In the low-temperature range, the Debye and brittle mechanisms may take place simultaneously causing disintegration of the bonds on both ends of the MB distribution (position 13) and provoking catastrophic, seemingly nonmotivated fracture of metal structures.[1-4] In the high-temperature range (position 15), the possibility exists of superposition of the Debye and ductile mechanisms leading to failure of parts and structures of thermal engines, blast furnaces, and other similar apparatus.[3,4] Separated by a large temperature interval, the brittle and ductile mechanisms cannot coincide.

Two-phase structure of solids and existence of a multivariant possibility of failure of the AM bonds make the "theoretical strength" notion meaningless from the physical standpoint. This notion was introduced at the outset of development of the atomistic approach to the problem of strength, played a stimulating role in its solution, and is still often unjustifiably used in the technical literature.[3,4,10]

Consider the process of formation of the deformation diagram (Figure 2.10) from the other standpoint. Specimen 1 (Figure 3.2) is loaded by an external pressure σ (either compressive or tensile) induced by a force F. When deformation takes place, it changes size and shape of the specimen displacing its surface L in space. The specimen can be regarded as a dynamic system consisting of N rotoses and mobile wall L. The latter will have $(3N + 1)$ degrees of freedom. In accordance with the definition of the MB distribution (Equation 2.3), the state of such a system is characterized by coordinates and pulses of all the rotoses, as well as by a position and pulse of the mobile surface.

Deformation causes displacement of energy levels in the system of N rotoses. We can represent the energy of the deformed solid as

$$U_i(P) = U_i + [U(r) + U(T)], \tag{3.36}$$

where U_i is the energy of the system of N rotoses and $U(r) + U(T)$ are the kinetic and potential energies of the surface of the solid. Nozdriov and Senkevick[49] have reported that the work performed over the AM system by changing external pressure σ by a value of $d\sigma$ is equal to $d\Omega_V = -VdP$. Hence, $\Omega = PV$.

Compared with the kinetic energy of the system of rotoses, the energy of thermal movement of surface L can be neglected because much smaller numbers of rotoses are on the surface than those in the volume. Therefore, Equation 3.36 becomes

$$U_i(P) = U_i + PV.$$

The function of state of such a system can be written as

$$Z(P,T) = \int \sum_i \exp\left(-\frac{U_i + PV}{kT}\right) \Omega(U_i) dV,$$

where summation is done over all i levels of the energy (value U_i depends on V), and integration is done over the entire volume of the body. The last expression makes it possible to use Analogy 2.63 and write

$$G = kT \ln Z(P,V). \tag{3.37}$$

Substituting Equation 3.37 into Equation 1.111, we find that

$$\Omega = kT[\ln Z(T) - \ln Z(P,T)]$$

or, finally,

$$\Omega = kT \ln \frac{Z(T)}{Z(P,T)}. \tag{3.38}$$

That the kinetic and potential parts of internal energy depend on different variables suggests that one two-parameter function $Z(P,T)$ can be represented as a product of two independent one-parameter functions: one in a three-dimensional space of pulses $Z(T)$ and the other in a three-dimensional space of coordinates $Z(P)$, i.e., $Z(P,T) = Z(T)Z(P)$. Then Equation 3.38 takes the form of $\Omega = kT\ln Z(P)$ or, taking into account the expression of entropy for quasi-static processes (Equation 2.17),

$$\Omega = Ts(P)_T, \tag{3.39}$$

where $s(P)T$ is the entropy that depends only on the pressure, but not on the temperature. This expression is of fundamental importance. First, it shows that the Ω-potential is determined by the bound energy of a body. Second, it becomes clear that fracture under noncontrollable conditions of loading ($P \neq$ const) occurs because of a shortage of heat (as in Equation 3.39 $T =$ const). On the other hand, upon heating when $T \neq$ const, it is impossible to fracture (in a common sense of this word) the body.

Like the rest of the parameters of state (Ω, F, G, etc.), N is a function of two (P and T) independent variables, i.e., $N(P, T)$. For this reason, solidification of bodies results in the formation not only of flat paired bonds (Figure 1.7b; considered in detail in Chapter 1) but also of chaotically oriented space formations (Figure 2.7). The orientation processes occurring in the force fields first deactivate the interatomic bonds of the first kind, $N(P)$ (in elastic region *oa* of the deformation diagram in Figure 2.10), and then the bonds of the second kind, $N(T)$ (in plastic region *ab*).

The same materials at different temperatures may fracture by the compresson (brittle) or by the dilaton (ductile) mechanism. The former takes place at low temperatures, where the apex of the MB distribution is between critical temperatures T_c and θ_s. Such a situation is represented by curve 1 in the lower field of Figure 3.9. In this case the fracture process simultaneously covers both branches (that outside the cold shortness limit is represented as hatched).

The second is characterized by an extreme yield and loss in shape of the body because the major part of the interatomic bonds is in the D state; it is represented by curve 3 in the same figure. By varying external conditions, it is always possible to form any ratio of the C and D phases. Their concentration at a given time moment per unit volume of a material determines the character of its deformation and fracture. The CD mechanism of deformation and fracture was described for the first time by the author in 1986[55] and then further elaborated in following studies.[12,18,21,24]

The preceding notions make it possible to explain behavior of solids in quasi-static[24] and alternating, single and combined,[21] force, thermal, radiation, or other fields and also to offer efficient methods for elimination[12] or activation[16] of the deformation and fracture processes. In accordance with Equation 3.19, electric and magnetic fields exert a similar effect on solid structures. Direct and indirect evidence supporting this fact can be found in References 1, 9, 20, 23, 29, and 30.

The discussed physical nature of the resistance of materials to external fields and aggressive environments allows us to conclude that the system of assuring strength and reliability of commercial products used today in practice is passive, inertial, and noncontrollable and, therefore, of low efficiency. It does not allow correction of the state of materials and structures depending on variations in external conditions, which inevitably leads to fracture of materials and failure of seemingly reliably designed (by traditional standards) and manufactured engineering objects. This primary drawback lowers the level of technogenic safety.[8] There is no way to improve this level while remaining within the frames of traditional concepts of strength and mechanisms of its exhaustion.[1-3]

The level of technogenic safety can be improved only on the basis of developing materials and structures adaptable to various service conditions. In this case, the materials and parts made of them will differ from those used today. These new materials, parts, and structures should not just change their state under the effect of the applied loads following the law of Equation 2.134, but also immediately restore its controlling actions (Equation 3.19). Methodologies and fundamentals underlying such materials and structures are under development.[12,16,18,21,24]

3.6 Scale Effect: Causes of Initiation, Forms of Manifestation, and Dangerous Consequences

The scale effect (SE) remains an unclear issue in solid-state physics and strength of materials, implying dependence of physical–mechanical properties on the volume (to be more exact, mass) of a body. Aleksandrov and Zhurkov[97] were the first to pay attention to it as early as 1928. Recently it has become evident that this effect is of a very general and fundamental character. It shows up in solids with different structures and properties, such as glasses,[2,91] polymers,[98] and composite materials[99-102] under the most diverse conditions of testing and loading, as well as at different types of the stressed state in force, thermal, electric, and other fields.[2-4,13,15,19,29,36,45,48,68-83,92-95] Analysis of the accumulated experimental data confirms the general nature of this phenomenon, which is attributed not only to a particular material but also to a solid state of the matter as a whole.

However, causes of initiation, conditions of occurrence, and possible consequences of SE for practice are still insufficiently studied. Some investigators[105-110] think that it is of statistical nature while others[102] explain it by technological or procedural causes. There are other opinions as well.[105] Studies [3,101] provide the most comprehensive and systematic review of results of experimental and theoretical investigations into various manifestations of this effect.

To understand the amount of knowledge accumulated on SE, we have analyzed more than 100 publications dealing with the phenomenon. In most publications (56), preference is given to the statistical nature of SE, believing that it is one of the consequences of the statistical character of fracture; authors of 26 publications give different views of its intensity, causes, and consequences. In some publications (18), the phenomenon was not found while the authors of others (9) think that it manifests itself under some conditions but not under others. Some researchers (three publications) failed to make unambiguous conclusions of its presence because of a substantial scatter of available experimental data. There is no unanimity in opinions of investigators concerning the existence and causes of SE.

Specimens of engineering materials used to investigate SE ranged from tens of microns (glass fibers) to 3 mm (cast iron) in diameter (or thickness). Size of their working surfaces varied within five orders of magnitude, and volume within two orders of magnitude. Scale dependencies were plotted using the working volume, cross section area, limiting surface, diameter, and length of the samples. For building materials (primarily concrete), experiments were conducted on cubic specimens of heavy or light concrete. Most tests were carried out on cubes with an edge of 20, 15, and 10 cm, more rarely on cubes with an edge of 7 and 5 cm, and even more rarely on cubes with an edge of 30 cm. As a rule, empirical coefficients were found to correlate strength of different sizes of the cubes; their values vary over very wide ranges in the data of different authors as well as within the data of a single author.

All the analyzed experiments are characterized by the following drawbacks:

- SE was estimated on the basis of insufficiently representative samplings. For example, concrete was investigated as a rule by testing from three to five samples of the same diameter.
- Not much attention was paid to the similarity of structure of different sizes of the specimens. For example, of 26 publications dedicated to concrete, only one study used samples taken from the same batch.
- Loading conditions were neglected. Most studies do not provide the rate and character of loading; only two mention the effect of loading conditions on SE test results.
- A phenomenological approach to the problem was taken; not one study considered the physical nature of SE.
- No consideration was given to a deep internal relationship (possibly, not yet realized) among SE, efficiency of available calculation methods, reliability, and durability of parts and structures.

This section aims to reveal the physical nature of SE, first, by summarizing the known facts. It was observed for all tested materials and test conditions that an increase in length, diameter, cross section, or surface area, as well as working volume, leads to a decrease in strength of a material and to a

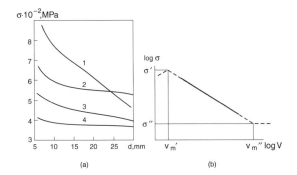

FIGURE 3.12
Scale dependencies observed in tension for cast iron (a) and fatigue limit for steel (b). (From Chechulin, B.B., *Scale Effect and Statistical Nature of Strength of Metals* (in Russian), Metallurgizdat, Moscow, 1963.)

corresponding change in its homogeneity.[3,102] This holds in static, dynamic, and fatigue tests and takes place in force,[3] thermal,[104] and electric[92,114] fields, as well as in tension, compression, bending, and other kinds of the stressed–deformed state. Typical scale dependencies plotted on the basis of analysis of numerous experimental data for cast iron in tensile tests are shown in Figure 3.12a[104] and for steel in fatigue tests in Figure 3.12b.[3] Numbers 1, 2, 3, and 4 designate different types of structure of cast iron, the specimens of which had diameter d mm.

Increase in scale parameters was accompanied not only by a decrease in a mean value of strength under different kinds of the stressed state, but also in its spread. Actual data on the scale homogeneity of different grades of steels, aluminum and magnesium alloys, polymers, and other materials can be found in Troshenko.[3]

Table 3.1 gives results of experiments conducted by the author and co-workers to study SE of concrete.[111] The study was carried out using specimens made of concrete of design grades 100, 300, 600, 700, and 800. Eighteen cubic specimens with an edge 20 cm in size and 126 specimens with an edge 10 cm in size were made of each composition of concrete. The compressive strength of concrete was determined at an age of 28 days. The same operator using the same equipment carried out all the tests.

It follows from Table 3.1 that the mean strength and standard deviation values obtained in testing cubic specimens with edges 20 and 10 cm in size do not coincide for all compositions. Also, whereas in the zone of low and medium strength (compositions 1 and 2) the impact of SE on the mean values is noticeable, it is very low for high-strength concrete (compositions 4 and 5). The scale dependence of spread of the strength values was observed as well in metals.[3,104]

Therefore, it follows from analysis of the entire set of experimental factors that SE has a general character. What is the root cause of its formation? Deviation of the value of any parameter in Equation 2.65 leads to a change in the

TABLE 3.1

Main Parameters of Statistic Distribution of Compressive Strength of Five
Compositions of Concrete

Composition No.	Statistic Characteristics of Strength of Concrete at Age of 28 Days						
	Cube Edge Size 20 cm			Cube Edge Size 10 cm			
	σ	g	s	s	g	s	K
1	12.0	1.7	0.51	13.2	0.58	0.65	0.90
2	30.4	7.5	1.57	31.6	1.71	1.92	0.96
3	57.3	10.0	3.01	58.5	2.97	3.33	0.97
4	68.7	12.2	3.27	69.8	3.04	3.41	0.98
5	81.1	12.7	3.81	81.4	3.76	4.22	0.99

Notes: σ is the mean compressive strength of concrete, MPa; γ is the mean error of mean strength, %; s is the standard deviation, MPa; and K is the scale factor.

state of a solid; this applies also to volume V. This symbol stands for the ability of a solid to change its size and shape in a preset volume dV and also for macroscopic variations of ΔV. The former is deformation (Equations 2.134 to 2.136) while the latter is the scale effect.[24,96] Under unchanged external conditions, a certain constant value of the Ω-potential is established in the same material, which means that a different resistance to external fields characterizes solids having different volumes. This becomes apparent if we rewrite Equation 3.39 as

$$\mathbf{P} = \frac{\Omega(T)}{V} \qquad (3.40)$$

and then differentiate

$$d\mathbf{P} = \frac{d\Omega(T)}{V} - \frac{\Omega(T)dV}{V^2}. \qquad (3.41)$$

As seen, Equation 3.41 coincides with the form set by Equation 1.53. It expresses the work done not by just one rotos but the entire set N in external fields. Figure 3.13 shows the graphical representation of this equation. Curve 1 represents the variation of the first term and curve 2 that of the second term. The resulting curve (solid line) shows that an increase in volume over a certain range of $V_0 - V_m$ leads to an increase in the resistance of the material, which reaches maximum at $V = V_m$ and then continuously decreases. It follows from Equation 3.41 that, in addition to $\Omega(T)$, the resistance of a solid is determined by the volume V and the rate of its variation, dV, during deformation.

At a constant and relatively low rate of deformation (set by the appropriate standards and usually kept constant during the tests), it can be considered that $dV = \text{const} \approx 0$. As such, Equation 3.41 is transformed into Equation 3.12.

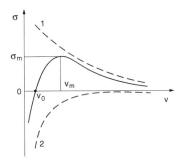

FIGURE 3.13
Generalized scale dependence of resistance of solids.

The curves shown in Figure 3.12 were plotted particularly for such conditions; they coincide in shape with curve 1 in Figure 3.13. If we take into account that, in accordance with Equation 2.153, the $\Omega(T)$ potential is equal to

$$\Omega(T) = \frac{1}{2\alpha}(\lambda - C_v),$$

we can conclude that the test results depend directly on the heat exchange parameters. Unfortunately, these parameters are not given due consideration in standard tests, leading to irregular spread of results of measurements and to a wide spectrum of opinions on different aspects of SE.

Volume V_m, at which the resistance to external effects is maximum, is referred to as the optimum volume. By equating (3.41) to zero, we find that

$$V_m = \frac{\Omega(T)}{d\Omega(T)}dV. \tag{3.42}$$

As seen, the optimum volume, V_m, is determined by the thermal–physical parameters of a material and depends on the deformation rate dV. Substitution of this value in Equation 3.40 yields

$$P_m = \frac{d\Omega(T)}{dV}. \tag{3.43}$$

As follows, the maximum possible resistance of a given material is determined by the volume gradient of the $\Omega(T)$ potential. It also follows from the relationship of equivalence (Equation 2.105) that

$$dV = \frac{dT}{T}V.$$

Substitution of this expression and the differential $d\Omega(T)$ from Equation 2.148.2 into Equation 3.40 yields

$$P_m = \frac{1}{12V}\left(C_v T + T^2 \frac{dC_v}{dT}\right).$$

Comparison with Equation 3.40 makes it possible to write

$$P_m = P_s + P_d, \tag{3.44}$$

where

$$P_s = \frac{\Omega(T)}{V} \quad \text{and} \quad P_d = \frac{1}{12}\frac{T^2}{V}\frac{dC_v}{dT}$$

are the resistance under static (or quasi-static) and dynamic loading. The first are realized under stationary and the second under nonstationary conditions of heat transfer.

Theoretical relationship 3.41 and Figure 3.13 obtained from the equation of state (2.65) have excellent experimental proof. For example, Sosnovsky[3] notes, based on the results of analysis of numerous experiments with engineering materials subjected to fatigue tests, that in the range of super-small sizes of specimens the $\sigma = f(V)$ dependence undergoes inversion and is transformed from an inverse to direct one (corresponding to value V_m' in Figure 3.12b). Fudzin and Dzako[108] concluded that some critical volume V_m' corresponds to the maximum of function $\sigma = f(V)$. The presence of inversion of SE (maximum of curve $\sigma = f(V)$) in the small-volume region under quasi-static loading of porous materials is reported in Troshenko.[109] The direct proof of existence of the scale dependence with a maximum, as applied to the ultimate stresses under axial loading of steel specimens, can be found in Troshenko,[3] while that in testing porous zirconia-based ceramics is shown in Figure 3.14.[109]

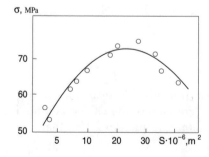

FIGURE 3.14
Manifestation of the scale effect in specimens of porous zirconia-base ceramics.

Scale phenomena in a small-volume region are of scientific rather than practical interest, as the volume and size corresponding to the inversion point are small. They can be estimated by Equation 3.42. For example, for porous ceramics (Figure 3.14), they were determined experimentally to be $V_m = 600$ mm^3 and $S_m = 20$ mm^2.[99] Size, volume, or mass of the overwhelming majority of actual objects belongs to the descending branch of plot $\sigma = f(V)$ (to the right from the maximum point in Figure 3.13). Investigators should pay particular attention to this region.

Based on the rapidly descending character of dependence $\sigma = f(V)$ (Figure 3.13), we can assume some critical volume V_m'' (Figure 3.12b) in terms of influence on the fatigue limit. If it is exceeded, the fatigue limit probably would not decrease,[110] an assumption mentioned also in Troshenko.[3] However, whereas inversion is confirmed theoretically (Equations 3.40 to 3.44) and experimentally (Figures 3.12 and 3.14), the latter assumption has no theoretical or experimental proof to confirm its existence. Moreover, it should be regarded as a dangerous delusion, evidenced by accidents and catastrophes at various engineering objects reported from time to time.[3,14,17] It is not excluded that one of their causes is underestimation of processes that occur in deformation and fracture of materials in the zone of the large-size end of scale dependence.

What is the physical nature of SE? To answer this question, let us subdivide a solid, with volume V_1 and made with special care, into a few standard specimens of smaller size V_2 (i.e., $V_1 > V_2$). We can write the equation of state at $T = $ const for each: $\mathbf{P}_1 V_1 = \mathbf{s}_1 T$; $\mathbf{P}_2 V_2 = \mathbf{s}_2 T$. Then we can find the relationship between them as

$$\frac{\mathbf{P}_1}{\mathbf{P}_2} = \frac{\mathbf{s}_1}{\mathbf{s}_2} \frac{V_1}{V_2}. \qquad (3.45)$$

Under specified conditions of specimen manufacture, it can be considered that each has the same meso- and macroscopic levels of entropy in Equation 2.112. Then, expressing macroscopic parameters V and $\mathbf{s}$ in accordance with Equations 2.97 and 2.109, in terms of microscopic variables, we have

$$\frac{\mathbf{P}_1}{\mathbf{P}_2} = \frac{h_1}{h_2}, \qquad (3.46)$$

where

$$h = \frac{\delta}{r^3}$$

is the indirect indicator of microhardness. It follows from Equation 3.46 that the resistance of a solid can be evaluated on the basis of indicators of hardness.[53,55,111,115]

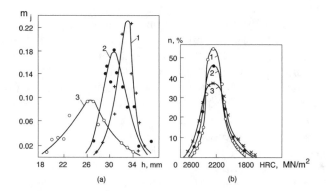

FIGURE 3.15
Empiric distribution of hardness of concrete (a) and steel (b) in specimens of different sizes.

Figure 3.15a shows empirical distributions of hardness of concrete determined by the author on cubic specimens with an edge 10 (1) and 20 (2) cm in size and on short columns $20 \times 30 \times 150$ cm (3), using the method of the height (h, mm) of rebound of the indenter.[115] Figure 3.15b shows variations in Rockwell hardness of steel (HRC, MN/m^2) measured on cylindrical specimens of 10 (1), 20 (2), and 60 (3) mm in diameter.[3] These data are comparable because the state of both materials can be described by Equation 3.65, with the only exception that, in concrete and other brittle materials (cast iron, ceramics, etc.), $\delta = $ const. This means that there is no way for the compresson–dilaton phase transitions to take place in them.[24] So, an increase in volume $V_1 < V_2 < V_3$ smoothes out the external potential field (Figure 1.8; movement along the axis of abscissas in a direction from r_{0_1} to r_{0_3}). In turn, this levels the energy state of interatomic bonds (change in positions from 1 to 3 in Figure 3.9). In Figure 3.15a, the recurrence of single changes is expressed in relative units, m_j, and in Figure 3.15b in percent, $n\%$.

The maximum energy of one of the atoms of a solid at an absolute zero temperature can be determined from formula[30]

$$u_0 = \frac{\hbar^2}{8m} \rho^{2/3}, \tag{3.47}$$

Because the mean distance between the neighboring particles is equal to the edge of a cube with volume $1/\rho$, Equation 3.47 can be rewritten as

$$u_0 = \frac{\hbar^2}{8mr^2}.$$

As seen, the maximum u_0 and mean u energies are very close to each other,[30] i.e.,

$$u = \frac{3}{5}u_0.$$

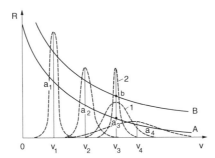

FIGURE 3.16
Diagram of smoothing of the internal potential field with an increase in volume of a body.

Thus, both are in inverse proportion to the square of distance between the neighboring atoms. This fact determines the character of Dependencies 3.40 and 3.41.

On the other hand, it is well known from quantum statistics[27] that the mean distance between neighboring permitted levels of energy, $R^{(u)}$ is determined as

$$R^{(u)} = ue^{-s(u)}, \tag{3.48}$$

where $s(u)$ is the entropy of a macroscopic system (Equation 2.112). The latter sets the density of levels of the spectrum energy (Figure 1.18). Since entropy is additive, the mean distance between the levels exponentially decreases with an increase in size of a solid, i.e., the quantity of particles in it or distance between them.

The plot of exponential dependence $R = f(V)$ is indicated by a solid line in Figure 3.16. The dotted line represents the transformation of the distribution of entropy with an increase in the volume of the solid: $V_1 < V_2 < V_3 < V_4$. Curves 1 and 2 are drawn in position V_3 and represent different density of material that fills out the same volume—V_3. In the case shown, the density of the second is higher than that of the first. An increase in density of the material causes the exponential curve to displace equidistantly from position A to position B, thus increasing the distance between the permitted levels of the energy by value a_3b.

At the same density of a material, the area delineated by curves $V_1 - V_4$ remains unchanged. An increase in volume smears distribution $s(u)$, changing its shape in the direction of V_1 to V_4. This means that, at a force, electric, or other excitation of rotoses, the value of energy step Δq (Figure 1.18), which they have to overcome during the process of transition from one state into the other, exponentially decreases (compare lengths V_1a_1, V_2a_2, V_3a_3, and V_4a_4). This has two consequences: first, this greatly facilitates the loss of a bond in each elementary event of its disintegration and, second, in a large volume, the number of the local zones, in which disintegration processes occur simultaneously, is always greater.

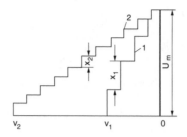

FIGURE 3.17
Different paths in overcoming the energy barrier in bodies of different volumes.

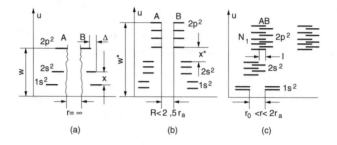

FIGURE 3.18
Diagram of formation of a paired bond in a carbon atom.

Figure 3.17 shows variation of a drift path passed at fracture by the atoms of specimens (parts and structures) with different volumes and made of the same material. The maximum energy barrier that atoms must overcome to lose mutual bond is designated as u_m. In Figure 3.16, the value of u_m corresponds to lengths V_1a_1, V_2a_2, ... The spread of the elementary energy step x in large volume V_2 is always smaller than that in small volume V_1, because Equation 3.48 sets inequality $x_2 < x_1$. Because overcoming a small barrier requires much less energy of external excitation, an atom reaches threshold u_m in small steps x_2 much faster and more easily than in large steps x_1. Moving over path 2, it has to pass a longer distance than that along path 1, forming larger fracture surface (determined by length of segments $0V_1$ and $0V_2$, respectively). Despite this, it particularly prefers this path as the most beneficial in terms of energy.

Because the fundamental significance and unexpected nature of these conclusions followed from analysis of the AM nature of SE, we provide a possibly more obvious and clear interpretation of the processes described. The energy of electrons in an isolated atom may have only specific values separated from each other by forbidden interval x. To illustrate, Figure 3.18a shows the energy spectrum of two isolated carbon atoms (A and B); two electrons are located at each level (Figure 1.1). Displacement of levels Δ relative to each other is indicative of their belonging to different circular orbits. So, let the nondisturbed atom have energy equal to W.

The mutual molecular bond is formed when atoms approach each other to a distance of $r < 2.5r_a$ (see Table 2.1). In this case each level is split into several sublevels that indicate the excitation of both atoms (Figure 3.18b). The excited state of atoms is marked by an asterisk. The depth of splitting of the levels ($2p^2$, $2s^2$, and closer to the nucleus) depends on the distance between atoms and the degree of overlapping of the wave functions of electrons. In accordance with Equation 3.47, drawing together causes an increase in the energy of electrons. Hence, $W^* > W$. Valent electrons experience maximum excitation (level $2p^2$), as they approach each other and excitation reduces the forbidden zone, i.e., $x^* < x$.

While approaching each other, the energy levels start overlapping (Figure 3.18c): the smaller the distance between the atoms, the larger the overlapping area of the valent zone and the greater the number of layers it penetrates. In the range of $2r_0 < r < 2.5r_a$, only the levels of valent electrons, $2p^2$, overlap each other, forming molecular elliptical orbits (see Figure 1.3) that tie together the formed ions, hampering their drawing apart. Not bound by deeper bonds (e.g., at level $2s^2$ or at level $1s^2$), ions can easily change their position inside valent orbits (Figure 1.3). This is the nature of the liquid state (see Table 2.1); there are no short-range bonds, which could add rigidity to the AM structure.

At $r_a < r < 2r_a$, the surface level $2p^2$ as well as the internal ones (e.g., $2s^2$ in Figure 3.18c) are overlapped, resulting in solidification of a body. Extension of the bond to deep layers adds rigidity, a specific feature of solids, to the AM structure. Their heating first causes destruction of the short-range bonds and then of the long-range bonds. Fracture always occurs under conditions of heat insufficiency; therefore, destruction of the short- and long-range bonds occurs simultaneously.

External effects (thermal, force, electric, etc.) cause a change in the natural energy of electrons (valent, first of all), and they transfer to one of the closest split-levels. Width of the overlapped area of excited levels l depends on the distance between the atoms, r. Density of splitting of the levels in accordance with Equation 3.48 is determined by the quantity of atoms per unit volume of a body, N/V (compare, for example, frequency of lines at level $2p^2$ at N_1 and N_2, where $N_2 > N_1$). The distance between the lines corresponds to the height of steps on lines 1 and 2 (Figure 3.17). The higher the frequency of the lines, the easier the transfer of an electron to more elongated elliptical orbits because it receives pulses of small magnitude but high frequency from the outside.

Eventually, the ellipse is transformed into a hyperbola (Figure 1.12a), and the electron leaves the AM bond. This escape not only weakens the molecular bond (Figure 1.3) that keeps atoms 1 and 2 together, but also violates the balance of the electric charges of ions A and B. Repelled from each other, positive ions complete the process of disintegration of the bond.

We investigated the large-size zone in concrete. All possible measures were taken to ensure equivalence of its structure: composition, as well as manufacturing conditions and storage of the specimens, their age, and moisture content, were maintained at constant levels. For this, the cubic specimens

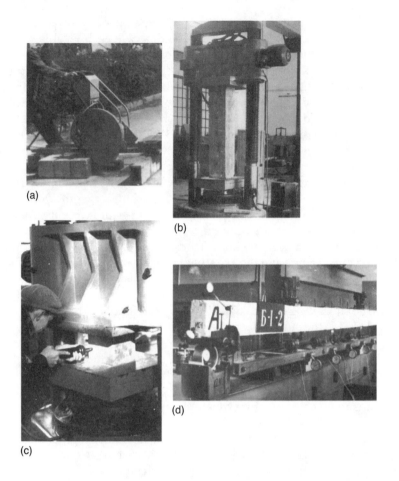

FIGURE 3.19
Investigation of the scale effect in concrete: (a) cutting of standard cubic specimens of different sizes from short columns, (b) testing of columns and (c) cubes in compression by preliminarily measuring their hardness, and (d) testing beams in bending.

were cut from massive prisms or short columns, measuring $20 \times 20 \times 80$, $20 \times 30 \times 120$, and $20 \times 30 \times 100$ cm (Figure 3.19a). The size of the cube edge varied from 20 to 5 cm with an interval of 5 cm. To decrease subjective effects, all the tests were conducted by the same operator using the same procedure and the same equipment (Figure 3.19b,c). After cutting and calibration, specimens of different sizes were combined into several groups. Each group comprised no fewer than 12 specimens of the same size. Specimens of each size subgroup were tested at a constant loading speed. The time of loading was varied from 4 to 5 s to 8 h.

For comparison, Figure 3.20 shows scale dependencies for metal rods (a)[103] and concrete (b). Ziobron[103] investigated SE in the variation of the length of high-strength bundle wire in a range from 5 to 70 cm at a constant diameter of wire equal to 5 mm (Figure 3.20a). In accordance with Equation 3.40, an

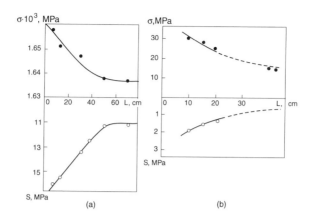

FIGURE 3.20
Scale dependencies for (a) steel and (b) concrete.

increase in size parameters of both materials led to a decrease in the mean strength σ (upper field) and increase in its homogeneity s (lower field).

In the tests, the size group was not limited to laboratory specimens (Figure 3.19c), but was brought to the scale of actual structures: prisms measuring $20 \times 20 \times 120$ cm, short columns measuring $20 \times 30 \times 150$ cm (Figure 3.19b) and beams of 4500 cm long (Figure 3.19d). Compressive strength of the prisms and columns was lower compared with that of the cubic specimens with the edge of 20 cm by 20 and 24%, respectively (extreme left points in the upper field of Figure 3.20b).

An approximately similar decrease was observed for bending strength of the beams. We did not obtain numerical values of the standard strength for large-size test pieces because of unavailability of representative sampling. However, we can state with a high level of confidence that increase in volume of a stressed material leads to decrease in spread of individual values of strength (dashed end of the plot in the lower field of Figure 3.20b).

Comparison of Figures 3.13 and 3.20b shows that, for concrete, the inversion point is in the specimens whose size is smaller than 5 cm, i.e., in the volumes, which are of no practical value. The volume of actual structures is located on a large-scale descending branch of curve $\sigma = f(V)$ (Figure 3.13). As the working volume of a material increases, a decrease in the mean strength takes place and the shape of its statistical distribution changes. A decrease in the spread is accompanied by a shift of its center to the zone of decreased strength indicative of an increase in the energy homogeneity of a structure (in Figure 3.17, line 1 transfers to 2). As reported in literature,[3] no attenuation of the scale phenomena in the large-size zone was detected.

Figure 3.21a shows the dependence of the load-carrying capacity of beams in bending F on their cross section area S. A similar result was obtained in a study[112] (Figure 3.21b). The authors used bending coefficient $A = \sigma_1/\sigma_2$ as the SE indicator, where σ_1 is the tensile strength of concrete in bending, σ_2

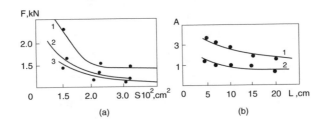

(a)   (b)

FIGURE 3.21
Scale characteristics of reinforced concrete (a) and concrete (b) beams in bending.

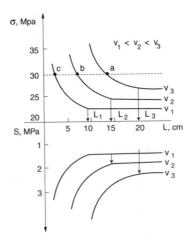

FIGURE 3.22
Effect of loading rate on scale characteristics.

is that in the central tension of the prisms measuring $10 \times 10 \times 40$ cm at standard loading rates, and L is the cross section height. Numbers in both plots stand for the following grades of concrete: (a) 1, 300; 2, 500; 3, 800; (b) 1, 300; 2, 600 MPa. Similar dependencies for cast iron and different types of steel can be found in Troshenko.[3] All of them can be approximated by Equation 3.40.

High cost of the experiments did not allow us to obtain a representative sampling and find other parameters of statistical distribution of strength in the large-size region. However, the discussed variation persists in this case as well: spread of single measurements of strength decreases with an increase in volume.

The variation of statistical parameters of the compressive strength of concrete with variations in sizes of the specimens and rates of loading is shown in Figure 3.22. It follows from this figure that the scale phenomena in concrete (as in any other material) are sensitive to loading conditions (see Equation 3.41) and, hence, correlate with dynamic effects. For each loading rate v is a

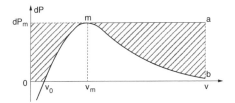

FIGURE 3.23
Character of manifestation of the scale and dynamic effects.

peculiar barrier L, starting from which its effect becomes especially pronounced. Thus, if $v_3 > v_2 > v_1$, then $L_1 < L_2 < L_3$. It is likely that the boundary between the smooth and sudden course of the scale dependence is the inflection point of the descending branch of curve $\sigma = f(V)$ (Figure 3.13). It is also seen that massive parts are more sensitive to SE under dynamic effects (see movement along *a–b–c* in Figure 3.22). For example, it is not observed in the cubic specimens having an edge size of 15 cm and at loading rate v_3, whereas it does show up in the specimens having an edge size of 20 cm.

Therefore, for one and the same material, under certain conditions SE may show up, and under other conditions may not, depending on specimen sizes and deformation rates. This process is controlled by density of the flow of the mechanical energy into a material from the outside. It varies with both the loading rate and the volume of a material, and characterizes the amount of the energy input in a unit volume of a material per unit time. SE is not observed at a constant density of the flow independently of the conditions. Experience shows that it is nearly impossible to detect SE in testing different-scale specimens by varying the loading rate proportionally to their volume (e.g., moving along line *a–b–c* from the right to the left; see Figure 3.22).

The selection of the scale row is always random. What causes changes in the scale dependencies (Figure 3.22) in small-size and, especially, large-size regions with variations in loading conditions? Rewriting Equation 3.41 as

$$dP = \frac{\Omega}{V}\left(\frac{d\Omega}{\Omega} - \frac{dV}{V}\right) \tag{3.49}$$

and equating the right-hand part to zero, we can find volume

$$V_0 = \frac{\Omega}{d\Omega}dV,$$

at which a body cannot resist external effects, and can just retain its shape (Figure 3.23).

Taking into account that the differential of Equation 3.40 is

$$d\Omega = \Omega\left(\frac{dN}{N} + \frac{dT}{T} + \frac{d\delta}{\delta}\right), \tag{3.50}$$

we can rewrite condition $dP = 0$ as

$$\frac{dN}{N} + \frac{dT}{T} + \frac{d\delta}{\delta} = \frac{dV}{V}. \tag{3.51}$$

Size parameter V accounts not only for variation of the scales of a body (as shape is preserved) but also for transformation of the shape at a constant volume. This follows also from Equation 2.137.

Comparison of Equations 3.51 and 2.137 shows that, if a body with volume V_0 has the ability to be deformed, then the right-hand side of Equation 3.51 would not be equal to dV/V but to $dV/V + V_0(\eta + \varepsilon)$, where $V_0 = $ const. In reality, $V_0(\eta + \varepsilon) = 0$. Hence, $\eta = \varepsilon$. Therefore, the amount of energy concentrated in volume V_0 is sufficient only for maintaining exchange processes $d\delta/\delta$, directed for compensation for random fluctuations of entropy (Equation 2.13) due to a change in the number of interatomic bonds dN/N — not enough for generation of resistance forces. The following condition should be met for a structure to resist the external effects:

$$\left(\frac{d\Omega}{\Omega} - \frac{dV}{V} \right) > 0.$$

In fact, if we take the second derivative of Equation 3.49

$$d^2P = \frac{\Omega}{V}\left(\frac{d\Omega}{\Omega} - \frac{dV}{V} \right)^2 + \frac{\Omega}{V}d\left(\frac{d\Omega}{\Omega} - \frac{dV}{V} \right)$$

and equate it to zero, $y^2 = -dy$, where

$$y = \frac{d\Omega}{\Omega} - \frac{dV}{V}.$$

Then, after integration, we obtain that

$$\frac{d\Omega}{\Omega} - \frac{dV}{V} = \text{const},$$

which determines coordinates of point m of the maximum of function dP (Figure 3.23). If

$$\frac{d\Omega_0}{\Omega_0} = \text{const},$$

then

$$V_m = \frac{dV}{\Omega(T)},$$

where

$$\Omega(T) = \frac{d\Omega_m}{\Omega_m} - \frac{d\Omega_0}{\Omega_0}$$

is the stored kinetic energy consumed during deformation. Substitution of V_m into Equation 3.49 yields

$$d\mathbf{P}_m = \frac{\Omega}{V} \frac{d\Omega_0}{\Omega_0}.$$

The Debye temperature θ determines individual properties of any solid (Equation 1.87). This temperature characterizes the ability of a body to accumulate internal energy (it is identified with the area of rectangle $0dP_mab$, Figure 3.23) and spend it, according to the law set by Equation 3.40, for an increase in volume (dashed zone) or for deformation (the figure limited on the top by curve V_0mb). In a range of $V_0 - V_m$, the first part decreases while the second increases. At $V > V_m$, on the other hand, its larger part goes to an increase in volume and the smaller is spent for deformation resistance. Comparison of Figures 3.20 to 3.21 with Figure 3.23 allows a conclusion that, in both cases, the size parameters were located on the right branch of the scale characteristic.

The scale dependence of resistance derived (Equation 3.41) from the thermodynamic equation of state (2.65) is well supported by numerous experiments.[3,101–104] Therefore, the efficiency of existing methods for design and calculation of engineering objects in which the SE is not considered is doubtful.

Investigation of SE is usually carried out by varying one of the size parameters (volume, diameter, cross section area, etc.) and by loading specimens at constant rates (as a rule, specified for a given material by an appropriate standard), i.e., by keeping to condition $L \neq$ const at $v =$ const. As such, the condition of a constant flow density is never met and thus SE is observed almost everywhere. Under service conditions, SE shows up for a different reason. In service, the size parameters are always constant, i.e., $L =$ const, but loading conditions are arbitrary, i.e., $v \neq$ const. This causes a nonconstant flow density and a more dangerous impact of SE. Danger grows because the dynamic loading conditions not only change the value of the mean strength (upper field in Figure 3.20) but also lead to an increase in energy homogeneity of a structure (compare plots v_1, v_2, and v_3 in the lower field of Figure 3.22).

An increase in homogeneity of the structure implies leveling of a local potential field, which is accompanied by a change in a drift path from position 1 to position 2 (Figure 3.17) and, as a result, by facilitation of the process of initiation and propagation of cracks. The only way of eliminating impact of SE and, therefore, preventing development of fracture processes due to altering external (environmental) conditions is to maintain the energy capacity of a material artificially within certain ranges. This can be implemented by supplying (or removing) additional energy from the outside to level the energy balance of the material.

It follows from Equation 3.43 that SE is determined by the ability of a structure to effectively redistribute internal reserves of kinetic energy $d\Omega(T)$ in deformation dV to achieve Ω = const when external factors vary. Interaction with the environment may lead to heat losses and thus to change in the Ω-potential according to Equation 2.153. In technical literature,[102–104] a wide variety of opinions concerning its impact and objective nature exists. Different opinions are also caused by different sensitivities with which each type of a structure made of the same material reacts to scale changes. The density of packing of the structure determines this sensitivity. Indeed, rewriting Equation 3.41 and accounting for the overall energy (Equation 2.142) related to one rotos yields

$$dP = \frac{1}{V}[d\Omega(t) - \omega\rho dV].$$ (3.52)

Figure 3.24 shows the dependence of the SE of concrete on the arrangement of a structure. The short-time strength σ (directly depending on the density; see Section 3.8) is taken as the integral indicator of a structure and the beginning of its impact is determined by scale parameter L (Figure 3.22). The shape of curve $L = f(\sigma)$ indicates a decrease in sensitivity of dense, high-strength concretes to scale changes. The zone of their scale sensitivity shifts toward small values of size parameters L. For this reason, specimens of standard sizes of these concretes sometimes do not exhibit SE (see Table 3.1). The noted trend is aggravated with a decrease in the loading rate (compare curves at $v_3 > v_2 > v_1$). A convincing proof of validity and versatility of these laws can be found in literature dedicated to studies of mechanical properties of high-strength concretes.[113–119]

Our considerations suggest that the governing laws at the level of the AM structure differ from those used in the known theories of strength,[3] which

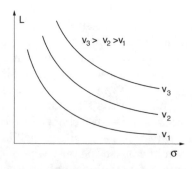

FIGURE 3.24
Scale sensitivity of concrete.

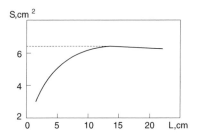

FIGURE 3.25
Variation of the specific fracture surface under scale effect conditions.

can easily lead to unpredictable consequences. For example, following the theory of strength of the weakest link[59] or fracture mechanics,[108] current design methods tend to ensure reliability and durability of structures by adding excessive material. It is thought that, the higher the content of a material in volume of a part and the higher its homogeneity, the stronger and more reliable the structure. However, as follows from Equation 3.41, this is not true. The effect of mass on the resistance of materials to external fields is ambiguous. On one hand, its increase causes a decrease in the level of internal stresses; this perception is in the basis of the existing methods of design of parts and structures (methods known earlier and currently under-developed). On the other hand, crack resistance decreases when the mass of material is increased.

As discussed, interatomic bonds of any solid exist in the internal potential field (Equation 1.26.2), which may have a different spatial configuration (Figure 1.8). In small volumes, this field has a clearly defined profile (line 1 in Figure 3.17). Propagation of cracks (if initiated, of course) is difficult in such a field. An increase in mass of a body smears the profile (line 2 in Figure 3.17). Thus, cracks are readily initiated, readily coalesce, and thus propagate, which is proven by the variation of the specific fracture surface under SE impact conditions (solid line in Figure 3.25). The fracture surface does not change with the adjustment of the loading rate, depending on the decrease in scale of a specimen (dotted line).

Figure 3.26 presents the results of stereolithography study of variation of a concrete structure in cubic specimens of different sizes subjected to differ-ent levels of stress.[120] The level of stress $\alpha_p = \sigma/\sigma_p$, where σ is the effective stress level and σ_p is the compressive strength, is plotted on the axis of abscissas. The total length of cracks, L, calculated per unit cross-sectional area, is plotted on the axis of ordinates. Numbers in the figure designate the following edge sizes of the cubes: 1, 30; 2, 25; 3, 20; 4, 15; and 5, 10 cm. At the first stage of loading, the destruction processes start developing only in specimen 1; the rest of the specimens exhibit no structural changes. At stage 2, in specimen 1 the total length of cracks reached a value of $L = 6$ cm, whereas in the second specimen they were only initiating. In specimen 5,

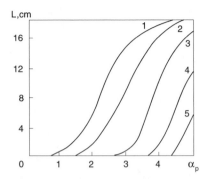

FIGURE 3.26
Scale dependence of crack resistance of concrete.

they become marked only at stage 5, i.e., when the cracking processes in specimen 1 have already reached saturation.

It becomes clear that sizes of a body have an ambiguous effect on its strength. Although their increase leads to a decrease in the effective stresses σ initiated by external fields (not to be confused with resistance **P** developed by the atomic-molecular structure), crack resistance of a structure decreases. Therefore, it is not by chance that parts and structures designed by these methods operate reliably under static and constant climatic conditions but unexpectedly fail under dynamic changes of the external factors.[3] These methods *a priori* cannot guarantee 100% safety of a structure, as their initial prerequisites are not adequate to the physical–mechanical processes occurring in the structure.

According to the data of the American Institute of Concrete,[121] 10% of load-carrying reinforced concrete members fail during the operation process, and 1% fail with catastrophic consequences. Reliability of metal structures is not better.[3,43] This is the level of technogenic safety achieved today and the price of the inadequacy of current methods for ensuring strength and reliability in the building industry. It cannot be radically increased if we remain within the frames of traditional notions of deformation and fracture processes.

The problem can be solved only on the basis of developing materials and structures adaptable to service conditions. The ability of a structure to generate resistance forces $d\mathbf{P}$ can be controlled by varying the value of energy parameter Ω in Equation 3.40. This means that, at any point of a large-size range $V > V_m$ (Figure 3.23), it is possible to use the permitted and also the potential range of resistance (the latter is dashed). As such, materials themselves and parts made of them will differ from those used today. Unlike the known ones, they not only should change their thermodynamic state under the effect of useful loads following the law set by Equation 3.65, but also should immediately restore this state under the impact of the governing effects (Equation 3.19). Development of a physical approach and principles of creation of such materials and structures is under development.[18,21,24]

These approaches make it possible to (see Equation 3.41 and Figure 3.17):

- Increase energy barrier u_m that can be done in design of a structure and in the process of manufacture of materials. Its value can be adjusted by technological methods at the stage of fabrication of parts and structures.
- Avoid smoothing energy steps preventing change of position 1 to position 2 (Figure 3.17) due to an increase in the efficiency of utilization of a unit volume of a material.
- Increase the height of energy step x due to proper utilization of the orientation effects.
- Decrease sensitivity of materials to dynamic loading conditions by supplying them with additional energy at the moment of action of those conditions. The energy methods for ensuring reliability and durability allow flexible reactions to variations in the stressed–deformed state by almost immediately lowering dangerous stresses. These methods can be used as the basis for fabrication of structures with a self-adjusting load-carrying capacity.[12,122]

3.7 Mechanism of Formation of the Maxwell-Boltzmann Factor

It can be seen from the equation of state written in the form of Equation 2.96 that the Ω-potential directly depends on the Maxwell-Boltzmann factor δ (Equation 2.94). The latter shows how densely, in which region of axis r_j (Figure 1.14), and at what energy the elements of set N are concentrated in a preset volume (positions 8 to 15). Let us consider $\Omega = f(\delta)$ at constant temperature (T = const) and density (N = const) of a material and in the absence of scale phenomena ($\Delta V = 0$), using Relationship 3.46. We conducted special preliminary investigations to study and upgrade known methods and develop new ones for quantitative estimation of this relationship for building materials and structures,[123–125] enhancing their metrological characteristics[126,127] and improving methods for interpretation of indirect measurements.[128,129]

It appeared that the integrated elasto-plastic method is the most suitable because it has the highest information potential for the preceding purpose,[125] combining two methods of hardness determination: rebound, or dynamic, hardness and indentation hardness (in the considered case the indenter had a spherical shape). The main peculiarity of the method is that measurement of both properties is performed simultaneously each time by using the same reference pulse of the kinetic energy of the indenter. The method and corresponding device are standardized.[130]

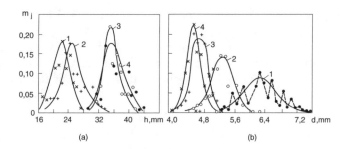

(a)   (b)

FIGURE 3.27

Empiric distribution of elastic (a) and plastic (b) hardness indicators obtained in testing concretes of design grades 100 (1), 300 (2), 500 (3), and 700 (4).

TABLE 3.2

Parameters of Statistical Distribution of Elastic (Columns 2 to 5) and Plastic (6 to 9) Parts of Indicators of Hardness

Designation	n, pcs	h, mm	S_h, mm	A_h	n, pcs	d, mm	S_d, mm	A_d
1	2	3	4	5	6	7	8	9
1	114	22.8	2.4	0.10	206	6.2	0.45	0.136
2	130	26.2	3.2	0.12	280	5.2	0.28	0.053
3	108	36.2	2.3	0.16	268	4.6	0.22	0.047
4	128	36.8	2.9	0.18	254	4.5	0.19	0.041

Although the experimental determination of hardness is simple compared with any other mechanical property, few studies of statistical principles of variations in elastic and plastic indicators of hardness are available.[3] At the same time, these parameters, determined experimentally, are most suitable to study function $\Omega = f(\delta)$.[131,132] Indeed, in accordance with Equation 3.46, the elastic indicator characterizes resistance of a body Pi to the action of a reference load at point i, while the plastic indicator characterizes its deformability dV_i.

Figure 3.27 shows variations in the empirical distribution of local elastic, h, and plastic, d, deformations determined in hardness testing of cubic samples with an edge size of 20 cm (Figure 3.19c) and made of concrete of design grades 100, 300, 500, and 700 at room temperature. The axis of ordinates shows repeatability of single measurements in relative units. Selected data on compressive strength of concrete are given in Table 3.1 (rows 1 to 4). The number of measurements n and basic parameters of statistical distribution of the indicators of hardness (mean values h and d, standard deviations S_h and S_d, and asymmetry A_h and A_d) are given in Table 3.2.

Reaction of a structure of concrete to the referenced pulsed mechanical effects fundamentally changes toward the high-strength zone. The displacement of centers of statistical distributions of elastic h and plastic d parts of overall local strains is opposite.[125] Under invariable experimental conditions and constant level of energy effects, the share of the former increases (positions 1 to 4 in Figure 3.27a) while that of the latter decreases (similar positions in Figure 3.27b). This means that the number of rigid bonds in a unit volume (grouped in Figure 3.9 on the left branches of curves 1 to 4) continuously grows, and that of plastic ones (on the right branches) decreases. The discussed displacement is accompanied by the transformation of the shape of distribution: the standard deviation of plastic strains S_d decreases (according to the data of Table 3.2, from 0.45 for concrete grade 100 to 0.19 for concrete grade 700), whereas the standard deviation of elastic strains S_h increases (see the lower field in Figure 1.14). This indicates the formation of substantial energy heterogeneity of the structure: the preset volume of concrete is increasingly saturated with the energy-extensive bonds with movement to the high-strength zone.

The apex of the MB distribution (Figure 1.14) approaches the θ_f-temperature (point d) and its homogeneity grows (compare positions 1 to 4 in Figure 3.27). Under loading, this leads to an increase in the concentration of centers of crack nucleation per unit volume. At the same time, the number of plastic bonds decreases, i.e., the energy zone near the end of curve 2 is released. Ductility of plastic traps (zones of crack arresting) decreases and their spatial field becomes more uniform. (In Figure 3.17, energy "staircase" 1 takes the form of 2.) As a result, a material becomes brittle and its fracture is explosive.

This is clearly seen in Figure 3.28, which shows the variation of the standard deviation of distribution of the compressive strength of concrete, S_σ with an increase in its value from 10 to 80 MPa (upper field). The lower field shows a corresponding variation of standard deviations of local elastic S_h and plastic S_d indicators of hardness.

To reveal the macroscopic consequences to which energy redistribution in elementary volumes of a material leads, measurements of the indicators of hardness were accompanied by investigation of the scatter of the compressive strength. Figure 3.29 shows ranges (solid line) and empirical distributions

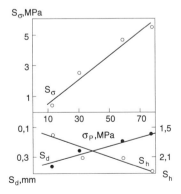

FIGURE 3.28
Variations in standard distributions of elastic, S_h, and plastic, S_d, hardness indicators, determining macroscopic heterogeneity of compression strength of concrete, S_σ.

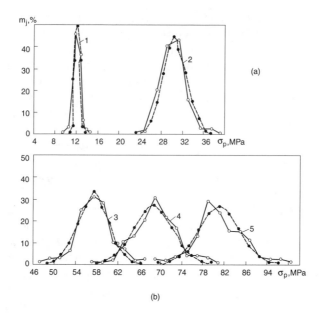

FIGURE 3.29

Empiric distributions of compression strength of concretes (parameters are given in Table 3.1).

of strength of concrete of compositions 1 to 5 (see Table 3.1) plotted using the results of 126 measurements.[111] The ranges are well approximated by a normal distribution curve (dotted line). The spread of the results increases with an increase in the mean strength and transition to the high-strength zone. Because all specimens were made of the same material using the same technology and because test conditions were identical, it is likely that this finding should be related to energy peculiarities of the structure of the tested material.[133,134] Many metals and alloys, rubber, and other materials exhibit similar variations in the homogeneity of strength characteristics not only in compression, but also in tension, bending, and other kinds of the stress state that indicate general character of the discussed finding.[3]

This is natural. For a given type of material, Ω = const. Tendency to an increase in resistance, which requires growth of $d\mathbf{P}$, inevitably leads to a decrease in deformability. (In this case, dV decreases.) This is most clearly proved by the variations of parameters S_d and S_h (lower field in Figure 3.28).

The correlation of the mean value of strength of concrete, σ and the standard deviation S for the compositions investigated is shown in Figure 3.30.[111] A straight line, $S = k_1\sigma$, where k_1 is the coefficient whose value was equal to 0.05, that passes through the origin of coordinates can approximate it. This expression is in good agreement with conclusions made by other investigators. For example, according to the data of Nevill, coefficient k_1 is equal to 0.04 and, according to the data of Ruch,[140] it is equal to 0.05. For normal distribution of the strength, dependence $S = f(\sigma)$ enables determination of the standard deviation using the known mean value found by testing a small

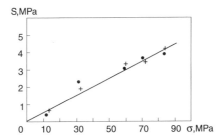

FIGURE 3.30
Dependence of standard distribution S of compression strength of concrete on its mean value σ.

number of reference specimens. This dependence is valid for low- and high strength zones.

Variation of the strength of concrete has been the subject of many studies,[133–142] mostly because of the necessity to improve the resistance regulation method or arrange statistical monitoring of the quality of products. Their results show that homogeneity of concrete is of exceptional importance for reliability and durability of structures.[137,138] However, the root cause of the formation of these properties and objective relationships between them have not been established yet. The previously mentioned studies were conducted using methodologies of passive noncontrollable experiment, i.e., by processing records of laboratory tests of standard specimens. This information characterizes the strength variation at the macroscopic level and is too subjective; therefore, in principle, it does not allow revealing the submicroscopic mechanism of formation of strength and deformability properties of materials in general and of concrete in particular.

Apparently, an insufficient number of publications are dedicated to investigation of microhomogeneity of structures. No information on interrelation of variability of structural–mechanical parameters at different scale levels is available either. Desov and Malinovsky,[138] Ruch,[140] and Conrad[141] studied the relationship between the standard deviation S and the mean strength of concrete, σ; regions of substantial data spread are reported. For example, Figure 3.31 shows dependence $S = f(\sigma)$ according to the data of NIIZhB;[138] no close relationship between these values was observed although there is a clearly defined trend to an increase in S with an increase in strength. The angular coefficient of the approximating straight line is equal to 0.57.

Figure 3.32 shows the results of statistical processing of 196 series of compression tests of cubic specimens of heavy concrete with an edge size of 20 cm made for research purposes. Depending on the particular purpose of the main experiment, the number of specimens of the same composition (each point in the field of the figure) varied from 3 to 12. In total, 2964 single tests were processed; the mean strength varied from 6.5 to 110 MPa. Super-high, early-strength cements were used to manufacture high-strength concretes for the specimens.

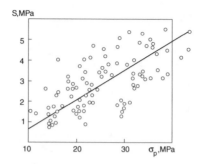

FIGURE 3.31
Dependence of standard deviation of compression strength of concrete according to the data of the Research Institute for Reinforced Concrete.

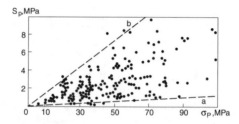

FIGURE 3.32
Variations in the standard short-time compression strength of concrete, depending on its mean value.

Analysis of Figure 3.32 allows the following conclusions: 95% of all experimental points lie between boundary straight lines *oa* and *ob*, whose angular coefficients are equal to 0.012 and 0.148, respectively. The stable trend is observed as increase in the standard deviation with an increase in the mean strength. In fact, the angle between the boundary straight lines indicates a direction and character of transformation of differential curves of distribution of the compressive strength with a distance to the high-strength zone (see Figure 3.29). Existing technology arranges the structure of concrete (as well as that of any other material) in such a way that a tendency to achieve the maximum possible value of strength unavoidably leads to an increase in the range of variation of this parameter at the macrolevel and, hence, to an increase in the probability of failure of the structure, especially under dynamic loading conditions.

As a useful conclusion, we can point to the direct analogy of this result with the temperature dependence of the Ω-potential (Figure 2.24). Indeed, one of the parameters of the Ω-potential, resistance **P**, can be increased using technological methods only through increasing parameter N in Equation 2.96. As such, $\Omega = f(N;\delta)$ at $T = $ const. As the Ω-potential is directly proportional to N, as well as to T, expression $\Omega = f(N;\delta)$ repeats the external form of function $\Omega = f(T;\delta)$ shown in Figure 2.24.

Now we can explain causes of the discussed behavior of mechanical properties of materials. The condition of equivalence (2.105) allows the equation of state (2.96) to be written as follows:[24]

$$\Omega = R\left(\frac{V}{V_\theta}\right)^3,$$ (3.53)

where $R = kNT$ is constant under the above specified conditions, and V_θ and V are the random microvolumes of a material at the Debye and any other local temperature. The Ω-potential (Equation 1.111) is a two-parameter function characterizing load-resisting, **P**, and deformation, V, properties of a solid. This also allows the right-hand side of Equation 3.53 to be represented as $\Omega = R(\delta_h \delta_d)$, where

$$\delta_h = \left(\frac{V_c}{V_\theta}\right) \quad \text{and} \quad \delta_d = \left(\frac{V_d}{V_\theta}\right)$$

are the elastic and plastic parts of the MB distribution. Converting this result into indirect indicators of hardness, we can write $\Omega = R[hd]$, where $h = \delta_h$ and $d = \delta_d$ are the statistical sets of elastic and plastic characteristics of hardness. As seen, they should repeat the inverse-proportional form of the relationship existing between parameters **P** and V of the Ω-potential.

Figure 3.33 shows how the elastic and plastic parts of response of a material to a local mechanical effect with constant energy are distributed depending on the mean compressive strength of concrete from 10 to 80 MPa.[127] The abilities of a material to resist such effects h and deform under their impact d are inversely proportional. Equation 3.53 predicts this form of the relationship.

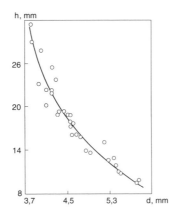

FIGURE 3.33
Redistribution of elastic, h, and plastic, d, parts of hardness indicators of concrete having mean strength ranging from 10 to 80 MPa.

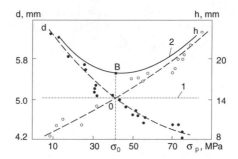

FIGURE 3.34
Formation of the Ω-potential in concretes of different strength.

Figure 3.34 shows the result of changes in elastic h and plastic d indicators of hardness with an increase in strength of concrete from 10 to 80 MPa. It can be seen that, at a constant number of interatomic bonds (V = const and N = const), an increase in strength occurs due to a continuous growth of the number of rigid bonds (h increases) and the corresponding decrease in plastic bonds (d decreases). Horizontal line 1 expresses equality Ω = const under the specified conditions. Curve 2 with a maximum at point B is the result of functions h and d and shows how parameters of the Ω-potential change within the strength range investigated, if Ω = const. As seen, within a zone of $0 - \sigma_0$, a decrease in resistance ($h \rightarrow 0$) is accompanied by an increase in deformability ($V \rightarrow \infty$). The trend observed in the $\sigma_0 - \sigma_{max}$ zone is reversed: an increase in h inevitably causes a decrease in d. There is a narrow zone (near vertical line $\sigma_0 B$) where a material has a relatively high load-carrying capacity $\mathbf{P}$ and sufficient deformability V; for concretes, it is equal to 40 MPa.

According to existing opinion, assignment of a maximum value of strength of concrete at the best homogeneity of its structure automatically ensures the required level of reliability and durability of reinforced concrete structures;[32] however, this is not true. As follows from Figure 3.34, the highest level of reliability and durability can be achieved only at a strictly specified combination of strength and deformability parameters. The material has sufficient resistance and good damping ability that prevent cracking only in the zone of 40 MPa.[14,17]

The inverse proportional form of relationship between resistance of a body $\mathbf{P}$ and its deformability V is determined by the nature of the Ω-potential (Equation 1.111) and reflected in the equation of state (Equation 2.65). It is embedded into the AM structure (see Figure 2.6 and Equation 1.53) and shows up in observations of behavior of all nonmetallic (Figure 3.34) and metallic (Figure 2.38) materials.

The most probable mechanism of transformation of the MB energy distribution of rotoses of concrete (using, for example, its dilaton phase) that controls strength and deformability (Figure 3.35) can be revealed using the results of analysis of the discussed hardness tests. In the region of low strength (up to a value of σ_0 = 40 MPa), in accordance with data given in

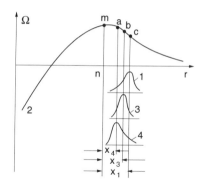

FIGURE 3.35
Transformation of MB distribution with variations in strength of concrete.

Figures 3.27 to 3.29 and 3.33, as well as in Table 3.2, it has the form of curve 1 with left-side asymmetry. As the strength increases (movement to the right along the axis of abscissas in Figure 3.34), the MB distribution moves to the left and takes the form of curve 4 with right-side asymmetry. The symmetrical shape of curve 3 in Figure 3.35 corresponds to point σ_0 in Figure 3.34. The apex of the MB distribution has locus *cba* on curve 2 of internal forces. Along the axis of abscissas, which determines sizes of local submicrostrains, this apex successively passes a path of x_1 to x_3 to x_4 approaching the Debye temperature (point *m*). To simplify our considerations, the mechanism of formation of the MB factor for concrete is shown in the zone of deceleration barrier cbq_1 and not in acceleration well q_2dc (Figure 1.14).

The physical meaning of the experimental dependence shown in Figure 3.33 is a macroscopic representation of dilaton part *mabc* of curve 2 in Figure 3.35. The shape of the MB distribution (curves 1, 3, and 4; the real representatives of which, on a macroscale, are the empirical dependencies shown in Figure 3.27) and its location on the axis of abscissas with respect to critical vertical line *mn* have a decisive effect on the process of deformation and fracture, i.e., on the structure of the Ω-potential.

To destroy a material characterized by type 1 of the MB distribution (a process that begins when its asymmetrical end reaches vertical line *mn* and ends at the moment when its apex approaches this line), it is necessary to pass path x_1, i.e., to accompany this process with substantial deformation. This way, materials located to the left from point σ_0 (Figure 3.34) react to loading. Such materials are characterized by a high degree of deformability (region *od* in curve *d* of this figure); because they have high deformability, in principle they cannot have substantial strength.

This conclusion follows from the fact that, according to the equation of state (2.96), materials whose structure is composed of the same chemical elements (ensured by θ = const) and made using the same technology (N = const) at T = const and V = const have Ω = const. This means that curves 1, 3, and 4 limit an equal area (Figure 3.35). The left-sided asymmetry of curve 1 increases the number of interatomic bonds with decreased resistance, at the same time lowering its maximum. The location (height) of the maximum (at the level of vertical lines passing through points *a*, *b*, and *c*) is responsible for the strength.

High-strength concretes exhibit the opposite character; in Figure 3.34, they are located more to the right from point σ_0 and correspond to distribution 4 in Figure 3.35. At high strength, they have much lower deformability, as can be seen from inequality $x_4 < x_1$ following from Figure 3.35 and from smoothing of the plastic part of the indicator of hardness in this zone (tendency of curve *d* to axis of abscissas to the right from point 0 in Figure 3.34). An increase in strength is ensured by a right-sided asymmetry of curve 4, which is indicated by an ascending branch of the elastic part of the indicator of hardness (curve *h* in Figure 3.34).

The mode of fracture of a material depends on the steepness of the left front of the MB distribution (curves 1, 3, and 4 in Figure 3.35). If the front is gently sloping (curve 1), fracture starts from a small fraction of bonds randomly dissipated in the volume in accordance with Equation 2.3. Nucleated simultaneously at many centers and propagated from them during deformation, cracks form a developed fracture surface. High-strength concretes, in contrast, enter the critical zones with a steep front (curve 4). This means that the process of fracture of these concretes is like an avalanche and thus the crack initiation centers are initially located close to each other. Although in low-strength concretes it takes time for the cracks to propagate from one center to another, in high-strength concretes the process of their coalescence takes place almost immediately. For this reason the former fail by ductile fracture[48] and the latter by brittle fracture.[142]

It is likely that transition from one type of fracture to the other should be accompanied by a corresponding change in the total fracture surface. This has experimental confirmation (Figure 3.36) and, at the same time, proves the validity of developed concepts of deformation and fracture. Curve $S = f(\sigma)$ (Figure 3.36) was plotted using the results of measurement of the actual fracture surface, determined by pulling inserted anchors of standard

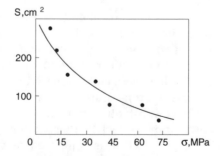

FIGURE 3.36
Variations in the total fracture surface of concretes of different compositions.

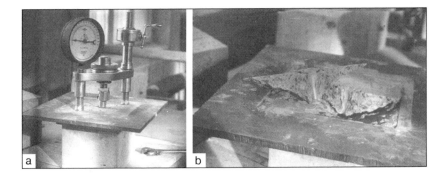

FIGURE 3.37
Pulling out of inserted anchors of standard sizes (a) for the determination of the total fracture surface (b).

sizes and shapes out of different compositions of concretes.[130] As might be expected, an increase in the compressive strength is accompanied by a decrease in the surface area of a cleavage cone (Figure 3.37).

Comprehensive theoretical (Chapter 2) and experimental studies (especially analysis of the data shown in Figure 3.34) indicate that, over the entire range of variations in the strength of concrete, the Ω-potential remains constant (horizontal line 1). Variations in the MB distribution (shown in Figures 3.27 and 3.35), occurring under the effect of variations in the quantitative composition of similar types of initial components and technological processing, cause a redistribution of its load-resistance component **P** (in Figure 3.34, represented by the elastic part of the indicator of hardness — curve *h*) and deformability component *V* (represented by the plastic part — curve *d*), leading to an inversely proportional dependence (Figure 3.33). Envelope 2 has the shape of a curve with a minimum (point *B*) and indicates mutual variations in elastic and plastic properties of a material (Figure 3.34).

The optimum proportion between strength and deformability characteristics is observed at this point. Deformability dominates in the low-strength zone and load-resistance dominates in the high-strength zone characteristics. This conclusion follows from the equation of state (2.96). Hence, it applies to all solids, not only to concretes, determining viability of a material under different loading conditions. Apparently, the highest efficiency under static and slowly changing loads will be exhibited by high-strength concretes.[143] Under dynamic conditions, the highest reliability should be expected from concretes with an optimal strength ($\sigma_0 = 40$ MPa) because they have a maximum damping ability at a sufficient strength.

The existence of optimum proportion between strength *E* and deformability ε characteristics of concrete is evidenced also by the shape of the relationship between its ultimate deformability and short-time strength σ (Figure 3.38) presented by Sytnik and Ivanov.[113] A similar relationship can be found in the study by Vasiliev.[144] Such a relationship is fundamental and holds for

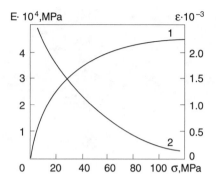

FIGURE 3.38
Variations in elasticity modulus 1 and ultimate tensile strength 2 of concrete.

TABLE 3.3

Parameters of Distribution of Short-Time Strength of Factory Compositions of Concrete

Concrete Grade	Designation	Quantity of Samples n, pcs.	Mean Strength σ, MPa	Standard Deviation S, MPa	Asymmetry, A
200	1	450	21.0	5.0	0.23
300	2	285	30.6	5.8	0.19
500	3	112	49.8	7.3	0.14

concretes and also for metals. Sufficient evidence of this fact can be found in Troshenko.[3]

The principles established were used to reveal causes of failures of parts produced by prefabricated reinforced concrete plants. In the mid-1980s, products manufactured in one of these plants were very frequently fractured. These products were manufactured from concrete of three design grades: 200, 300, and 600. Cubic specimens with an edge size of 20 cm were made from the commercial grades of concrete, at least nine pieces of each grade. Sampling of formation of the products was made at the stations. In addition, results of laboratory tests of the reference cubes (normally with an edge size of 10 cm, and more rarely with an edge size of 15 cm) were statistically processed. The representative sampling was formed for a period of 1.5 months, covering 6 years that preceded the tests. Specimens with an edge size of 20 cm were used for measurement of the indicators of hardness and after that they were tested (for compression) using a standard method. Results of the measurement of hardness were processed and analyzed using the preceding procedure. Table 3.3 and Figures 3.39 and 3.40 give results of statistical processing of the plant tests of cubic samples with an age of 28 days.

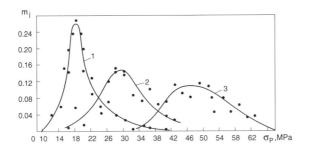

FIGURE 3.39
Empiric distributions of short-time compression strength of concrete of commercial grades 200 (1), 300 (2), and 500 (3).

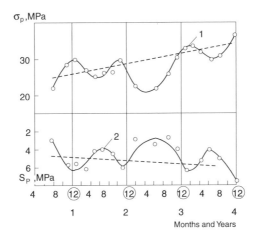

FIGURE 3.40
Analysis of consistency of the technological process of preparation of the concrete mixture of grade 300.

Figure 3.39 shows that statistical distributions of strength for all grades of concrete have a right-sided asymmetry, which predetermines brittle failure of the material, according to comments made for Figure 3.36. Figure 3.40 shows stable seasonal variations in statistical indicators σ_p and S_p. These variations occur in one phase: an increase in σ_p unavoidably leads to an increase in S_p, and vice versa (curves 1 and 2). Also, σ_p and S_p were found to be higher for parts made during winter months than for those made in summer months. For the parts made in autumn and spring, these parameters have intermediate values.

A characteristic feature is that seasonal variations do not occur in design grade 300, but relative to a straight line that tends to increase (dotted line). The obtained data show that the manufacturing process at the considered plant is inconsistent in time and that the plant has no flexible readjustment system. The combined effect of continuous growth of the mean strength (curve 1 in Figure 3.40), seasonal variations of the standard deviation (curve 2), and presence of the right-sided asymmetry of empirical distributions (Figure 3.39) lead to failure of reinforced concrete parts at stages of transportation and installation, and at an early period of service.

The results of analysis of the extent to which the MB distribution affects the processes of deformation and resistance of materials to external fields allow development of technological approaches to the control of mechanical properties. Their description can be found in the following chapters. Although no similar investigations for metals have been conducted, nothing indicates any substantial deviations from the conclusions made in this section.

3.8 Dependence of Mechanical Properties on the Packing Density of a Structure

Section 2.8 shows that the arrangement of different chemical elements in a structure changes density and affects the value of the Ω-potential. In a formed body, the number of interatomic bonds does not remain unchanged; it may increase due to continuation of the structurization processes or decrease under the effect of external factors.

Comparison of Equations 2.36, 2.51, and 2.52 allows us to write

$$\Omega = G = \omega N. \tag{3.54}$$

If the ω-potential determines the shape of the curve of potential energy of an individual rotos (Figure 1.14), then the G-function characterizes its average value for the entire set N in a given volume V. Figure 3.41 shows

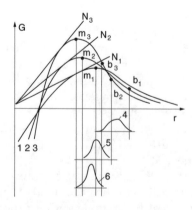

FIGURE 3.41
Dependence of the G-potential on the number of interatomic bonds N in volume V.

formation of the interaction function G, depending on the number of rotoses N arranged in volume V. Because $G = \omega N$, an increase in the concentration of the bonds N (Figure 3.41, $N_1 < N_2 < N_3$) is accompanied by an increase in steepness of the ω-potential. Its shape changes in the direction from position 1 to position 3.

An increase in the steepness of the ω-potential causes a decrease in the spread of the compresson and dilaton phases. This follows from a decrease in the distance between vertical lines passing through the following pairs of the points: $m_1 - b_1$, $m_2 b_2$, and $m_3 b_3$. A decrease in the spread of the D-phase is accompanied by transformation of the MB distribution in the direction of 4 to 5 to 6 from wide in position 4 to acute apex shape in position 6. This variation of the ω-potential (curves 1, 2, and 3) should lead to a rise in the mean strength of a material and transformation of the MB distribution in the direction of 4 to 6 to an increase in its spread.

How can we check this statement? To do this, consider a body in which the number of interatomic bonds varies with time. Let its resistance at the moment of time τ_0 be equal to $\mathbf{P}_0$. Solidification or destruction changes $\mathbf{P}_0$ and by moment τ it becomes equal to $\mathbf{P}$. Designate the number of rotoses that formed or failed per unit time as η. Then $\tau_1 = 1/\eta$ expresses the period of time during which the resistance changes by a dimensional unit (e.g., 1 MPa). The number of bonds that cause the change of $d\mathbf{P}$ in resistance is equal to

$$\frac{1}{\eta} d\mathbf{P}.$$

The number of bonds that formed again (or failed) during time $d\tau$ is proportional to time τ, i.e.,

$$d\tau = \frac{1}{\eta} \tau d\mathbf{P}.$$

By dividing variables

$$\frac{d\tau}{\tau} = \frac{1}{\eta} d\mathbf{P},$$

integrating, and accounting for the initial conditions, we find that

$$\mathbf{P} = \mathbf{P}_0 \pm \eta \ln \frac{\tau}{\tau_0}.$$

Assuming (accounting for Equation 3.54) that $P_0 = \omega \rho_0$, where $\rho_0 = N_0/V$, we can write

$$\mathbf{P} = \omega \rho_0 \pm \eta \ln \frac{\tau}{\tau_0},$$

where the plus sign refers to structurization and minus to destruction processes. The last formula expresses the law of strengthening on solidification or loss in strength under the effect of external factors. We can express it as

$$\sigma = \sigma_0 \pm \eta \ln \frac{\tau}{\tau_0},$$ (3.55)

where σ_0 is the initial value and σ is the current value of strength (in compression or tension). Equation 3.55 allows experimental verification of the effect of density (the number of interatomic bonds per unit volume of a material) on the resistance to external effects.

Concrete is a unique investigational object for these purposes because the process of its solidification can be expanded in time; it is a heterogeneous conglomerate with structure parameters varying with time. A variable element of the structure of concrete is a crystalline agglomerate of new hydrate formations.[145] It develops due to nonhydrated cement grains, which are conserved in the bulk of a material at the manufacturing stage to support structurization processes, thus increasing N.

The results of systematic investigations into the principles of time variation of strength of conventional and high-strength concretes with more than 40 compositions of concrete mixture are now considered.[146–148] Cubic specimens and prisms of standard sizes were made of each composition of concrete. Aged from 3 days to 6 years, these specimens were tested to compression and tension. Elastic and plastic indicators of hardness were measured prior to the tests. No fewer than nine samples of each age were subjected to the tests.

A total of 1300 samples were made of all compositions and tested during this period. Some were solidified under natural conditions while others were subjected to heat treatment. Main statistical characteristics of parameters studied were calculated from the test results. Figure 3.42 shows experimental dependencies of the family of those defined by Equation 3.55.[146–148] Curves 1 to 5 were plotted for concretes prepared on the base of super-high, early-strength cements of grade 800 and the rest on the base of cements of grades 500 to 600.

Analysis of experimental data shows that an increment in strength and its statistical spread in hydration of residual cement grains under natural temperature–humidity conditions are really described by the function given by Equation 3.55. The structurization processes obeyed the logarithmic law also at early stages of solidification (Figure 3.43).[149] This proves the order of formation of a structure of solids (Figure 3.41) predicted by the equation of state (3.54).

Note that the same logarithmic law (Equation 3.55) governs the structurization and mechanical-destruction processes, linking the processes of increase in strength (solidification) and its exhaustion (durability). These should be regarded as two sides of one and the same objective principle governed by the MB statistics (Equation 2.3). This has identical

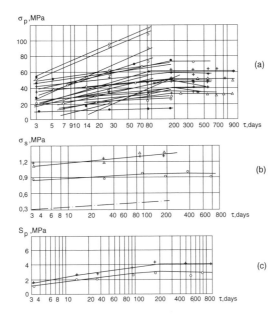

FIGURE 3.42
Logarithmic dependence of the mean strength of concrete in compression (a) and tension (b), following the diagram in Figure 3.37, and standard deviations (c).

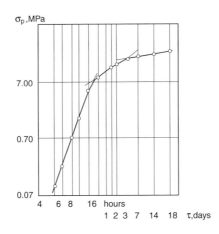

FIGURE 3.43
Change of strength with time at early stages of solidification of concrete.

logarithmic expression describing increase and loss in strength of metallic[10] and nonmetallic (Figures 3.42 to 3.43) materials.

Potentiality of conserved cement grains is not unlimited. As their hydration occurs, the structurization processes attenuate. Figure 3.44 shows dependence of hydration of cement $g(\tau)$ on the age of concrete τ according to Gansen.[149] As seen, hydration activity of both types of cement (1 and 2)

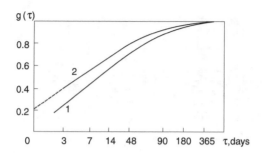

FIGURE 3.44
Dependence of the degree of hydration of Portland (1) and fast-hardening (2) cements on the age of concrete.

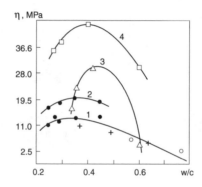

FIGURE 3.45
Dependence of parameter η in Equation 3.55 depending on the water–cement ratio, cement activity, and initial solidification conditions: 1: heat-humid treatment (HT) and AC 50 MPa; 2: natural solidification conditions (SC) and AC, 50 MPa; 3: SC and AC, 60 MPa; 4: SC and AC, 85 MPa.

becomes almost completely exhausted after 180 days. Before that time a continuous increase occurs in the short-time strength. The angle of inclination of the curves in this time interval determines the value of parameter η in Equation 3.55. The value of η depends on the water–cement ratio, cement activity, and initial solidification conditions (shown in Figure 3.45). As seen, any curve $\eta = f$ (water–cement ratio, cement activity, initial solidification conditions) has a maximum. Position of this maximum on the axis of abscissas depends on the cement activity, i.e., the higher the cement activity, the higher the values to which this maximum corresponds.

Deviation of water–cement ratio to both sides from the maximum is accompanied by a dramatic decrease in η, attributable to the fact that a decrease in water–cement ratio leads to consolidation of structure and, hence, to deterioration of conditions of migration of moisture in the bulk of the material and its access to nonhydrated cement grains. As the water–cement ratio increases, the overwhelming majority of cement grains enter into hydration

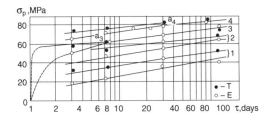

FIGURE 3.46
Effect of solidification conditions on the process of increase in strength of concretes of the following water–cement ratios: 1, 0.77; 2, 0.57; 3, 0.32; 4, 0.26.

reaction during setting and concrete intensively picks up strength at early stages of solidification. During subsequent periods, the hydration processes slow considerably. At the same water–cement ratio and cement activity, the value of η depends on initial temperature–humidity conditions. In concretes subjected to any heat–humid treatment, η is lower than in those solidified under natural conditions (curves 1 and 2 in Figure 3.45) because most cement particles experience the hydration reaction during a heat–humid treatment.

Figure 3.46 shows the effect of solidification conditions on the process of increase in strength for four compositions of concrete differing in water–cement ratio. Some specimens were solidified under natural conditions (circles and letter E) and others were subjected to heat treatment (solid points and letter T). Intensification of the process of hydration of cement at the initial period leads to the fact that $\sigma_T > \sigma_E$ up to a certain age of any composition of concrete. This difference gradually levels until equality $\sigma_T = \sigma_E$ is reached. The lower the water–cement ratio, the earlier the moment of leveling of the intensity of the structurization processes. For example, for water–cement ratio = 0.26 to 0.32, it begins after 7 to 28 days, whereas for water–cement ratio > 0.32, the difference between σ_T and σ_E was observed up to 180 days.

Therefore, the resistance of reinforced concrete structures to external effects differs in different periods of their life (storage, transportation, installation, and service). It changes because a part of cement is conserved in the structure and also depending on the temperature–humidity conditions of solidification of concrete. At the initial period, it is higher for the structures subjected to heat–humid treatment, while during subsequent periods it is higher for those structures that gained their strength under natural conditions.

For this reason, the processes of internal structure stabilization occurring under stresses (during transportation and after installation) lead to a mismatch of durability. This results in premature failures occurring at early periods of service and, hence, to a growth of cost of repair of the entire building structure. To ensure the durability of a building constructed from structures fabricated at different periods, it is necessary to differentiate their stressed state depending on the age of concrete. Arrangements and

TABLE 3.4

Values of Parameter η for Compressive Strength of Concrete

Strength at Age of 28 days, σ, MPa	10–18	20–28	30–38	40–48	50–90
η_s	4.3	12.8	16.3	15.2	12.1
η_s	0.51	2.27	2.53	3.01	3.77

TABLE 3.5

Parameters of Distribution of Plastic Deformations, d, and Elastic Recoil of a Striker, h, Obtained in Solidification of Grade 300 Concrete

Age, days	n_d, pcs	d, mm	S_d, mm	A_d	n_h, pcs	h, mm	S_h, mm	A_h
3	220	5.73	0.33	0.057	—	—	—	—
28	280	5.26	0.28	0.053	130	26.2	3.29	0.125
90	260	4.90	0.25	0.051	119	30.7	2.34	0.076
180	260	4.70	0.18	0.038	111	32.8	1.59	0.048

coordination of the time of manufacturing of structures and priorities of their assembly in the framework of a building should decrease operating costs.[128]

Values of parameter η_σ for the mean value of the compressive strength and the standard deviation η_s calculated from the results of tests of 42 compositions of concrete are given in Table 3.4.[148]

For an entire period of investigation of the structurization processes, compressive (Figure 3.19b,c) or tensile (Figure 3.37) tests of the specimens were preceded by measurements of indirect indicators of hardness by an integrated elasto-plastic method.[125] Figure 3.47 shows the time dependence of empirical distributions of the indirect hardness indicators for concrete of grade 300. The major parameters of distributions (statistical mean h and d, standard deviations S_h and S_d, and asymmetry A_h and A_d) as well as other information required for identification and analysis are given in Table 3.5.[147]

The actual character of distribution of energy in microvolumes of a material during solidification (Figure 3.47) coincides with that predicted by the equation of state (Figure 3.41 and positions 8 to 10 in Figure 1.14). It follows from comparison of Figures 3.27 and 3.47 that the mechanism of formation of the Ω-potential under the effect of technological factors and that caused by natural hydration of cement particles conserved in the volume are the same. They are governed by the quantity of interatomic bonds N and by the MB distribution of the internal energy between them. This is evidenced by the manner of variation in statistical parameters h and d. Distribution of h shifts to the zone of increasing values of elastic rebound (apex h moves to the right along axis h), and its standard deviation S_h decreases (Figure 3.47 and Table 3.5). This means that the reference pulse of a tentative load is taken up by an increasing part of more rigid and elastic bonds, i.e., the left end of distribution G, while moving in Figure 3.41 to the left in the direction of 4, 5, and 6 increases its steepness.

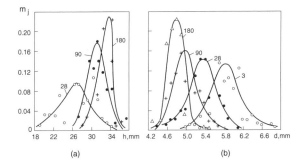

(a)   (b)

FIGURE 3.47
Empiric distribution of indirect indicators of hardness: elastic rebound of indenter h and plastic deformations d, obtained on concrete of grade 300 during solidification (age in days is indicated by numbers over the plots).

Movement of distribution d toward lower values of residual strains with a simultaneous decrease in its spread indicates that it is accompanied by a decreasing number of elementary bonds. The opposite variations in hardness indicators point to an increase in homogeneity of the structure of concrete at the microscopic level. As discussed earlier, it has significant influence on the processes of fracture of specimens and structures, particularly in early periods of loading, because of simultaneous formation of a large number of pre-fracture centers and thus propagation of an avalanche crack.

Scillard[121] presented data on the technical state of prestressed reinforced concrete structures at an early period of service of engineering objects in the U.S., Canada, and in a number of Pacific and Far East countries. He noted that the total number of defects (from minor to catastrophic) approached 10% of the total number of structures examined. According to the estimate made by the American Institute for Reinforced Concrete, the number of cases of serious defects reaches 1%. The same applies to early periods of service of metal structures,[3,43] thus indicating the versatility of the effect considered. Understanding its nature allows development of preventive methods that are the subject of the following chapters.

Long-time service of structures sometimes leads to a substantial increase in strength at the expense of ductility. (The technical state shifts to the right from the optimal state designated as σ_0 in Figure 3.34.) For example, cases in which concrete serviced for more than 30 years in hydrotechnical structures showed an increase of 3.5 times in the compressive strength, compared with that aged 28 days, have been reported.[150] Results of an investigation of the statistical distribution of hardness indicators on steel samples cut out from a pipe in service for 100,000 hours in the first loop of a nuclear power station[151] show similar behavior.

If we assume hypothetically that the process of hydration of cement conserved in manufacturing continued for 30 years, then, by making elementary calculations using Equation 3.55 and data from Table 3.4, we find that the

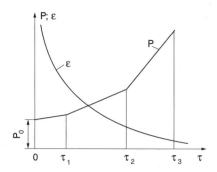

FIGURE 3.48

Diagram of variations in the parameters of the Ω-potential (resistance **P** and deformability ε) of materials during long-time operation in slowly varying external fields.

mean increase in strength for the entire range of all possible compositions of concrete in the most favorable case is equal to no more than 3.1 times, compared with that aged 28 days, but not 3.5 times.

Some of the compositions whose time characteristics are shown in Figure 3.42 solidified under conditions of 100% humidity (i.e., under conditions similar to those described in Gnutov and Osipov[150]), but they exhibited no marked increase in strength after 180 to 200 days of solidification. This means that the hydration process under any solidification conditions (moreover, under high humidity conditions) does end by that time. This conclusion is supported by data from Figure 3.44, while direct proof is given in Table 3.4. Thus, the processes of time variations of the strength of concrete influence other factors, in addition to the hydration factors. They include the orientation effect (see Section 3.4 and Figures 2.7 to 2.9), which leads to an extra increment (on top of the hydration effect) in the strength.

Now we can reconstruct, with a high degree of confidence, the variations in the technical state of a material during long-time service in constant force fields (Figure 3.48). At the initial period of operation, the resistance is estimated by P_0. In time interval of $0 - \tau_1$, the solidification process continues in concrete while the aging processes take place in metals and alloys. The latter are associated with redistribution of energy inside the AM set accompanied by a change in shape of the MB distribution (positions 11 to 15 in Figure 1.14). Occurring in solids, these processes are long-term and associated with adaptation of a structure formed under certain technological conditions to new, i.e., service, conditions. As an example, Figure 3.49 shows the plot of aging (variation of Brinell hardness HB) of aluminum alloy D16T at different temperatures.[152]

Both processes (solidification and aging) lead to increased strength (compare Figures 3.49 and 3.42). Region $\tau_1 - \tau_2$ (Figure 3.48) determines the orientation effect whose unavoidable consequence is an increase in function **P** in the direction of action of an external load. The moment of time τ_2

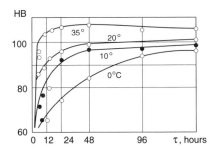

FIGURE 3.49
Variations of hardness of aluminum alloy D16T in aging.

corresponds to the completion of orientation restructuring and the beginning of the destruction process. The latter ends at point τ_3 with fracture of a material.

Experience shows[3] that failure of parts and structures has highest probability at the initial and final periods of service. At this point we could give an exhaustive explanation of this known fact. Failures at the initial period of service occur because of an insufficient resistance of a material, P_0, or excessive deformability, $dV \approx \varepsilon$, whereas those at final period of service are caused by exhaustion of ductility resource. Function P (schematically shown in Figure 3.48) is conditional. To reveal the nature of the phenomena that take place here, we separated individual processes in time. In reality, they occur simultaneously from the beginning of operation to its end. However, this simplification does not distort the general picture and has no effect on the results of the analysis.

3.9 Variation of State in Compression and Tension

While deforming a body according to the diagram shown in Figure 2.8, external fields influence its ability to generate volumetric resistance forces **P** and change its technical state. We can rewrite Equation 2.140 at $\Delta V = 0$ and $T = \text{const}$ in the form of

$$\mathbf{P}\left(\frac{d\mathbf{P}}{\mathbf{P}} + \frac{dV}{V}\right) = kT\rho\delta\left(\frac{dN}{N} + \frac{d\delta}{\delta}\right), \tag{3.56}$$

where

$$\rho = \frac{N(P,dV,\delta)}{V}$$

is the density of the deformed body, which depends on the dipole pressure **P**, the change in volume dV, and the MB factor δ (positions 8 to 15 in Figure 1.14).

Let parameter **P** change only its direction. As such, dipoles can change only their orientation in space but not geometrical dimensions (positions I to IV in Figure 2.8a). This makes phase transitions impossible ($d\delta = 0$), and Equation 3.56 transforms into

$$\mathbf{P}\frac{dV}{V} = kT\rho(P, dV)\delta\frac{dN}{N} . \tag{3.57}$$

Substitution of Equation 2.137 into Equation 3.57 yields an expression describing the variation of the state at the elastic stage of deformation in strength testing of standard specimens in an invariable thermal field

$$\mathbf{P}(\varepsilon + \eta) = kT\rho(P, dV)\delta\frac{dN}{N} . \tag{3.58}$$

Because η is normally not measured in technical tests,

$$\mathbf{P}\varepsilon = kT\rho\delta\frac{dN}{N} . \tag{3.59}$$

Comparison of Equation 3.59 with the empirical Hooke's law Equation (2.139) makes it possible to represent the three-dimensional parameter **P** in a linear case (corresponding to uniaxial tension) as the elasticity modulus E and thus write

$$\sigma = kT\rho\delta\frac{dN}{N} . \tag{3.60}$$

The theoretical analogue of the empirical Hooke's law, describing deformation processes in electric and magnetic fields, is the Langevin formula (Equation 3.27).

Expression 3.56 is an analytical representation of the deformation diagram (Figure 2.10) whose elastic part, having an orientational nature, is described by Equation 3.60. The plastic part of the diagram is determined by a change in the value of parameter P during the process of phase transitions (Figure 1.15) $d\delta/\delta$ occurring in deformation dV/V.

Deformation (as well as resistance) of solids is initially a thermal-activation process (Equation 2.140). The system of notions developed in this book allows identification of an elementary deformation event with an increase in the size of the EM dipole (an increment in sizes of horizontal sections of the potential well in Figure 1.7a at levels t_1, t_2, etc.). As such, the dipoles of brittle materials tend to horizontal line q_2c (Figure 1.14), limited by the critical states (positions 1 and 4 in Figure 1.15), one of which corresponds to cold-shortness temperature T_c. Having moved into this position and received no thermal support for transition into a new stable state, the dipoles disintegrate.

The structure of any solid is made of cold and hot EM dipoles. The former exist near the θ_f-temperature while the latter are grouped near θ_s (Figure 1.14). Positions 8 to 10 and 11 to 15 show different variants of their distribution in the same set N. Parameters of state, $\mathbf{P}$, V, and T, following the law in Equation 2.65, fix them in a specific structure to form entropy s. External fields (force, thermal, radiation, etc.) violate this dynamic equilibrium (Equation 3.19), activating the orientation processes (elastic part of the diagram in Figure 2.10) and phase transitions (its inelastic part). All solids, without any exception, have a dipole structure. The only difference between them is that some can perform phase transitions (metals and alloys), and others (cast irons, stones, ceramics, etc.) cannot.

At $T = $ const a force field changes only the shape of the MB distribution (for ductile materials in the direction of 11 to 12 to 13 in Figure 1.14). Affecting the entire set N, a thermal field at the same time causes its major part to move in the direction of 8 to 9 to 10 (for brittle materials) and 12 to 14 to 15 (for ductile materials).[12,24] Cooling changes positions in the reverse direction. Tension of ductile materials (change in positions in the direction of 11 to 12 to 13) brings the hot end closer to the θ_s-temperature (vertical line *brb*), and the cold end closer to T_c (identified with vertical line *crc*).

In contrast, compression, by increasing steepness of both fronts (the reverse change in positions 13 to 12 to 11), causes its end to move out of the dangerous zones. Periodic compression–tension changes in the stressed state increase the probability that the end regions of the MB factor will exit the critical temperatures, thus activating the process of accumulation of damages in the structure (dashed regions in position 13) and initiating fatigue of materials.[24]

If a body of dimensions $V = SL$ experiences a uniaxial σ stressed state, and this stress affects area S in direction L, the Ω-potential can be written as

$$\Omega = \frac{F}{S}V = FL = \frac{dT}{dr}NL. \tag{3.61}$$

As follows, all members of set N perform resistance work due to formation of the temperature gradient. Those whose location r coincides with direction L exert resistance through changing their size and shape (Figure 2.8b), while the rest are subjected to the action of moment $M(\sigma)$, which causes their reorientation in the direction L (Figure 2.9).

Consider a body in an external uniaxial force field with intensity σ. In this case, the equation of interaction (Equation 3.19) takes the form

$$V(\mathbf{P} \pm \sigma) = Ts(P)_T. \tag{3.62}$$

The plus sign in front of σ relates to compression and the minus sign to tension. It is seen that, at $T = \text{const}$, the external force field, while changing entropy $s(P)_T$, violates the initial order in the AM system according to Equation 3.61. How can we check that this change takes place in reality?

It follows from Equation 2.18 that, under isothermal conditions, the force field performs the work over a solid through changing the logarithm of the state function Z. Substitution of Equation 2.17 into Equation 3.62 yields

$$\Omega(\sigma) = kT \ln Z, \tag{3.63}$$

where $\Omega(\sigma) = V(\mathbf{P} \pm \sigma)$ is the value of the Ω-potential in external field σ. Comparing Equation 3.63 with Equation 3.53 yields

$$\Omega(\sigma) = N[hd]. \tag{3.64}$$

Expression 3.64 shows that a change of the Ω-potential in the force field finds reflection in the domain of the indirect indicators of hardness. This finding allows experimental investigation of Function 3.62.

Cubic samples with a 20-cm edge (Figure 3.19c) and short prismatic concrete columns measuring $20 \times 30 \times 120$ and $20 \times 30 \times 150$ cm (Figure 3.19b) were used as specimens in our experimental study. Two kinds of stressed–deformed state were created in the specimens: compressive and tensile. The former was achieved by central compression of the specimens while the latter was achieved by their off-center compression. The load was applied stepwise up to fracture. Four batches of column and cubic specimens of four different compositions of concrete with design strength of 10, 30, 60, and 80 MPa were tested.

At each step, the process of loading was accompanied by measurements of the indicators of hardness, which were determined by the integrated elastoplastic method.[125] The number of measurements on each specimen was in the range from 60 to 110. The results thus obtained were processed to calculate the quantitative estimates of statistical parameters of distribution of hardness.

Figure 3.50 shows the variation in the mean value of elastic rebound of the indenter. In this picture: h = curve 1, standard deviation S = curve 2, and asymmetry A = curve 3 in tension (a) and compression (b) of concrete. In tension, the longitudinal strains ε and, in compression, relative levels of compressive stressed state

$$\alpha_p = \frac{\sigma}{\sigma_p},$$

where σ is the current value of external stresses and σ_p is the maximum value corresponding to fracture, were plotted on the axis of abscissas.

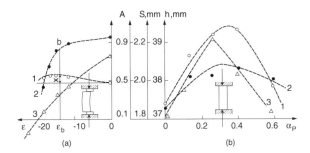

FIGURE 3.50
Variations in the parameters of statistic distribution of elastic hardness indicator of concrete in tension (a) and compression (b).

As seen from Figure 3.50, the transformation of distributions has a fundamentally different character in tension and compression. Note that in compression all the dependencies have maximum near $\alpha_p = 0.3 \div 0.4$, whereas in tension all of them continuously decrease (to be exact, for h this trend is expressed very slightly). The shape of curve 1 is indicative of power and depth of the processes of redistribution of the internal energy in structure of a material under loading.

These processes have two phases in compression. The first is associated with the orientation effect, occupying a range of the stressed state from $\alpha_p = 0$ to $\alpha_p = 0.3$ to 0.4. The second describes development of mechanical–destructive decomposition of structure under the effect of transverse deformation. Here the manner of variations of all parameters repeats that taking place in longitudinal tension (Figure 3.50a). This phase is located in a range of compressive stressed state from $\alpha_p = 0.3 \div 0.4$ to $\alpha_p = 1$.

These data allow the internal state of a material to be described in the domain of the indirect indicators of hardness.[54,132] Figure 3.51 shows transformation of the MB distribution of rotoses of concrete in tension, (a), and in compression, (b). As in Figure 1.14, region 2 of the constitutional diagram, which determines the ability of rotoses to generate volume resistance forces **P** counteracting external one-, two- or three-axial stresses σ, is shown in its upper part. Let the MB distribution of concrete have form 1 (Figure 3.51b) in an unstressed state (distribution 14 in Figure 1.14 corresponds to unstressed state for ductile materials). For convenience, the locations (in the D-zone) of these distributions in tension and compression are separated: point *a* in curve 2 corresponds to the first kind of stressed state while point *c* corresponds to the second kind.

As seen, in tension (Figure 3.51a), the position of maximum distribution (line *ab*) remains almost fixed with respect to the θ_f-temperature (vertical line *mn*) (positions 11 to 12 to 13 for ductile materials, Figure 1.14), whereas in compression (Figure 3.51b) the maximum of MB distribution successively takes positions from 1 to 5 (corresponding to an increase in the stressed level

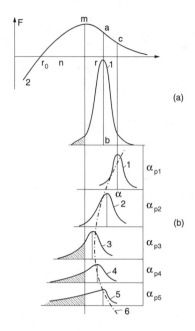

FIGURE 3.51

Representation of the stressed–deformed state of concrete in space of indirect indicators of hardness in tension (a) and compression (b).

from α_{p_1} to α_{p_5}), describing dashed–dotted locus 6 in the energy domain. Its analogue in the domain of the indirect measurements of hardness is curve 1 in Figure 3.50b. For ductile materials it successively takes positions 13 to 12 to 11 (Figure 1.14).

Increment in function h in tension by a value of ε_b (shown in Figure 3.50a by length x) indicates an insignificant shift of vertical line ab to the side of the critical zone mn due to orientation of rotoses in the direction of the external force field σ_s. This shift, however, is sufficient for the end of the left branch of distribution 1 to be in the C-phase (dashed zone). Rotoses of brittle materials are incapable of performing CD transitions (see Section 2.6). This means that internal dilaton pressure **P**, summing up with external tensile stress $(-\sigma_s)$, leads to an immediate failure of bonds located in the dashed zone and, as a consequence, to the nucleation of numerous prefracture centers in the bulk of a material.

This hypothesis is supported by a small decrease in the standard deviation S (curve 2) and asymmetry A (curve 3) to the right from vertical line $b\varepsilon_b$. Subsequent deformation (to the left from vertical line $b\varepsilon_b$) causes a catastrophic decrease in spread of distribution of S and its asymmetry A (dashed zone increases to cause failure of an increasingly large number of rotoses located in the left part of distribution 1). The fracture process develops at an increased rate due to a progressive increase in tensile dilaton pressure, which is a consequence of the orientation process (Figure 2.8a). In this case, the deformation diagram (Figure 2.10) includes only elastic region *oa*. Line 1 in Figure 3.50a is its reflection in the domain of the indirect indicators of hardness.

The compressive stressed–deformed state is formed following a more complicated law. At the beginning, the compressive forces cause a decrease in distance *r* between all the AM particles. This shifts the MB distribution to the left. (In Figure 3.51b, position 1 changes into position 2.) The left-side asymmetry *A* causes an increase in spread of *S*. This is indicated by an increase in all parameters of distribution of hardness (Figure 3.50b) at the initial stages of loading. A small escape of the left end of the distribution to the C-phase (position 2 in Figure 3.51b) cannot cause multiple disintegrations of the *D*-rotoses because this is prevented by the external compressive pressure $(+\sigma_p)$.

The process of reorientation of rotoses in the transverse direction takes place simultaneously with compacting along the compression axis. It ends when the compressive stressed state level reaches a value equal to $\alpha_p = 0.3 \div 0.4$. By this moment, the MB distribution takes position 3 (Figure 3.51b), characterized by the escape of a substantial part of the distribution into the C-phase. With further increase in the compressive stress level, the reoriented rotoses start deforming in the transverse direction, which initiates the destruction process whose mechanism is described in analysis of Figures 3.50a and 3.51a The only difference is that the presence of the external compressive pressure leads to forcing out of the distribution to the opposite side (see the sequence of positions 3, 4, and 5 in Figure 5.51b).

Graphic representation of function $h = f(\alpha_p)$ in compression (curve 1 in Figure 3.50b) coincides with a typical compression diagram of concrete $\sigma_p = f(\varepsilon)$ (Figure 3.11). This evidences, first, the adequate description of structural processes in the domain of the indirect indicators of hardness and, second, successful selection of the investigation methods and, most important, the reliable description of the stressed–deformed state by Equation 2.96.

Special experiments were conducted to obtain a direct proof that stressed–deformed state develops particularly in the direction predicted by Equation 2.96. The programs of those experiments provided for stereology investigations,[159] microscopy examinations, and hardness tests of slices of the specimens subjected to different stress levels up to complete fracture. For these purposes three to four sections with different orientations with respect to the direction of loading were made from the specimens subjected to different degrees of structural damage.

Structural changes were observed using a polarization microscope; fluorescent coloring of the slice surface revealed microcracks. Microhardness of each structural component was determined on the same sections by pressing the diamond pyramid into its surface and measuring the size of indentation. Special consideration was given to a contact layer of the cement stone and filler, and to an intergranular space. From 30 to 50 measurements were taken

TABLE 3.6

Microhardness of Structural Components of Two Compositions of Concrete

Water–Cement Ratio (WC)	Solidification Conditions	Mean-Statistical Microhardness of Samples, H, kg/mm²									
		Solidification Time, days									
		3		7		28		90		After HT	
		H_c	H_o	H_c	H_o	H_c	H_o	H_c	H_o	H_c	H_o
0.26	Air	92	86	108	95	114	109	173	141	—	—
	Water	109	97	124	111	140	131	175	158	—	—
	HT	—	—	—	—	—	—	—	—	140	115
0.42	Air	89	84	103	96	108	104	165	139	—	—
	Water	96	91	129	110	132	120	170	152	—	—
	HT	—	—	—	—	—	—	—	—	121	106

Note: HT = heat–humid treatment of the samples, H_c = microhardness in the zone of filler and cement stone contact, and H_o = same in the non-contact regions.

in each zone. Table 3.6 gives results of the microhardness tests of the slices of two concrete compositions prior to loading.

For all compositions investigated and all varieties of solidification conditions, hardness of the contact layers was always found to be higher than that of the cement stone that fills the intergranular space — a fact noted previously.[154] This indicates location in the volume of the left and right ends of the MB distribution exactly. In particular, the left-sided dilatons are located in the zone of contact of the filler grains and cement stone, and the right-sided ones are located in the intergranular space. The validity of the previously described mechanism of the development of compressive stressed state can be judged based on observations on in which of these regions and at what stage of loading local microdefects of structure are initiated in the tests.

The results of our micro-optical examinations show that, starting from stresses equal to $\alpha_p = 0.15/0.25$ and up to fracture stress, readily seen cracks emerge in the contact layers, whereas no such cracks were observed in the intergranular space. The number and size of these cracks continuously grow with the stress level in specimens. Under high loads, some regions exhibit coalescence of neighboring contact cracks through narrow intergranular bridges. This proves the fact, known long ago in concrete science, that structure starts fracturing from the contract regions.[48] It indicates that fracture propagates from a strong (but brittle) to weak (but ductile) D-link, i.e., the MB distribution approaches the Debye temperature with its left end.

At high stress levels, development of the stressed–deformed state is associated with the course of the destruction processes described by Equation 2.109. It was investigated using stereology methods,[120] according to which the spatial arrangement of a three-dimensional structure is reproduced using the study of flat sections of the test specimens. These methods make it possible to establish quantitative relationships between structural parameters and external effects, reveal the physical nature of these relationships, and find or check the laws governing them.

Stereology methods are based on the fact that relative volumes of components of heterogeneous systems are proportional to relative areas in the cross section of corresponding specimens, as well as to projections of the areas onto an arbitrarily selected line. The special internal structure of a body can be reconstructed using the results of investigation of the distribution of structural components on the flat sections of this body. These methods are extensively used in metals science[120] and have been used for determination of the content of cement and filler in concrete.[155] However, these methods have not been applied to any study of destruction processes occurring in structural materials under loading.

The science of concrete[48] regards concrete as a conglomerate consisting of inert fillers (quartz sand, crushed granite, etc.) strongly bonded to each other on contact surfaces by a solidified cement layer. From the geometrical point of view, its structural components can be classified, using a dimensional criterion, as linear, laminated, and equiaxial.

> *Linear* components have the dominating dimension in one direction. They include hairline cracks, pore channels, and other defects and their orientation in the bulk of a material is random. Therefore, the probability of coincidence of their axial line with a plane of section is very low.
>
> *Laminated* components have dimensions developed in two directions. Their third dimension is small compared with the two others. They include the majority of the experimentally observed micro- and macrocracks. After relieving the external load, the cracks close to take the shape of complicated broken surfaces. In a section plane, they have the form of thin laminae or extended lines.
>
> *Equiaxial* components include those having commensurable sizes in three mutually perpendicular directions. In concrete, this group includes fine and coarse particles of the filler.

The adequacy of the information on mechanical-destructive changes of a structure depends to a considerable degree on the proper choice of location of the section and its orientation with respect to the direction of action of the external force. The position of the section plane with respect to the direction of the load should be selected to enhance appearance of defects having the form of broken lines. Normally such a position is found experimentally. In the considered case, it turns out to be equal to 30 to 40°. Detection of cracks on flat concrete sections was carried out by deposition and subsequent removal of a thin layer of a bright, easily penetrating paint.

The methodology of the experimental study was as follows: flat plates 50 to 10 mm thick were cut from cubic specimens (Figure 3.19c) or prisms (Figure 3.19b) by a diamond tool after the required compressive or tensile level of the stressed–deformed state was achieved. The plates were dried until their weight remained unchanged. After that, a thin layer of fluorescent paint was deposited on the cut surfaces. After 3 to 5 min, the paint was removed to expose open pores, cracks, and other structural defects. Sections thus prepared were investigated by stereology methods. The initial background created by an inherent imperfection of the structure of concrete was preliminarily estimated and accounted for in subsequent calculations.

The most important indicator of the spatial system of cracks is their total area per unit volume of a material. It characterizes the degree of destruction and is expressed in cm^2/cm^3 or cm^{-1}. The system of flat lines, i.e., traces, of cracks is estimated by their total length per unit area of the section and measured in cm/cm^2 or cm^{-1}. Another characteristic, of no less importance, is the spatial orientation of cracks. In a flat section plane, it shows up as a certain orientation of their traces and can be used as a criterion for estimation of the direction of development of the cracking process. Measurement of the total surface of cracks per unit volume and estimation of their orientation with respect to the direction of action of the external force were carried out using Saltykov's method of directed secants.[120]

The essence of the method is as follows: if parallel straight lines (or a rigid transparent matrix contour) are superimposed in a random manner on the plane of a section, the total surface of cracks, L, can be estimated by the number of intersections of the matrix contour with the system of their traces, m,

$$\Sigma L = \frac{\pi}{2} m. \tag{3.65}$$

This formula constitutes the basic relationship used in the method of random secants and enables determination of the total length of cracks per unit volume of a material.

Any partially oriented system of cracks can be represented as two combined subsystems: one fully oriented in a certain direction, and the other isometric. Then the degree of orientation can be expressed as

$$\beta = \frac{\Sigma L_o}{\Sigma L_o + \Sigma L_i}, \tag{3.66}$$

where ΣL_o and ΣL_i are the lengths of the oriented and isometric parts of cracks, respectively, per centimeter. In the isometric system, the mean number of intersections per unit length of the secant does not depend on its orientation in the plane of a section.

A clear picture of the degree of orientation of the system of cracks in the section plane can be obtained from the diagram of intersections, which expresses the relationship between the mean number of intersections per unit length of the secant and its direction in the polar coordinate system. Information required for its construction can be obtained as follows. A plate of

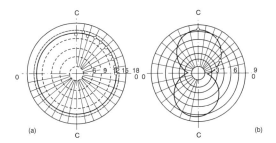

FIGURE 3.52
Number of intersections for the isometric (a) and partially oriented (b) systems of cracks.

a transparent material with a mesh of parallel lines scribed on it is placed on a section of concrete specimen preliminarily treated with paint. The lines are randomly oriented and uniformly distributed over the entire section surface.

The mean number of intersections of the secants with crack traces in the section plane is counted per unit length of a secant (e.g., per centimeter) in a given direction m_β. Then the mesh is turned to a certain angle β and again the mean number of intersections is counted. This operation is repeated until the mesh makes a full revolution in the section plane with respect to the assumed orientation axis. The diagram of intersections is plotted using the series of numbers m_β in an angle range of $0° < \beta < 360°$.

Figure 3.52 shows examples of the diagram of intersections for isometric (a) and partially oriented (b) systems of cracks. Direction of the axis 0–0 coincides with the direction of orientation. The mean number of intersections per unit length of the secants in this direction is referred to as $m_\|$ while that in the normal direction (along axis C–C) as $m_\perp$. Assuming that the partially oriented system can be divided into two subsystems, i.e., isometric and fully oriented, we can write in accordance with Equation 3.65 that

$$\Sigma L_i = \frac{\pi}{2} m_\| .$$

It is apparent that the $m_\perp$ number is made up of the number of intersections of both systems: oriented and isometric. The number of intersections only with the oriented part of cracks on the secant normal to the orientation axis is equal to the difference $m_o = m_\| - m_\perp$. Then

$$\Sigma L_o = \frac{\pi}{2} m_o = \frac{\pi}{2} (m_\perp - m_\|). \tag{3.67}$$

The total length of all the traces of the partially oriented system of cracks on the section plane is determined from the following formula:

$$\Sigma L = \Sigma L_i + \Sigma L_o = \frac{\pi}{2} m_\perp . \tag{3.68}$$

Having determined two mean numbers of intersections on secants ($m_\parallel$ and $m_\perp$), parallel and normal to the orientation axis, we can calculate the degree of orientation of the system of cracks using Equation 3.66. Accounting for Equations 3.67 and 3.68, we can express it as

$$\beta = 1 - \frac{m_\parallel}{m_\perp}. \tag{3.69}$$

Until now consideration was given to determination of destructive parameters in a plane. Similar correspondence between the mean number of intersections on secant m and the value of an absolute specific surface of the spatial system of cracks holds also for the volume. As shown in Saltykov,[120] it is established by the relationship

$$\Sigma S = 2m \cdot cm^{-1}. \tag{3.70}$$

Rewriting Equation 3.70 accounting for Equation 3.65, we obtain the basic formula of the space method of random secants:

$$\Sigma S = \frac{4}{\pi} \Sigma L. \tag{3.71}$$

It is applicable only to the isometric system of cracks.

Fracture of a material is the orientation process (see Section 3.4). Its direction is set by external conditions, i.e., support plates of the press used in tests of the specimens or boundary interfaces in actual parts and structures. The preferred orientation (axis C–C in Figure 3.52) points to the most probable direction of development of an avalanche crack. The complete picture of the intensity and character of fracture can be obtained only by simultaneously determining specific surface ΣS, shape, direction, and degree of orientation of the system of cracks. To obtain these characteristics, it is necessary to cut a sample into a certain number of sections whose planes are oriented randomly in space. As shown by experience of metals science,[120] the number of flat sections should be not fewer than six to achieve satisfactory accuracy of analysis.

The degree of orientation of cracks was studied at several levels of stresses. Plotting several flat diagrams of intersections using the results of analysis of differently oriented sections (Figure 3.53), one can construct dependence of the type of $\beta = 1 - \gamma$, where

$$\gamma = \frac{m_\parallel}{m_\perp}. \tag{3.72}$$

Changes in the kind of the stressed state (tension or compression) and its intensity affect the shape of flat diagrams of intersections. Figure 3.54 shows the variations in the degree of orientation of cracks depending on the level of a compressive, α_p and tensile, α_s stressed state.

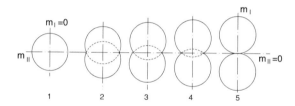

FIGURE 3.53
Sequence of flat distributions of the number of intersections obtained on five differently oriented cuts at a constant level of the stressed state.

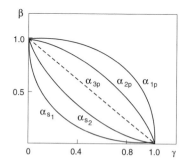

FIGURE 3.54
Variations in the degree of orientation, β, depending on the level of the stressed state in compression α_p and tension α_s.

Experimental studies show that in uniaxial compression the spatial system of cracks has a cylindrical symmetry. This makes investigation of the cracking process at a given stressed–deformed state much simpler because it is enough to make just two sections: one in the plane that passes through the axis of symmetry and the other in the direction normal to this plane. In the axial section, the cracks appear as broken lines, and in the normal section they appear as a system of spots.

Figure 3.55 shows typical dependencies of the total specific surface of cracks per unit volume of a specimen ΣS and the degree of their orientation β on the level of the compressive stressed state. As seen, up to a certain level of the stressed state, α_m, the concentration of cracks in the volume increases slowly, but the degree of orientation rapidly reaches its saturation level. It was established for central compression $\alpha_m = 0.25/0.3$, i.e., close to maximum of curves 1, 2, and 3 in Figure 3.50b. The process of coalescence of microdefects and growth of cracks starts on reaching this value, continuing to the level corresponding to $\alpha_{max} = 0.8/0.85$. At this moment, avalanche coalescence of the system of individual cracks and the development of avalanche cracks occur.

Figure 3.56 shows the character of growth of cracks in the volume of a material in tests of the prism-shaped specimens subjected to central compression (a)

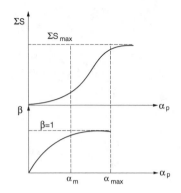

FIGURE 3.55
Dependence of the intensity of the destructive processes, ΣS, and degree of orientation of cracks, β, on the stressed state level α_p.

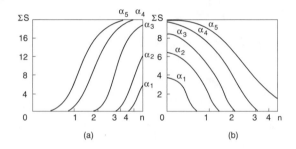

FIGURE 3.56
Accumulation of damage in the cross section of a concrete specimen in tension (a) and compression (b) at different levels of loading α.

and tension (b) (off-center compression). Specific surface of cracks per unit volume of the material, ΣS, at different levels of the stressed state ($\alpha_1 < \alpha_2 < \cdots \alpha_s$) is plotted on the axis of ordinates, and number of the concentric zone counted from the symmetry axis of a specimen to its periphery, n, is plotted on the axis of abscissas. As seen, the fracture process in compression starts from the center of the specimen and propagates to its periphery (Figure 3.56a). The picture observed in tension is opposite, i.e., accumulation of damages begins from the external surface and then propagates to the center (Figure 3.56b).

Similar processes in force fields take place in metallic bodies. The only difference is that CD phase transformations add certain peculiarities. An indirect proof of this postulate can be found in Sosnovsky.[151] Table 3.7 presents the results of statistical analysis of hardness of steel specimens made of a stainless steel (0.08%C, 18%Cr, 12%Ni, 1%Ti) cut from a pipe having nominal diameter of 550 mm and wall thickness of 25 mm. These specimens were cut from the pipe at its initial state and after 100,000 hours of service in the first loop of a nuclear power reactor (pressure 10 MPa, $T = 543$ K).

TABLE 3.7

Parameters of Empirical Functions of
Distribution of Vickers Hardness of
Steel before and after Service[151]

Parameter	Metal State	
	Initial	After Service
d_1, mm	0.482	0.422
d_2, mm	0.479	0.417
S_d, mm	0.0108	0.0077
HV	160.3	207.5
K_a	0.994	0.981

Note: S_d = standard deviation; HV = Vickers
hardness; and $K_a = d_2/d_1$ = asymmetry factor.

The spread of sizes d_1 and d_2 of indentations was studied using a Vickers
hardness tester in the circumferential and axial directions, respectively.

As Table 3.7 shows, the hardness of the tube as well as its spread and
asymmetry were increased during service life. It should be noted that statistical measurements of hardness allow revealing even a slight anisotropy
of properties of a material (the last row in Table 3.7).

In the context of these CD concepts of the structure of solids, these data
can be interpreted as follows. Service caused displacement of the MB distribution inside the *C*-phase, *cabkm* (Figure 2.36) closer to the Debye temperature (critical boundary *mn*). This is evidenced by an increase in hardness
from HV 160.3 to 207.5 with a simultaneous increase in steepness of its fronts.
(There was a decrease in spread of the indicators of hardness.) Formation of
anisotropy in the circumferential direction (d_1 changed more than d_2) is
attributable to a long-time effect on structure by the service pressure of 10
MPa. Naturally, it caused the orientation transformation of this structure.

The mechanism of this transformation is considered in Section 3.4 and was
experimentally confirmed for concrete (Figure 3.50b). As a result, ductility
of the steel degraded and its strength increased. There was a redistribution
of energy between parameters *P* and *V* of the Ω-potential, which caused the
displacement of the technical state of the material to the right from optimal
value σ_0 in Figure 3.34.

3.10 Complex Stressed States: Mechanism of Formation and Prospects of Application for Fracture Prevention

The internal resistance **P** has a tensor nature (Figure 2.7) and is distributed
in the volume of a body in accordance with the Boltzmann statistics (Equation 2.3). Its external manifestation, i.e., strength σ, is strictly determined by
the direction of action of the force fields. Therefore, it is of great interest to

understand how a material behaves in complex stressed–deformed states where it is simultaneously subjected to compressive and tensile forces.

To reveal the behavior of materials under such stressed states, we can rewrite Equation 3.19 for volumetric stressed–deformed states:

$$P_o \pm \Delta(\sigma_1 + \sigma_2 + \sigma_3) = kT\rho\delta, \qquad (3.73)$$

where σ_1, σ_2, and σ_3 are mutually perpendicular components of the external stresses. These stresses can be induced by each of the fields mentioned in the derivation of Equation 3.19. Equation 3.73 allows description of any kind of stressed–deformed state, starting from the simplest tension (Equation 3.60). If a specific material, whose θ, P_o, and ρ are constants, is in a biaxial tensile force field under conditions $T = \text{const}$, then Equation 3.73 becomes

$$\sigma_1 = \alpha B - \sigma_2, \qquad (3.74)$$

where $B = P_o - kT\rho$ is a constant and α is the thermal expansion coefficient according to Equations 2.154 and 3.20.

Figure 3.57 shows the results of a study where a biaxial tensile stressed state was applied to high-strength aluminum alloy 24S-T81;[44] this study also contains similar dependencies for other steels and alloys. Hollow cylindrical specimens were used. The longitudinal axial stress was created by the application of tensile ($-\sigma_1$) or compressive ($+\sigma_1$) loads and the stresses on the periphery ($-\sigma_2$) were created by the internal pressure in the hollow of specimens. Because the wall thickness of the specimens was small, the radial stress σ_3 was ignored. Different shapes of experimental points in Figure 3.57 are used to characterize different orientations of the fracture surface: triangles

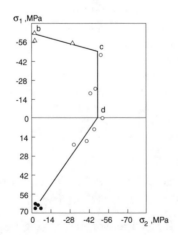

FIGURE 3.57
Biaxial stress–strain state of high-strength aluminum alloy 245-T81 induced by tension.

indicate that it was parallel to the transverse section, circles, along the axial section, and dots, at an angle of 45° to the axis of species.

As predicted by the theory (Equation 3.74), σ_1 and σ_2 are really related to each other by a directly proportional dependence with an angular coefficient α. Its fundamentally different character in the upper and lower coordinate fields is attributable to the fact that, in tension–compression (lower field), the continuous trend of the MB factor to transformation of its shape in a direction of 11 to 12 to 13 (Figure 1.14) during the entire deformation process is hampered by the stresses induced due to axial compression ($+\sigma_1$). As such, the θ-mechanism of fracture is dominant (right dashed zone in position 13 of Figure 1.14).

In the tension–tension mode (upper field), no such a counteraction exists. Therefore, Dependence 3.74 shows up only in line bc, i.e., until the cold end of the MB factor approaches the T_c temperature. The simultaneous occurrence of the T_c and θ_s mechanisms of disintegration of bonds leads to catastrophic fracture of the specimens (along vertical line cd) and is proven by different orientation of the fracture surface.

If, under the same conditions, a complex stressed–deformed state is created in a material by axial compression ($+\sigma_1$) and transverse tension ($-\sigma_3$), Equation 3.73 assumes the form

$$\sigma_3 = \alpha B + \sigma_1. \tag{3.75}$$

The indirect indicators of hardness were measured only in the direction normal to that of action of compressive forces. Figure 3.50 shows the variation of their distribution parameters. The shape of curves in this figure shows that, in this direction, one might expect an increase in tension resistance compared with the stress-free state. This expectation is confirmed also by Equation 3.75.

Special experiments were carried out to check this hypothesis. Compressive loads ($+\sigma_1$) were created in concrete columns measuring $20 \times 30 \times 120$ cm (Figure 3.19b) having the steel anchors from 30 to 50 mm long with threaded ends installed on their lateral faces (clearly seen in Figures 3.19b and 3.37b). A portable press-pump pulled out these anchors (Figure 3.37a).

At least two tests were carried out at each level of reduction $\alpha_p = \sigma_1/\sigma_p$, where σ_p is the compressive ultimate strength of concrete. Figure 3.58 shows a typical dependence of the type of $\sigma_3 = f(\alpha_p)$ having a maximum. This type of the dependence holds for all investigated classes of concrete (heavy, light, cellular, etc.). It was observed by the authors[156] and also by other investigators[157,158] and can be detected not only by direct methods[156–158] but also by ultrasonic methods[159–160] (especially by the hardness measurement).[53,54] Moreover, it is so pronounced that it is mentioned in regulatory documents.[161]

The ascending branch corresponds to the course of the orientation processes; it is equivalent to elastic region oa of the deformation diagrams (Figures 2.10 and 2.22). Orientation saturation in the zone of maximum leads

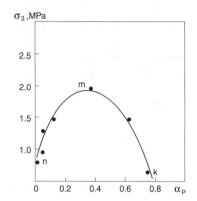

FIGURE 3.58
Complex stress–strain state induced by longitudinal compression and transverse tension.

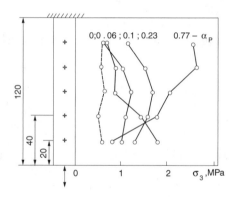

FIGURE 3.59
Development of the stressed–deformed state in centrally compressed structural members.

to the tension of the material in the transverse direction and reaching the θ-temperature by the hot end of the MB distribution, and, as a consequence, to beginning of the degradation process. An increase in the concentration of microdefects in the volume of a material leads to a decrease in the pulling out force.[55] Depending on a particular value of σ_p, an increase in σ_3 is estimated to be equal to 30 to 60%. Maximum of the curve is located in the range $\alpha_p = 0.2$ to 0.45. It was observed that the higher σ_p, the higher the α_p.

Figure 3.59 is instructive to understanding the formation of the stressed–deformed state (Equation 3.75) in the bulk of structural elements. This figure shows the variation of σ_3 along the height of the columns with variation of α_p from 0 to 0.77. Each point in the plots is the mean value of σ_3 measured in a given section on several columns made from the same composition of concrete. The numbers in the upper part of the coordinate field show the values of the compressive stressed state level at which they were obtained.

In the stress-free state, σ_3 is almost the same along the entire height of the column. Small deviations are attributable to local heterogeneities of structure of the material. At the first stage of loading ($\alpha_p = 0.06$), the stressed–deformed state covers the zone between the lower plate of the press (shown by a vertical arrow) and center of the column height. The largest variation of the state is observed near the base. Subsequent stages correspond to plots having maximum (levels 0.1 and 0.23). This means that the orientation maximum (indicated by point m in Figure 3.58) moves upward with loading.

Indeed, at a level of the lower anchor, as α_p varies from 0 to 0.77, σ_3 increases (from point 0 to point 1), reaches maximum (at point 1), and then decreases (sequence of points 2, 3, and 4) to a value close to the initial state (points 4 and 0 almost coincide). Here the stressed state passes the whole cycle of development (Figure 3.58): from point n up the ascending branch of curve nmk and through maximum (point m), approaching at a level of $\alpha_p = 0.77$, the zone of failure of the column (point k). During this time, at a height of the fifth anchor, the orientation process approaches its maximum. The nonhydrostatic nature of the stressed state is a result of the effect of the proximity of interatomic bonds[20] and should be accounted for in application of the active methods of prevention or retardation of fracture.[18]

The nonhydrostatic nature shows up not only in the longitudinal direction (along the axis of action of compressive forces) but also in the transverse direction. Transformation of the MB distribution starts at early stages of loading in the subsurface layers and propagates deep into a material. This is explained not so much by geometrical differences as by specific conditions of the subsurface layers.[162] This influences their CD composition and, therefore, the development of a stressed state.

This is supported by the results of direct examinations and indirect measurements. Figure 3.60 shows the variation of the pulling force in pulling out inserted anchors of different length from concrete subjected to different compressive stressed states. As seen, curves of type $\sigma_s = f(\alpha_p)$ for anchors of different lengths have an equidistant displacement. This can be explained by:

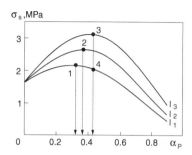

FIGURE 3.60

Variations of rupture strength σ_s relative to the level of axial reduction α_p for inserted anchors of different length (1 = 30 mm, 2 = 35 mm, 3 = 48 mm).

- Methodological causes, implying that a higher pulling force is required for a greater length of the rod at the same level of stressed state, which leads to displacement of curves upward.

- Physical causes, i.e., consolidation, orientation saturation, and subsequent loosening of the material propagate from the periphery of a section to the center of a sample, which leads to a shift of apexes toward higher values of α_p (sequence of points 1, 2, and 3).

Orientation strengthening of a structure at a depth of 30 mm ends at a level of stressed state equal to 0.32 (point 1). At $\alpha_p = 0.43$ it ends at a depth of 48 mm (point 3). At this moment the loosening process has already been occurring in the subsurface layers (point 4 in curve l_1). In loading of concrete by compression with subsequent tension, Kublin[158] also observed a dramatic inflection in curves of stresses and strains. The nonhydrostatic nature of the stressed–deformed state requires reconsideration of the basic principles of the existing methods of calculating strength and reliability in the design of parts and structures.

The formation of compressive and tensile stressed states was evaluated also by a change in parameters of a narrow superhigh-frequency (SHF) radio-wave beam.[123,163,164] Figure 3.61 shows the test of specimens using the SHF electromagnetic method in the nonloaded state (a) and results of measurement of attenuation of the radiation amplitude at different levels of compressive stressed state (b). The amplitude u was measured at three levels of compressive stressed state: 0.12, 0.23, and 0.87 (curves 1, 2, and 3, respectively). At the initial stages of loading (dotted line 1), a uniform consolidation of the material occurs over the entire cross section of the specimen. Oscillation of the amplitude of the radio signal that passed is caused by an inherent heterogeneity of the structure in the section investigated.

As compressive stresses increase (dashed curve 2), forced reorientation of dilaton rotoses (Figure 2.8a) takes place. This reorientation is accompanied by a deeper consolidation of the material (amplitude of the signal received decreases, and curve 2 goes down to a level lower than curve 1). The shape of curve 2 varies in the peripheral layers (extreme left and right zones of the

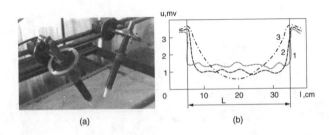

(a)   (b)

FIGURE 3.61
Investigation of concrete specimens using the SHF electromagnetic method in the unloaded state (a) and results of measurement at three levels of reduction (b).

Figure 3.61b). Here the amplitude of the signal increases, indicating the beginning of loosening of the material. This process is most clearly seen in curve 3. In the central region, it is lower, and in the peripheral regions it is higher than curves 1 and 2. Therefore, the consolidation process continues at the center of the sample; loosening, while propagating from the periphery to the center, goes deeper and deeper into the material.

Concrete is a dielectric material.[165] Dielectric bodies in the nonloaded state have inherent structural anisotropy (anisotropy of molecules, crystalline lattice, macrostructural heterogeneity), which leads to heterogeneity of dielectric properties. First, orientation (region *oa*), and then mechanical-destructive (region *ab* of the diagram shown in Figure 2.10) anisotropy form under the effect of external force fields. Rudakov et al.[166] report that, in heterogeneous environments, the mechanical types of anisotropy generate anisotropy of electric properties. For example, formation of a stressed state in a material causes an increment in the phase of electromagnetic radiation.[167]

$$æ = \frac{2\pi L v}{\lambda} \sigma, \tag{3.76}$$

where ϑ is the relative radio-polarization coefficient of mechanical stress σ, λ is the electromagnetic wavelength, and L is the specimen thickness.

Sloushch[168] substantiates correlation between variations in the electromagnetic radiation phase and dielectric permittivity of the environment. It is shown that an increment in the phase of the electromagnetic wave that passed through the material is determined by the expression

$$æ = \frac{2\pi L}{\lambda} (\sqrt{\varepsilon} - 1), \tag{3.77}$$

where $\varepsilon = \varepsilon_p + \varepsilon_o$ is the sum of inherent, ε_p, and force field induced, ε_o, dielectric permittivity.

Comparing Equations 3.76 and 3.77, we can write

$$\varepsilon = \left(v\sigma + \sqrt{\varepsilon_p} \right)^2. \tag{3.78}$$

Figure 3.62 shows the variation in dielectric permittivity ε in one of the column sections (Figure 3.19b), depending on the stress level. Theoretical dependence (3.78) is shown by a solid line and points indicate experimental values for one of compositions of concrete. Formula 3.78 is a parabolic equation; its apex has the following coordinates: $\varepsilon_{max} = 2\varepsilon_p$ is the maximum possible value of dielectric permittivity for a given composition of concrete under the effect of compressive stresses

$$\sigma_{max} = \frac{\sqrt{\varepsilon_p}}{v}. \tag{3.79}$$

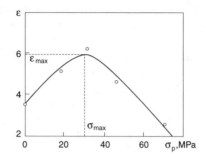

FIGURE 3.62
Dependence of dielectric permittivity ε of concrete on compressive stresses σ_p.

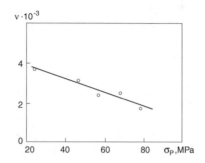

FIGURE 3.63
Value of relative radio-polarization coefficient v for concretes of different compositions.

It is seen that dependencies $\varepsilon = f(\sigma)$ and $\alpha_p = f(\sigma_s)$ (Figures 3.62 and 3.60, respectively) have similar shapes. This is natural, as they are based on the same physical phenomena occurring at the AM level and exerting the same effect on parameters ε and α_p. They describe reaction of a structure to external compressive forces. Relationship 3.78 allows values of relative radio-polarization coefficient v to be calculated from ε_p and σ_p measured experimentally. These values are shown in Figure 3.63 for the investigated compositions of concrete; this dependence characterizes the degree of polarizability of different compositions of concrete in electric fields and might have practical consequences. Their essence is as follows.

It is shown in Section 3.4 that elastic deformation at the AM level is the orientation process. If mechanical loading of dielectric materials is accompanied by a simultaneous effect of the electric field, which creates AM polarization of the opposite sign (Figure 2.8), this allows resistance and deformation to be controlled and the fracture process to be hindered or avoided.

Is it possible to use force fields to create artificial anisotropy of mechanical properties? The answer to this question is of scientific and high practical interest because, if it is possible, it allows development of methods of stimulation or retardation of fracture. We can show now that by purposeful

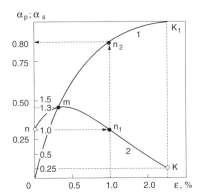

FIGURE 3.64
Variations in the relative level of rupture strength α_s on the relative level of reduction α_p.

formation of the stressed state (within the elastic orientation stage), it is possible to increase the resistance of materials to deformation in one direction (in this case, in the direction of compression) and to fracture in the other direction.

Figure 3.64 presents (in the combined coordinates) the correlation between the level of compressive stressed state α_p (curve 1) and the corresponding level of the rupture strength, α_s, in the direction normal to the action of the compressive force (curve 2). Relative values of strain $\varepsilon\%$ are plotted along the axis of abscissas.

Combination of curves on the same coordinates allows the mechanism of formation of the stressed–deformed state to be identified in time and space. At the absence of reduction (strain), the relative value of the rupture strength is determined by point n ($\alpha_s = 1$). As the compressive load increases (i.e., ε grows), the ability of the material to resist compressive (in the direction of action of the force) and tensile (normal to the force direction) stresses also increases (α_p and α_s increase). This relationship persists to point m on curve 2. Maximum resistance to rupture is observed at $\alpha_p = 0.25$ to 0.5. In this case, strengthening initiated by the external load amounts to 20 to 40% ($\alpha_s = 1.2$ to 1.4). Its particular value depends on the design composition of concrete. A quantitative measure of strengthening in tension can be the maximum value of α_s. Figure 3.65 shows the dependence of $\alpha_{s_{max}}$ on the composition of concrete, whose indirect indicator is compressive strength σ_p.

As the compressive load increases, the strengthening process propagates from the periphery to the center of the section (region nm of curve 2 in Figure 3.64). Coalescence of localized microdefects and formation of avalanche cracks occur in the peripheral zones, starting from point m. The rupture strength begins to decrease (the left branch of curve 2 in region mk); at a level of compressive stresses equal to 0.7 to 0.8 (point n_2 in curve 1), rupture strength reaches the initial value (points n and n_1 in curve 2). By the moment when the maximum compression resistance is achieved (point k_1 in curve 1), concrete still can withstand some transverse tensile forces (point k in curve 2).

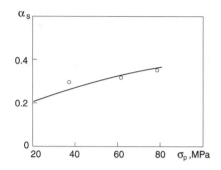

FIGURE 3.65
Degree of orientation strengthening for concretes of different compositions.

FIGURE 3.66
Evaluation of the level of the compressive stress–strain state of foundations of a highly critical object (a) and columns of an emergency structure (b).

At the depth at which the anchors are inserted, about 15% of the bonds are capable of bearing the load. However, concrete in the bulk is so loose and so penetrated by macrocracks that the process of their coalescence acquires catastrophic character. The descending branch of curve 1 (located to the right from point k_1 and not shown in Figure 3.64) characterizes development of the fracture process deep into the material, formation of avalanche cracks, and fracture.

There are no known direct methods for evaluation of the compressive stressed state level of concrete (as well as of the rest of structural materials) in the science and practice of civil engineering.[169] The effect of orientation strengthening can be used for evaluating it in massive structures. The method is based on small-volume local fracture of concrete caused by pulling out of the inserted anchors (Figure 3.37b) and utilization of the calibration dependence whose analogue is shown in Figure 3.58.[156] Ambiguity caused by the two-branch character of the calibration dependence is avoided if the anchors are pulled out at three levels of stresses: zero σ_o, effective σ_e, and auxiliary σ_a. As such, σ_e stresses are determined from the magnitude value and variations of pulling force F at the above levels. This method was suggested in 1981[129] and still finds application for evaluation of compressive stressed state in building of critical structures (Figure 3.66a) and in emergency situations (Figure 3.66b).

3.11 Mechanical, Thermal, Ultrasonic, Electron, Chemical, and Other Effects in Deformation and Fracture

External fields cause mechanical, thermal, ultrasonic, chemical, and other effects in solids. Mechanical effects are described by parameters of resistance, P, σ and deformation, dV, dV/V, η, and ε, and are obvious. All reflect different aspects of the Ω-potential and are interrelated to thermal effects. Generated at the AM level (Equations 1.51, 1.53, and 1.69), this relationship penetrates through all scale levels as reflected in the macroscopic thermodynamic equation of state (2.65).

The adequacy of mechanical and thermal processes for a rotos can be proved another way. Taking into account that the difference $(u - u_1)$ in Equation 2.176, in accordance with Equation 2.105, is always equivalent to a change in kinetic energy $dU(T)$, and using the second-order infinitesimal terms of the expansion, we can write the quadratic equation with respect to deformability x

$$\frac{1}{2} \Delta x^2 + Fx \pm du(T) = y, \tag{3.80}$$

where the sign in the front of the kinetic energy depends on to which side from equilibrium x_1 the rotos changes its size. This equation has four variables, three of which (F, x, and $dU(T)$) correlate. It has two solutions with respect to x:

$$x_{1,2} = \frac{1}{\Delta}\{-F \pm [F^2 - 2\Delta du(T)]^{1/2}\}. \tag{3.81}$$

They mean that, whatever the phase of the rotos (compresson or dilaton) in deformation to both sides from the equilibrium position, sooner or later its energy state becomes critical (T_c or θ for compresson, or θ and ∞ for dilaton; Figure 1.14).

The plot of Function 3.80, which is a parabola, is shown in Figure 3.67. Coordinates of the apex m of this parabola are determined by solving this quadratic trinomial and are equal to

$$x_0 = \frac{F}{\Delta} \quad \text{and} \quad y_0 = \frac{1}{2}\frac{F^2}{\Delta} + du(T);$$

x_1 and x_2 are its roots (Equation 3.81). The steepness of branches of the parabola depends on the rigidity coefficient Δ (Equation 3.20). Rigid rotoses (e.g., 1 in Figure 1.10b) have steep branches, and ductile rotoses (positions 2 and 3 in the same figure) have gently sloping branches. It can

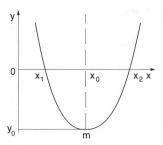

FIGURE 3.67
Function of deformation of a rotos.

be seen from Figure 3.67 that deformation of any rotos involves overcoming resistance forces F and is accompanied by thermal effects (radiation or absorption of heat), $dU(T)$, as in Equation 3.80, or formation of temperature gradient in Equation 3.61.

However, the processes of deformation under the effect of external thermal fields should not be absolutely identified with those of heat release in mechanical deformation. The thermal diffusivity of materials is anisotropic,[20,23] determined by the space component of the MB distribution (Equation 2.2). During the process of uniaxial loading, the concentration of heat flow in the direction of the force field increases (Equation 3.61), which leads to anisotropy of the internal thermal field in different directions. Excess of the thermal energy in one of them is not always compensated for by its shortage in the other. As a result, some directions are characterized by radiation of heat and others by its absorption.

Nonidentity of the processes of thermal expansion and heat release in mechanical compression, as well as thermal compression (decrease in linear sizes during cooling) and absorption of heat in mechanical tension, is proved by facts obtained in the study of the energy state of a deformed material.[35] A simple calculation shows that the amount of heat Q that leads to a change in temperature of a unit volume by dT is much greater than the energy of deformation A accompanied by the same value dT of the same amount of material.

For example, uniaxial compression of a steel specimen to stress $\sigma = 100$ MPa leads to an increase in its temperature by $dT = 0.1°C$.[44] This is equal to the supply of the following amount of heat to a unit volume of the specimen:

$$Q = \rho C_p dT = 7.8 \cdot 0.11 \approx 0.1 \text{ cal} \approx 3.8 \text{ kg} \cdot \text{cm},$$

where ρ is the density and C_p is the heat capacity of the metal.

At the same time,

$$A = \frac{\sigma^2}{2E} = \frac{10^4}{2 \cdot 2.1 \cdot 10^4} \approx 0.24 \text{ kg} \cdot \text{cm},$$

where E = $2.1 \cdot 10^4$ MPa is the elasticity modulus of the metal. As seen, A is lower than Q by an order of magnitude. This calculation is true under the condition of equality of heat capacities $C_p = C_p(\sigma)$ in nonloaded and stressed states. So, in compression, the amount of the external energy A spent for the deformation process controls the transfer of the amount of internal energy significantly greater than A.

The change of dT in temperature of a body, or a transfer of the above amount of heat Q to it, leads to thermal expansion or compression of the body by a value much smaller than deformation ε that causes the same change in temperature dT. Indeed, the uniaxial stress $\sigma = 100$ MPa leads to strain of the considered steel specimen equal to $\varepsilon(\sigma) = 0.15 \cdot 10^{-6}$, and a change in temperature of dT causes relative strains

$$\varepsilon(T) = \alpha dT = 10^{-6},$$

where α is the coefficient of volume thermal expansion (Equation 2.152), i.e., 6.5 times greater than in the deformation. This is natural because heating or cooling is an isotropic process (see Section 2.7), while deformation is an anisotropic process (Equation 2.136 and Figures 2.8 to 2.10 and 3.59, and their explanations). Therefore, it is clear that the mechanism of transfer of the amount of heat Q to a body through heating by dT is not identical to the process of a change in its temperature by dT caused by deformation.

Assuming that $x \gg x^2$ and ignoring the second-order infinitesimal effects associated with the rigidity coefficient Δ (Equations 2.154 and 2.177), which determines steepness of the leading front of the force characteristic of a rotos (compare curves 1 and 3 in Figure 1.10), we can rewrite Trinomial 3.80) in a simplified form:

$$\omega = Fx = \pm du(T). \tag{3.82}$$

Dividing its left and right parts by an initial interatomic spacing x_0 and designating the relative variation in sizes of the rotos in the direction of x as

$$\varepsilon = \frac{x}{x_0},$$

we have

$$F\varepsilon = \frac{du(T)}{x_0}. \tag{3.83}$$

Taking into account that $E = F/x_0$ can be interpreted as the initial elasticity modulus of the rotos, write that

$$\varepsilon E = \frac{du(T)}{x_0^2}. \tag{3.84}$$

In accordance with Equation 1.90,

$$du(r) = du(T) = \frac{\partial u(r)}{\partial x_0} = F_0 \quad \text{and} \quad \sigma = \frac{F_0}{x_0^2} = \frac{F_0}{S_0}.$$

Then, accounting for Equation 3.84, we obtain $\sigma = \varepsilon E$, i.e., Hooke's law for an elementary bond, which determines the elastic or reversible deformation in the direction normal to site S_0. As seen, this is an orientation law, forms at the AM level, and is readily detected by macroexaminations (see Section 3.4 and Figure 2.10). Therefore, Equation 3.84 can be expressed as

$$S_0 = \pm \frac{du(T)}{\sigma}. \tag{3.85}$$

This determines the share of internal energy transformed into heat in formation of a new surface unit through microcracking at stress σ, or that value of the thermal energy that must be introduced into a material (or removed from it) to stop time development of the fracture process. Although some studies have pointed out the correlation of deformation and fracture with thermal effects,[35,44] we provide the physical explanation and quantitative estimation to this correlation.

It follows from Equation 3.84 that the higher the value of E of the rotos, the higher the thermal reaction observed in its deformation. In other words, the steeper the fronts of the force characteristic of rotoses (see Figure 1.10) of a material, the larger the amount of heat radiated or absorbed. For example, E of metal is five times higher that that of grade 600 concrete. Therefore, heat release in deforming of a metal is by an order of magnitude higher than that of concrete.

Relationship 3.84 has a general character. The rotos always reacts to a deviation of temperature (either fed from the outside or released in forced deformation) by a change in product εE (while a solid, naturally, reacts by change in the Ω-potential) rather than by a change in one of these parameters separately. Figures 3.33 and 3.34 provide experimental proofs of this fact. Structural materials react differently to an increase in temperature. These differences are attributable to a different intensity of variations in parameters ε and E.

There are bodies in which an increment in ε is much greater than that in E. As a result, ductility grows and resistance falls—changes observed in metals and alloys. Some types of concrete and rock have the opposite character of changes in these properties. Within a certain temperature range, they are characterized by a marked increase in strength with an insignificant increase in deformability.[40]

Thermal effects accompany elastic (Equation 3.84) and destructive (region ab of diagram $\sigma = f(\varepsilon)$ in Figure 2.10) deformation. If we replace $Nd\omega = d(\omega N) - \omega dN$ in Differential Relationship 2.60, then

$$d(\Omega + G) = -sdT + kTdN, \tag{3.86}$$

where $G = \omega N$ and $\omega \approx kT$, according to Equations 2.51 and 2.146. Taking T out of brackets in the right-hand part, and noting that $ds = dT/T$ from the definition of entropy (Equation 2.13), we find that

$$\frac{1}{T}d(\Omega + G) + \mathbf{s}ds = kdN.$$

Using the equation of state written in the form of Equation 1.111, we obtain

$$\frac{1}{T}d(2\Omega - F) + \mathbf{s}ds = kdN. \tag{3.87}$$

Substitution of the differential form of the G and F potentials for systems with a variable number of interacting particles (Equation 2.48.III and IV) changes Equation 3.87:

$$\mathbf{s}ds + \frac{1}{T}\left[2(\mathbf{P}dV + Vd\mathbf{P})\right] = kdN.$$

Hence,

$$T\mathbf{s}\frac{ds}{dN} + 2\frac{d\Omega}{dN} = kT. \tag{3.88}$$

As seen, a change in the number of interatomic bonds, dN, leads to a decrease in the bound part of the internal energy $T\mathbf{s}$ and the Ω-potential. This process is inseparable from a change in thermal conditions of a body (the right-hand part of the equality).

It follows from Equation 3.84 that the εE product decreases with a decrease in temperature. In this case, a material might fail due to the cold shortness phenomenon (see Section 1.10). Let physical–mechanical characteristics of a material be determined at temperature T_1 and represented by diagram $\sigma = f(\varepsilon)$ (Figure 3.68). Let a structure fabricated from this material be subjected to a certain stressed–deformed state (compression, tension, or any other kind) at the level of point a_1 and be operated under conditions of dramatic temperature gradients. With a decrease in temperature, point a_1 does not remain in place, but moves in the coordinate system under the effect of two opposite trends. A decrease in ε moves it to the left, i.e., closer to the origin of coordinates, while a decrease in E causes it to slide along a line tangent to curve $\sigma = f(\varepsilon)$ in the direction from point a_1 to maximum point a_{max}.

If the intensity of movement from E_{a_1} to E_{max} is in excess of the intensity of movement to zero, under a constant external load $\sigma_{a_1} = \text{const}$, point a_{max}

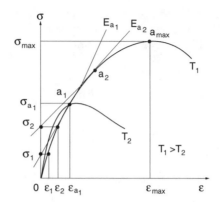

FIGURE 3.68
Variation in the shape of the deformation diagram on cooling of a material.

may coincide with a_1. As a result, fracture occurs not at σ_{max} but at σ_{a_i}. The trend of E_{a_1} to E_{max} is none other than

$$\frac{d\sigma}{d\varepsilon} \to 0$$

that leads to reshaping of curve T_1, which will eventually take the position of curve T_2. The described manner of temperature variations of the deformation curves is observed in practice.[3]

If loading of a specimen is carried out at a constant temperature, $T = const$, the slope of the tangent continuously changes so that it cuts off larger and larger lengths along the σ axis: σ_1, σ_2, etc. They correspond to points ε_1, ε_2, etc. along the axis of abscissas. These points determine the residual strains in unloading from the stressed–deformed levels: σ_{a_1}, σ_{a_2}, etc. and characterize the intensity of accumulation of damage, i.e., that share of the volume excluded from the work at each level of stresses: σ_{a_1}, σ_{a_2}, etc.

By definition, $E = d\sigma/d\varepsilon$ at $T = const$. Substituting this in Equation 3.84 and changing from AM level to macroscale, we find that

$$\frac{d\varepsilon}{\varepsilon} = \frac{d\sigma}{kT},$$

where

$$kT = \sum_{i=1}^{i=N} \frac{du(T)}{S_o}.$$

After integration and substitution of the initial conditions, we obtain the equation of deformability:

$$\varepsilon = \varepsilon_{max} \exp\left(-\frac{\sigma}{kT}\right). \tag{3.89}$$

This establishes the correlation of strain with the external stress and also with temperature T. The higher the temperature, the lower the rigidity of rotoses Δ, i.e., the smaller the number of electron levels collectivized by atoms that are part of structure of a rotos (see Figure 1.3), the more ductile the material. This relationship is valid for orientation region *oa* of diagram $\sigma = f(\varepsilon)$ (Figure 2.10); in region *ab*, the destruction processes start affecting it.

During the deformation process, a rotos may be in one of the critical states. These states are determined by cold shortness temperature T_c and Debye temperature θ (Figure 1.14). The Debye temperatures (Equations 1.87 and 1.95) are related to each other by Equation 1.101 and can be expressed in terms of the velocity of ultrasound v in a given material[49] as

$$\theta = \left(\frac{\hbar v}{2kr_a}\right)^{1/2}, \tag{3.90}$$

where r_a is the crystalline lattice parameter (Equation 1.48.I).

If we raise both parts of Equation 3.90 to power of 2 and multiply them by mass of atoms that make a rotos, m, we obtain

$$g = \frac{2k}{\hbar} m\theta^2 r_a, \tag{3.91}$$

where $g = mv$ is the ultrasonic pulse released in separation of atoms of the rotos following the diagram shown in Figure 1.15.

The frequency of harmonic oscillations of an atom having mass m in a flat elliptical motion as part of the rotos is found from Equation 1.50 or the condition[30]

$$v = \frac{1}{2\pi}\left(\frac{\Delta}{m}\right)^{1/2}. \tag{3.92}$$

On the other hand, it is related to the θ-temperature by Equation 3.90. Taking into account that, in Equation 3.90, $v = v/2r_o$ is the frequency of oscillations of atoms of the rotos at the θ-temperature, and substituting Equation 3.92 into Equation 3.90, we find a relationship between the coefficient of rigidity of the bond Δ (Equation 2.154) and the θ-temperature

$$\Delta = Bm\theta^2, \tag{3.93}$$

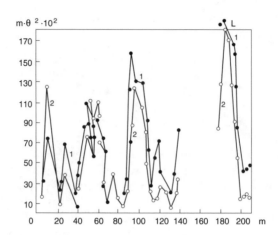

FIGURE 3.69
Variations in evaporation heat L (curve 1) and rigidity coefficient of interatomic bond (2) depending on the position of a chemical element in the periodic system.

where

$$B = \left(\frac{2\pi k}{\hbar}\right)^2$$

is a constant included in the Boltzmann, k, and Planck, η, constants.

Rigidity of the bond Δ depends on the same AM parameters as the Debye temperature θ (Equation 1.87) does. Therefore, it must repeat the manner of its variations (Figure 1.21) in the periodic system of chemical elements. Figure 3.69 shows the dependence of evaporation heat of simple solids, $L = f(m, z, e)$ (curve 1) and rigidity coefficient $\Delta = f(m, z, e)$ (curve 2) on the position of a chemical element in Mendeleev's table.[92] As seen, these parameters, while repeating the manner of variations in θ, do change periodically in the transition from one element to the other.

Rewriting Equation 3.91 and accounting for Equation 3.93, we have

$$g_v = \frac{1}{\pi} B^{1/2} \Delta r_a, \tag{3.94}$$

Acquiring the phase transition temperature θ during the process of internal heat transfer (Figure 2.24) and failing, rotoses radiate internal energy u (Equations 1.22 and 1.23) not only in the form of a thermal phonon (Equation 3.35) but also in the form of an ultrasonic pulse (Equation 3.94).

The phenomenon of producing ultrasonic waves in the disruption of the AM bonds in a solid during the process of energy restructuring is known as acoustic emission. Studies on this phenomenon began recently and at present it is intensively developed.[170–179] It is thought to be one of the most promising areas of investigation of the mechanism of fracture of structural materials.[1,3] The acoustic emission method is based on registration of elastic waves emitted by solids during deformation and fracture, followed by decoding of the information contained in them. Acoustic emission sources are located at micro- and macroscopic scale levels. At the first level, the acoustic emission pulse is generated due to an elementary failure of the AM bond, while at the second, it is generated due to coalescence of microcracks in a cluster. The major advantage of this method is that it offers a possibility for investigation of the fracture process in real time.

A particular shape of an acoustic emission pulse is determined by individual acoustic properties of the radiation source (Equation 3.94) and environment; to the first approximation it can be thought of as depending on peculiarities of occurrence of two successive processes. The first process is associated with a sudden disruption of the bond and instantaneous redistribution of internal forces in its surrounding. It determines the leading front of the pulse. The character of its variations can be approximated by a function of the type of

$$\sigma_i \exp\left(\frac{\tau}{a}\right),$$

where σ_i is the local stress at moment τ of disruption of the bond, and a is the time constant that depends on acoustic properties of the source of the signal and the environment. The second determines the shape of the trailing part of an acoustic emission pulse and is associated with redistribution of stresses in the surrounding of a rotos that failed.

Major parameters of an acoustic emission pulse are amplitude, energy, total duration, duration of the leading and trailing parts, and frequency spectrum. Acoustic emission is a stochastic pulsed process whose parameters are assigned by the set of individual sources chaotically located in the bulk of an object. Regularity of their operation is determined by the MB distribution (Equation 2.2) and by the character of its entering into the range of the Debye temperatures (Figure 1.14). In the force fields, it depends on the kind of stressed state (tension, compression, bending, etc.). The parameters of acoustic emission signal carry valuable information on mechanical-destruction processes.

For example, the amplitude and energy of acoustic emission are associated with the energy of a single acoustic emission event; the energy of an acoustic emission pulse is proportional to its area. A single acoustic emission pulse, while propagating in the bulk of the object, reaches the acoustic emission transducer and then, in the form of a single electric signal, arrives at the input of the recording (measuring) device where it is transformed. On the readout of this device, the signal has the form of an attenuating pulse of high-frequency oscillations. The typical form of such signals is shown in Figure 3.70, where U (volt) is the electric voltage of the acoustic emission

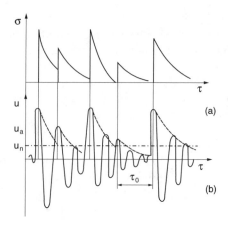

FIGURE 3.70
Original (a) and registered (a) acoustic emission signals.

signal, U_a is its amplitude, U_n is the threshold level, and τ_0 is the pulse duration.

The measuring device records the number of events equal to the number of primary spikes of the acoustic emission pulses and the number of oscillations equal to the number of spikes of the electric signal over the threshold level U_k. It is seen that the number of excesses is associated with the amplitude and energy of the primary signal. Therefore, the energy of the acoustic emission source can be evaluated from this number. Reliability of recording of the number of acoustic emission events depends on conditions of propagation of acoustic signals and resolution of the measuring equipment used.

A no less important parameter is the activity or frequency of occurrence of the acoustic emission pulses, which characterizes the rate of development of damage. Typical dependencies of activity $n(\tau)$, quantity of pulses n, and intensity U of acoustic emission, derived in pulling inserted anchors out of concrete mass (Figure 3.37), are shown in Figure 3.71. Similar results were obtained in compression testing of concrete.[180]

These characteristics allow us to determine the distribution function of single acoustic emission sources by energy $f(F)$. Indeed,

$$f(F) = \frac{dn(F)}{n}$$

represents the total number of rotoses N whose interaction energy is very close to the required one within a range from $F_i + dF$ to F_i, where $n(F)$ is their quantity that immediately fails after the external load reaches F, and n is the number of bonds involved in the work from the beginning of loading and

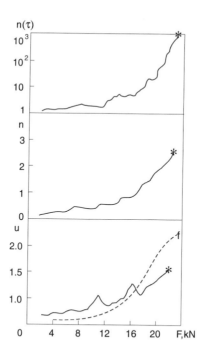

FIGURE 3.71

Variations in the acoustic emission parameters in pulling out of inserted anchors from concrete.

that have already failed by the moment of examination. An increment in the acoustic emission level at a given time moment is equal to

$$dU = U_a dn(F),\qquad(3.95)$$

where U_a is the mean amplitude of the acoustic emission signal under load F_i, calculated as a mathematical expectation of random process[175]

$$U_a = n \int_0^\tau \beta_{a,r}(\tau - \tau_i)d\tau,$$

where β_r is the characteristic of the shape of a single pulse of one of the two types (acceleration or relaxation), τ is the time, and τ_i are the moments of emergence of the acoustic emission pulses.

As proved by experience,[180,181] concrete is characterized by the relaxation type of acoustic signal. Therefore,

$$U_a = n(\tau) \int_0^\tau \beta_2(\tau - \tau_i)d\tau.$$

Assuming that the shape of the signal remains unchanged during the entire period of loading and that the sequence of occurrence of the pulses is continuous, we find that $U_a = n(\tau)D$, where

$$D = \int_0^\tau \beta_r(\tau - \tau_i)d\tau \qquad (3.96)$$

is the constant value for a given material and test conditions. After the external load reaches F, the total acoustic emission level can be found using Equation 3.95:

$$U = \int_0^F U_a dn(F),$$

or accounting for Equation 3.96:

$$U = \int_0^F n(\tau)Dnf(F)dF.$$

Differentiating this expression with respect to the upper limit, and considering P to be a current coordinate, we find $f(F)$ as

$$\frac{dU}{dF} = Dn(\tau)nf(F),$$

or, finally,

$$f(F) = [Dn(\tau)n]^{-1}\frac{dU}{dF}. \qquad (3.97)$$

That is, to obtain the function of distribution of rotoses by energy $f(F)$, it is necessary to write the character of variations of the acoustic emission parameters n, $n(\tau)$, and U during the process of loading a material and differentiate dependence $U = f(F)$ (dotted line in the lower field of Figure 3.71).

In Expression 3.97, the first term $A = [Dn(\tau)n]^{-1}$ is the normalization factor. It can be stated that, accounting for the assumption made earlier that $1/D =$ const, the character of its variations is determined by the type of dependencies $n(\tau) = f(F)$ and $n = f(F)$ obtained from the experiment (upper and medium fields in Figure 3.71). Figure 3.72 shows the general form of function $A = f(F)$.

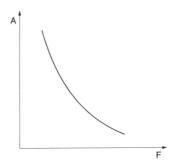

FIGURE 3.72
Variation of normalization factor A depending on the level of stressed state F.

From the formal standpoint, Figure 3.72 indicates that A in Expression 3.97 changes direction of the gradient of function

$$f = \frac{dU}{dF}$$

to opposite (starting from the point of inflection of the dotted curve in the lower field of Figure 3.71). From the physical standpoint, this indicates the objective nature of the rigid-bond diagram of fracture of solids that originates in their AM structure. In fact, when $F \to 0$, then $A \to \infty$, and at a small but finite value dU/dF and $f(F) \to \infty$ at an early stage of loading.

In addition to mechanical, thermal, and ultrasonic effects, fracture of solids is accompanied by electron emission. This phenomenon was registered as a discovery in 1984[38] and is explained by the fact that atoms, while losing bonds between themselves at the θ-temperature, immediately move to the potential wells (Figure 1.7) of their closest neighbors (according to the diagram of Figure 1.15). As such, common molecular electron shells (e.g., 3 and 4 in Figure 1.3) disintegrate. Having lost centers Fe and C around which they previously moved along the closed paths, valent electrons become free and thus the fracture surface becomes electrically charged.

It follows from Equation 3.51 that deformation dV/V not only decreases the number of interatomic bonds dN/N in volume V developing destruction processes and decreases entropy (Equation 2.13) dT/T stimulating orientation processes, but also affects the chemical composition of a solid $a\delta/\delta$, while changing parameter θ following the law set by Equation 2.141. Ample evidence of these effects can be found in texts on materials science[6,7,60] and resistance of materials.[2,3] All of the preceding effects find practical application in studies of stressed–deformed state of materials, parts, and structures, as well as for other purposes.

4

Variations of State under Dynamic and Quasi-Static Loading Conditions

4.1 Introduction

In addition to the parameters considered in Chapter 3, another important parameter, namely time, affects the state of bodies. Its effect is accounted for in the differential form of the equation of state (Equation 2.140). Chapter 3 considers the processes of variations of the state that take place during time intervals achieved on modern testing equipment and specified by appropriate standards. If a variation of state takes place within a time interval shorter than the standard one, the dynamic effect should be added into the consideration (Section 4.2); if this time is much longer, then durability is under consideration (Section 4.3).

Section 4.2.1 discusses the extent to which the dynamic effect has been investigated. Section 4.2.2 deals with the physical nature of this effect; in experiments, it becomes apparent through strengthening, embrittlement, and reduction in the total fracture surface area. Theoretical conclusions are confirmed by testing standard specimens subjected to compression and tension (Section 4.2.3), and by testing large-size structural members in bending (Section 4.2.4).

The differential form of the equation of state (Section 2.5) allows deriving the equation of durability in force and thermal fields (Section 4.3.1). Good agreement between theory and experiment was achieved using compression, tension, and bending tests (Section 4.3.2). Section 4.4.2 presents a way to achieve durability of parts and structures within the limits of existing design methods. Section 4.3.4 traces links between basic characteristics of long-time structural processes: durability, creep, and fatigue life. An insight into the nature of these processes makes it possible to suggest principles of the physical theory of reliability and a method for calculation of durability (Section 4.3.5).

4.2 Dynamic Effect (DE)

As shown in Section 1.11, the crystalline structure of a solid is not indifferent to the rate of application of a mechanical load (Equation 1.104 and Figure 1.26). Its increase leads to a certain distortion of the internal potential field (Figure 1.8) to enhance the effect of resistance to external effects.

4.2.1 The Known Results

Depending on the functional application, parts and structures are subjected to a wide variety of external loads, from static to dynamic and with different intensity and duration.[1-3] Moreover, independently of service loading conditions, many parts and structures are subjected to random dynamic effects (technological, transportation, erection, etc.) in earlier periods of their existence. Engineering experience supports the theoretical estimates (Equation 1.104) of sensitivity of many materials to the rate of deformation,[42-48] which is natural, because time is one of the basic parameters of the equation of state (2.138).

The modern mechanics of materials used in design tries to compensate for the lack of knowledge about reaction of materials to expected and random dynamic effects by assigning increased safety factors to the results of theoretical calculations. However, as shown in Section 3.6, this does not solve the problem of safety. Moreover, it aggravates the situation by increasing the probability of fracture of parts and structures due to the scale effect. Only knowledge of the objective principles of processes occurring in a structure under the effect of various loading conditions and their arbitrary alteration allows development of efficient methods for controlling reliability and durability of materials and structures. This is the only way to create conditions for rational utilization of material resources and elimination of accidents and disasters due to materials failures.

The influence of the deformation rate on mechanical properties of metals is described in References 2 and 3, that for building materials in References 182 through 190, and for natural rocks in References 191 through 193. For example, Kvirikadze[182] used specimens made of heavy and light concrete having $10 \times 10 \times 40$ mm dimensions in tension and compression tests, varying the rate of loading from $2.5 \cdot 10^{-4}$ to $4.75 \cdot 10^{-5}$ MPa/sec. He concluded that an increase in the rate of loading leads to an increase in strength in both compression and tension.

Similar experiments were carried out at the Bucharest Civil Engineering Research Institute.[183] Cubic concrete specimens were subjected to compression tests. The rate of loading was varied from 0.7 to 4.6 MPa/sec, resulting in increase in strength from 5 to 15%.

Verker[184] described the results of Watstein's experiments on the effect of rate of loading on strength and deformation properties of concrete. According to his data, variation of the rate of deformation by six orders of magnitude leads to an increase in strength and elasticity modulus of light and heavy concrete by a factor of 1.85. In experiments conducted by Sheikin and Nikolaeva,[186] variation of the load rate from $4.3 \cdot 10^{-4}$ to $2.6 \cdot 10^{-5}$ MPa/sec, i.e., by a factor of 160, led to a variation of 10% in strength properties of concrete. In experiments conducted by Bazhenov, strength of cement concretes increased by 25 to 30% with a five-fold increase in the rate of loading.[187] Grushko, Glushchenko, and Ilyin investigated the effect of dynamic loads in bending on a structure of concrete.[188] Fine-grained concretes and concretes based on granite and limestone aggregates were tested; variation of the load rate by five orders of magnitude led to a variation in strength by 20 to 40%.

Results of experimental studies on the effect of the rate of loading on the stressed–deformed state of rock are given in References 192 and 193. The tests were conducted on cylindrical limestone samples 30 mm in diameter and 45 mm high by varying the load rate from 20^2 to 10^{-4} MPa/sec under different supporting conditions: e.g., specimens of one group were tested with their supporting surfaces lubricated with paraffin. The test time ranged from 0.3 sec to 110 h. Variation of the load rate by six orders of magnitude led to a deviation of the value of mean elasticity modulus by a factor of 4 to 6, depending on the stress level.

Analysis of these experimental data allows the following conclusions:

- All materials, without exception, react to an increase in the rate of loading by increasing resistance and decreasing deformability, i.e., through changing the structure of the Ω-potential.

- The intensity of strengthening depends on the peculiarities of the structure of a specific material.

- Most studies discuss the existence of the dynamic effect and provide its estimates; only a few consider its physical nature. However, no one widely accepted concept of the depth and causes of the dynamic effect exists. Some investigators identify it with temperature effects,[105] concluding that, for steels, an increase in temperature by 61°C is equivalent to a decrease of about 1000 times in the rate of deformation.

We should point out another methodological discrepancy in the considerations of parts and structure made of several materials essentially differing in their physical nature. On the one hand, results of the fundamental theoretical–experimental studies on physics of strength[6,7,10,48] and systematized experimental data[3,19,45] indicate that the time of loading is a basic parameter of deformation and strength of structural materials. On the other hand, different standards disregard this fact in imposing requirements on the rate of application of load on the determination of physical–mechanical characteristics of materials[194–198] and in the estimation of load-carrying and deformation properties of finished structures.[199] Under actual service conditions, the intensity of loading of a structure varies over wide ranges; however, such different materials as steel and concrete contained in one structure are always subjected to external effects at the same rate. Unfortunately, this fact is ignored.

For example, in determination of resistance of cement to bending, the mean load rate is specified to be equal to (50 ± 10) N/sec. In determination of resistance to compression, this rate is equal to (2 ± 0.5) MPa/sec or, accounting for the standard dimensions of specimens, it is specified to be equal to (5 ± 1.25) kN/sec, i.e., higher by two orders of magnitude.[195] When testing concrete to compression, the load on a specimen should be gradually increased at a constant rate equal to 0.6 to 0.4 MPa/sec up to fracture.[194] The rate of load application in the evaluation of strength of concrete in tension and cleavage was set at (0.05 ± 0.02) MPa/sec,[194] i.e., by a lower order of magnitude than in compression. The load rate of the reinforcement steel samples in tensile tests should not exceed 10 MPa/sec.[198]

Other investigators also indicate drawbacks of the regulatory documents. For example, Krol and Krasnovsky[189] are of the opinion that these documents should only specify general conditions and requirements to tests rather than stringently set a particular procedure regardless of the test objectives. The use of the same procedure inevitably leads to losses associated with missing or depreciation of useful information.

This is equally applicable to the standard on methods for estimation of strength, rigidity, and crack resistance of assembled reinforced concrete structures (the standard for CIS countries). This standard specifies the sequence of steps of application of the load, but does not regulate the rate of its application at each step.[199] According to this standard, the load is applied in incremental steps in fractions of a control load. At each step, the load is kept invariable during no less than 10 min up to the control load and not less than 30 min under the control load. Then, the archived load is kept invariable during 10 min in the further incremental steps up to fracture. Experience shows that, with the available means of loading and those of measurement of deformations, the time needed to complete a test of one specimen is 24 h or even longer if the preparatory operations are included. Such intensity and sequence of operations are very far from that wide variety of loads, which occur under actual service conditions, and not in agreement with those loading conditions under which strength properties of concrete and steel were estimated.

Discrepancies in recommendations for the rates of loading are associated with lack of systematic studies on the effect of the intensity of loading on the stressed–deformed state of materials and structures. Nothing can justify ignoring this fact because loading conditions have substantial impact on the reliability and durability of parts and structures.

Investigation of the fracture principles under dynamic effects becomes especially important in connection with the necessity to develop practical methods and techniques for deceleration or activation of destructive processes. The former will find application in fabrication of structures with guaranteed failure-free operation and minimum material consumption. The

latter will promote decreases in power and labor consumption of technological processes employed in production and processing of natural resources,
and forming operations in fabrication of parts and structures.

4.2.2 Physical Nature of the Dynamic Effect and Its Manifestation through Strengthening, Embrittlement, and Localization of Fracture

We can express the volume of a body, V, from Equation 3.4 in explicit form
and differentiate it in analogy with Equation 3.41, which yields

$$dV = \frac{d\Omega(T)}{\mathbf{P}} - \frac{\Omega(T)d\mathbf{P}}{\mathbf{P}^2}. \tag{4.1}$$

It follows from comparison of Equations 3.40 and 4.1 that rates of variation
of resistance, $d\mathbf{P}$, and deformation, dV, are reciprocals. Graphical representation of both has the form of a curve with a maximum (Figure 3.23). Rewriting
Equation 4.1 and accounting for Equations 3.40 and 2.137 yields

$$\eta + \varepsilon = \frac{d\Omega(T)}{\Omega(T)} - \frac{d\mathbf{P}}{\mathbf{P}}. \tag{4.2}$$

It can be seen that, at any CD transformation of a structure, $d\Omega(T)/\Omega(T)$,
the higher the relative rate of variations in resistance $(\eta + \varepsilon)$, the lower the
deformability $d\mathbf{P}/\mathbf{P}$. So, what is the cause of this relationship?

To answer this question, consider again the equation of state in the form
of Equation 2.65. Parameters V and T can be conditionally regarded as the
external parameters, and $\mathbf{s}$ and $\mathbf{P}$ as the internal parameters. The latter cannot
change on their own; they are forced to change under the effect of external
conditions. Volume V is a potential variable. To change it, a given body must
come in direct contact with other objects.

Unlike a potential interaction, the thermal interaction can take place without
contact. The majority of engineering objects used in practice have no regulated
thermal insulation so they continuously exchange heat with their surroundings. Normally, these objects are small compared with the environment and
thus they can be regarded as bodies placed in contact with a heat tank.
Although this fact is natural and thus hardly ever noticed, it plays a decisive
role in structural CD transformations, which, in turn, may lead to catastrophic
consequences for a material at a sudden decrease in the ambient temperature.

Differentiation of the left- and right-hand parts of Equation 2.65 yields
Equation 2.134. We can regroup its terms as

$$V\dot{\mathbf{P}} - \mathbf{s}\dot{T} = T\dot{\mathbf{s}} - \mathbf{P}\dot{V}, \tag{4.3}$$

where, in analogy with Equation 1.23, parameters of the type of $\dot{\mathbf{P}}$ are
differentials with respect to time, and compare it with the first law of

thermodynamics written as Equation 2.24. The latter, relating the right-hand part of Equation 4.3 to variations in the overall internal energy of a body, $\dot{U}$,

$$\dot{U} = T\dot{s} - \mathbf{P}\dot{V}, \tag{4.4}$$

is the law of conservation of energy in the deformation process of a body subjected to the action of external objects (direct work). At $\mathbf{P}$ = const, T = const, and $\dot{V} \neq 0$, it results in $\dot{s} \neq 0$. Therefore, the left part of Equation 4.4 can be considered the law of conservation of energy for the body to perform work over external objects (reverse work) due to variations in temperature $\dot{T} \neq 0$ through $\dot{\mathbf{P}} \neq 0$ at V = const and s = const. This means that the differential form of the equation of state (4.3) expresses the generalized law of conservation of energy for any thermodynamic processes. With opposite variations, the direct and reverse works are mutually compensated for and, under specially created conditions, can have no effect on a structure. We suggest using this phenomenon for justification of thermodynamic methods to control deformation and fracture of engineering objects.[12,18,21]

A very important practical conclusion follows from Equation 4.3: reliability and durability of engineering objects are determined not as much by their physical–mechanical characteristics as by their thermal–physical ones. This fact should always be borne in mind in selection of materials for parts and structures operating in dynamic external fields.

It can be seen from Equations 3.41 and 4.1 and Figure 3.23 that the resistance of a body increases with an increase in the deformation rate, reaches maximum σ_m at point m under optimal loading conditions for a given material, V_m, and gradually decreases to zero with further intensification of loading. Existence of this relationship between resistance and deformation rate is not an exotic feature or exception to the rule, but rather the rule that follows from the equation of state (2.65).

We can rewrite Equation 3.41, accounting for the definition of the thermal expansion coefficient α (Equation 2.152) and providing that dV = const and

$$\left(\frac{dV}{dT} \right)_P = \text{const:}$$

$$D = \frac{\dot{\Omega}(T)}{C_1 \alpha} - \frac{\Omega(T)}{C_2 \alpha^2}, \tag{4.5}$$

where $D = \dot{P} / \dot{V}$ is the increment in the resistance of the AM structure on high-rate deformation, which is a quantitative measure of the dynamic effect.

Figure 4.1 shows the dependence of D on parameter α. As seen, with an increase in the rate of volume deformation, $\dot{V}$ (which corresponds to a decrease in parameter α), dynamic resistance D increases (movement from the right to the left along the axis of abscissas), reaches maximum D_d at point

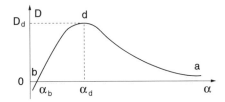

FIGURE 4.1
Thermodynamic characteristic of solid bodies.

d, and then dramatically decreases with further acceleration of loading (trend from α_d to α_b).

Theoretically obtained Equation 4.5 finds excellent confirmation in practice. As concluded in Troshenko[3] using the results of numerous experiments, dynamic strengthening and embrittlement of materials occur at low and moderate deformation rates (up to 10^3 s^{-1}) (in regions $\alpha > \alpha_d$ of the diagram in Figure 4.1). Inversion of the dynamic effect occurs at the deformation rates close to $5 \cdot 10^4$ s^{-1}: instead of increasing, the resistance decreases, accompanied by an increase in ductility. In the dynamic diagram, this region is located to the left from inversion point d. Hereafter, the dynamic effect to the right of inversion point d (in region ad) is referred to as the direct dynamic effect, and that to the left (in region db) is referred to as the inverse dynamic effect. The direct dynamic effect is of highest practical interest.

Equality $D = 0$ is possible only when one of two conditions is met: either $\dot{P} = 0$ or $\dot{V} \to \infty$. As such,

$$1/\alpha_b = \dot{\Omega}/\Omega. \tag{4.6}$$

Differentiation of Equation 4.5 yields

$$\dot{D} = \frac{\ddot{\Omega}}{\alpha} - \frac{\dot{\Omega}(\dot{\alpha}+1)}{\alpha^2} + \frac{2\Omega\dot{\alpha}}{\alpha^3}. \tag{4.7}$$

At inversion point $d : \dot{D} = 0$. Equating (4.7) to zero yields a quadratic equation with respect to $1/\alpha$:

$$\frac{2\Omega\dot{\alpha}}{\alpha^2} - \frac{\dot{\Omega}(\dot{\alpha}+1)}{\alpha} + \ddot{\Omega} = 0.$$

Assuming that $\alpha \gg 1$, we find its solution in the form

$$\left(\frac{1}{\alpha_d}\right)_{1,2} = \frac{1}{4}\frac{\dot{\Omega}}{\Omega} \pm \left[\frac{1}{16}\left(\frac{\dot{\Omega}}{\Omega}\right)^2 - \frac{\ddot{\Omega}}{2\Omega\dot{\alpha}}\right]^{1/2}. \tag{4.8}$$

As $\alpha_b < \alpha_d$ (Figure 4.1), $1/\alpha_b > 1/\alpha_d$. Comparison of Equations 4.6 and 4.8 shows that expression in square brackets cannot exceed $\frac{3}{4}\frac{\dot\Omega}{\Omega}$. In the limit,

$$\frac{3}{4}\frac{\dot\Omega}{\Omega} = \left[\frac{1}{16}\left(\frac{\dot\Omega}{\Omega}\right)^2 - \frac{\ddot\Omega}{2\Omega\dot\alpha}\right]^{1/2};$$

Hence

$$\dot\alpha = -\frac{\Omega\ddot\Omega}{\dot\Omega^2} \qquad\qquad (4.9)$$

or, after integration,

$$\alpha = \ln\left(\frac{\dot\Omega}{\Omega} - \frac{\ddot\Omega}{\dot\Omega}\right) + \text{const}. \qquad\qquad (4.10)$$

We can express Equation 4.8 accounting for Equations 4.6 and 4.10:

$$\left(\frac{1}{\alpha_d}\right)_{1,2} = \frac{1}{4\alpha_b} \pm \frac{1}{\alpha} + \text{const}.$$

This equation shows that a change in rigidity (or deformability, which is the same, accounting for their inverse Relationship 2.154) to one side from the inversion point d leads to an increase in the resistance of the body, and that to the other side leads to its decrease. The capability of the kinetic part of internal energy Ω of redistributing (first, $\dot\Omega$, and second, $\ddot\Omega$ derivatives) in a given volume V is responsible for all these changes.

In the region of the direct dynamic effect we can ignore the higher-order efects. Thus, ignoring the second quadratic term in Equation 4.5, we can write

$$\dot{\mathbf{P}} = \mathbf{P}_0(\dot\eta + \dot\varepsilon), \qquad\qquad (4.11)$$

where $\mathbf{P}_0 = \Omega\dot V$ is the resistance of a body under quasi-static loading conditions. As seen, the higher the transverse, $\dot\eta$, or longitudinal, $\dot\varepsilon$, (or both) deformation rate, the higher the increment $\dot{\mathbf{P}}$ in quasi-static resistance $\mathbf{P}_0$. This rule is universally valid in practice. The meaning of Equation 4.11 becomes apparent from consideration of deformation diagrams $\sigma = f(\varepsilon)$. Figure 4.2 shows such diagrams for concrete (a) and limestone (b)[193] obtained in compression, σ_p, and steel (c)[3] in tension, σ_s.

Numbers in the plots stand for the following deformation rates: in Figure 4.2a, in kN/sec, 1 = 2.0, 2 = 1.2, and in Figure 4.2b in MPa/sec, 1 = 10^2, 2 = 10, 3 = 1, 4 = 10^{-1}, 5 = 10^{-2}, 6 = 10^{-3}, 7 = 10^{-4}. Relative deformations expressed in arbitrary units are plotted on the abscissa in Figure 4.2a and those expressed in percent are plotted in Figure 4.2b. Figure 4.2c shows the effect of the longitudinal

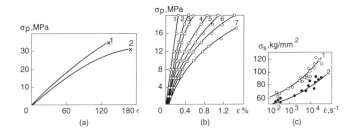

FIGURE 4.2
Manifestation of the dynamic effect in concrete (a) and limestone (b) in compression and in steel 1045 (c) in tension.

deformation rate $\dot{\varepsilon}$ in s^{-1} on the upper, 1, and lower, 2, yield limits of steel 1045. If we take into account that α in Definition 2.152 is a value inverse to ε, it becomes clear that, for steel, all the experimental points lie in region ad of the direct dynamic effect (Figure 4.1).

Similar diagrams can be found in other publications on dynamic strength of various materials.[182–193] For example, Troshenko[3] discusses a substantial effect of the deformation rate on strength and deformation properties of polymers. For nylon, its increase from 5 to 15 mm/sec leads to an increase in the yield and tensile strength, as well as in elasticity modulus by 1.6, 1.13, and 1.94 times, respectively, while strains at fracture decrease by a factor of 2.3. Analysis of typical deformation diagrams for many other materials given in Troshenko[3] shows that all of them repeat that shown in Figure 4.1. An equidistant shift of the temperature dependence of the yield strength of steel 10G2FB at different deformation rates (Figure 2.31) also supports this fact.

Maximum of function $\mathbf{P} = f(\dot{V})$ is determined from the condition that both parts of Equation 4.2 are equal to zero, i.e.,

$$\eta = \varepsilon \quad \text{(I)}, \qquad \frac{d\Omega(T)}{\Omega(T)} = \frac{d\mathbf{P}}{\mathbf{P}} \quad \text{(II)}. \tag{4.12}$$

Taking into account the kinetic nature of deformation (Equation 2.105) and resistance (Equations 2.152 and 2.153), it can be concluded that the dynamic transformation of the MB distribution (Figure 2.36) up to the point of maximum m leads to utilization of the kinetic part of material strength (Figure 2.37). At point m it becomes fully exhausted (Equation 4.12.I). Equality of longitudinal, ε, and transverse, η, strains (Equation 4.12.I) indicates that the body can no longer resist a change in its shape. Upon passing through point m, where

$$\frac{d\Omega(T)}{\Omega(T)} > \frac{d\mathbf{P}}{\mathbf{P}},$$

the resistance decreases and an excess heat goes to an increase in ductility.

The definition of the differential of entropy (Equation 2.13) and the relationship of equivalence of parameters T and V (Equation 2.105) allow us to express Equation 4.3 as

$$PV\left(\frac{\dot{V}}{V}+\frac{\dot{P}}{P}\right)=-\dot{T}[1+s(P)_T].$$ (4.13)

By definition (see Section 2.8), $s(P)_T = k\ln Z(P)$ is a value that, by many orders of magnitude, is less than 1. This allows the left-hand part of Equation 4.13 to be simplified by approximation $1 + s(P)_T \approx 1$. Then,

$$P\left(\frac{\dot{V}}{V}+\frac{\dot{P}}{P}\right)=\frac{\dot{T}}{V}.$$ (4.14)

The relationship of Equivalence 2.105 suggests that

$$\frac{\dot{V}}{V}=\frac{\dot{T}}{V}=d\mathbf{s}.$$ (4.15)

Substitution of Equation 4.15 into Equation 4.14 yields

$$\dot{P}=d\mathbf{s}(1+P).$$

Assuming that $P \gg 1$, write that

$$\dot{P}=Pd\mathbf{s}$$ (4.16)

or, in accordance with Equation 4.15, again coming back to strains, we can write

$$\dot{P}=P\frac{\dot{V}}{V}.$$ (4.17)

It can be seen that a material responds to any attempt to change its geometrical size and shape by an adequate increase in resistance. Our consideration resolves differing opinions on the terminology associated with the dynamic effect.[189] It shows that the terms "rate of deformation" and "rate of loading" are adequate.

Now we are ready to consider the question about the processes occurring in the structure of a body under dynamic loading conditions. To answer this question, we must find differentials of both parts of Equation 2.96:

$$P=\left(\frac{\dot{P}}{P}+\frac{\dot{V}}{V}\right)=G\delta\left(\frac{\dot{G}}{G}+\frac{\dot{\delta}}{\delta}\right),$$

where $G = kNT$. Under quasi-static conditions, it always holds that

$$PV = G\delta. \tag{4.18}$$

Then,

$$\frac{\dot{P}}{P} + \frac{\dot{V}}{V} = \frac{\dot{G}}{G} + \frac{\dot{\delta}}{\delta} \tag{4.19}$$

Equation 4.18 is feasible only if $P = G$ and $V = \delta$. When $\dot{P} = 0$ and $\dot{G} = 0$, Equation 4.18 becomes

$$\left(\frac{\dot{V}}{V}\right)_P = \left(\frac{\dot{\delta}}{\delta}\right)_G.$$

Subscripts P and G show that this equality is valid only if $P = $ const and $G = $ const. Accounting for Equation 2.137 makes it possible to write that

$$\dot{\eta} + \dot{\varepsilon} = \frac{\dot{\delta}}{\delta}. \tag{4.20}$$

By Definition 2.141, it can be considered that

$$\delta = \left(e\frac{T}{\theta}\right)^3,$$

so then

$$\dot{\delta} = 3\delta\left(\frac{\dot{T}}{T} - \frac{\dot{\theta}}{\theta}\right), \tag{4.21}$$

where e is the base of natural logarithms.

Substitution of Equation 4.21 into Equation 4.20 yields

$$\dot{\eta} + \dot{\varepsilon} = 3\left(\frac{\dot{T}}{T} - \frac{\dot{\theta}}{\theta}\right). \tag{4.22}$$

Expressions 4.20 and 4.22 reveal the physical nature of energy restructuring that leads to the dynamic effect. It can be seen that transverse, $\dot{\eta}$, and longitudinal, $\dot{\varepsilon}$, deformation rates change the initial order of the atomic–molecular (AM) system of the body, the quantitative measure of which is entropy ds (Equation 2.13) and the Debye temperature θ (Equation 1.87). We can say that

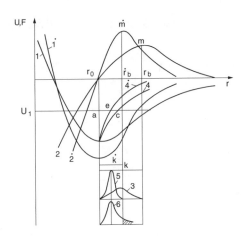

FIGURE 4.3
Explanation of the physical nature of dynamic strengthening and embrittlement of materials.

the dynamic effect changes the configuration of the local potential field of individual rotos (Figure 1.8) and of the entire set N (positions 8 to 15 in Figure 1.14). Deviation of entropy ds from an equilibrium value changes the structure of a body according to the diagram shown in Figure 2.8, and $d\theta$ leads to its chemical regeneration, whose phenomenological description can be found in any course of metals science.[6,7,60] Graphical interpretation of the process set by Equation 4.22 is shown in Figure 4.3.

In the equilibrium state (Equation 4.18), let the flat section of the potential field of the interatomic bond in an arbitrarily chosen direction r (plot of the potential energy) have form 1, corresponding to force characteristic 2. The initial value of the Debye temperature θ_0 is determined by vertical line mrb,[24] which divides the energy state of each bond into two phases: compresson $(r_0 - r_b)$ and dilaton $(r_b - r)$. We assume that, at the initial time moment, the Maxwell-Boltzmann distribution 3 is in the compresson phase, as is shown in the lower field of Figure 4.3. When the energy state of the bond changes in depth of potential well 1, its geometrical size is set by curve 4. At level u_1, let the size of the bond in direction r be characterized by length ac.

High-rate deformation leads to transformations of the internal potential field: now it is defined by curve $\dot{1}$, instead of curve 1. This causes a corresponding change in the force (from 2 into $\dot{2}$) and deformation (from 4 into $\dot{4}$) curves and, thus, leads to a shift of vertical line mrb to position $\dot{m}\dot{r}_b$, which is equivalent to a change of the Debye temperature from θ_0 to θ. As such, the rigidity of the bond also changes. Its deformability at level u_1 is now determined by length ae, instead of ac. The compresson phase is reduced from $k = r_0 r_b$ to $\dot{k} = \dot{r}_0 \dot{r}_b$. Steepness of the supporting compresson front $r_0 \dot{m}$ of curve $\dot{2}$ grows, compared with previous position 2. The described processes inevitably lead to a change in location and shape of the Maxwell-Boltzmann distribution; it moves from position 3 to position 5.

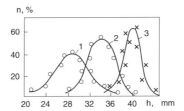

FIGURE 4.4
Empirical distribution of elastic rebound of the indenter at three rates of loading in cubic samples.

Increase in steepness of the supporting compresson front $r_0\dot{m}$, accompanied by narrowing of the Maxwell-Boltzmann distribution, leads to an increase in strength of the material and, simultaneously, to its embrittlement because of an increase in the rigidity of the bonds. Steps of energy restructuring may not coincide in time. For example, if the discussed shift of vertical line $\dot{m}\dot{r}_b$ outstrips the process of transformation of the distribution, it may lead to failure of the right-hand part of the bonds (dashed zone on curve 6).

Experiments prove the described mechanism of structural rearrangement. Concrete cubic specimens of 20 cm edge size were subjected to compression with the same reduction at different load rates. Then, within a very short period of time (not until the equilibrium state was achieved), the values of hardness were measured using the method shown in Figure 3.19c. Figure 4.4 shows the obtained distributions of hardness at the following loading rates (kN/sec): 1 = 1.2, 2 = 2.5, and 3 = 3.5. As seen, an increase in the rate of reduction is accompanied by an increase in mean hardness (the distribution shifts to the right) and a simultaneous increase in homogeneity (steepness of the distribution fronts grows).

Recurring to previous designations for longitudinal, $\dot{\varepsilon}$, and transverse, $\dot{\eta}$, strains (Equation 2.137), and accounting for the relationship of equivalence Equation 2.105[24] (which shows that deformation can have no nature other than thermal), we can express Equation 4.22 as

$$\frac{ds}{s} = 2\frac{dL}{L} - 3\frac{d\theta}{\theta},$$

where differentiation with respect to time is omitted. Integration and accounting for the initial conditions yields

$$\frac{S_1}{S_{0_1}} = \frac{S_2}{S_{02}}\left(\frac{\theta_0}{\theta}\right)^3,$$ (4.23)

where S_{0_1}, S_1 are the normal (to the direction of action of the dynamic forces) cross-sectional areas of a body before and after fracture, respectively, and

$$S_{0_2} = L_0^2; \quad S_2 = L^2$$

are the parallel (to the direction of action of the dynamic forces) cross-sectional areas of a body before and after fracture, respectively.

Equation 4.23 can be rewritten as

$$S_1 = S_{0_1} \frac{S_2}{S_{0_2}} \left(\frac{\theta_0}{\theta} \right)^3. \tag{4.24}$$

Assuming that, when quasi-static loading is the case, the cross-sectional area of a body after fracture, S_1, hardly differs from the initial one, S_{0_1}, we can see that dynamic loading leads to decrease of S_1, which is caused by apparent inequalities

$$S_{0_2} > S_2 \quad \text{and} \quad \theta > \theta_0.$$

Indeed, Relationship 4.24 is valid in practice.

Figure 4.5a shows variations in fracture surface S under dynamic conditions of loading of concrete having rate v in tension (1) and compression (2). Curve 1 was plotted using the actual surface of fracture (Figure 3.37b) in the test where the inserted anchors (3) were pulled out of block (4) at different load rates as shown in Figure 4.5b. Deformations of the cone of cleavage (5) were fixed by linear displacement sensor (6). The loading device was attached to the threaded part of the anchor (3) (Figure 3.37a). The data for curve 2 were obtained from stereological investigations[120] of internal sections

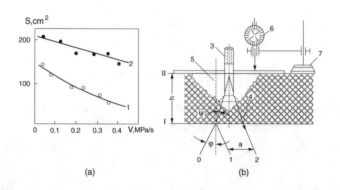

(a)   (b)

FIGURE 4.5
Variation in the fracture area (a) in pulling inserted anchors out of concrete (b) under dynamic loading by tension (1) and compression (2).

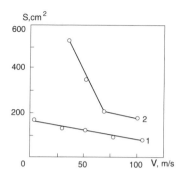

FIGURE 4.6
Impact of the dynamic effect on the fracture surface for marble (1) and limestone (2).

of cubic samples brought up to a certain degree of reduction at different rates (Figure 3.19c).

Studying the effect of the loading rate on the efficiency of rock fracture, Voblikov and Tedder[192] observed a similar trend (Figure 4.6). They fractured marble (1) and limestone (2) using a striker with the chisel angle of 90°. The striker was moved at different speeds v while its stored kinetic energy was kept constant. Variations in the loading rate within two orders of magnitude (from 1 to 100 m/sec) led to a decrease in fracture surface S cm^2 of the marble samples by a factor of 2, and that of the limestone samples by a factor of 3. A similar effect is observed in metals and alloys. For example, it was shown in experiments on the dynamic resistance to cracking of steels X70 and 1012FB that the cross-sectional area of crack propagation decreased by approximately 60% when the load rate increased.[3]

In Equation 4.22, the conditions of dynamic deformation are deliberately not specified. Thus, the dynamic effect can take place in forward (loading at an increasing rate) and in reverse (unloading) directions. This assumption can be proved experimentally. In the previously described experiment, the level of reduction was kept constant for a certain time period for some cubic specimens. Then the specimens were unloaded at different rates. The time of unloading varied from several seconds to several hours. After this, the specific length of cracks, ΣL cm^{-1} per unit surface of a cut was measured as discussed in Saltykov.[120] The relationship between the specific length of cracks and the rate of unloading obtained this way is shown in Figure 4.7. As seen, exceeding a certain unloading rate leads to the intensification of the cracking process. For a given composition of concrete, the ultimate rate of unloading turned out to be equal to 0.8 kN/sec.

A similar effect was observed by Rozhkov[200] after unloading the specimens kept under load for a long time. He pointed out that high-rate unloading of concrete from any level of stresses does affect its structure, inducing extra internal stresses that cause formation of new microfractures chaotically oriented with respect to the direction of action of the external forces.

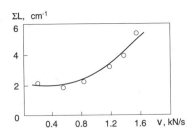

FIGURE 4.7
Effect of unloading rate on the intensity of cracking of concrete.

Therefore, an increase in the intensity of the input of mechanical energy into a body causes an increase in homogeneity of the internal potential field (compare the shape of curves 3 and 5 in Figure 4.3), which leads to increased resistance. In the homogeneous fields, this favors the processes of initiation, development, and propagation of cracks (see comments on Figure 3.17), thus decreasing the total fracture surface.

4.2.3   Investigation of the Effect in Tension and Compression

Not many publications are dedicated to investigation of the extent to which the rate of deformation affects physical–mechanical properties of the same material at different stressed–deformed states. Here, we present results of manifestation of the dynamic effect in tension (rupture), compression, and bending of specimens made of heavy concretes of different compositions. The rate of loading was varied from $1.6 \cdot 10^{-3}$ to 0.4 MPa/sec. The compressive stressed–deformed state was created in specimens of cubic shape with 10 and 20 cm edge size (Figure 3.19c) and in prisms (Figure 3.19b). At a given loading rate, the testing time of one specimen ranged from 40 sec to 8 h.

Dynamic tensile tests were conducted by pulling out inserted anchors from concrete (Figure 3.37). The actual fracture surface was determined after each removal. During loading, the measurement of dielectric characteristics of the stressed material was carried out using the SHF electromagnetic method (Figure 3.61a), acoustic emission, and by absolute strains.

The scheme of the experiment is shown in Figure 4.5b. The narrow electromagnetic beam (designated by 0) was used to scan at an angle to the rear surface (I) of short concrete columns (4) having thickness h. Inserted anchors (3) were installed on front surface II during concreting. The process of loading and pulling out resulted in the formation of cleavage cone 5 (Figure 3.37b). Incident beam 0 penetrates through block h to reach the metal plate located on surface II, then is reflected from it and goes out from concrete in direction 2, at distance a from beam 1 reflected from surface I. Strains were measured using remote displacement sensor 6 installed on frame 7.

For the test, 120 cubic specimens with 10 cm edge size were made from each of the three compositions of concrete to calculate parameters of statistical

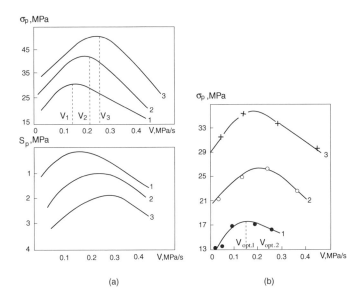

FIGURE 4.8
Variation in strength of heavy concrete in compression under dynamic conditions of loading of cubic specimens (a) and prisms (b).

distributions of the compression strength. Specimens of each series were divided into three groups with 40 pieces each, then the specimens of each group were tested under three different loading conditions. Prior to loading, the hardness of all the specimens (cubes, prisms, and columns) was tested using a dynamic-hardness tester (the Shore scleroscope) at three values of impact energy (Figure 3.19c).

Figures 4.8 and 4.9 show variations in the major parameters of statistical distribution of strength of heavy concretes in compression and tension. Specimens were made from cement 500, and their water–cement ratio was 1 = 0.57, 2 = 0.48, and 3 = 0.39, respectively. Standard deviation for the tensile strength was calculated from nine measurements for each concrete composition under all loading conditions. Attempts to obtain dependence of the $S_p = f(v)$ type for strength of prisms failed because it was impossible to obtain a representative statistical sampling.

Function $\sigma_p = f(v)$ has a maximum for all of the investigated compositions. Its formation is associated with a different degree of deformation rearrangement of the structure and confirms theoretical dependence (4.11) shown graphically in Figure 3.23. Unlike metals and alloys, whose maximum is observed at deformation rates of $5 \cdot 10^4$ s^{-1}, [3,201,202] this maximum for concrete was found to be within a range of 0.15 to 0.20 MPa/sec in compression and 0.35 to 0.40 MPa/sec in tension (rupture).

Some investigators also observed a maximum on the dynamic characteristic.[183,193] For example, Figure 4.10a shows mean compression strength and

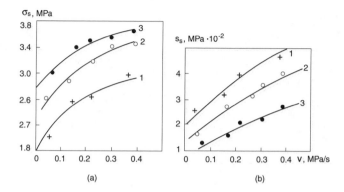

FIGURE 4.9

Variation in the mean value of σ_s and standard strength S_s of heavy concrete in rupture. Grade 600 cement, water–cement ratio: $1 = 0.51$, $2 = 0.43$, $3 = 0.32$.

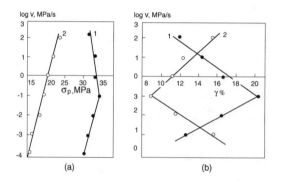

FIGURE 4.10

Variation in strength of limestone in compression (a) and its variation coefficient, $\gamma\%$ (b) with the rate of loading in the presence (curves 1) and absence (curves 2) of friction on the supporting surface.

Figure 4.10b the variation coefficient for limestone, plotted according to the data of Kuntysh and Tedder.[193] Dependence 1 was obtained for the case where the friction forces were acting on the supporting surfaces, and dependence 2 for the case where these forces were removed (or significantly reduced) by placing paraffin lubricant on these surfaces. As seen, the loading rate exerts a substantial effect on parameters of statistical distribution of the compression strength independently of the test conditions for concrete and limestone. Variations in the loading rate within six orders of magnitude lead to variations in the strength of limestone from 30% (at the presence of friction) to 45% (at its absence). It is also seen that test conditions affect the results significantly. At the presence of friction, there is a maximum on the $S_p = f(v)$ curve whereas no maximum was observed in the absence of friction. The variation coefficient has an opposite character of variation for the discussed

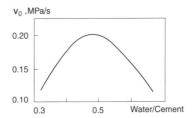

FIGURE 4.11
Dependence of the optimal rate of dynamic loading on the composition of concrete.

cases. Special experiments will determine whether or not these differences are accounted for by procedural reasons.

It follows from Figure 4.8 that each concrete composition is distinguished by its own dynamic characteristic. The position of the maximum is determined by the initial MB distribution (evident from the results of hardness measurement shown in Figure 3.27). During manufacturing, it is set by the water–cement ratio. A decrease in this ratio leads to an increase in the density of the structure and, therefore, in the number of "cold" dilatons (located closer to the Debye temperature) with simultaneous decreases in the number of "hot" dilatons. For this reason the maximum first shifts to the zone of high loading rates (shift of the curve from position 1 to position 2) and then, with an increase in density, shifts in the opposite direction (from position 2 to position 3). Each concrete composition has its optimum loading rate: v_1 for the first composition, v_2 for the second, and so on. Generalization of the discussed dependence is shown in Figure 4.11.

An important practical conclusion follows from analysis of Figures 4.8 and 4.11: each concrete composition has its optimal dynamic mode of operation. That is, at given physical–mechanical properties of the source components and manufacture technology, there is the best composition in terms of dynamic properties (corresponds to the maximum of the curve in Figure 4.11).

Each concrete composition reacts differently to high-rate load application. This statement is supported by different angles of inclination of tangents to the ascending and descending branches of the curves and noncoincidence of the points of maximum on the coordinate plane in Figure 4.8. Quantitative measures of the response of a concrete to dynamic effects may be the range of variation in the strength, $\sigma\%$, and strain rate sensitivity μ (Figure 4.12), expressed in relative units.

It can be seen that, within the range of the loading rates and concrete compositions used in the tests, the higher the standard cubic strength σ_p (Figure 4.12a) or the lower the water–cement ratio (Figure 4.12b), the narrower is the range of dynamic strengthening. This means that, with an increase in the standard strength of concrete (within a range of $\sigma > \sigma_0$ in Figure 3.34), its resistance to peak dynamic loads decreases. By varying the quality of source materials and the manufacturing technology, it is possible to obtain the microstructure of concrete that has the best adaptability to varying external

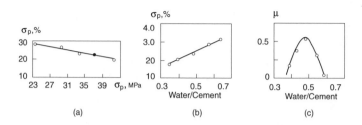

FIGURE 4.12
Correlations of the loading rate range with the standard strength (a), composition of concrete (b), and its dynamic sensitivity (c).

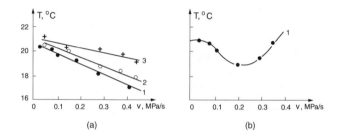

FIGURE 4.13
Variations in temperature of concrete depending on the rate of inducing the stress–strain state by tension (rupture) (a) and compression (b).

force effects (the zone of maximum sensitivity in Figure 4.12c). In Figure 3.34 it is located near point σ_0.

To utilize the resistance of concrete efficiently and thus avoid accidents and catastrophes, design of the composition, selection of the manufacturing technology, and selection of the resistance characteristics used in the design calculations should be carried out accounting for the dynamics of service force effects. Remember that an increase in short-time strength is accompanied by an increase in the probability of failure of concrete under dynamic loading conditions. Therefore, high-strength concretes should not be used under such conditions.[14]

The dynamic characteristic is formed in the depth of the AM structure. It follows from the equation of state (2.96) that the stressed state and temperature mode of a material are interrelated; a change in one affects the value of the other. To reveal this relationship, high-rate tests of concrete were accompanied by measurement of its temperature within the fracture zone. Figure 4.13 shows the variation of temperature with the load rate in tension (a) and compression (b). Numbers designate the same compositions as in Figures 4.8 and 4.9.

Variation in the temperature of the stressed material is observed in tension (rupture) and in compression with an increase in density of the flow of energy input. However, the character of this variation is fundamentally different.

In tension, a gradual decrease in the fracture temperature takes place as the loading rate is increased, whereas in compression the temperature jump ranges from 0.15 to 0.20 MPa/sec. In both cases the intensity of temperature variations depends on the density of packing of the structure: the higher the water–cement ratio, the higher the temperature gradient, other conditions being equal (compare curves 1 and 3).

A similar character of variation in the heat of the fracture process was observed earlier.[35,92] For example, while considering the fracture process from energy standpoints, Kuzmenko[35] formulated the general principle of temperature variations in deformation of a solid: any increase in volume of a certain quantity of a material causes a decrease in its temperature, and vice versa. It is significant that the temperature of a deformed material always changes to the side, which provides compensation for any change in volume. For example, a solid is heated in compression and cooled in tension.

The first statement for concrete causes no objections. It is predicted by the equation of state (2.96), and is always met in practice. The second statement, however, requires additional explanation. The external one-, two-, or three-axial pressure disturbs the intrinsic energy equilibrium of the space-oriented interatomic bonds (Figure 2.7). Application of the external tensile load to a body causes expansion of the MB distribution and transformation of part of the rotoses that constitute it, from the dilaton phase into the compresson phase, or from the compresson phase into the dilaton phase. This process may occur only due to conversion of some part of kinetic energy into potential energy.

A decrease in the kinetic part of the internal energy means cooling of the material. Cooling of concrete starts from cold dilatons due to their approaching the Debye temperature and is accompanied by extra embrittlement of the material (Figure 4.9a). Unlike the heated dilatons, which may only deform (due to a change in orientation or dimensions), the cooled dilatons are capable of bearing the external load (see the right-hand branch of curve 2 in Figure 3.51a). This leads to a continuous increase in dynamic strength (Formula 4.11 and Figure 4.9) and embrittlement. Without exception, all investigators notice a brittle character of fracture of concrete under dynamic loading.[182–194]

In compression, the fracture process consists of two stages, i.e., from longitudinal deformation and densification to transverse loosening and disintegration of the structure. Longitudinal deformation implies a decrease in the amplitude of the Boltzmann distribution due to transformation of dilatons into the compresson phase. Lowering of the energy level in the potential well (Figure 1.7) is accompanied by thermal radiation. As a result, the temperature of the solid increases.

The saturation of the compresson phase leads to transverse deformation and fracture; this process requires an inflow of heat. So, at high rates of loading, the destructive processes occur under conditions of increased temperatures caused by preceding longitudinal compression. Therefore, the dynamic strength increases first (Figure 4.8), reaches maximum in the zone of minimum heat radiation (Figure 4.13b), and then decreases because of an increase in local temperatures. The effect of a dramatic decrease in dynamic strength

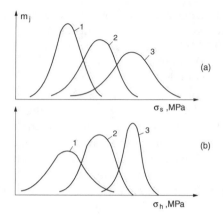

FIGURE 4.14
Transformation of macro- and microdistributions: tensile strength under dynamic loading (a) and hardness (microstrength) (b) at three values of the impact energy.

in compression has long been practically applied for crushing brittle materials.[192,193] Finally, the dynamic strengthening allows a natural analogy with the "cold-shortness" phenomenon (see Section 1.10). This analogy is attributed to similarity of the processes of formation of the resistance of materials at the AM level and its manifestation under variable external conditions.

The dynamic processes occurring in the structure of concrete in rupture can be evaluated on the basis of variation in statistic macro- (Figure 4.9) and microscopic distributions of elasto-plastic parameters of hardness with variation of the impact energy (Figure 4.14). Macroscopic tensile strength increases (Figure 4.9a) with an increase in the loading rate due to the displacement of the distribution to the right (positions 1, 2, and 3 in Figure 4.14a). Simultaneously, its amplitude increases (in accordance with Figure 4.9b).

This change in the macroscopic distribution is caused by an improvement in homogeneity of the structure at the microlevel S_h (positions 1, 2, and 3 in Figure 1.14b). This is attributable to embrittlement of the material with an increase in the loading rate. Embrittlement leads to weakening of the process of deceleration of crack nucleation. Interaction of microcracks results in the development of major cracks without branching. As a result, the total fracture area decreases (Figures 4.4 and 4.5). It should be noted in conclusion that statistical parameters constituting the probability methods of design of parts and structures[3,204] naturally follow from the differential form of the equation of state (Equations 4.1 to 4.5).

4.2.4 Dynamics of Bent Structural Members

Studies on the dynamic effect in metals[3] and nonmetals[41,192] are normally conducted using laboratory specimens or samples whose dimensions are much smaller than those of real parts. The processes of their dynamic restructuring

(Figure 4.3) develop simultaneously in the entire volume of a material. At the same time, as shown in Section 3.10, compressive stressed–deformed state and subsequent destructive processes in massive structural elements (short columns) even under quasi-static loading conditions develop with time in a nonhydrostatic manner (Figure 3.59).

In this connection, it seems reasonable to investigate the formation of stressed–deformed state in parts and components of real dimensions. Experimenting with real parts is associated with considerable labor, material, and energy costs. No doubt, this money will be well spent if we manage to obtain new information or confirm, refine, or, possibly, disprove old information. For example, we might be able to determine whether processes of formation of stressed–deformed state on standard specimens and actual structures differ (and, if so, to what extent), how a part made of fundamentally different materials behaves at different service conditions, whether the kind of stressed–deformed state affects the mechanism of formation of the dynamic effect, and other important issues.

To be able to answer these and other questions, it was decided to investigate the behavior of materials under other than standard stress–deformation, namely, in bending. Reinforced concrete members working in bending, i.e., beams, served as specimens. They were made from two different materials— concrete and steel (reinforcement bars). Concrete is a material of dilaton type and steel is a material of compresson type. Concrete performs effectively in compression, and steel in tension. The beams had a rectangular shape with a cross section of 100×200 mm and length of 4150 mm. Such sizes were chosen on the basis of capabilities of the experimental equipment used.

Grade 300 concrete with a water–cement ratio of 0.47 was used. The concrete mixture had the following weight composition per cubic meter: cement — 380 kg, sand — 65.5 kg, crushed rock with a fraction of 5 to 15 mm — 1300 kg, and water — 179 kg. The beams were reinforced with welded space frames. Two bars of the die-rolled section, with a diameter of 14 mm, made of steel of class A-III were used as the principal reinforcement. Distribution reinforcement was made of steel A-I bars having a diameter of 6 mm. Design of the experimental beams, their reinforcement, and test matrix are shown in Figure 4.15.

The beams were tested as simply supported and the load was applied as two concentrated forces acting on quarters of the beam span. The zone of pure bending extended to 2000 mm and was sufficient for installation of the required number of devices for measurement and detection of initiation and propagation of cracks. The beams were placed on two supports: hinged-immovable self-adjusting, and rolling-contact in the form of ring dynamometer. The latter was used also to measure the value of the support reaction. The supports were made as metal plates covered by rubberized cord fabric. The beams were loaded by a hydraulic machine (Figure 3.19d). The value of the loads was set by a force meter of the machine with the smallest scale division of 12.5 kg (125 N), and controlled by the ring dynamometer with an accuracy of ±5 kg (±5 N).

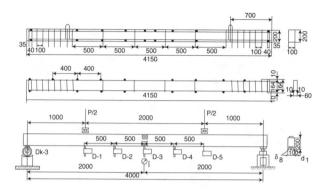

FIGURE 4.15

Design of testing experimental reinforcing bars: *D*, linear displacement meter; Dk-3, ring dynamometer; *δ*, strain gauges.

Strains in concrete and reinforcing bars as well as deflections of the beams were measured during the tests. Deflections were measured at several points simultaneously, accounting for the displacements of the supports. The moment of initiation of cracks was determined using an ultrasonic device, while the width of their opening was measured with a microscope.

In total, 36 beams of the same type were tested under 10 modes of loading, including the maximum loading developed by the machine and the mode where the load was applied stepwise holding the beam for 10 min at each step and for 30 min when the control load was reached. The time of loading to fracture was varied from 1 min to 400 h. No fewer than three beams were tested in each mode. In doing so, the rate of increase in stresses in a compressed zone of concrete was varied within a range of three orders of magnitude: from 0.001 to 3 MPa/sec.

Mechanical characteristics of the reinforcing-bar steel were determined at the standard rate of loading. Yield strength was 409.7 MPa and tensile strength was 653.2 MPa. The recording devices were calibrated during the tensile tests of the bars, then used for measurement of strains in the reinforcing bars. Values of fiber strains of concrete in the compressed zone and those of the reinforcing bars in the tensile zone were measured using strain gauges with a nominal length of 20 and 50 mm.

The age of concrete by the time of the tests was over 6 months. By that time, the processes of hydration of residual grains of cement had been completed and the structure of the material had been stabilized. Independently of the rate of loading, the parameters of stressed–deformed state of the bent members and structural materials were recorded on reaching the same values of external forces.

Figure 4.16 shows the dynamics of the stressed–deformed state of concrete in the compressed zone (a) and stretched reinforcing bars (b) near the central section, as well as deflection of the beams in the middle of the span (c). These data correspond to 2500 kgf (25 kN), which is 64% of the average fracture

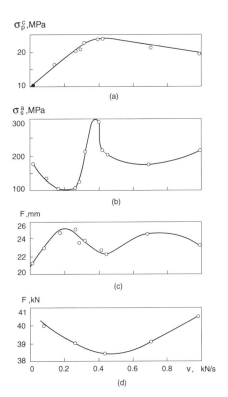

FIGURE 4.16
Variation in the stressed state of concrete in the compressed zone (a) and stretched reinforcing bars (b) near the central section, as well as deflection at the center of the span (c) and strength (d) of bent reinforced concrete members with variation in the loading rate.

load. The load-carrying capacity of the beams in bending (d) is placed in the same coordinate system.

The first two characteristics (Figure 4.16a,b) are local, whereas the rest (Figure 4.16c,d) are integral. Comparison of the theoretical curve in Figure 3.23 with the experimental curves, determined in central compression of specimens (Figure 4.8), pulling out of anchors (Figure 4.9a), and the dynamic characteristic of concrete in the compressed zone of the structure (Figure 4.16a) proves a versatility of the proposed mechanism of structural transformations occurring under high-rate loading conditions. It does not depend on whether the kind of stressed–deformed state is tension, compression, or bending. A two-branch dependence of resistance of concrete on the rate of input of mechanical energy from the outside is observed in all the cases. The stressed–deformed state in bending is described also by Equation 3.75; the only difference is that compressive, σ_1, and tension, σ_3, stresses are formed in the same bending plane.

As follows from Figure 4.16, the integral parameters of bent reinforced concrete members, deformability (c) and load-carrying capacity (d), are similar

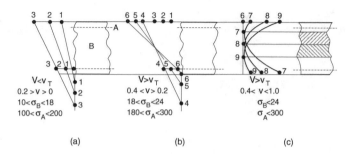

FIGURE 4.17
Explanation of dynamic processes developing in structure of the bent reinforced concrete members.

to those of the source materials, stretched bars (b) and concrete (a), although their variations go in opposite directions. This indicates that, with variations in loading conditions, their deformability is determined by the damping properties of reinforcing bars of the tensile zone, whereas the load-carrying capacity is determined by the efficiency of processes of dynamic restructuring of concrete.

Reinforced concrete is a combined structure, consisting of materials that resist compression and tension differently (see Section 2.6). Such a structure can adapt itself to dynamic loading conditions through redistribution of stresses between its components, depending on their individual peculiarities, relative location in the volume, and the relationship between the rate of application of an external load, v, and the relative rate of redistribution of the kinetic part of the internal energy, $v_T = d\Omega(T)/\Omega(T)$ (Equation 4.2). Under actual service conditions, the loading rate could vary widely, but the processes of deformation and restructuring are determined by parameter v_T.

The discussed character of restructuring of concrete in compression at any loading rate is versatile and determined by experimental dependencies shown in Figures 3.50 and 3.58. At low loading rates (within a range of $0.2 > v > 0$ kN/s), at $v < v_T$ the stress wave, taking a successive series of positions 1, 2, and 3, covers the entire area of central sections (Figure 4.17a). In Figure 3.58 this process corresponds to the displacement from initial point n to that of orientation saturation m.

Nonuniform compression of concrete occurs at the central sections, where the entire member as a whole is deformed. As a result, tensile stresses in reinforcing bars of the lower chord decrease by 100 MPa (Figure 4.16b). Further increase in the loading rate leads to $v > v_T$ and the time needed for complete restructuring of concrete in the stretched zone is not sufficient. When this happens, despite a continuing densification of the compressed zone (vector positions 1, 2 to 5, and 6 in the upper horizontal axis in Figure 4.17b), compression of the lower zone begins to decrease (see positions 4, 5, and 6 on the lower horizontal axis in Figure 4.17b). This leads to a drastic increase in tensile stresses in the reinforcing bars of the lower chord (Figure 4.17b in region $0.4 > v > 0.2$). This process continues until the dynamic characteristic of concrete (Figure 4.16a) reaches maximum at $v = 0.4$ kN/s. This moment corresponds to point m of maximum of the quasi-static characteristic (Figure 3.58). Then destructive processes control all parameters.

Further increase in the loading rate leads to an increase in density of energy flux from the outside to the upper zone; the process of densification of the concrete structure changes to its fracture. This process associates with the motion along the descending branch of the quasi-static characteristic from point m to point k in Figure 3.58. Its maximum, taking a successive series of positions 7, 8, and 9, moves from the upper chord to the lower chord, increasing density of the deeper layers of concrete and causing failure of the upper ones (dashed zone in Figure 4.17c). As a result, tensile stresses in reinforcing bars of the lower chord again decrease, whereas dynamic characteristics of steel and concrete transfer to descending regions (Figure 4.16a,b). The described mechanism of redistribution of internal forces is fully determined deformability and strength of the bent members.

Relief of stresses in the stretched reinforcing bars leads to a decrease in rigidity of the entire system. The deflection of the beams increases when the loading rate is in the ranges of $0.2 > v > 0$ and $0.7 > v > 0.4$ (Figure 4.16c). It decreases when the reinforcing bars assume a larger portion of the external load (ranges $0.4 > v > 0.2$ and $1.0 > v > 0.7$).

Gvozdev[203] described a similar dependence. Analyzing the mechanism of impact of the loading rate on deformability of the bent reinforced concrete structures, he showed that the stretched concrete sections between cracks successively ceased resistance to loading as the loading rate decreases. The compressed zone of concrete becomes shorter, and deflections of the bent members increase. It was noted that concrete behaves as an elastic body under very rapid loading. As such, the formation of plastic strains does not lead to decrease in the strength of concrete; rather, it leads to its growth. Intensification of loading of the bent members leads to a change in shape of the stress diagram and displacement of the neutral axis.

In these tests, a rate of 0.4 kN/sec was found to be critical. Starting from this value, the dynamic strengthening of concrete stops and function $\sigma_p = f(v)$ changes from increasing to decreasing. As such, the character of fracture also changes.

Figure 4.18 shows similar-type beams after they exhausted their load-carrying capacity at different loading rates. The time of loading to fracture of the first beam (a) was 88 sec, and that of the second beam (b) was 2 h, 38 min. The beam in its initial state is shown in Figure 3.19d. Comparing Figures 4.18a and b, one can notice the difference in the total fracture surfaces (different concentration of cracks per unit length of the beam). In the first beam (a), the number of cracks formed in the zone of pure bending was $n = 14$ and their total length was $L = 873$ cm, while in the second beam (b), these values were $n = 32$ and $L = 1664$ cm. Failure of the first beam was caused by formation of several through-cracks under the supporting surfaces (cracks across the section are seen well in these locations in Figure 4.18a).

FIGURE 4.18
Appearance of the beams after fracture in loading for 88 sec (a) and 2 h, 38 min (b).

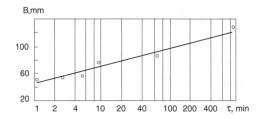

FIGURE 4.19
Dependence of the specific length of crack B in the bent reinforced concrete structures on the time of loading, τ.

Exhaustion of the load-carrying capacity of the second beam resulted from crushing the compressed zone of concrete in the middle of the span (cleavage of concrete is seen in this zone in Figure 4.18b).

The first case is attributable to activation of destructive processes occurring in concrete at high loading rates (Figure 4.17c), and the second can be explained by high deformability of the entire structure (Figure 4.16c). Superposition of these processes at a rate close to critical (0.4 kN/s) leads to a decrease in the load-carrying capacity. For this reason, plot $F = f(v)$ has a minimum here (Figure 4.16d). From that moment, high-rate fracture becomes brittle.

Dependence of the mean length of one crack B on loading time τ where $B = L/n$, is shown in Figure 4.19 in the semi-logarithmic coordinates. It proves the theoretically predicted (Equation 4.5) and experimentally observed (Figures 4.6 and 4.7) correlation between loading mode and fractured area.

Increase in the loading rate leads also to a change in the appearance of structures after relieving the load. An elastic aftereffect leads to restoration of the initial shape of the beams subjected to high-rate loading. After quasistatic loading, the shape of the beams undergoes substantial changes, acquiring residual deflection.

The dynamic effect is of great scientific significance and practical interest. Based on analysis of the cracking process occurring under different external conditions, Finkel[36] concludes that dynamic impact in the direction, which does not coincide with the principal load, hinders cracks propagation. Equation 4.5 and Figure 4.1 present the physical interpretation of this fact.

A change in the shape of a local potential field from position 1 to position 1̇ (Figure 4.3) increases energy consumption of the cracking process. Theoretical principles of deceleration of fracture in the force fields are considered in Komarovsky.[12] Preliminary results of utilization of the dynamic effect for improvement of workability of metals and alloys, as well as decreasing the energy consumption of the related processes, are discussed in Komarovsky and Astakhov.[16]

4.3 Durability

Efficiency of production depends to a substantial degree on reliability of industrial products; methods of ensuring reliability determine the level of technogenic safety. Insufficient reliability of engineering objects not only causes economic, social, or moral losses,[8] but also affects the fates of new technical ideas and technical progress as a whole.[204] Improvement of reliability increases competitiveness of products and efficiency of production. Therefore, all efforts to improve reliability are critical.

The semi-probability calculation method assures reliability by adding excessive material.[205] The uncertainty elements, inevitably introduced together with the empiric safety factors, make it necessary to change to calculation methods developed using the statistical theory of reliability.[2,3] However, the latter operates with incomplete knowledge of the initial structure of materials and probable consequences of the effects of various combinations of external factors. Such incomplete knowledge makes it impossible to predict the results of structural processes exactly. Instead, only a certain estimation of these results having probabilistic nature can be made. The amount and scope of knowledge about the external factors, parameters of environment, and mechanism of their effect on structural materials must be widened and increased to make such predictions adequate.[206]

According to Gnedenko,[204] experimental approaches that allow only phenomenological conclusions are not sufficient to solve such an integrated problem as reliability. The unjustified belief in the omnipotence of mathematics and its auxiliary tool, computer modeling, is not of any help either. Any fracture regarded outwardly as accidental is not without a cause. Any event that appears accidental has its causes and obeys its own peculiar laws. The accidental nature of fracture lies only in the fact that its causes and the laws it obeys are not yet known and, thus, unclear.[9]

Only a deep insight into the essence of a phenomenon, investigation of the mechanism that governs it, and cognition of the earlier hidden cause-and-effect relationships and laws make this event predictable rather than accidental. Only this approach helps to reduce the role of randomness in the behavior of materials and structures, and can create conditions for the development of a theory of design and reliability on a basis other than probability.

Sorin[9] and Melomedov[207] have suggested that the theory and practice of reliability should be developed on the basis of investigation of those physical phenomena that occur in materials under the effect of external factors, service conditions, and time. This shift of interests toward physics of reliability is natural and justifiable and reflects understanding that only revealing the nature of the destructive processes will open a way for an objective estimation of reliability of engineering objects. In the preface to Sorin's book,[9] Gnedenko states that "this direction presents a rich soil which should be cultivated with a special care because it nourishes all the branches of science related to the improvement of reliability with the required information."

4.3.1 Equation of Durability

Equation 2.96 determines the state of a solid in thermal and force fields. It follows from this equation that one could hardly be distinguished from the other. Whether the solid is affected by the force field or not, it is always in the thermal field (at least under the effect of fluctuations of the ambient temperature).

The equilibrium state of a solid is characterized by the equality of mechanical, Ω, and kinetic, $\Omega(T)$, potentials (Equation 2.175). In turn, each of them also depends on several parameters: the first depends on the ability of the solid to resist various external effects, P, or deform under their effect, V, and the second depends on the number of interatomic bonds, N, and absolute, T, and Debye, θ, temperatures (Equation 2.96). Any variation in one of these parameters inevitably affects the value or direction of the others; this causes a solid to transform from one state into the other. (The independent influence of these parameters on the state of a body is considered in Chapter 3.)

Immediately after solidification, a body is characterized by certain initial values of the Ω-potentials $\Omega_0 = \Omega_0(T)$ with corresponding initial state parameters: P_0, V_0, T_0, N_0, and θ_0. During the lifetime of the body these parameters do not remain constant because the processes of adaptation or, to be more exact, correspondence of the AM structure to external fields (force, thermal, radiation, and others) or active environments always occur inside this body. The state parameters can be conditionally subdivided into two groups: internal, i.e., N and θ, and external, i.e., P, V, and T. The body interacts with external objects, fields, or environments by a contact (P, V) or noncontact (T) means.

As a rule, solid bodies in the form of structural members or machine parts exist in certain (natural or artificial) force, σ, and thermal, T_b, fields, and their deformability is limited by external constrains V_b. When a body performs useful functions, its state parameters P, V, and T should not exceed the limits of technical restrictions on σ, V_b, and T_b. If the limit is exceeded, however, the body can no longer function in a normal way and thus fails. This occurs because of an insufficient strength or brittleness or because of an excessive deformability (see comments on Figures 2.6 and 3.34), and very rarely because of temperature restrictions T_b. Let us call the period of time from the beginning of functioning to reaching one of these limitations "durability."

Based on experience accumulated in the theory and practice of reliability, Sedyakin[208] formulated a fundamental principle of reliability of complex technical systems, which can be regarded as a physical law. It is based on the notion of the resource s spent by a system during time τ:

$$s(\tau / v) = \int_0^\tau f(\tau / v) d\tau, \tag{4.25}$$

where $f(\tau/v)$ is the rate of failures of similar-type elements under conditions v. Because the probability of faultless operation during time τ is

$$\omega(\tau / v) = \exp\left[-\int_0^\tau f(\tau / v) d\tau\right],$$

then $s(\tau/v) = -\ln\omega(\tau/v)$.

Therefore, it is likely that the notion of system resource (or life) defined by Equation 4.25 is identical to entropy, according to Boltzmann (Equations 2.14 to 2.17). Entropy (Equation 2.17) is a natural part of the equation of state (2.65) and, as shown earlier, determines the orientation, phase, and destructive components of the resource (or service life) of structural materials (Equation 2.113). Expanding the results obtained in Section 2.5, we can add more sense to the physical nature of entropy; however, for structural materials, it should be regarded only as resource (service life).

The destructive character of consuming resources (Section 3.11) is accompanied by the release of the internal energy in the form of thermal,[35,36] ultrasonic,[37] electron, and x-ray[38] radiations. For materials found in invariable or slowly varying force and thermal fields, Equation 3.19 assumes the following form:

$$\frac{\Omega_0 \pm V\sigma}{kT} - s(\tau) = g, \tag{4.26}$$

where the plus sign corresponds to compression and the minus sign to tension, Ω_0 is the initial value of the Ω-potential, $V = \Delta V_0$ is the working volume that has transformed (according to the diagram shown in Figure 2.8) under the effect of the σ and T fields, g is the energy flux due to resource exhausted during time τ, and $s(\tau)$ is its residual part of this resource.

Using the definition of flux g as the amount of energy Q transferred through unit surface S per unit time τ,[31] i.e., $g = Q/S\tau$, and integrating the right-hand part of Equation 4.26 at $Q = $ const with respect to time from τ_0 to τ and with respect to surface area from S_0 to S, we derive the equation of durability[209,210] in the following form:

$$\tau = \tau_0 \frac{S_0}{S} \exp \frac{1}{Q} \left[\frac{\Omega_0 \pm V\sigma}{kT} - s(\tau) \right],$$
(4.27)

where $\tau_0 = 10^{-13}$ s is the time of a complete revolution of dipole D on the orbital ring (Figure 1.7b).

The processes of microcracking result in an increase in the total surface of a body from S_0 to S.[62] One part of energy Q released in this case is spent to maintain the deformation process (differentiation of dipoles according to the diagram shown in Figure 2.8b) and to heat the body, while the other part is emitted into the environment (see Section 3.11). If the body is brought up to fracture (in this case $s(\tau) = 0$), with no regard to variations in S and Q (as is usually done in technical strength tests of materials), Equation 4.27 for this case is transformed into

$$\tau = \tau_0 \exp \frac{\Omega_0 \pm V\sigma}{kT}.$$
(4.28)

A similar expression for durability was derived experimentally, based on analysis and generalization of results of extensive and time-consuming experiments conducted at the Laboratory of Physics of Strength of the A.F. Ioffe Physical–Technical University of the Russian Academy of Sciences (St. Petersburg) under the leadership of academician Zhurkov. These experiments began in 1952, and the results have been continuously published (see, for example, Regel and Slutsker's review article[62]). Therefore, we can state that the validity of the equation of durability (4.28) theoretically derived here is proven experimentally.

For example, various sources[10,62,211] present results of systematic investigations of durability of loaded solid bodies. Durability of materials τ was determined in tensile tests of specimens at fixed temperature T and at given stresses σ maintained at a constant level up to fracture. Experiments were conducted by varying temperatures and stresses over wide ranges. Correspondingly, durability was varied from thousandths of a second to many months. More than 100 different materials with different structure were tested. We can note, however, that extensive literature sources are dedicated to durability in tension while data on other types of simple and combined loading, especially those in actual structures, are very scanty. Figure 4.20 shows time dependencies of tensile, compressive, and bending strength of different types of materials tested on standard specimens and structural members.

As seen, durability monotonically decreases with an increase in stresses — a tendency observed for all materials tested. Similar investigations carried out on different types of concrete loaded by tension (pulling out) (Figure 4.5b) are in full agreement with these results.

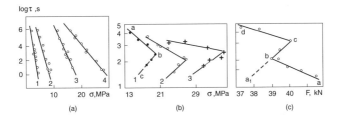

(a)   (b)   (c)

FIGURE 4.20

Time dependence of strength in tension (a), compression (b), and bending (c) of different classes of materials in testing of specimens (a, b) and structural members (c): (a): 1, paper; 2, wood; 3, cement stone; 4, glass-reinforced plastic; (b): concretes of the same composition but with a different water–cement ratio: 1, 0.37; 2, 0.43; 3, 0.59; (c): reinforced concrete beams 100 × 300 × 4150 mm in size with a welded bar frame and concrete with water–cement ratio of 0.43.[212]

These dependencies appear to be linear on the semi-logarithmic coordinates. Analysis of a large amount of experimental data allowed the authors of the kinetic concept of strength to derive a general formula to relate durability to stresses and temperature:

$$\tau = \tau_0 \exp\left(\frac{U_0 - \gamma\sigma}{kT}\right). \tag{4.29}$$

Analysis of this equation resulted in a conclusion that processes of gradual accumulation of structural damages occur in loaded solid bodies. This conclusion is confirmed by results of experiments with interrupted loadings.[214] It was found that, if a specimen is kept under a load for a certain period of time, it loses strength, i.e., its durability decreases.

At $T = $ const, tension changes the shape of the MB factor in a direction of 11 to 12 to 13 (Figure 1.14). The shape of the MB factor tends to position 13 by increasing the level of tension stress. As such, a greater part of set N remains beyond the critical temperature T_c and θ_s (the dashed zone in Figure 1.14). As a result, the service life (resource) $s(\tau)$ is shorter and durability τ is lower (Figure 4.20a). Compressive stressed–deformed state is formed in a different way.[53] It takes the end of the MB factor out of the dangerous zones, thus increasing life and durability due to elongation of elastic region oa of the deformation diagram (Figure 2.10).[16] As a result, region cb is formed on plots 1, 2, and 3 (Figure 4.20b).

After orientation saturation (Figure 2.8a) at point b (corresponding to point a in Figure 2.10), the degradation processes, developing in the transverse direction, add an inclination as in tension (region ba in Figure 4.20b) to the durability plot. Measurement of hardness during loading allows us to give the previously discussed interpretation to processes of exhausting the service life (resource) in tension and compression.[54]

To explain durability in bending in accordance with definition of the Ω-potential (see Section 2.3), the numerator of the exponent (Equations 4.27 and 4.28) should include $\sigma = \sigma_1 - \sigma_2$, where σ_1 is the stress in the compression zone and $(-\sigma_2)$ is that in the tension zone of a bent member. The durability plot for reinforced concrete beams (Figure 4.20c) is represented by broken line *abcd*. At low (branch *ab*) and high (*cd*) durability, the dominating contribution to the exhaustion of the service life (resource) is made by the tensile stress component $(-\sigma_2)$ of the stressed–deformed state, whereas in region *bc* it is made by the compression stress component, σ_1. Tensometric measurements of the compressed zone of concrete and stretched reinforcing bars in testing of durability of structural members in bending proves the validity of this sequence of changes in stresses (Figure 4.16).

The described temperature dependence of resistance of materials to deformation and fracture (see Section 2.7) allows us to conclude that an increase in temperature decreases durability under loading conditions. To illustrate, Figure 3.18 shows results of investigations of the effect of temperature on durability of aluminum alloy under a tension load.[211] As seen, for each temperature the $\log\tau = f(\sigma)$ dependence remains linear, but its angle of inclination to the axis of ordinates changes accordingly. With a decrease in *T*, the inclination angle grows, approaching the vertical axis. This explains the catastrophic character of fracture of metals at low negative temperatures.

In this case the MB distribution takes position 11 in Figure 1.14. An increase in temperature, in contrast, causes a decrease in the inclination angle and, therefore, leads to a decrease in durability, although this time for a different reason, i.e., due to weakening of the structure and an increase in ductility. Also, while shifting to the right, the MB distribution transforms part of the interatomic bonds into the dilaton phase (position 15 in Figure 1.14).

The discussed approach of the durability graph to the vertical axis with lowering temperature (position 1 in Figure 4.21) is used by the authors of the kinetic concept of strength[10] to explain the introduction of the "ultimate strength" concept. Analysis of variation in durability for different materials over a wide range of temperatures allows us to conclude with a high degree of confidence that ultimate strength is not a physical characteristic of a body.

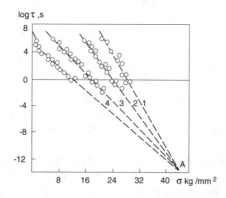

FIGURE 4.21
Dependence of durability on the temperature for aluminum alloy: 1, 18; 2, 100; 3, 200; 4, 300°C.

The significance of the kinetic concept of strength[10,62,64,211] lies primarily in revealing the decisive role of the kinetic part of the internal energy in deformation and fracture processes. It pays major attention to the right-hand part of the equation of state (2.96) and ignores its natural link with the Ω-potential, which determines mechanical properties. For this reason, this theory appears as a particular case of more general and versatile thermodynamic concepts elaborated in this book.

Supporters of the kinetic theory think that thermal fluctuations play the decisive role in breaking interatomic bonds. They also believe that, in the absence of an external load, thermal fluctuations cause arbitrary breaks of the bonds. After some time, these breaks recover because their formation and restoration of a bond require identical values of the fluctuations.

An external tensile load decreases the energy of disintegration of a bond and thus increases the frequency of breaks; what is most important, by facilitating the thermal fluctuation disintegration of bonds, it hampers their recombination. As such, the rate of disintegration of bonds is much higher than that of their restoration. This initiates accumulation and coalescence of elementary fractures. The fracture energy is often taken from the stored thermal energy of a body, rather than from the work of an external force. The role of the external force is in facilitation and direction of the fracture effect of thermal fluctuations.

Although the role of thermal fluctuations in fracture of solids revealed in the kinetic theory is correct, it is necessary to clarify some important points. Over the entire thermal range available to interatomic bonds of a given body, thermal fluctuations lead to deformation of this body rather than to its fracture. They cause disintegration of the bonds or CD phase transition only at brittle failure temperature T_c (at the bottom of the potential well, Figure 1.7) and Debye temperature θ. As such, it is very important that phase transitions are accompanied by a change of sign of internal stresses.

The authors of the kinetic concept of fracture were not able to obtain a full analogue of Equation 4.28; they also failed to disclose in full the physical nature of the main parameters of this equation. This imposes substantial limitations on the practical implementation of the kinetic concept according to which the parameter that stands ahead of σ has a decisive effect on strength. It is designated as γ and, in the discussion of its physical meaning, it is called "index of the concentration of stresses in a loaded body." As stated, this parameter determines the known substantial difference between the actual and theoretical strength of a solid as well as causes an increase in strength during technological processing. The authors of the kinetic concept of fracture do not give any quantitative measure of this parameter or method for controlling its value. In our opinion, this can hardly be done within the frame of the kinetic concept.

In deriving Equation 4.28, we did not formally introduce the parameter at σ. It was naturally determined as the deformation part of the Ω-potential (Equation 1.111), characterizing the effect of the quantity of a material on the MB distribution of the internal energy in the bulk of a body. It has dimension of volume in theory (Equation 2.96) and in experiments.[10] This parameter serves as a functional that distributes external load σ in accordance with the internal needs of each local region of a material. We emphasize here that it distributes this load between local regions rather than elementary bonds, because interaction between atoms determined by the short order does not change in a directed effect on a structure.[211] As such, a change occurs in superatomic, rather than atomic, structure of a body, with dimensions of tens and hundreds of atomic dimensions.

Finally, the most important point is that papers dedicated to the kinetic concept of strength do not discuss the content of term U_0 in Equation 4.29. Calling it "initial activation energy," the authors of these works are of the opinion that clarification of the physical meaning of this parameter is of great practical importance, in addition to the scientific importance. It is stated that only data on the physical nature of U_0 can serve as a reliable basis for a purposeful solution of one of the most important problems of materials science — increase in strength.

Chapters 1 and 2 discuss the physical nature of this parameter and the terms of the corresponding equations indicate the methods for controlling its value. This makes it possible to solve not only a particular problem of materials science, i.e., increase in strength of materials, but also a general engineering problem, i.e., development of efficient methods to control deformation and fracture. Once developed, these methods ensure 100% reliability and durability of highly critical engineering objects and decrease energy consumption of many technological processes.

It can be considered that the rate of variation in the MB factor, $\dot{Z}$ (i.e., the rate of change in the energy states of rotoses, Equation 2.17), should not differ much in order of magnitude from the mean rate of thermal fluctuations of rotoses that constitute them.[30] This assumption makes it possible to write that

$$\tau_0 = 1 / \dot{Z} \approx 10^{-13} \text{ sec.}$$

Equations 4.28 and 4.29 are simplified. The first suggests that $\tau_0 = 1/\dot{Z} =$ const, and the second follows from analysis of fracture of solids under the simplest kind of loading, i.e., the uniaxial static tension. This simplification is justifiable in terms of revealing physical principles of fracture. However, it is of high scientific and practical interest to reveal more general cases closer to actual working conditions. Attempts to extend principles established within frames of the kinetic concepts to more general cases lead to the known difficulties.[62]

For example, even in uniaxial tension, widening the temperature range or increasing the intensity of stressed–deformed state causes deviation from the exponential dependence set by Equation 4.29. Such deviations arise also in other loading conditions and kinds of stressed–deformed state; they no longer adequately describe experimental data in the radiation fields and under the combined kinds of loading. Regel et al.[10] describe various cases of such deviations and attempt to explain their causes. The approximate character of Equation 4.29, associated with constancy of pre-exponential factor τ_0, is indicated as one of these causes. In reality (and this will be shown later), this is the main cause, rather than one in a series of causes.

4.3.2 Structural Transformations in Sustained Force and Thermal Fields

When no other external fields but thermal ones affect a body, the first law of thermodynamics (Equation 4.4) that describes variations of the state in such fields has the following form:

$$\dot{U} = \lambda \dot{T},\tag{4.30}$$

where λ is the thermal conductivity. Assuming that overall energy U consists of potential $U(r)$ and kinetic $U(T)$ parts and that the first of them does not depend on the temperature, yields

$$\frac{\dot{U}(T)}{\dot{T}} = \lambda.$$

Taking into account the expression for kinetic energy (2.145), differentiating it with respect to T, and accounting for heat capacity C_v (Equation 2.143), we obtain

$$\lambda = C_v \dot{T}.\tag{4.31}$$

Comparison of Equations 4.30 and 4.31 shows that variations in the thermal fields result in transfer of heat from the environment to a material (at $\lambda > C_v$) or from a material into the environment (if $C_v > \lambda$), due to continuous transformation of the MB distribution (change in positions from 8 to 10 and from 11 to 15 in Figure 1.14).

We can express the overall internal energy in terms of the sum of the potential and kinetic parts:

$$\lambda \dot{T} = \dot{U}(T) + \dot{U}(r).\tag{4.32}$$

Primitive function of $U(T)$ is determined by Equation 2.142.
The potential part can be found from the formula

$$U(r) = \omega \exp\left(-\frac{\omega}{kT}\right),\tag{4.33}$$

where ω is determined by Equation 1.53 and sets the shape of the potential well (Figure 1.7) depending on physical–chemical parameters A and α (Equation 1.29) of a given material, and

$$\exp\left(-\frac{\omega}{kT}\right)$$

characterizes the probability of its filling in accordance with the MB factor (Equation 2.5).

As seen from Equations 1.53, 4.33, and 2.142, the overall internal energy U, as well as its components $U(r)$ and $U(T)$, depends on two parameters: volume V and temperature T. Taking partial differentials of Equation 4.33, we write

$$\left[\frac{\partial U(r)}{\partial V}\right]_T = \frac{\partial U(r)}{\partial V}\exp\left(-\frac{\omega}{kT}\right)\left[1+\frac{\omega}{kT}\right] \quad \text{(I)};$$

$$\left[\frac{\partial U(r)}{\partial T}\right]_V = \left(-\frac{\omega}{kT}\right)^2 \exp\left(-\frac{\omega}{kT}\right) \quad \text{(II)}. \tag{4.34}$$

The extreme of the Ω-potential occurs when both partial differentials become zero. It is clear that, under any conditions,

$$\exp\left(-\frac{\omega}{kT}\right) \neq 0.$$

Then three equalities are possible:

$$1)\ \frac{\partial U(r)}{\partial V} = 0, \quad 2)\ \left(-\frac{\omega}{kT}\right) = 0 \text{ and } \quad 3)\ 1+\frac{\omega}{kT} = 0.$$

The first two are of no interest, whereas the third can be written as follows, accounting for Equations 1.26 and 1.28:

$$kT = \frac{\alpha}{V} - \frac{A}{V^2}. \tag{4.35}$$

A graphical solution of the equation having two variables T and V (Equation 4.35) is shown in Figure 4.22. Its left part is a straight line originating from the coordinate origin and having a constant angular coefficient k. It is versatile and common for all materials. The right part is shown as a two-branch dependence with maximum at point θ. Its shape is determined by

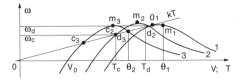

FIGURE 4.22
Variation in the ω-potential under the deformation, V, and thermal, T, effects.

physical–chemical parameters A and α of a specific material and is set in the coordinate system and, therefore, in the crystalline structure by two temperatures: cold shortness temperature T_c and Debye temperature θ.

The energy parameters of the entire diversity of solids are arranged to the left of the extreme position designated by number 1. In general, Equation 4.35 has two solutions (see points of intersection of straight line kT with curves 2 and 3), located in the compresson, T_c and ωc, and dilaton, T_d and ω_d, phases. An exception is the extreme case (position 1), where the plots contact each other at one point O_1. Location of potentials ω to the right from position 1 is unnatural.

As seen from Equation 4.33, the value of the potential energy, which determines stability of the equilibrium state of a body, is affected by a superposition of two functions: ω (curves 1, 2, and 3 in Figure 4.22) and

$$\exp\left(-\frac{\omega}{kT}\right)$$

(positions 1, 2, and 3 in the lower field of Figure 3.9), each with one maximum. The maximum of function ω is fixed on solidification by the Debye temperature θ (positions θ_1 and θ_2 in Figure 4.22) or by the crystalline lattice parameter r_a in Figure 1.7 (which is the same). At a fixed position of function ω (positions 1, 2, and 3), the ω-potential may have two values: in position 2 they correspond to points c_2 and d_2 of intersection of plots ω and kT, and in position 3 they correspond to points c_3 and d_3. These points determine coordinates of the most probable or maximum values of function defined by Equation 4.33. The plot of the latter is a two-branch dependence with a maximum.

It can be seen from Figure 4.22 that such dependencies cannot exist at points c_3 and d_2 because, in the first case, their maximum lies near critical point V_0 (in Figure 1.7 in a linear form, the coordinate r_1, equivalent to temperature T_c corresponds to this point), and in the second case, near the Debye vertical axis θ_2. In both cases the formation of left (cold) branches of the MB distribution is impossible because of a high probability of failure of rotoses at temperatures T_c and θ.

Affected by the thermal force fields, characteristic temperatures T_c and Θ often vary in service, so (see comments on Figure 1.22) the ω-function, while shifting to the right, may land in critical position 1, where C and D phases degenerate, becoming uncertain. As a result, a structure made of these phases fails suddenly, which is regarded as an accidental and seemingly nonmotivated process. To be able to show that nothing is accidental in this process we should answer a question: what is the mechanism of this process?

We can write partial derivatives of Functions 4.33 and 2.142 recalling remarks made considering Equation 4.35:

$$C_V = \left[\frac{\partial U(T)}{\partial T} \right]_V = \frac{12\pi^4}{5} k\delta V + 3kV \qquad \text{(I)}$$

$$C_T = \left[\frac{\partial U(T)}{\partial T} \right]_T = \frac{3\pi^4}{k} \delta\omega_T + 3\omega_T \qquad \text{(II)}$$

$$P_V = \left[\frac{\partial U(r)}{\partial T} \right]_V = -\left(\frac{\omega_p}{\omega_T} \right)^2 \exp\left(\frac{\omega_p}{\omega_T} \right) \qquad \text{(III)} \qquad (4.36)$$

$$P_T = \left[\frac{\partial U(r)}{\partial T} \right]_T = \frac{\partial \omega_p}{\partial V} \exp\left(\frac{\omega_p}{\omega_T} \right) \left(1 + \frac{\omega_p}{\omega_T} \right) \qquad \text{(IV)},$$

where C_V and C_T are the heat capacities and P_V and P_T are the internal resistances developed by the AM structure of a body at a constant volume and temperature, $\omega_p = \omega$ and $\omega_T = kT$ are the thermodynamic potentials of a rotos at constant resistance **P** and temperature T (Equation 4.35).

In constant or slowly varying force ($\dot{P} = 0$) and thermal ($\dot{T} = 0$) fields, the equation of state (4.4) assumes the following form:

$$\dot{U} + (\mathbf{P} \pm \sigma)\dot{V} = 0, \qquad (4.37)$$

where σ is the external mechanical stress of tension (minus sign) or compression (plus sign). It is always necessary to distinguish one-, two-, or three-axial external stress σ and internal volume resistance **P**, which has a plus sign in the compresson phase and a minus sign in the dilaton phase. Taking into account Equations 4.36.I and 4.36.III, in analogy with Equation 2.153, we can write

$$\sigma = \frac{dT}{dV}(\lambda - C_V - P_V). \qquad (4.38)$$

As seen, the resistance of materials is always determined by a relationship among three thermal flows: one external flow,

$$Q = \lambda \frac{dT}{dV},$$

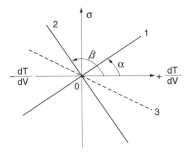

FIGURE 4.23
Dependence of resistance σ on the value and sign of the external (1) and internal (2) heat flows.

and two internal flows,

$$g_1 = -C_V \frac{dT}{dV} \quad \text{and} \quad g_2 = -P_V \frac{dT}{dV}.$$

The internal flows characterize the processes of conversion of the energy from kinetic to potential, or vice versa. The type of external field sets their direction: compression $(+\sigma)$, tension $(-\sigma)$, or their combination. The external flow can activate or weaken these processes.

Figure 4.23 shows formation of resistance σ depending on the value and sign of the total heat flow. Its components are shown as straight lines with angular coefficients

$$Q = \tan\alpha \frac{dT}{dV} \text{ (line 1)} \quad \text{and} \quad g_2 = -\tan\beta \frac{dT}{dV} \text{ (line 2)},$$

where

$$\lambda = \tan\alpha \quad \text{and} \quad C_v = -\tan\beta.$$

The relative positions of straight lines 1 and 2 set the direction of a resulting straight line

$$g = (\lambda - C_V) \frac{dT}{dV} \text{ (line 3)}.$$

It determines the response of the material depending on the value and direction of the resulting heat flow. Heat transferred to the material from the outside,

$$\left(+\frac{dT}{dV} \right),$$

decreases resistance, whereas that released from it,

$$\left(-\frac{dT}{dV}\right),$$

increases resistance. This variation in strength is characteristic of metals and alloys because their structure consists mostly of C rotoses. Supplied heat transforms part of the hot rotoses from the C to D phase, thus decreasing tensile strength (see Figure 2.36 and comments on it).

Cooling, on the contrary, increases the compresson concentration and, thus, increases the tensile strength (transformation of positions 12 to 11 in Figure 1.14). Acting for a long period of time, such flows move the structure of the material to a new position adapted to these conditions, but it takes some time to change it. Therefore, any sudden change in the external effects leads to catastrophic consequences (see Figure 4.22 and comments about it).

The moment of a change in direction of the heat flow is determined by difference $(\lambda - C_V)$. It depends on the relationship between the internal thermal–physical parameters (λ and C_V) and the external parameters, i.e., the intensity of the force field $(+\sigma)$. Thermal conductivity λ characterizes the capability of rotoses to exchange kinetic energy between themselves and the environment, while heat capacity C_V characterizes their ability to accumulate and preserve it.

Figure 4.24 shows the mechanism of formation of resistance of a body,

$$\sigma = tg\gamma \frac{dT}{dV},$$

corresponding to thermodynamic equation (4.38), where $tg\gamma = (\lambda - C_V - P_V)$ is the angular coefficient. The resistance to compression $(+\sigma)$ is due to dilatons and that to tension $(-\sigma)$ is due to compressons; the angular coefficients of these phases are different (γ_c and γ_d). Depending on the thermal conditions of deformation (value of and C_V), a straight line in the coordinate system can take different positions (e.g., positions 1 and 2). This means that the strength of a material may have only that value permitted by its heat content at a given moment of time; therefore, heat content is a strength regulator.

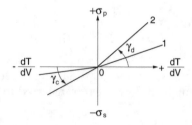

FIGURE 4.24
Resistance to compression $(+\sigma)$ and tension $(-\sigma)$ at different heat contents.

Different combinations of the effect exerted on it by service conditions are given in Table 3.7 and in Figure 1.22.

As a mechanical analogue of heat flow

$$Q = \lambda \frac{dT}{dx},^{49}$$

thermodynamic Equation 4.38 reveals the physical (to be more precise, orientation–destruction) nature of the mechanism of resistance of bodies to external force fields σ (Figure 2.10). It can be seen that the natural reaction to these fields is the formation of a temperature gradient in the direction of external stresses in the bulk of a material. Because they are oriented along the internal heat flow, rotoses create an internal mechanical moment, that counteracts the external moment (see Equation 3.31). The inverse is also true: resistance of a material to external force effects can be increased and fracture of the structure can be avoided if the properly oriented heat flow is formed in this material. The expression in brackets is a mechanical equivalent of the coefficient of thermal conductivity λ or its inverse value, i.e., heat resistance χ.

In analogy with heat flow Q, mechanical stress σ can be regarded as a flow of the internal potential energy that affects the material. Equation 4.38 shows very clearly that thermal and mechanical processes occurring in a structure under the effect of the force fields are mutually related. It cannot be otherwise, as they both characterize different sides of the same process, i.e., the process of orientation and deformation transformation of rotoses (Figure 2.8) under the effect of variations in the kinetic or potential components of the overall internal energy of a body. The analogy between mechanical and thermal phenomena becomes natural and clear, if we recall that the fundamental law established by the equation of equivalence (2.105) is in its basis.

Let us show that the third term in the left-hand part of Equation 4.37 is the differential of the energy of the external potential field Π with respect to the volume of a working media, V (Π should not be confused with the internal potential field $U(r)$). By expressing the volume in terms of the mean size of a rotos, r, from Equation 2.109, we find its derivative $\dot{V} = \beta N r^2 \dot{r} = S\dot{r}$, where $S = \beta N r^2$ is the cross-sectional area of the body in the direction normal to loading and $\dot{r} = dr/d\tau$ is the rate of deformation of rotoses in the r direction. Then,

$$\sigma dV = \frac{F}{S} dV = \frac{\partial \Pi}{S \partial r} dV = \frac{\partial \Pi}{\partial V} = d\Pi. \tag{4.39}$$

Expressing the overall internal energy U in terms of the sum of the kinetic, $U(T)$, and potential, $U(r)$, parts, we can express Equation 4.37, accounting for Equation 4.39, as

$$\dot{U}(r) + P\dot{V} \pm \dot{\Pi} = \dot{U}(T). \tag{4.40}$$

The presence of derivative $\dot{\Pi}$ in the left-hand part of this equation violates thermal balance of a material. Consider now how the left-hand, i.e., potential, part varies and the consequences of this variation.

As seen from Equations 4.36.III and 4.36.IV, the potential part of the internal energy, $U(r)$, depends on two external parameters: volume V and temperature T. The external potential field Π affects $U(r)$ particularly through these parameters. As such, any change in the volume due to deformation affects $U(r)$ directly, and that in temperature affects $U(r)$ through derivative dV (see Equations 4.36.I and 4.36.II). We can write the auxiliary relationship,

$$dU(r) + d\Pi = \frac{\partial U(r)}{\partial V}dV + \frac{\partial \Pi}{\partial V}dV = d[U(r) + \Pi],$$

which allows one to express Equation 4.33 for the potential part of the internal energy, $U(r)$, accounting for the external potential field Π applied to the body, as

$$U(\Pi) = U(r) + \Pi = [U(r) \pm \Pi]\exp\left(\frac{U(r) \pm \Pi}{kT}\right). \tag{4.41}$$

The sign of Π accounts for the direction of action of the external field: plus is for compression and minus is for tension. As seen, the external field changes the depth of the potential well (Equation 1.26) and the shape of the MB distribution. The latter is transformed under the effect of the potential component (the numerator of the exponent factor) and the kinetic component (denominator).

Figure 4.25 shows the variation in the internal potential energy of a rotos, $U(r)$ (curve 1), in its differential $F = dU(r)/dr$ (curves 2, 2p, and 2c), and in

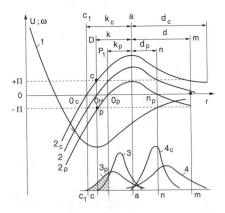

FIGURE 4.25
Variation of the internal potential field, Ur, and the resistance (curves 2, 2c, and 2p), as well as the MB distribution in the compresson (curves 3 and 3p) and dilaton (4 and 4c) phases in the material affected by external tension ($-\Pi$) and compression ($+\Pi$) force fields.

the MB distribution in the compresson (curves 3 and 3*p*) and dilaton (curves 4 and 4*c*) phases of a body affected by the external tensile (–Π) and compressive (+Π) potential fields. A number of conclusions can be drawn looking at this figure:

> The external tensile load lowers the zero energy level *or* (Figure 4.25) to position (–Π). This causes a decrease of O_HP in the potential well depth.
>
> This load causes shift of curve 2, which determines the resistance forces, to position 2*p*.
>
> It reduces the compresson zone from *k* to k_p and the dilaton zone from *d* to d_p.
>
> Due to a decrease in the numerator in the exponent factor and formation of the internal heat flow (Equation 4.38), the MB distribution, while moving to the left, changes its shape from that in position 3 to that in position 3*p*.

Reduction of the *C* phase with a simultaneous increase in the spread of the MB distribution leads to failure of part of the rotoses (dashed region on curve 3). The only way of preventing fracture is to resist development of these processes. Our analysis shows that deceleration of fracture based on these considerations is feasible; methods intended for such deceleration are substantiated in the next chapters.

In compression (zero level occupies position [+Π]), the structural processes occur in the opposite direction: depth of the potential well increases by a value of O_HC, the *C* and *D* zones expand to k_c and d_c, respectively, and MB distribution 4, while moving to the left, reduces its spread (curve 4*c*). Distribution 3 corresponds to metals and distribution 4 corresponds to stone materials, as the former are subjected mostly to tension and the latter to compression. In the stone materials only the ends of the left branches in the initial state are in the *C* phase (to the left from vertical axis *a–a*), which is why they have low tensile strength.

Transformation of the shape of the distribution from 4 to 4*c* at the macroscopic scale shows up as a volumetric reduction (compression) accompanied by improvement of the structure; it increases resistance to all types of external effects. This phenomenon is considered in detail in Section 3.10 for quasi-static loading of concrete. The favorable effect of reduction on resistance of metals and alloys to dynamic loading conditions has been discussed in Troshenko[3] and Ekobori.[44] As loading intensifies, the distribution gradually moves to the left, and the left-hand bonds increasingly transfer into the *C* phase. As such, mutual repulsive forces (differential curve 2*c* changes its concavity in an opposite direction), rather than attractive ones, become dominant. An increase in the concentration of the *C* rotoses leads to squeezing the material in the transverse direction. For this reason, stone materials fracture in compression due to transverse tension[48] and metallic ones change their initial shape.[2,3]

In service, the processes of redistribution of the internal energy between its potential and kinetic parts occur continuously under the effect of variations in the external force and thermal fields. As such, the end of the cold branch of the MB distribution may periodically find itself in the zone of brittle failure of rotoses T_c (corresponding in Figure 1.7a to point r_a), while that of the hot branch may cross the Debye temperature θ (vertical line propagating through point b) to activate the CD transitions. All this causes active thermal-destructive processes to occur in a material, inevitably leading to accumulation of structural damage and decrease in durability.

In addition to durability under quasi-static conditions (Equations 4.27 and 4.28), the equation of state (2.96) also makes it possible to explain the physical nature of fatigue of materials under various external fields (force, thermal, radiation, and others),[24] derive expressions for cyclic strength and durability under similar and combined types of loading, and substantiate the idea of its prevention and elimination.[215] An equation of the type of Equation 4.28, which is used as the basis of the kinetic concept of strength,[10] is a particular case of Equation 4.27 and does not allow for the CD nature of fracture (Figure 1.15); therefore, it has numerous deviations from the logarithmic dependence.[62]

4.3.3 Durability in Tension (Rupture), Compression, and Bending

In the equations of durability (4.28 and 4.27), one of the parameters (resistance P) is a vector value and, therefore, is sensitive to a particular direction of action of external fields. In other words, durability depends on the absolute value of external stresses, σ, and also on their direction. For this reason, durability of materials should differ depending on a particular stressed–deformed state. The authors of the kinetic concept of strength noticed this fact[62] and this point of view does not contradict engineering experience.[2,3]

If in Equation 4.28 the stress changes its sign, durability increases instead of decreases. Apparently, this can be confirmed if we conduct durability tests using the same test material in tension and compression. Unfortunately, not many experimental investigations on durability of materials under different stressed–deformed states are available in the literature; however, such studies are very important for practice. They are especially important because they help reveal the physical mechanism of exhaustion of strength and deformability and, possibly, lead to finding methods to control it.

Concrete was the best subject for such investigations due to the relative simplicity of formation of different types of stressed–deformed state. Durability in tension (rupture) was studied by pulling inserted anchors out from the bulk of materials and measuring the actual fracture surface (Figures 3.37 and 4.5b) at room temperatures ($T = \text{const}$). The time of loading was varied from several seconds to many days. Before any rupture test, the measurement of hardness by a dynamic-hardness tester at three levels of the impact energy (Figure 3.19c) should be conducted. A portable hydraulic press-pump (Figure 3.37a) was

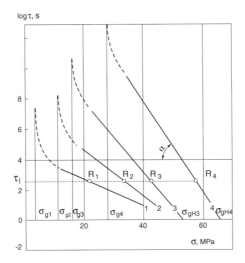

FIGURE 4.26
Durability in rupture (tension) of concretes of different compositions: water–cement ratio = 0.57 (1); 0.42 (2); 0.35 (3); 0.27 (4).

used for the short-time tests and gravitation load created long-time conditions. Typical rupture (tensile) plots of durability of different compositions of concrete are shown in Figure 4.26.

Dependence $\log \tau = f(\sigma)$ was found to be linear for all investigated compositions of concrete. The current procedure of design of the composition of concrete and the traditional technology lead to a nonequidistant shift of durability curves. The strength corresponding to a standard mode of loading is designated by R and located at the τ_H level in Figure 4.26. As R shifts to the zone of high values, the angle of inclination of the durability plot, α to the axis of abscissas continuously grows. As a result, even a slight variation in the stressed–deformed state leads to an appreciable decrease in durability. In transition to the high-strength region, the damping ability of concrete gradually deteriorates (Figure 3.34). This means that the rational region of practical application of high-strength concretes should be limited to static loading conditions.[14,17]

Figure 4.20b shows typical plots of durability of different compositions of concrete subjected to compression. All the compositions were manufactured using 500 grade cement: number 1 corresponds to the water–cement ratio of 0.57, 2 to 0.48, and 3 to 0.39. Unlike in tension, function $\log \tau = f(\sigma)$ appears to be a broken line for compression. The presence of the deflection point on this function indicates deep changes in the mechanism of formation of the stressed–deformed state. The physical nature of such changes is considered in detail in Section 3.9.

As loading conditions change (from dynamic to quasi-static and then to sustained), durability gradually grows. However, in region *cb* this is accompanied by an increase in resistance and, after inflection point *b*, it is accompanied by a decrease in resistance. This means that short-time compression has a favorable effect on the structure of a material. It exerts a unique restoration effect, the essence of which is in transformation of the elliptical shape of polyatomic molecules (Figure 2.8) into a spheroidal shape (Figure 2.7). This effect is not peculiar only to concretes; it is of universal nature.

Other investigators have also paid attention to this effect in studies of low-cycle fatigue.[215–218] Particularly, they note that holding the specimen under tension causes a decrease in durability, whereas holding it under compression may lead to an increase in durability under certain conditions or even to compensation for a damaging impact of tension. Moreover, low-cycle loading with holding only in a half-cycle of compression may even lead to an increase in the number of cycles to fracture. This effect can find application for restoration of performance of parts and units of machines and mechanisms (see the next chapters).

Location of inflection point b (Figure 4.20) in the coordinate system is entirely determined by peculiarities of structure of a specific type of concrete. Long- and short-time strength is a function of the Ω-potential. The latter is set in the structure of a material at the stage of its manufacture. Therefore, only at this stage is it possible to regulate the combination of properties that determine reliability and durability of a structure purposefully.

Also note that the homogeneity of the structure determines the steepness of the branches of the durability plots. Figure 4.27 shows dependencies of the slope of the quasi-static and dynamic regions of the plot to axis σ on the homogeneity index. When moving to the high-strength region (horizontal section of plots at $\tau = $ const), dynamic sensitivity of concrete to compressive effects decreases (the lower branch flattens). Therefore, reliability and durability of high-strength concretes have the highest values in operation under static and quasi-static loading.[55]

Determination of the boundary and degree of dynamic weakening of concrete in compression is of a high practical importance. (This equally applies to other artificial and natural stone materials.) This is required for correct estimations of reliability and durability of structures operating under random peak loads and finds application in development of energy-saving and waste-free technologies for production and processing of natural resources and for other purposes.

Figure 4.20c shows the plot of durability of bent reinforced concrete members. Their testing procedure is described in Section 4.2.4 and illustrated in Figure 4.18. In the semi-logarithmic coordinates it is shown by broken line *abcd*. At low durability (branch *ab*), there is no time for individual peculiarities of concrete and steel reinforcing bars to manifest themselves to a complete extent. Therefore, the load-carrying capacity of such a combined system is determined by averaged properties of both materials.

During long service of a structure under applied load, the resistance of concrete of the compressed zone has the dominating effect on the load-carrying capacity—a conclusion prompted by the presence of a characteristic deflection of function $\log \tau = f(\sigma)$ in region *bcd*. A conditional continuation of branch *cb*

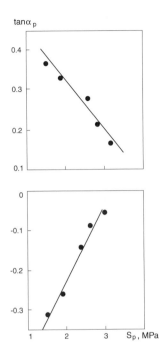

FIGURE 4.27
Dependence of the steepness of branches of
the durability plot for concrete in compression
on the homogeneity index.

in the direction of point a_1 and comparison of broken line a_1bcd with the curve
of durability of concrete in central compression (Figure 4.20b) further support
this conclusion.

At present, strength of reinforced concrete members in bending (as in other
kinds of stressed state) is determined under arbitrary loading conditions.
The term "arbitrary" is used here with respect to the plot of durability, in which
the rated conditions are set by one of the horizontal sections (e.g., level τ_H
in Figure 4.26). So far, the choice and substantiation of these conditions are
explained in various terms (level of perfection of the force inducing methods,
level of perfection of recording equipment, etc.) but not by a wide variety
of actual service conditions. Reliable information on the combined work of
steel and concrete within other time ranges is evidently not sufficient.

In this connection, the selection of structural materials on the basis of their
physical–mechanical properties investigated over a narrow time range, as
well as their positions in the bulk of a structural member, as related to the
entire time spectrum, looks arbitrary. This means that the durability plot
(Figure 4.20b) is a particular case out of a wide variety of other possibilities.
This realization may not be the best for given service conditions.

In estimation of durability of a particular material one should always know
whether the estimated region is on a linear part of dependence $\log\tau = f(\sigma)$
or covers the curvilinear zone, i.e., whether this dependence obeys in principle
the linear law. The authors of the kinetic theory of strength[10,62,63] are of the
opinion that deflections and singularities of this dependence are related as

a rule to instability of structure of a given material. This consideration assumes variability of the coefficients (V, γ) of σ in Equations 4.28 and 4.29 due to redistribution of stresses in interatomic bonds in a loaded body.

Multiple time tensile tests with recording absolute deformations of concrete (according to the diagram in Figure 4.5b), together with dynamic-hardness measurements,[125] led to realization that the linearity of the durability plot is a measure of compliance of the actual microscopic distribution of resistance of elementary volumes of a material (represented by a measure of dynamic hardness, Figure 4.4), with the logarithmic law (Equation 2.5).

The inverse statement is also true. Linearity of the plot is met by satisfying two conditions. The first is that the initial distribution of the hardness measures, measured, for example, at the τ_H level, should obey the logarithmic law. The second is that this principle should be stable in time, which means that mechanical-destructive degeneration of structure with a variation in the loading mode should not change the original distribution law, i.e. the process of microcracking should not be accompanied by irreversible relaxation phenomena. In other words, in any region of plots 1 to 4 (Figure 4.26), the distributions of hardness measures should be logarithmic.

The discussed experimental results allow suggesting accelerated nondestructive testing methods for estimation of durability of concrete (and any other material) in compression and tension (rupture). The time rupture tests and measurements of current forces and absolute values of deformations result in the family of cumulative curves (Figure 4.28), where F, kN, is the rupture force and f, mm, is the absolute deformation. The necessary and sufficient conditions of linearity of the durability plot are met when the cumulative curves are described by the logarithmic law at any value of durability of a stressed material $(\tau_1, \tau_2, \ldots \tau_n)$.

To verify this, it would be sufficient to pull the inserted anchors out of the bulk of a material (Figure 3.37) at two considerably different rates of loading. Repeating the tests several times at each rate and measuring the maximum force and actual fracture surface simultaneously with proving the linearity,

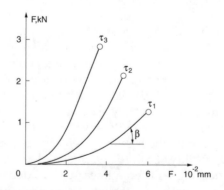

FIGURE 4.28
Family of the cumulative curves plotted in testing concrete on durability in tension.

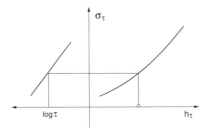

FIGURE 4.29
Nomogram for estimation of durability of concrete from the results of nondestructive tests.

we obtain two groups of points in the coordinate plane ($\log \tau - \sigma$). Using averaging of these points, we can draw a line of durability (designated as τ_H in Figure 4.26). The obtained line is used to determine the parameters of Equation 4.28.

To conduct an accelerated durability test in compression, the nondestructive method, in which the height of rebound of the indenter is measured, can be used.[129] The essence of the method suggested is that statistical arrays h_τ and d_τ of the indirect hardness indicators are obtained by testing cubic samples or prisms at several values of the indenter impact energy. Then the samples are brought to fracture at a different loading rate to obtain σ_τ. The degree of compliance of arrays h_τ and d_τ with the logarithmic law is estimated; the better this compliance is, the more linear the durability plot. Correlation dependencies of the $\sigma_\tau = f(h_\tau d_\tau)$ type (Figure 4.29) are plotted in the combined coordinate system.

When durability of concrete is to be estimated in the complete structures, the reverse sequence is used utilizing the obtained graph. Of special benefit for this purpose are the nondestructive methods and facilities that use electrical signals created by transducers placed in the structure in the process of dynamic impact. Such facilities, simultaneously with automatic recording of measurement results, allow their statistical processing and evaluation.[219,220]

4.3.4 Dynamic and Sustained Strength. Assuring Durability by Traditional Design Methods

Studies of durability of various materials revealed semi-logarithmic nonlinearity in regions of low values of σ, which is especially appreciable in metals at high temperatures (Figure 4.30).[62] In Betekhtin[211] and Zhurkov,[221] this nonlinearity is attributed to the fact that Equation 4.29 was derived studying disintegration of interatomic bonds, with no accounting for the possibility of their recombination. These processes, however, occur simultaneously in a material. At $\sigma \rightarrow 0$, the intensity of the process of restoration of broken bonds increases, which prevents accumulation of structural damages and development of fracture.

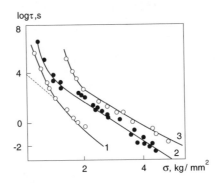

FIGURE 4.30

Manifestation of nonlinearity of the logarithmic dependence of durability in tension in the field of low loads for the following materials: 1, silver; 2, aluminum; 3, polymer.

Some investigators[222] are of the opinion that curve $\log\tau = f(\sigma)$ asymptotically approaches a certain straight line that cuts off on the axis of abscissas (a segment taken as "safe load"), rather than the axis of ordinates. Others[221] think that this assumption is arguable because it is impossible to verify it using a real-time experimental check, although they do not deny the existence of such a possibility.

In our opinion, the asymptotic trend of the durability plot to a certain limit (levels σ_{g1}, σ_{g1}, ... in Figure 4.26) is not only possible but also imperative. Experimental facts (Figure 4.30) as well as theoretical estimations and service experience on metal structures (see Figure 1.22 and comments on it) provide sufficient evidence in favor of this statement.

Figure 4.31 shows Function 4.28 at different values of exponent

$$A = \frac{\Omega_0 \pm V\sigma}{kT}.$$

Condition $A_1 > A_2 > A_3$ corresponds to different levels of the intensity of the external field, $\sigma_1 < \sigma_2 < \sigma_3$. In quasi-static external fields, where $dV = \text{const}$, variation in the position of the curve is determined only by internal energy U or, to be more exact, by the heat content of a material, H (Equation 2.29). As H decreases (for example, when temperature of the environment dramatically decreases), the durability curve moves from A_1 to A_3. Even at an invariable level of effective stresses, $\sigma = \text{const}$, and relatively high resistance P_0, this leads to a decrease in durability from τ_1 to τ_3. In such situations (which worsen if accompanied by dynamic loading), the probability of failure of the structure grows substantially. In fact, this is the case observed in reality (Figure 1.22).

Decrease in the temperature of cold shortness T_c in force and thermal fields with time (lower part of Figure 1.22) means a change in the spatial shape of the internal potential field (Figure 1.8). In accordance with Equation 1.68, this is possible only with an increase in the crystalline lattice parameter r_a,

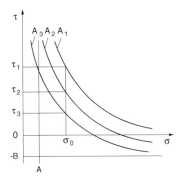

FIGURE 4.31
Variations of durability of materials, τ, in variable force and thermal fields.

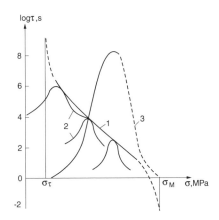

FIGURE 4.32
Substantiation of existence of dynamic, σ_M, and sustained, σ_v, strength.

with a simultaneous decrease in density of AM of the electrostatic field, α (Equation 1.17). This variation in the parameters of the local field leads to its leveling over the entire working volume in the direction of r_{a_1} to r_{a_3} (Figure 1.10b). In such a field, the resistance of a body, **P**, decreases and conditions become more favorable for crack initiation and propagation (see comments on Figure 3.17).

The trend of durability plots A_1, A_2, and A_3 (Figure 4.31) in the field of high values of durability to their asymptotes (e.g., to AA_2) is predetermined by parameter A in Equation 4.28. In addition, the durability plot is a locus of critical points corresponding to the ultimate resistance of a material that is achieved at different rates (curve 1 in Figure 4.32). The transition to the critical state at any rate is characterized by a peculiar cumulative curve of the type of $\sigma = f(\varepsilon)$. The set of such curves for concrete, limestone, and steel is shown in Figure 4.2; curves for metals and different alloys can be found in Troshenko.[3] The moment of fracture coincides with the point of inflection of the cumulative curve or with the maximum of the differential distribution.[223] It appears that the durability plot itself is an envelope of the instantaneous MB distributions (curves 2) and, in reality, a super-logarithmic integral law developed in time. Its differential form can be obtained by graphical differentiation (curve 3).

The semi-logarithmic time law of durability (curve 1), as well as any cumulative curve, is always characterized by the asymptotic trend to a certain limit.[223] Moreover, it makes sense to consider the asymptotic approach not only in the long-time region (the upper dotted region of curve 1), but also in the super-short zone (its lower dotted end). For structural materials, these asymptotes have actual physical meaning and are, perhaps, the only constants: dynamic, σ_M, and sustained, σ_τ, strength. As such, σ_M is determined by Equation 4.11.

The shape of the durability plot of any material is governed by the process of mechanodestruction (see explanations to Figure 2.10). To find out to which stage of the destruction corresponds the moment of transition from level σ_τ to a linear region, it is necessary to carry out time-consuming observations and measurements of the temporary mechanodestruction transformations.[128,129] However, we can point out that the achievement of the limit of sustained strength coincides with the transition of the initial incubation period of fracture (Figure 2.8a) to the second period, i.e., formation of multiple microcracks (Figure 2.8b).

Berg[48] came to the same conclusion; in the afterward to his book,[149] he presented distinctive features of sequence σ_{g_1}, σ_{g_2}, ... σ_{g_n} similar to those shown in Figure 4.26: the boundary of formation of microcracks, i.e., structural parameter A in Equation 4.28, that determines the position of asymptote σ_g is a variable that depends on the rated strength of a material (i.e., points lying on horizontal line τ_H). He pointed out that, despite the spread of data associated with nonhomogeneity of concrete, these data clearly show that the boundary of formation of microcracks, σ_p, increases with an increase in the rated strength σ_T, i.e., sequence $\sigma_{g_1} < \sigma_{g_2} <$... is satisfied (Figure 4.33).

Parameter A is of principal significance for many important applications of the theory of strength and deformability of concrete. For example, as established by direct experiments,[128,129] A corresponds to the absolute fatigue limit of concrete. Methods of acoustic emission, super-high-frequency radio waves, and others reliably distinguish transition periods of the mechanodestruction process (see Figures 3.50 and 3.51). Therefore, they can be employed for determining the sustained strength limit under forced loading conditions. This limit corresponds to a sudden change in the direction on the plots shown in Figures 3.50, 3.58, and 3.62.

Existence of the sustained strength is evidenced by the multi-century practice of building and failure-free operation of engineering objects of the most diverse application. The failure-free operation originates from the stage of design. Results of investigations make it possible to reveal and describe the

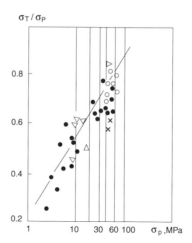

FIGURE 4.33
Dependence of the limit of cracking σ_T on the rated strength σ_p of concrete.

mechanism of ensuring reliability and durability by using up-to-date design methods.

As follows from Figure 4.26, the source information for these methods is obtained from accelerated tests of materials in a strictly fixed time range of loading (e.g., at τ_H level). As a result, only particular and shortened statistical samplings (one of plots 2 in Figure 4.32) become available to the designer from a wide variety of those possible in service. In Figure 4.26, the average values of these samplings are designated as $R_1, R_2 \dots$. They characterize no more than behavior of a material under given loading conditions. Regulatory documents enforce the described procedure of the tests,[194–199] which, in our opinion, has substantial drawbacks. Limitations included in the regulatory base do not allow obtaining comprehensive information on the behavior of materials under various operation conditions and do not guarantee 100% certainty in failure-free operation of parts and structures.

Working in the zone of sustained strength is assured by assigning design resistances. The available procedure for their regulation is based on a hypothesis of brittle fracture, which, although logically faultless, is questionable from the physical standpoint (see Chapters 1 and 2) and has no experimental proof. According to this hypothesis, the failure of the entire system occurs after the ultimate state is achieved in the weakest link. Nevertheless, the expected result is achieved because the design resistance is assumed to be lower than the lowest value of the actual resistance, i.e., at the extreme point of the left end of distribution 2 (Figure 4.32).

This approach allows the service stressed–deformed state to be in the zone of sustained strength σ_τ, but does not guarantee the exact coincidence of R_{min} and σ_τ. Therefore, at $R_{min} > \sigma_\tau$, the reliability of a part decreases, whereas at $R_{min} < \sigma_\tau$, it becomes economically faulty. Such a design system provides a high degree of reliability under static loading conditions, but fails to guarantee the safe operation of parts and structures working in random force, thermal, or other fields, especially at their unfavorable combinations.

New knowledge of the physical nature of deformation and fracture (Chapters 1 and 2) suggests alternative methods for ensuring reliability and durability. In accordance with Equation 4.27, the trend of τ to ∞ can be achieved two ways: either at $\mathbf{P} \to 0$ or at $A \to 0$. The first condition contradicts physical sense; current design methods provide meeting of the second condition due to an increase in internal parameters of a material, i.e., N and U (see explanations to Equation 4.27). A can be made infinite also by the other method, i.e., by creating conditions to force $dV \to 0$. Assuming that $dV = 0$ (Equation 4.14), we will have

$$\frac{\dot{V}}{V} + \frac{\dot{\mathbf{P}}}{\mathbf{P}} = 4\frac{\dot{T}}{T} - 3\frac{\dot{\theta}}{\theta}. \tag{4.42}$$

This equality substantiates a fundamentally different, but so far unknown in world practice, method for ensuring reliability and durability, i.e., the energy method. It ensures the constancy of mechanical parameters of the Ω-potential ($\mathbf{P}$ and V) through variation of the kinetic ones (T and θ), preventing departure of the material structure from the equilibrium state. Information on this method can be found in Komarovsky[12] and is also described in the following chapters.

4.3.5 Relationship of Durability, Creep, and Fatigue Limit

The durability plots (Figures 4.20 to 4.26) can be considered the stressed–deformed state diagram developed in time. In the force fields, the reorientation of rotoses chaotically located in the bulk of a material takes place in the direction of this field (according to the diagram shown in Figures 2.8 and 3.7). Naturally, this process is accompanied by transformation of the MB distribution (Figures 4.3 and 4.25). The reorientation takes place in a reversible manner until the ends of MB distribution reach one of the critical temperatures: T_c or θ (Figure 1.14). At this moment, according to the diagram in Figure 1.15, the CD phase transitions start developing into the mechanical-destructive process under thermal insufficiency. The physical nature of this process has been described. Loading conditions and character of interaction of a material with the environment and other fields have a decisive effect on the CD phase transitions.

Vertical sections of a durability plot at $\sigma = \text{const}$ make it possible to obtain rheological characteristics, while horizontal sections at $\tau = \text{const}$ allow estimation of resistance. The first, called creep curves, are plotted for long-time loading conditions, and the second are deformation diagrams $\sigma = f(\varepsilon)$ (Figure 2.22) obtained in short-time strength tests. The first are characterized by a long-time effect of the constant level of stresses ($\tau \neq \text{const}$, $\sigma = \text{const}$), and the second by gradual (sometimes quite long) variations in the force parameter (τ = const, $\sigma \neq$ const). Rheological and strength tests are usually conducted separately; results of these tests are rarely related.[2,10]

Physical processes occurring at the disintegration of a structure under short- and long-time loading have been studied to a different extent of completeness. The development of the tests in time ($\tau \neq$ const), and measurement of the current and ultimate values of strains ($\varepsilon \neq$ const) by intensifying the load ($\sigma \neq$ const), yield a family of stress–strain diagrams $\sigma = f(\varepsilon)$ (Figure 4.2). At present, this is the most common, although not the best, method for graphical representation of results of destructive tests. These diagrams can be used for plotting the time development of the stressed–deformed state.

Relating the values of maxima of the curves shown in Figure 4.2 to the time of dwelling of a material under a load, the durability plot $\log \tau = f(\sigma)$ (Figure 4.26) is a part of the constitutional diagram. Including only the dynamics of resistance variations, it characterizes only one aspect of the fracture process, namely, the work of stressed bonds of a section. The other aspect, i.e., the principle of accumulation of damages in a structure, escapes investigators' attention. It can be evaluated by variations in deformability because the latter is controlled by the flow of failures of interatomic bonds.

Comparing the ultimate values of deformability at each time of holding under a load, ε_τ, and the associated resistance, σ_τ, we can obtain dependence of the type of $\varepsilon_\tau = f(\sigma_\tau)$. This establishes the relationship between the number of supporting and failed bonds at any rate of loading. To be more exact, it allows evaluating the relationship between compresson and dilaton bonds, depending on the density of flow of the external mechanical energy.

Such dependence is not hard to obtain at high rates of loading. To do this, the experimental setup should be equipped with linear displacement sensors (proximity probes) for recording parameters. In the zone of long-time strength, obtaining the source data involves considerable difficulties associated with the long duration of the experiment. However, our analysis of the mechanism of exhaustion of strength (Chapters 1 and 2), as well as the available experimental data,[3,202,224–227] allows the shape of dependence, $\varepsilon_\tau = f(\sigma_\tau)$, for the entire strength range to be predicted with a high level of confidence. Figure 4.34 shows the general shape of such dependence.

It has maximum ε_τ, whose position is determined by sustained strength σ_τ or by an asymptotic approximation of the durability plot (Figures 4.26 and 4.32). The shape of the left branch of the curve (especially its initial part) results from the dynamic loading conditions, whereas the right branch can be predicted from the results of structural transformations in long-time tests.[202]

Opinions of investigators concerning the dependence of ultimate deformability of concrete on its short-time strength differ. The issue is about the possible shift of the curve (Figure 4.34) in the coordinate system, depending on the composition of concrete (to be more exact, on its structural peculiarities). Some investigators believe that ε_τ hardly varies with a variation in the structure and has the same value over the entire practically achievable range of the short-time strength. The second group presents evidence of an

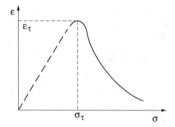

FIGURE 4.34
General form of the relationship between ultimate deformability ε_τ and resistance σ_τ of the same material in durability tests.

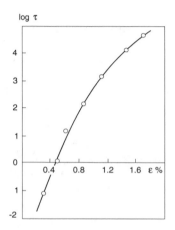

FIGURE 4.35
Time dependence of deformability of limestone.

increase in ε_τ as the strength increases, whereas the third group maintains that ε_τ decreases with growth of the strength.

For example, Pisanko[118] concluded that $\varepsilon_\tau = 1.47 \cdot 10^{-3}$ = const within a range of decrease in strength from 35 to 65 MPa. Pinus[119] thinks that ε_τ hardly differs within concrete grades 500 to 700, and is equal to $1.51 \cdot 10^{-3}$, on the average. Experiments conducted by Sytnik and Ivanov[113] show the presence of a definite dependence of the type $\varepsilon_\tau = f(\sigma_\tau)$ (Figure 3.42).

Curves characterizing variations of deformability of a material depending on the time of loading can be plotted in a similar way. To illustrate, Figure 4.35 shows such dependence for limestone. The source data for it were taken from Kuntysh and Tedder.[193] Similar data for concrete can be found in the paper by Vasiliev.[144]

This dependence resembles very much the durability curve (Figure 4.26). In our opinion, it should be so because both characterize behavior, under loading, of different parts (failed and supporting) of the same CD set that determines macroscopic properties of a solid.

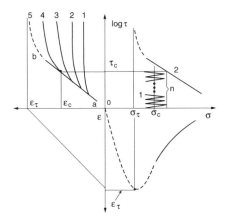

FIGURE 4.36
Relationship between durability τ, creep ε, and fatigue limit n.

Combining the plots shown in Figures 4.26, 4.34, and 4.35 in the common coordinate system, we obtain a complex diagram (Figure 4.36) that simultaneously characterizes supporting and deformative properties of a material in the stress–deformed state. Dependence $\varepsilon_i = f(\sigma_i)$ serves as a link between these properties.

The resistance of a material with stable structure has a single-value dependence on the time (durability plot 2 in the right-hand upper field of Figure 4.36). Relationship between resistance σ_τ and ultimate deformability ε_τ is now characterized by the two-branch dependence (right-hand lower field). Development of the structural-deformation processes in time results in a multivalue dependence (left-hand upper field). The deformation ambiguity is an inherent property of any stable structure and can be attributed to a fundamentally different character of occurrence of the CD phase transitions and destructive processes, depending on the level of a destructive factor. At a low value of external loads, the recombination processes, which always accompany CD transformations, balance the degradation changes and thus the structure eventually stabilizes and adapts to new external conditions. This possibility exists at stresses ranging from zero to a value of sustained strength σ_τ. Development of the deformation process within this range is determined by sequence of creep curves (1, 2, ... 5).

Upon exceeding the σ_τ level, mechanical-destructive processes do not attenuate. They are no longer limited to local microvolumes; instead, they progress and affect more and more regions of a material. Deformability does not depend now on the relaxation capability of the structure because it is limited by the time effects. As durability decreases (to the right from σ_τ, in the right-hand upper field), deformability is determined by region ab of the boundary deformation curve (upper left-hand field).

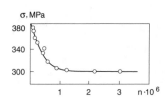

FIGURE 4.37
Fatigue limit diagram.

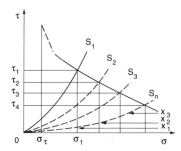

FIGURE 4.38
Showing exhaustion of strength during the service time under arbitrary loading conditions.

The stressed–deformed diagram (Figure 4.36) allows us to establish the relationship between durability, creep, and fatigue limit. Let the external forces induce stresses in a material, which vary following the periodic law with respect to the mean level of σ_c. Periodic variations (1) fill out the time interval of the durability plot at the σ_c level. The point of their intersection with durability plot (2) determines cyclic life τ_c and deformability ε_c. Fatigue limit n at a preset level of the stressed state is calculated by dividing durability τ_c by a period of the τ_n cycle.

Varying conditions of cyclic loading (shift of the σ_c level along axis σ), we obtain a different number of cycles, n. This number characterizes the fatigue limit of a material and depends on the shape of plot 2. Plotting the fatigue limit diagram, i.e., dependence of stress σ_c on the number of cycles n (Figure 4.37), is the usual practice in fatigue tests.[3] In the rectangular coordinates, it is a curve asymptotically approaching the straight line that cuts off stresses equal to fatigue limit on the axis of ordinates. As seen, this curve repeats the shape of the durability plot and the fatigue limit coincides with the value of sustained strength σ_r.

Using these results, we can describe (with a high degree of confidence) the mechanism of exhaustion of resistance of a stressed material under the effect of accidental loads of different intensity and duration. Assume that a constant load that induces stresses σ_1 is applied to a body having cross section S_1 (Figure 4.38). This stressed–deformed state corresponds to durability τ_1. Let stresses σ_1 be in excess of the value of sustained strength, which means that irreversible structural changes will accumulate in the body with time. Their macroscopic measure is deformability (the left-hand upper field in Figure 4.36). The flow of failures of interatomic bonds leads to a gradual decrease in the load-carrying section area.

As a consequence of this, curve S_1 shifts to positions S_2, S_3, etc. Physically, this shift means a steady decrease in the safety factor for dynamic resistance (Figure 4.3). If, at the beginning of loading it was equal to x_1, later on it decreases to x_2, x_3, etc. If, at the beginning of service, the material could withstand substantial random overloads (in a range of $\sigma_1 + x_1$) within the corresponding time interval $(0 - \tau_1)$, after some time both parameters will decrease to form strength $(\sigma_1 + x_1)$, $(\sigma_1 + x_1)$, etc., time $(0 - \tau_2)$, $(0 - \tau_3)$, etc., rows that tend to zero. This leads to the fact that effective stresses near the expected durability are much in excess of the initial level, σ_1, and any excess of it, even for a short time, will lead to catastrophic failure of parts and structures.

4.3.6 Method for Estimation of Durability in Design

In different periods of the existence of parts and structures (manufacture, storage, transportation, assembly, operation) a material is subjected to different external effects. With respect to the structure, the latter may have a passive or an active nature. The external effects are active if they influence the performance of a material during the entire period of functioning of parts: humidity, temperature, pressure, chemical composition, and aggressiveness of the environment such as presence of radiation, physical fields, etc. Condition and physical–chemical properties of the working environment have an activating or decelerating impact on structural processes and, as a consequence, on strength, deformability, stability, and durability of the material. The passive effects include those associated with preparation of a part for performing the consumer's functions (transportation, assembly), or those that arise at the stage of their active operation.

Differentiation of the equation of state (2.134) yields the entire spectrum of stressed–deformed state, which is possible for a material in force and thermal fields. If we take dT in the right-hand part and PV in the left-hand part out of brackets and replace $ds = dV/V$ (this can be done using the relationship of equivalence (2.105) which establishes adequacy of the processes of mechanical deformation and thermal expansion), we obtain

$$\mathbf{P}V(\alpha + v) = \alpha T - (\varepsilon + \eta), \tag{4.43}$$

where ε and η are the longitudinal and transverse strains, α is the thermal expansion coefficient (2.152), and

$$v = \frac{1}{\mathbf{P}}\left(\frac{\partial \mathbf{P}}{\partial V}\right)_V$$

is the thermal pressure coefficient.[23] (Subscripts indicate the fixed parameters.) Then, following the principle of equivalence and assuming that $(\alpha + v)_p = (\varepsilon + \eta)_T$, we can write

$$PV = T\frac{\alpha}{\alpha + v} - 1. \qquad (4.44)$$

The equation of state is expressed in terms of the parameters, which can be determined experimentally. It allows one to determine the Ω-potential, and, if temperature T and volume V are given, the resistance **P**.

The rest of the parameters can be calculated using the following method.[23] First we determine enthalpy $dH = Tds = C_p dT$. Integration of this expression in a given thermal field yields

$$H = \int_{T_{min}}^{T_{max}} C_p dT + H_0. \qquad (4.45)$$

Because heat capacities of a solid body at a constant pressure and volume are equal, $C_p \approx C_v$, C_p can always be replaced by C_v. To find H experimentally, it is necessary to measure C_p or C_v in a service range of temperatures $T_{max} - T_{min}$ and to know H_0 at a certain fixed temperature T_0. Behavior of a material in external fields does not depend on the specific value of heat content H, but on its change during service time interval. If so, then

$$\Delta H = H - H_0 = \int_{T_{min}}^{T_{max}} C_p dT,$$

which eliminates the need of finding constant H_0. Entropy **s** is found from the same values of heat capacity,

$$\mathbf{s} = \int_{T_{min}}^{T_{max}} \frac{C_p dT}{T} + \mathbf{s}_0 = \int_{T_{min}}^{T_{max}} C_p d\ln T + \mathbf{s}_0,$$

by graphical integration of experimental dependencies $C_p f(\ln T)$ or $CV_f(\ln T)$. Variation in entropy in a given thermal field is equal to

$$\Delta \mathbf{s} = \mathbf{s}(T) - \mathbf{s}(T_0) = \int_{T_{min}}^{T_{max}} C_p d\ln T.$$

In addition to T, force field σ also has influence. This influence is determined as[23]

$$\left(\frac{\partial \mathbf{s}}{\partial \sigma}\right)_T = -\left(\frac{\partial V}{\partial T}\right)_\sigma = -\alpha V. \qquad (4.46)$$

By integrating, we find

$$s(T,\sigma) = -\int_{\sigma_{min}}^{\sigma_{max}} \alpha V d\sigma + s(T_0,\sigma_0).$$

Hence, variation in entropy at the simultaneous effects of force and thermal fields is

$$s(T_0,\sigma_0) - s(T,\sigma_{max}) = -\alpha V \Delta\sigma, \tag{4.47}$$

where $\Delta\sigma = \sigma - \sigma_0$ is the range of variation of the external stress. The working volume is found from Equation 4.47:

$$V = \frac{\Delta s}{\beta \Delta\sigma}, \tag{4.48}$$

where

$$\Delta s = -[s(T,\sigma_{max}) - s(T_0,\sigma_0)]$$

is the required safety factor for strength to counteract external stresses in the $\Delta\sigma$ range.

This method of calculation is based on fundamental physical laws and thus is universal. It allows principles of deformation and fracture of materials in variable force and thermal fields (and, naturally, in each of them separately) to be described from unified positions. Also, it indicates the principal direction for further improvements of the strength design procedures and reduction of experimental parts of studies necessary to substantiate reliability and residual life of machines and mechanisms. Compared with traditional approaches, this method requires extra costs associated with an increase in volume of the source information and updating the regulatory procedures. This is a natural compensation for precise prediction of behavior of materials, parts, and structures during operation and for trustworthy conclusions on their reliability and durability.

In this connection, the collection of reliable information on service loads and thermal fields is of primary importance. For example, multiple loads acting on civil engineering objects can be conditionally divided into two types, deterministic and random, depending on the character of their initiation and variation in time. The former, on initiation, remain constant or gradually vary following the known laws. The latter are random functions of time. For example, every earthquake is accidental, while the time of action of seismic loads is determined by the velocity of propagation of an elastic wave ($(1$ to $5) \cdot 10^5$ cm/s); in some technological processes (cutting, deforming) the rate of load application reaches 1 s^{-1} or higher. During construction of

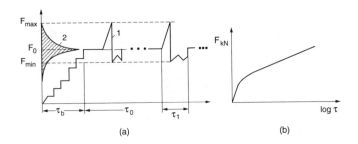

FIGURE 4.39
Examples of (a) variations in service loads with time (1), and differential (2), and integral (b) laws of their distribution.

civil engineering structures, the loads increase continuously within dozens of days, whereas during operation, some of them remain invariable for decades.

Therefore, under actual operation conditions, the rate of load application as well as the time of their impact varies within 15 to 20 orders of magnitudes. The rate of deformation of concrete specimens and reinforcing bars in testing to various kinds of the stressed state (tension, compression, bending, etc.) is regulated stringently within a range of $1 \cdot 10^{-4}$ to $2 \cdot 10^{-2}$ s^{-1}.[194–199] Matching strain rates in testing and those expected in actual service would provide reliable information on behavior of materials under specific operation conditions and a substantial reserve for decreasing material consumption of parts and structures.

Variations in loads can be represented as a time spectrum (Figure 4.39). Loads having instantaneous character are located at one end and those determining long-time rheological processes at the other. At the initial period of operation, the loads monotonically increase during time τ_b, then they stabilize and retain their value at certain level F_0 within the time of their active operation, τ_0. During this active time of operation, the applied load may fluctuate by certain value 1 (during short periods of time τ_1) and the frequency and amplitude $F_{min} - F_{max}$ of these fluctuations are set by sequence and intensity of the technological cycle.

Summing up the time of dwelling of the material under the same load during a preset period, we obtain the differential law of distribution 2. Although it does not account for the sequence of changing in the stressed–deformed state, it can be considered to the first approximation and represents the convolution of the time stochastic process of loading 1. Its graphical integration yields the mirror reflection of the cumulative curve (Figure 4.39b).[223] Superimposed with the durability plots (Figure 4.26), it yields a graph that can be used to solve a set of problems on the achievement of the required level of reliability and durability (Figure 4.40).

The shape of an integral curve of loads 1, 2, and 3 in solving any specific problem remains unchanged. However, varying the volume of a material that participates in resistance can control its location in the coordinate field. Frequency of filling the coordinate grid and shape of curves 4, 5, and 6

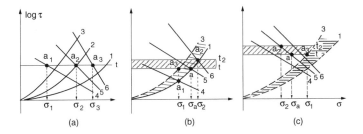

FIGURE 4.40
Nomograms for solving the following prediction-type problems at the stage of design:
(a) estimation of durability; (b) prediction of the probable damage due to gross design errors
and decrease in the level of technological discipline; (c) selection of a rational type of structural
material.

depend on many parameters. Among them the following are important: a
particular kind of stressed–deformed state, value of the Ω-potential of a
specific material, kind of packing of dipoles in its structure (Figure 2.17),
and extent of perfection of the technology employed. In other words, it
depends on those factors that determine life $s(\tau)$ (Figure 4.27). The levels of
durability τ are set on the basis of the functional purpose of a part under
consideration and vary, as a rule, within narrow ranges.

The discussed graph allows evaluation of various design solutions, analysis of cost efficiency of acceptance of any of them at the stage of design,
estimation of consequences of violation of technological discipline, unfore-
seen deviations in operation conditions, etc. The type of engineering problem depends on the combination and possibility of regulation of these
variables.

When the level of durability, $\tau = $ const is set strictly but there are no
limitations on the working volume of a material, $V \neq$ const, and its resource
$s \neq$ const, (e.g., due to varying physical–mechanical properties of source
elements, methods of their interlocation in the bulk of a part, kind, sequence,
and intensity of technological effects, and other factors), the so-called uncon-
strained or direct problem is the case. At the first stage of its solution, the
value of micro- and macrolevels of the life (Equation 2.112) is selected (plots
4, 5, or 6) (Figure 4.40a). Then, the required level of macroscopic entropy
(Equation 2.113) is ensured by varying the shape and size of a structural
element (1, 2, and 3), formulating requirements for its manufacturing technology, and assigning rational production methods.

Thus, working point a on horizontal line t is determined, taking into consideration the dynamic loading conditions and critical cost efficiency. For
example, the assigned level of reliability can be ensured using different
combinations of quality s and quantity V of the material (sequences of points
a_1, a_2, and a_3). Such variants allow the action of different constant loads ($\sigma_1 <
\sigma_2 < \sigma_3$) at substantially different levels of dynamic stability (compare areas
of figures $\alpha_1\sigma_14 > \alpha_2\sigma_25 > \alpha_3\sigma_36$). This permits control of the degree of dynamic
strengthening of a material[24] at high-rate application of random loads.

The working point may not be in the optimal zone due to significant design errors, misuse of manufacturing technology, low-quality assembly, or deviations in service conditions. For example, dramatic variations in manufacturing quality of a material or inconsistency of subsequent technological operations lead to a spread of the working point in fields 4 to 6. Inadequate knowledge of the laws of exhaustion of resource leads to the same consequences.

Figure 4.40b shows the displacement of the working points at $\tau = $ const, $V = $ const, and $s \neq $ const (significant design errors, inconsistency of technological process, etc.). Displacement of point a to position a_2 leads to material over-expenditure, increase in power and labor consumption in part manufacturing, decrease in the safety factor for dynamic strength, and deterioration of reliability.

The excessive safety factor for durability $(a - a_2)$ increases nonproductive costs for reconstruction (horizontal region between straight lines t and t_2). Shift of a to position a_1 is inevitably accompanied by a decrease in durability of the considered component. To avoid failure of the whole assemblage (whose durability may be much in excess of the local level) and maintain performance of the object, it is necessary to restore its resource. This involves an increase in reparation costs (displacement of point a_1 to position a_3). In both cases, an increase in operating costs is caused by excessive amount for materials, power, and labor (dashed region between curves 1 and 3).

Design and manufacturing errors, as well as aging and thus wear of technological equipment, lead to variation in parameter $V \neq $ const at $\tau = $ const and $s = $ const (Figure 4.40c). In this case, point a, having one degree of freedom and moving, for example, along curve 5, may leave the optimal zone, coming to position a_1 or a_2. Such a shift causes similar consequences: position a_1 requires repair operations (to move it to position a_3), and point a_2 corresponds to an excessive durability and nonproductive costs.

Other problems may occur with constraints; however, their occurrence is not that frequent. For example, conditions $\tau \neq $ const, $V = $ const, and $s = $ const occur in manufacturing of a batch of similar-type parts of the same components made using different technologies or when the parameters of the existent technology substantially vary. It characterizes a requirement for restoration of durability in optimization or updating of a part.

In design and manufacturing of similar-type parts intended for operation under conditions with fundamentally different environment parameters, it is necessary to solve a problem having three variables $\tau \neq $ const, $V \neq $ const, and $s \neq $ const. Condition $V = $ const, $\tau \neq $ const and $s \neq $ const holds at limited material resources and in the absence of flexible control of the technological process. The necessity to operate a part in the presence of stringent stability limitations requires carrying out analysis of the following condition: $V = $ const, $\tau \neq $ const, and $s = $ const. All of these discussed problems can be considered and illustrated in a similar way.

5

Solids in Active Media

5.1 Introduction

The initial condition of materials in parts and structures is formed during solidification[41] and then is corrected by technological procedures.[3,44] During service, the external fields and aggressive media influence it through one or several parameters of Equation 3.19 and often lead to premature failure of engineering objects.[1-3,43] This is attributable[1-4] to insufficient accuracy of design methods, incomplete knowledge of material properties, poorly substantiated fracture criteria and strength margins, imperfection of materials and technological processes, deviation from normal operating conditions, untimely and low-quality maintenance and repair, etc.

The chief and not yet completely realized cause of this failure, however, is change of the internal condition of materials in time (Equations 4.27 and 4.28) because electrodynamic processes initiated by external factors are proceeding in the structure of solids (Figure 1.15 and Section 3.2). Among these processes, thermal energy has a special place because it is included in the equation of state (2.96) — directly through parameter T and indirectly through δ parameter. It is capable of concentrating in local volumes t and easily converting into other kinds of energy, and characterized by high velocities of transfer from one object to another. In addition to an equiprobable direction of distribution, it is capable of forming internal heat flows, etc.

The result of these processes is referred to using different terms, namely aging (heat, radiation, etc.), embrittlement (radiation, thermal, etc.), and damage (force, chemical, hydrogen, moisture, etc.). The AM mechanism of these processes is determined by the features of CD arrangement of the structure (Figure 1.14), while the rate at which they proceed is determined by external conditions, namely loading mode, service temperature, impact of active media and energy flows, etc. Unfortunately, the evaluation of durability and strength analysis is currently carried out using rather simple procedures accounting for only partial characteristics of static strength.[3]

This chapter is devoted to analysis of the influencing mechanisms of specific factors on the structure condition, namely aging in heat and force fields (Section 5.2), embrittlement in hydrogen-containing media (Section 5.3), structural damage in radiation fields (Section 5.4), and moisture-induced softening (Section 5.5). Moisture impact is most essential for porous materials (natural and artificial stones, ceramic, polymers, composites, etc.).

Internal processes, proceeding under the influence of diverse external factors, change durability of structures; Section 5.6 presents the schematic of predicting such an impact. Structural transformations can be reversible and irreversible. Reversible processes can be controlled in order to bring the structure into its initial condition. Thermal, magnetic, electrical, chemical, and even force fields will find an application for such a purpose. Recovery is possible during repair, standby mode, or in operation; these methods are outlined in Section 5.7.[21–24]

5.2 Aging

It follows from the equation of state in its parametric form (1.111) that body resistance to external impacts P and its ability to deform, while preserving its dimensions and shape V, are entirely determined by the ratio of free energy F and Gibbs thermodynamic potential G. Free energy F creates a positive (tensile or dilaton) component and Gibbs thermodynamic potential G is responsible for the negative (compressive or compresson) component of internal pressure. Dynamic equilibrium of the interacting system of rotoses, is characterized by the equality of the compresson and dilaton pressures in local volumes. This is possible only when condition $F = G$ is satisfied. Equality $\Omega = 0$ at $V \neq 0$ means that, in the equilibrium state, $\mathbf{P} = 0$.

Structural materials solidified under certain conditions are then processed under other conditions and used under conditions that may substantially differ from the previous ones. The material always tries to match its structure to the conditions. In publications such processes are called aging.[3,206] Aging of metals and alloys is usually understood to be the processes of changing their physico-mechanical properties, proceeding in time by the schematic given in Figure 1.15. This process is manifested as phase transformations of the structure in the solid state. This is due to the gradual transition of the structure from the metastable condition (with respect to the external conditions) into the stable or equilibrium condition (in Figure 1.14, identified with the change of positions in the 11 to 15 range).

Aging may lead to both improvement and deterioration of the mechanical properties (again, only with respect to the specified service conditions). As has been discussed, improvement of some properties invariably causes deterioration of others (Figure 3.34). Due to the generality of the equation of state (2.65), this takes place everywhere in practice.[2,3,19,44] For instance, strengthening in aging (MB factor tending to position 11) of many heat-treated alloys on iron, aluminum, copper, magnesium, nickel, and cobalt base is accompanied by simultaneous decrease of ductility.[228] This is natural because in this case the absolute value of the Ω-potential does not change. In order to eliminate such an influence, it is recommended traditionally to use special nonaging steels, alloyed with titanium, aluminum, and zirconium.

The aging processes are controllable; their rate can be regulated by changing the appropriate parameters of state (Equation 2.96). They underlie many modern approaches of technological processing of parts and structures that are believed to be complicated and insufficiently studied so far.[207,228]

The main kinds of phase transformations[11,60] are thought to be the polymorphous transformation, martensite transformation with martensitic structure decomposition, solid-state dissolution and decomposition of oversaturated solid solutions, their ordering and disordering, and formation of eutectoid mixture and its decomposition. These transformations are divided into two groups. The first covers those that proceed without any change of phase composition and are associated only with the rearrangement of the crystalline structure. After electrodynamic rearrangement of the structure by the schematic shown in Figure 1.15, they preserve their molecular configuration (Figure 1.3). In the second group, new phases form with the change of composition. The spatial configuration of the molecules is not recovered (Figure 2.8).

Both groups can proceed in a broad temperature range, including the most often observed in practice temperatures. A profound CD rearrangement of the structure requires either very high (positions 10 and 15 in Figure 1.14) or quite low temperatures (positions 8 and 11). These are achieved with special kinds of heat treatment and also during operation of many parts and structures (gas turbines, space vehicles, etc.).

Figure 2.24 shows the influence of macro- and microscopic temperature parameters T and δ on the Ω-potential at $N = const$ (Equation 2.96). Number 1 designates the direct proportionality (in keeping with Figure 1.14, θ_s temperature is located on it), and 2 and 3 limit the range of MB distribution (positions 11 to 15) on the opposite branches of the braking barrier dbq_1 at the specified temperature. It is seen that the higher the body temperature, the higher the Ω-potential and the greater its range. (In Figure 1.14, it varies by 11 to 12 to 14 to 15 sequence.) For instance, at temperature T_1 the MB factor has a sharp-peak shape that determines the small range of Ω_1 along the abscissa axis, and at T_3 it flattens out, causing an appropriate increase of the range of Ω-potential up to the value of Ω_3. Deformation in force fields at $T = const$ yields similar results. In this case, the straight line 1 takes a horizontal position, and the shape of MB factor changes by 11 to 12 to 13 sequence (Figure 1.14).

Of great practical interest are the processes related to decomposition of martensite structure and oversaturated solid solutions (precipitation processes) because aging of alloys is quite often understood to mean exactly these processes. They arise from an unstable (metastable) structure of an alloy, acquired as a result of technological processing (for instance, quenching or work hardening) and associated with the formation of a CD relationship uncharacteristic of this alloy under these conditions (see explanations to Figure 1.13). As seen from Equation 1.111, compared to the equilibrium condition, the metastable condition is characterized by a higher level of free energy F. In such a condition, the structure is oversaturated with D-bonds. Higher dilaton pressure initiates and sustains a spontaneous transition from the metastable into the equilibrium condition, which is characterized by a lower level of free energy and new CD ratio.

The dilaton part of the volume can be found as (Equation 4.36.I)

$$V_d = \frac{C_V}{16\pi\delta + 3k} \tag{5.1}$$

and thus the compresson part is

$$V_c = V - V_d, \tag{5.2}$$

where V is the total volume of a material.

As discussed in the previous chapter, the volume of any material always consists of two phases, namely dilaton and compresson. For this reason, attempts to obtain the design "theoretical strength," assuming a single-phase composition of the AM structure of solids, are doomed to failure.[3,10,22] It is rational to consider just a certain degree of approximation to it, due to formation of the most beneficial compresson concentration (in tension) or dilatons (in compression).

Maximal excess pressure is found from the Equation 4.36.III if we assume that $T = T_n$ in Equation 2.172. The minus sign in Equation 4.36.III means that the internal pressure is always directed to counteract its cause. For instance, an interacting system of rotoses, while resisting tension, gives rise to compressive forces but, while expanding, it resists external compression. Figure 5.1 shows the mechanism of formation of Equation 4.36.III for various CD concentrations. The upper field (a) shows temperature dependence of internal pressure **P** in compresson oversaturation of the volume, and the lower field (b) at dilaton. Numbers designate the following graphs:

$$(1) \ y_1 = \frac{U(T)}{kT}; \qquad (2) \ y_2 = \left[\frac{U(T)}{kT}\right]^2; \qquad (3) \ y_3 = \exp\frac{U(T)}{kT}.$$

Dotted lines 4 designate superpositions of the pair dependencies (1 to 3) and (1 to 2). They indicate the zones of the maxima of tensile P_s and compressive P_p pressures. Certain temperatures, T_k and T_d, correspond to these maxima. Mismatch of pressures and temperatures initiates CD transitions in the first case (in upper field $P_s > P_p$ and $T_k < T_d$) and DC transitions in the second case (here $P_p > P_s$ and $T_k > T_d$). In both cases, the equilibrium sets in

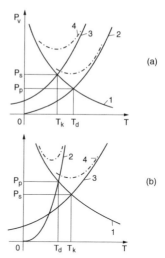

FIGURE 5.1
Nature of variation of internal pressure P_V at volume saturation with (a) compressons and (b) dilatons.

when $P_p = P_s$ and $T_k = T_d$. External force and thermal fields can activate or slow these processes, depending on the plus or minus sign in the left-hand part of Equation 3.19.

As follows from the phase diagrams (Figure 1.13), the specific volumes of compresson V_c and dilaton V_d phases are functions of internal pressure **P** and temperature T. Indeed, at certain temperatures, *edabf* curve limiting the two-phase condition of the body and having a bell-like shape (assigned by the sequence of curves 4 to 6 and changing with temperature) changes its shape (corresponding to curve 4) and becomes one point (point a). This change corresponds to the moment when one of the flat rotoses (Figure 1.7b) of the polyatomic molecule of ellipsoid shape (Figure 2.7) reaches the Debye temperature. Length of sections q_2c and cq_1, determining CD ratio, changes continuously.

External pressure has a similar influence. In Figures 2.36 and 3.51, *mn* is the vertical line that separates specific volumes V_c and V_d at the appropriate level of compressive stressed–deformed state. Different areas under distributions 1, 2, and 3 to the left and right of it evidently correspond to different relative amounts of compressons and dilatons. Let us designate their specific concentrations in the initial state (curve 1) through x and $(1 - x)$, respectively. Then the total volume of material, referred to as the unit of mass, is equal to:

$$V = xV_c + (1-x)V_d.$$

From this it follows that

$$x = \frac{V - V_d}{V_c - V_d}, \quad \text{and} \quad (1-x) = \frac{V_c - V}{V_c - V_d}.$$

The ratio

$$\frac{x}{1-x} = \frac{V_c - V_d}{V_c - V}$$

determines the CD relationship. It is obvious that the specific concentration of compressons and dilatons is inversely proportional to the areas, into which MB distribution is divided by the *nm* vertical. By analogy with the rule of the lever[31] known in the theory of phase transformations of gases and liquids, this relationship for solids can be referred to as the rule of areas. It determines the CD ratio in a given volume of material at a given moment of time. Its particular value can be found from the difference in the strength of the same material obtained in testing specimens under the same conditions, or from the ratio of the elastic and plastic components of the hardness indicators (Figure 3.33).

CD transition processes govern all the kinds of aging and appear as the decomposition of oversaturated solid solutions. Let us consider an alloy consisting of base metal *A* and alloying component *B*. On heating, the rotoses change their energy condition in direction 6 to 9 (Figure 2.47) and lower the bond energy by increasing free energy. Indeed, according to Equation 2.39, when $U = $ const, T_s reduction leads to increase in F. This facilitates the solubility of the *B* component in the *A* matrix, thus increasing the saturation limit of the solid solution.

Transforming as shown in the lower field in Figure 2.47, MB distribution switches most of the rotoses into D-phase (position 5 in Figure 1.15), thus bringing the alloy into a single-phase state. Solidification during abrupt cooling in quenching fixes this single-phase condition of the structure. Because it is metastable, the structure decomposes during aging into two phases: compresson and dilaton (see Figure 2.17). The compresson phase, evolving from D-solution, can be a metal compound or a new solid solution with a different content of the dissolved component.

Figure 5.2 shows the equilibrium diagram of the system of *AB* components that form the aging alloys. After heating to the temperature of T_3 or higher, and rapid cooling to room temperature T_1, *AB* solid solution turns out to be oversaturated and, hence, metastable. Remaining dissolved in *A* component is not the amount *m* of element *B* that corresponds to limit solubility of *B* in *A* at room temperature T_1, but a greater amount, *n*, that dissolves in *A* at temperature T_3. After quenching, the alloy is a single-phase oversaturated (i.e., having a high dilaton concentration) *AB* solid solution, whereas its equilibrium condition at room temperature T_1 (and at higher temperatures up to T_3) should have two phases (compresson–dilaton) with lower content of the *B* component in the *AB* solid solution.

Because of this state, the processes of decomposition and depletion of oversaturated solid solution and its transition into a more stable two-phase dilaton–compresson state, in which the free energy of the phases has the

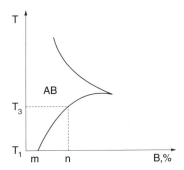

FIGURE 5.2
Equilibrium diagram of aging.

lowest value, take place in the alloy. In some overcooled solid solutions individual stages of decomposition can start and proceed at room temperature while in others the triggering of these processes requires heating to higher temperature, i.e., can start only after an appropriate energy rearrangement of the rotoses (see Figure 1.15).

The mechanism of the process of aging with decomposition of oversaturated (or dilaton) solid solutions can be qualitatively described: at the first stage, the diffusion of B component atoms and their clustering in the grain periphery proceed inside AB solution under the influence of excess D-pressure (directed away from D-phase). At the second stage, the D-phase sections, which are most enriched with B component, develop very small areas with new configuration of rotoses (already compresson) characteristic either of B component, or of its chemical compound with A component, or of their transition phase. The structures of these rotoses and those of the initial solid solution are still not very different from each other; therefore, their regular conjugation takes place.

At the third stage, the rotoses of a new phase are rearranged to such an extent that they already generate K-pressure (opposite direction to that of dilatons). Under these opposite pressures, D- and K-rotoses separate and move away from each other; this results in the formation of independent dispersed particles of the new compresson phase (so-called dispersion hardening) and in reduction of D-phase concentration. This process is accompanied by reduction of the amount of dissolved element in the initial solid solution to the equilibrium value. The fourth stage begins when the coarsening (coagulation) of dispersed particles takes place and finishes with the transition of the metastable modification into a stable phase characterized by equality of D and K pressures. A more detailed description of the processes of structure rearrangement during aging can be found in References 11, 60, and 61.

The nature of the aging process depends on many factors,[61,228] such as temperature (greater rate of structure rearrangement at higher temperatures), chemical composition (at a particular temperature, slower aging of refractory alloys than that of the others), structural prehistory (acceleration of the stabilization process by preliminary irradiation or cold deformation), etc. Considered together, these factors determine the main service parameter, namely, the time of aging with the structure transition into the equilibrium state.

Modern science and engineering practice focus on the processes of deformation and cracking, believing that these are responsible for reliability and durability of engineering objects.[22] This means that the consequence rather than the cause of failures is considered. The latter is hidden from researchers attention thus far, lying in phase transformations of the structure (Figure 1.15). Reversal of the sign of internal stress is often not accompanied by any change of material composition; therefore, it cannot be distinguished by modern instrumentation or by visual observations. The main drawback of traditional concepts of the nature of strength lies in underestimation of the role of phase transformations, which fundamentally change the nature of materials resistance and deformation.

5.2.1 Thermal Aging

Many parts and components of machines and mechanisms in service are exposed to long-term thermal force impacts. In this connection, it is interesting to analyze the general laws of thermal aging. For this purpose, let us consider the simplified temperature dependence of the Ω-potential that follows from Equation 2.96 and take into account Equation 2.141

$$\Omega = kNT\left(e\frac{t_i}{\theta}\right)^3,\qquad (5.3)$$

where $e = 2.71$ is the base of the natural logarithms, its cube being equal to $\pi^4/5$. This dependence is shown in Figure 5.3 using the assumption that the

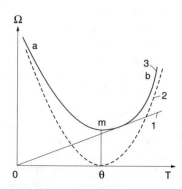

FIGURE 5.3
Principle of formation of temperature dependence of the Ω-potential.

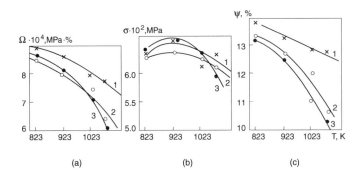

FIGURE 5.4
Temperature dependence of mechanical properties of steel 12Kh18N9T (0.05% C, 12% Cr, 18% W, 9% Ni, 1% Ti) on aging duration.

number of interatomic bonds does not change during aging, i.e., $N = $ const (curve 3). It is derived using graphic multiplication of the linear function I ($y_1 = kNT$; Figure 2.24) and cubic parabola 2 ($y_2 = \delta^3$; Figure 2.23). The latter consists of two branches, *am* and *mb*, with Debye temperature, θ, the boundary between them.

Refractory alloys are metastable[3] because, during aging, they undergo decomposition of the solid solutions with the formation of secondary carbide and intermetallic phases. In this connection, structural changes must be taken into account in determining the intensity of lowering the load-carrying and ductile properties in time and depending on temperature.[3,229]

Figure 5.4 presents the nature of the change of mechanical properties of stainless steel 12Kh18N9T (0.05% C, 12% Cr, 18% W, 9% Ni, 1% Ti), depending on aging duration at different temperatures.[3] Graphs marked by numbers 1, 2, and 3 correspond to aging duration of 500, 3000, and 5000 h, respectively, at the temperatures indicated on the abscissa axis. Figure 5.4a shows the nature of Ω-potential variation, depending on the aging temperature. The product of resistance σ and bulk deformations ψ, expressed in percent, is taken as its quantitative measure; Figures 5.4b and c show the concurrent change of resistance σ and bulk deformability ψ.

Comparison of Figures 5.3 and 5.4a shows that, first, the temperature dependence of Ω-potential for the given alloy in the studied temperature interval falls on the descending portion *am* of curve 3. Second, the equation of state (2.96) correctly describes the general nature of Ω-potential variation. The latter, in turn, correctly determines the direction of the change of the material resistance to fracture (Figure 5.4b) and deformation (Figure 5.4c) in aging. Analyzing the experimental data[3,229] in the same way can provide further evidence of the validity of this statement.

It follows from Figure 5.4 that the higher the aging temperature and the longer its duration, the smaller the Ω-potential value. (Compare the relative position and trend of graphs 1, 2, and 3.) Why does its change have this form? To answer the question, we should consider the equation of state written as

Equation 2.179. Its right-hand side incorporates the kinetic parameters, which change comparatively easily under the impact of the external, primarily thermal, fields.

In order to achieve high strength properties, the high level of the kinetic part of internal energy is fixed in the metastable structure by thermomechanical treatments. Its essence is in shifting of the MB distribution from position T_1 into position T_3 in Figure 2.24. In the following period, adapting to the service conditions, the material invariably loses the achieved state (and thus properties) in coming back from metastable condition T_3 to stable condition T_1. It is logical that the intensity of such a loss is higher at higher process temperature and its duration. This is exactly why $\Omega = f(T)$, $\sigma = f(T)$, $\psi = f(T)$ dependencies in Figure 5.4, having an actually inversely proportional nature, are equidistantly shifted relative to each other with the increase of soaking time. (Compare the relative position of curves 1, 2, and 3.)

Change of the physico-mechanical properties of the alloys (strength, hardness, electrical resistance, corrosion resistance, etc.) is due to the decomposition of oversaturated solid solutions in aging. The decomposition process is often accompanied by change of volume (approximately by 3%[207]), which may lead to stresses developing in the grains. These stresses can sometimes slow the decomposition process. It follows from the equation of state written in thermodynamic potentials (Equation 1.111) that phase transformations are controlled by the energy relationship, equal to the difference between free energy F and G-potential. Let us rewrite Equation 1.111 in the following form:

$$\mathbf{P} = \frac{F}{V} - \frac{G}{V}. \tag{5.4}$$

The graph of Function 5.4 is shown in Figure 5.5 (curve 3). It is formed by two inversely proportional functions,

$$y_1 = \frac{F}{V} \quad \text{(curve 1)} \quad \text{and} \quad y_2 = \frac{G}{V} \quad \text{(curve 2)},$$

with different directions and variation intensity at the same conditions of the alloy existence. (G is determined by the potential and F by the kinetic components of full internal energy U.)

Emergence of a new phase (elementary CD transition) proceeds predominantly on the matrix grain boundaries where the free energy of the atomic clusters is higher and, therefore, the frequency of the nuclei formation is higher than that inside the crystal. In addition, the nucleation process is facilitated by solute atoms penetrating faster in this region because of a greater difference between the compresson and the dilaton pressures than that inside the grain. Stresses arising from the pressure difference and preventing nucleation usually relax faster at this regions than in other parts of

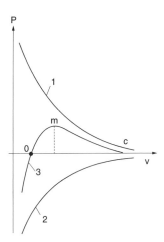

FIGURE 5.5
Redistribution of internal stresses at phase transformations.

the crystal. Grain boundaries are favorable zones for formation of abnormal concentration of the inclusions.

The processes proceeding at the initial stages of aging (appearance of submicron heterogeneity in the distribution of the atoms of solute component in the oversaturated solid solution, coherent bond between the dilaton and compresson rotoses, highly disperse particle precipitation) result in alloy strengthening, increase of its hardness, and resistance to plastic deformation (*om* section of curve 3, Figure 5.5). However, subsequent stages related to disperse particle coagulation are always accompanied by a loss of strength (*mc* section of curve 3). Softening is due to a loss of coherence of the bond between the new compresson phase and the solid solution, and its depletion of the alloying component during precipitation of the latter. When sufficiently long-time intervals are considered, the strength decreases to that of the alloy before aging or even smaller (final portion of *mc* section). Ostreikovsky[229] reconstructed the aging process from elementary subprocesses. The result is $P = f(V)$ dependence, which is similar to that following from the equation of state (1.111) and given in Figure 5.5.

The aging process is often characterized by curves with two or more intermediate maxima of hardness. In multicomponent alloys, such as Co–Fe–V, precipitation of more than one K-phase can result in the appearance of two hardness peaks, if these are formed with different rates. This is attributable to the mismatch in time or space of the maxima (point *b* on CD curve, Figure 1.14) of paired combinations of different chemical elements present in this alloy.

At least three fundamentally important conclusions follow from the equation of state (1.98) and its graphic interpretation (Figure 5.5). The first concerns a fundamental inversely proportional relationship between the deformation and strength properties; another pertains to the effectiveness of the traditional design and calculation procedures to ensure (as commonly accepted) reliability and durability of parts and structures by incorporating strength

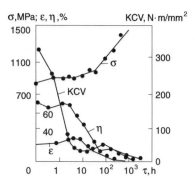

FIGURE 5.6
Change of the major mechanical properties of the β-alloy of titanium, depending on aging duration at 773 K.

and ductility margins through adding excessive material. The third follows from comparison of the equilibrium diagram (Figure 1.14) that describes the mechanism of CD transformations of the rotoses and curve 3 (Figure 5.5), characterizing the relationship of compresson and dilaton pressures (Equation 5.2) in volume V of the actual body. Their shape coincidence is indicative that CD heterogeneity, initiating in the depth of the AM structure, is manifested in the equation of state (2.96) and describes the behavior of actual solids. The latter statement needs no comment, but because the first and second have fundamental importance, they will be discussed in detail.

According to Equation 2.137, the change of parameter dV in the specimens of constant dimensions is deformation. It follows from Equation 5.4 and Figure 5.5 that strength and deformation properties are connected by an inversely proportional relationship. As experience shows, it is common for compresson materials, including metals and alloys (Figure 2.38).[3] As another example, Figure 5.6 shows the nature of variation of the major mechanical properties of the β-alloy of titanium depending on aging duration at 773°C. With its extension approximately up to 500 h, a continuously increasing strengthening of the alloy is observed, accompanied by lowering of elongation, ε, and reduction in area, η, as well as impact toughness KCV. With further extension of aging duration, the alloy completely loses its ability to undergo plastic deformation.

Such a nature of strength and deformation property variations is inherent in the bulk of AM structure (Figure 2.6) and, at the macroscale, it is manifested in the metallic and in nonmetallic materials (Figure 3.34). The strengthening maximum is found in the temperature range (673 to 773 K) where the most intensive embrittlement is observed.[231] Ample evidence of a similar nature of variation of the strength and deformation properties of metals and alloys is provided by Troshenko.[3]

It can also be seen from Equation 5.4 and Figure 5.5 that, with increase of the mass of material concentrated in the volume, its resistance (with other conditions being equal) to external effects decreases (see the descending *mc* section of curve 3). The causes for development and the possible consequences of this phenomenon are analyzed in detail in Section 3.6. Considering its extreme importance, let us reiterate the major conclusion: the greater the volume *V*, the higher the thermal lag of the structure that leads to increase of free energy *F*. As result, G-potential spreads over the volume, lowering all kinds of structure resistance (see Figure 3.17 and comments on it).

Thus, existing design methods do not guarantee 100% reliability of parts and structures through material redundancy because they do not take into account the actual physical processes in solids. On the contrary, increase of material content increases the probability of failure of the parts and structures.[1,3,8]

The estimation of the time or rate of the aging process can be considered as follows. Determination of absolute temperature (1.24.2) allows writing the equation of state (2.65) in the following form:

$$PV = \frac{mv^2}{2} s_{P,T},$$

where $s_{P,T} = s_P + s_T$ is complex entropy, in which components depend only on temperature (at $P = $ const) and pressure (at $T = $ const). The time of a stable existence of rotoses τ can be considered as the value inversely proportional to the averaged rate of their decomposition process, i.e., $\tau = 1/v$. Then,

$$\tau = K / P^{1/2}, \tag{5.5}$$

where

$$K = \left(\frac{ms_{P,T}}{2V} \right)^{1/2} = \text{const}$$

at $N = $ const and $V = $ const. The plot of Function 5.5 is given in Figure 4.31; this dependence is proven experimentally in extensive durability studies of various materials at constant load.[10]

Ostreikovsky[229] has noted that decomposition of oversaturated solid solutions lowers the alloy corrosion resistance. The particles of the precipitating new phase differ in their composition and structure from the main solution. Having different dissolution potentials, they develop an anomalous electrochemical activity; anode particles of the precipitates dissolve in the presence of the electrolyte while the cathode particles lead to the matrix dissolution around them. Stress corrosion cracking is often found in aging alloys and is related to localized precipitation along the grain boundaries. Presence of even small amounts of such precipitates may lead to cracking along the grain boundaries in the highest stresses of the sections.

5.2.2 Strain Aging

A martensitic structure is created in steels to achieve their high hardness and strength. This structure is a specific kind of acicular microstructure of metal alloys Cu–Al, Cu–Zn, Cu–Sn steels, some metals, and even nonmetallic materials formed as a result of martensite transformations. This type of aging also covers polymorphous transformations in cobalt, titanium, zirconium, and titanium- and zirconium-based compounds; noncarbon iron alloys with chromium, nickel, manganese; etc.[207]

According to the general theory of phase transformations,[60] martensitic structure forms in an alloy heated to a certain temperature and then cooled at a high enough rate. Both parameters depend on the composition of a specific alloy. Martensitic structure can be formed in a small amount at constant temperature; it is a metastable single-phase solid solution of carbon in iron with a cubic body-centered lattice. As a result of excess carbon atom penetration into α-iron lattice, it is distorted and becomes tetragonal.

Multiple studies of the processes of strain aging of unstable austenitic steels have demonstrated that deformation is also accompanied by marten-sitic transformations. (Analysis of these studies can be found in Troshenko.[3]) Special heat treatment yields in the structure having austenitic phase com-position in which the required strength and ductility properties are provided by controlling the decomposition kinetics at static deformation.

In terms of CD concepts, martensitic transformations (also at deformation) can be presented as follows. Let the MB distribution of the austenitic struc-ture be represented by curve 1, located in the compresson area $r_o m$ of force curve 2 in Figure 5.7. In Figure 1.14 its analogue is designated as cbq_1. The initial value of the Ω_i-potential of the austenitic structure can be graphically represented by the product of areas: one is limited by MB distribution 1 and the second is *acfd*.

In was shown in Section 2.2 (Equation 2.18) that the force fields have no influence on the probability distribution inside the MB factor (i.e., do not change the shape of curve 1). However, they cause change of values of the admissible energy levels (i.e., shift distribution I into position 3). As discussed

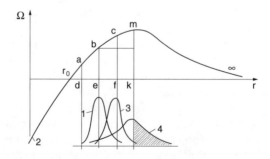

FIGURE 5.7
Martensite transformations in deformation.

earlier, such a shifting is achievable not only by deformation but also by heat treatment. As such, in accordance with Equation 2.15 and Figure 2.2, the shape of curve 1 is somewhat changed but remains within the limits of C-phase. If MB distribution is shifted so much that it occupies a part of D-phase (hatched area under curve 4), the aging process proceeds not by a compresson–compresson, but by a compresson–dilaton mechanism (Figure 1.15).

The solid solution transformation processes considered in the previous section correspond to this mechanism. CD transitions through Debye temperature require much more energy than interphase compresson–compresson transitions (see Section 2.7). Therefore, the nucleation energy in martensitic transformation is much smaller than that at the structure decomposition in the solid state. This is exactly the feature responsible for high rates of martensitic transformations.

Induced by shifting the MB distribution from position 1 into position 3, deformation changes the Ω-potential. It is now equal to the product of areas limited by distribution 3 and curvilinear figure *ebmk*. The latter is larger than that of initial figure *acfd* by the area of curvilinear triangle *bmn*. It can be identified with the increment of free energy F. Change of MB distribution position from 1 to 3 leads to higher admissible energy levels (compare the length of verticals *eb* and *fc* in the zone of maximum MB distribution) and is accompanied by simultaneous lowering of deformability. (Distance to critical point *m*, where a high probability of failure exists, is shortened from *ek* to *fk* when the rotos goes through the θ-temperature.) Such transformations lead to an increase in the load-carrying parameter **P** of Ω-potential growing and a decrease in its deformation parameter **V**. It should be anticipated that, at deformation aging, strength properties increase and deformation properties decrease.

Experiments confirm these expectations. Figure 5.8 presents a typical example of experimental data on the influence of tensile prestrain, φ, at room

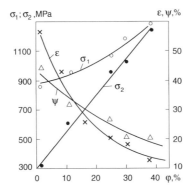

FIGURE 5.8
Influence of tensile prestrain on mechanical properties of austenitic steel 40Kh4G18 (0.4% C, 4% Cr, 18% Mg) at room temperature.

temperature on the mechanical properties of austenitic steel 40Kh4G18 (0.4% C, 4% Cr, 18% Mg; similar dependencies are found for a wide variety of steels[3]). Compared to the austenitic state (corresponding to point 0 on the abscissa axis), the states after deformation are characterized by increased ultimate strength σ_1 and yield strength σ_2. Increase in prestrain φ leads to higher limit stresses. On the other hand, ductility properties decrease abruptly; they are lower, the higher the degree of work hardening.

Compared to other kinds, martensite features the highest hardness and also higher brittleness. At comparatively low temperatures of heating of martensite (up to 100 to 250°C), the right end of distribution 3 goes into D-phase (Figure 5.7). CD transformations lead to carbon precipitation from α-iron lattice and to the formation of ultrafine particles of iron carbide, which decrease the degree of lattice distortion. The diffusion process of martensite decomposition is completed with the equilibrium state and, thus, MB distribution 3 coming back to initial position 1 (Figure 5.7). With temperature increase, decomposition rate rises significantly, shortening the time required to reach the equilibrium state. For instance, at 100°C, the initial state is recovered within 1 h and at 20°C, it takes 10 years.[207]

A semi-empirical equation of temperature dependence of martensite decomposition rate according to Umansky[60] is

$$nV = 10^{19} \exp\left(-\frac{U}{kT}\right),$$
(5.6)

where n is the number of carbide crystals (and, hence, the number of depleted sections of the solid solution) formed in a unit of time in a unit of volume, V is the average volume of the depleted section, U is the activation energy of the decomposition process equal to $3.3 \cdot 10^4$ cal/mol. Similar dependencies, theoretically derived from the equation of state, are given in Chapter 4.

Table 5.1 presents the results of calculation of the period of martensite half-life using Equation 5.6. Experiments have demonstrated[60] that martensite decomposition kinetics and aging processes in oversaturated solid solutions obey certain general law that have so far been hidden from researchers' attention. CD concepts make their nature understandable and theoretically well grounded.

TABLE 5.1

Martensite Half-Lives at Different Temperatures

	Temperature, °C						
	0	20	40	60	80	100	120
Half-life	340 years	6.4 years	2.5 months	3 days	8 hours	50 min	8 min

The presence of alloying elements starts essentially influencing martensite decomposition kinetics only at temperatures above +150°C. At lower temperatures, this proceeds at almost the same speed as in a carbon steel. Carbide-forming elements (chromium, titanium, vanadium, molybdenum, and tungsten) significantly slow the carbide phase coagulation and decomposition process at temperatures above 150°C at the expense of lowering the carbon diffusion rate.[60]

Deformation (especially, cyclic deformation) leads to destabilization of MB distribution 3 fixed in the metastable condition (Figure 5.7) and, as a consequence, promotes the change of the initial martensitic structure. It is natural that, depending on the deformation conditions, such changes may proceed on both sides of the metastable condition, i.e., MB distribution 3 can be shifted to the left or to the right from line *fc* (Figure 5.7) with an equal degree of probability.

Shifting direction determines the nature of change of the strength and deformation properties of steels graphically represented in Figure 5.8. Troshenko[3] provides a sufficient number of examples demonstrating the increase of σ_1 and σ_2 (MB distribution 3 approaching the Debye temperature) and their lowering (for instance, at high-frequency vibration) to indicate that distribution 3 is returning to initial position 1.

In addition to martensite transformations, deformation is quite often accompanied by allotropic transformations. Their essence consists of the transformation of one (as a rule, dilaton) modification of elements into another (compresson); they proceed at Debye temperature θ. In Figure 5.7, it corresponds to critical point *m*. Such transformations are related to the possibility of formation of different crystalline structures by the same element. These are formed by rotoses of different spatial configuration (see Chapter 1). The allotropic transformations of a number of pure metals:

$$Fe_\gamma \rightarrow Fe_\alpha; \quad Co_\beta \rightarrow Co_\alpha; \quad Ti_\beta \rightarrow Ti_\alpha; \quad Zr_\beta \rightarrow Zr_\alpha, \quad etc.$$

are well studied.[230] These are characterized by a noticeable change of volume; for instance, in tin it increases by 25.6%.

Allotropic transformations take place in those alloys in which the end of the MB distribution penetrates deeply into D-phase (curve 4 in Figure 5.7). This is indicated by the rate of such transformations becoming noticeable only at considerable overcooling of the alloy.[229] In the majority of the metals, the new phase crystals turn out to be oriented in relation to the initial matrix. This is natural because reduction of the rotos energy changes their shape from the volumetric (formed as a result of the increase of the number of azimuthal planes) into a shape elongated in the direction assigned by the cooling heat flow (see Figures 1.4 to 1.6). A regular orientation of crystals, formed in the isotropic environment, indicates a tendency of free energy to minimum. The processes of aging of other materials — for instance, polymer — can be considered along similar lines.

5.3 Hydrogen Embrittlement

Metals subjected to various kinds of stressed–deformed state crack in the most diverse corroding media. For instance, fasteners made of martensite steel with 12% Cr (cathode) contacting the aluminum roof (anode) fail in a humid environment after several hours, while many high-strength steel parts fail within several days or weeks in the stratal waters of oil or natural gas wells containing hydrogen sulfide. In these and other similar cases, hydrogen is the cause of cracking, so this kind of failure is called hydrogen cracking. It results from hydrogen embrittlement of metals.[3]

Moreover, Reference 232 reports the existence of a new state of the matter that differs by a considerable value of the elasticity modulus characteristic for the solid state of the matter and an extremely low value of the shear modulus that is characteristic of the liquid state. Appearance of this new state is the basis of mechanical instability of crystalline and amorphous alloys exposed to the combined action of stress field and highly intensive diffusion flow of hydrogen and deuterium. Concepts of CD nature of the strength of solids developed in this book provide an interpretation of this phenomenon.

This is based on the change of the innate rotoses (Figure 1.7b) due to implantation of a hydrogen atom. As an example, Figure 5.9 gives the schematic of an initial (a) and hydrogen-containing (b) rotos of iron. Let us assume that, in the initial condition, the two iron atoms Fe_1 and Fe_2, forming a rotos, are bonded by outer incompletely filled electronic levels 3 and 4 containing 14 and 2 electrons, respectively. The internal shells (1 and 2) are completely filled and stainless, and their electrons do not interact with each other. Appearance of hydrogen atoms H in the metal structure essentially changes the situation.

Consisting of one proton and one electron, hydrogen has small size compared to other atoms and is a chemically active element. Small dimensions of the proton allow it, first, to easily penetrate between the matrix molecules,

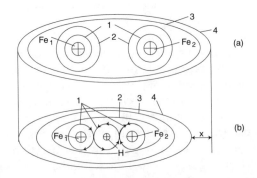

FIGURE 5.9
Schematic of formation of rotoses of hydrogen-containing iron.

and, second, to have not more than two atoms as its closest neighbors. An electron of hydrogen atom H, finding itself between two iron atoms, Fe_1 and Fe_2 (Figure 5.9b), not only starts servicing its own proton but also participates in the movement along the first energy levels about the nearest matrix neighbors. (These trajectories are schematically represented in the form of eight-shaped curves indicated by arrows.) The extra (hydrogen) electron pushed out to higher, emptier levels turns the circular trajectories of the electrons moving there into elliptical trajectories (2, 3, and 4).

While in the initial condition the iron atoms were connected by two common elliptical orbits 3 and 4, now they are connected by four orbits 1, 2, 3, and 4. Appearance of additional molecular orbits increases the rotos rigidity while its dimensions are reduced by distance x. This is not just an abstract assumption; on the contrary, it is based on reliably established experimental facts. For instance, the distance between the atoms in an oxygen molecule is 1.7 Å and that of hydrogen bond in a water molecule is 1.1 Å. Similarly, the size of the fluorine molecule is 1.6 Å and that of hydrogen fluoride is 0.9 Å.[26]

This is exactly the reason hydrogen-containing bonds have a high coefficient of rigidity (2.177), determined by the Δ parameter (see Figure 2.47). Appearance of such bonds in the material structure changes its condition by increasing Debye temperature θ. Crystalline lattice parameter r_a is shifted from position r_{a_3} in the r_{a_1} direction with appropriate change of the potential curve shape (Figure 1.10b). Such a nature of its change is further confirmed by substances with hydrogen bonds having higher melting, boiling, and evaporation temperatures than their isomers. (Molecules of both have the same composition.)[26,34]

This results in change of configuration parameter α (1.17), dynamics of the localized motion of the atoms (Figure 1.6), form of potential well (Figures 1.7 to 1.8), and of the rotos equilibrium diagram (Figure 1.14). Consequently, the characteristic temperature θ_s and electrodynamics of phase transformations (Figure 1.15) also change.

Figure 5.10 presents a graphic interpretation of such changes. In the initial condition (Figure 5.9a), let the potential function (Equation 3.2) for the metal (in Figure 1.14 in the zone of θ_s-temperature) have shape 1 in the r direction. In the equilibrium state it forms braking barrier 2. The distance between verticals r_d0 and mH-BH where initial MB distribution 3 is located gives the range of the K-phase. Linear dimensions of initial EM dipoles at any energy level (for instance, ω_1), are assigned by the length of horizontal section ac between vertical r_d0 and curve 4, which is the locus of points determining characteristic temperature T_c.

Hydrogen appearance in the structure (Figure 5.9b) changes the shape of the Ω-potential from 1 to 1H and the shape of the equilibrium diagram from 2 to 2H. The C-phase right boundary mb shifts to the left to position mH to bH, thus reducing its range. This leads to the transformation of initial gently sloping MB factor 3 into peaked 3H. Dimensions of hydrogen-containing dipole are now assigned by curve 4H and not by curve 4. If they were equal

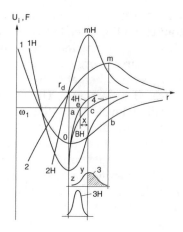

FIGURE 5.10
Physical essence of hydrogen embrittlement of metals and alloys.

to *ac* section in the *r* direction at the ω_1 level initially, now they are equal to *ae*, i.e., they were reduced by distance *x*. This distance determines the size of the submicroscopic defect appearing in the material crystalline structure as a result of transformation of the matrix rotos into hydrogen-containing one (Figure 5.9). The hatched zone under curve 3 (Figure 5.10) indicates that part of the rotoses exposed to hydrogen impact. This means that the material structure turned out to be filled with a chaotically arranged system of microscopic cracks having size *x*. This is the physical essence of hydrogen cracking.

Appearance of hydrogen-containing bonds in the bulk of material changes mechanical[3] as well as thermophysical, electrical, and other properties.[29] This is natural because a profound connection exists between them.[24] When hydrogen embrittlement of the metal is discussed, the deterioration of any of the many mechanical properties is meant, namely elongation in tension, reduction in area, ultimate strength and durability, crack propagation rate, etc.[232]

As an illustration, Figure 5.11 shows the influence of hydrogen on the change of the reduction of area at fracture, ψ, in tension for nickel (a) and aluminum (b) alloys.[3] Similar trends are observed for steel. Reduction of area ψ can change three- to tenfold or even more. Embrittlement is manifested only slightly at low (of the order of 73 K) as well as at elevated temperatures; ranges differ for different materials. In the former case, formation of hydrogen-containing rotoses is difficult because the matrix atom spacing is reduced so much that, even in the absence of hydrogen, electrons in the second and first energy levels start interacting. In the latter, these atoms are so far away from each other that triple interaction (Figure 5.9b) becomes impossible.

Hydrogen cracking results in three- to fourfold or even greater decrease of the long-time strength, as well as a considerable deterioration of fatigue characteristics of high-strength steels and thus in intensifying fatigue crack initiation and propagation.[3,215] Any method of preventing hydrogen penetration

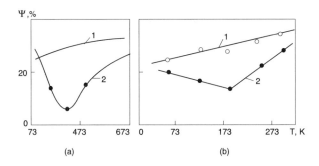

FIGURE 5.11
Temperature dependence of elongation in rupture for (a) annealed alloy Inconel 718, testing in air (1) and in the atmosphere of dry gaseous hydrogen at the pressure of 50 MPa (2) and (b) aluminum alloy 7075-T6651 before I and after 2 cathode hydrogenation.

into the material structure and slowing its migration leads to increased resistance in hydrogen-containing media.

5.4 Radiation Damage

Structural materials of the major units of thermal and fast fission reactors, as well as nuclear fusion reactors, operate under extremely complicated conditions. A fast neutron flow in some of them can be up to $1 \cdot 10^{16}$ n/cm$^2 \cdot$ s at the irradiation temperatures of intrareactor components ranging from 973 to 1272 K. Only a comparatively small number of materials can stand such conditions.[3,233,234]

Unlike the force or thermal impact (nature of their influence is explained using curves 11 to 15 in Figure 1.14), radiation factor is manifested as a result of a strictly localized application of a relatively high energy accompanied by a noticeable change of the crystalline lattice.[234] Atoms 1, 2, 3, etc. of a solid are excellent targets for neutron n (Figure 5.12). For instance, atoms of iron, zirconium, and niobium are 55, 91, and 93 times larger, respectively (depending on the number of the chemical element in the periodic system), than the neutron weight and size. Beginning from the energy of several electron-volts, the neutron has a pulse sufficient to induce the target atom excitation when colliding with it. High-energy neutrons with energy much higher than 14 MeV form in the nuclear fusion reactions. After the first collision, they preserve an even greater energy and thus are capable of meeting up to 12 and more atoms in their path before being entrapped by any of their nuclei.

For instance, at the first collision of neutron n with atoms 1 and 3 of a structural material K (Figure 5.12), they are drawn closer to each other and, at the second collision, the distance between atoms 4 and 2, 5 and 2, and 7 and 2 becomes greater. At the third collision, atom 3 moves away from

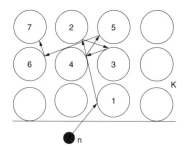

FIGURE 5.12
Schematic of neutron n penetration into the structure K of a material.

atom 5, coming closer to atom 1, etc. A series of collisions results in the change of the relative arrangement of atoms in the rotoses (Figure 1.6). Violation of the previous order changes the dimensions and electrodynamics of the rotoses (Figure 1.7b), spatial configuration of the molecules (Figure 2.8), and the local potential field (Equation 1.26). Therefore, the shape of the equilibrium diagram of EM dipole (Figure 1.14) and entropy (Equation 2.97) are also changed.

One part of the rotoses lowers the frequency of phonon radiation of T_i and moves in the energy space from the initial MB distribution 3 to the "cold" (left) end (Figure 5.10). As a result, the plastic condition characterized by curve 2 changes into brittle characterized by curve 2H. The other part changes its condition in the opposite direction (atoms 4, 2, and similar ones in Figure 5.12), which results in the reverse transition (brittle to plastic). The resultant condition of the material is determined by superposition of these two competing processes, namely strengthening of some local volumes and loss of strength and increase of ductility in others.

Therefore, neutron irradiation changes the initial function of condition Z_o to Z_n (Equation 2.64). Curves 3 and 3H in Figure 5.10, for instance, can serve as their graphic interpretation. As seen, MB function of Z_n determines all the physicomechanical properties of the material in the irradiated condition (Equation 2.66).

5.4.1 Strengthening and Softening

In general, the concepts developed in this book allow extending knowledge on structural transformations in solids in the thermal and force fields (Sections 2.7, 3.4, and 3.5) to radiation fields. Figure 5.13 presents stress–strain diagrams of zirconium alloys obtained at different radiation doses (a) and testing temperatures (b).[3] In the first case, testing was conducted at a temperature of 573 K. Curve condition 1 shows a strain–stress diagram for the initial state and curves 2 to 4 after irradiation by the following doses— 2, $6.3 \cdot 10^{18}$; 3, $6 \cdot 10^{19}$; 4, 10^{20} n/cm²—with more than 1 MeV energy. As seen, the influences of radiation and thermal fields on the material are similar;

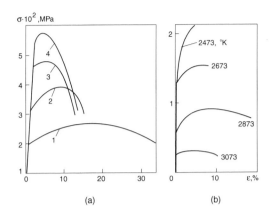

FIGURE 5.13
Stress–strain curves of zirconium alloys under the impact of (a) radiation and (b) thermal fields.

increasing the level of radiation is similar to lowering the testing temperature that leads to increase of limit stresses and to ductility drop. This is natural because the consequences of both impacts on the AM set are the same (Sections 2.7 and 5.2).

The difference is as follows. The radiation field has selectively local influence on the AM structure with gradual coverage of its greater part, while the temperature field acts on it as on a single whole. In the first case, the MB factor, as it transforms, stays in C-phase all the time (for instance, in 3H position in Figure 5.10); in the second case, shifting completely to the right, it transfers a certain part of C-rotoses into D-phase (hatched section in position 3). This also affects the deformation behavior of the material: at irradiation the modulus of normal elasticity remains practically unchanged while it changes substantially at thermal impact (compare Figures 5.13 and 2.39).

An integral link among radiation, thermal, and mechanical effects (theoretical basis discussed in Section 2.5) is manifested in their reversibility. Zirconium materials, namely alloys with small (up to 2.5%) additives of tin, niobium, iron, chromium, and copper, are widely used in the nuclear industry. If zirconium-2 is subjected to thermomechanical pretreatment,[235,236] when the extent of deformation increases up to 70%, its yield point rises to 755 MPa and subsequent irradiation increases it by just 6%. In the annealed condition, the same radiation field results in 31% increase of the yield point. If the degree of work hardening is lowered, then, with increase of the annealing temperature, the relative effect of radiation strengthening is increased.

Understanding these processes points to a possibility of recovery of radiation damage of the structure through thermal, mechanical, or other competitive impacts; Troshenko[3] provides experimental evidence. According to the data provided in this study, the irradiation of a titanium–zirconium alloy at 293 K increased σ by 30% on average and, at testing temperature above 573 K, the effect of radiation strengthening is gradually relieved. This effect

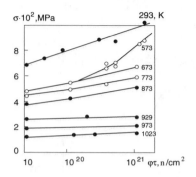

FIGURE 5.14
Dependence of limit stresses for steel 348 at different testing temperatures (numbers at curves) on radiation dose.

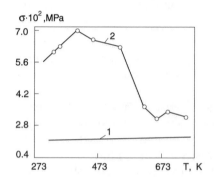

FIGURE 5.15
Dose-temperature dependence of the yield strength of steel A304 ($7 \cdot 10^{20}$ n/cm² dose) at room temperature (1, without radiation; 2, after irradiation).

is attributable to partial annealing of radiation defects during testing. A typical exposure-temperature dependence of limit stresses for steel 348 (Figure 5.14), irradiated at the temperature of 563 K, and tested at various temperatures (designated by numbers on the curves), provides additional proof. Radiation dose $\varphi\tau$ is plotted on the abscissa axis in n/cm².

Irradiation temperature has an essential influence on the change of strength properties determined at room temperature (Figure 5.15). As seen, the maximal increment of ultimate strength corresponding to points b_1 and b_2 in the deformation diagram (Figure 2.22) is found in the temperature range of 373 to 473 K. At temperatures above 673 K the difference in σ for the irradiated 2 and nonirradiated 1 steel becomes practically unnoticeable. This is attributable to the fact that, at not very high temperatures (in the range of 273 to 473 K), the rotos atom paths (Figure 17b) become so close that the extremes in the equilibrium diagram (Figure 1.14) located in the potential well (Equation 3.2) become more and more peaked.[24] (In Figure 5.10, cooling is illustrated by transition of position 2 into 2H.) This is why the MB factor in Figure 1.14 in the considered temperature range does not have shape 15, or even 14, but rather the shape of curve 12. Dynamic impact of the neutrons invariably causes its change in the direction 12 to 11, taking the end beyond critical temperatures T_c and θ_s and, hence, increasing the tensile strength.

At irradiation temperatures higher than 673 K, the shapes of the equilibrium diagram and the MB factor (positions 14 and 15) become blurred and the atoms (in their orbital motion) move away from each other to such a distance that the rotos is formally transformed from rigid into ductile. The energy of the neutron collision with one of them is no longer sufficient to produce an essential change of the dimensions of an individual rotos or of the shape of the MB factor as a whole.

As shown by experience, some AM systems, under the influence of neutron irradiation, go through CD phase transitions (Figure 1.15) with decomposition of rotoses. This leads to softening, which is observed in tensile testing in a broad temperature range including room temperatures. Troshenko[3] gives numerous examples of such behavior of materials.

5.4.2 Embrittlement

Under the influence of any external fields (including radiation fields), the condition of solids always changes so that the left-hand potential and right-hand kinetic parts of Equation 3.19 were always equal to each other and remained constant under the equilibrium conditions. This means that, in order to satisfy condition $\Omega = $ const, strengthening (characterized by the increase of P parameter) was always accompanied by embrittlement (resulting from lowering of parameter V being conjugate with P parameter). A distinction is made between high-temperature radiation embrittlement (HTRE) and low-temperature radiation embrittlement (LTRE).[3] HTRE is understood to be a more intensive and profound lowering of ductility properties of materials exposed to neutron radiation at short-term and long-term testing at high temperatures, compared to LTRE and to simple temperature aging. LTRE is found in the temperature range between 293 and 723 K when the increase of neutron irradiation dose is accompanied by a rather abrupt reduction of elongation.[3,237–245] As follows from Equation 2.134 and Figure 2.6, no matter what the reasons for strengthening, namely denser structure (Figure 3.34), thermal (Figure 5.6), force (Figure 5.8), or other fields, it is always accompanied by embrittlement. And neutron irradiation is not an exception to this list (Figure 5.13a). Figure 5.16 presents the change of elongation in tensile testing of a high-nickel, fine-grained alloy Kh20N45M4B (20% Cr, 45% Ni, 4% Mo, 1% B), depending on testing temperature and radiation dose.[3]

After austenization the alloy has high ductility in a broad range of temperatures (curve 1). Specimens were exposed to the following doses: 2, $1.6 \cdot 10^{21}$ n/cm² at $T = 623$ K and energy $E \geq 0.85$ MeV; 3, $5.5 \cdot 10^{21}$ n/cm², $T = 573$ K and $E \geq 0.85$ MeV; 4, $1 \cdot 10^{22}$ n/cm², $T = 573$ K and $E \geq 0.85$ MeV; 5, $7.7 \cdot 10^{22}$ n/cm², $T = 543$ to 608 K and $E = 0.5$ MeV. As seen, beginning

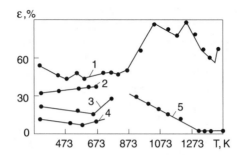

FIGURE 5.16
Temperature dependence of elongation at rupture of specimens of high-nickel alloy Kh20N45M4B (20% Cr, 45% Ni, 4% Mo, 1% B) in the initial 1 and irradiated 2–5 conditions.

from the dose of $1.6 \cdot 10^{21}$ n/cm^2, elongation of the alloy drops by 20 to 30% (curve 2) and, as the dose increases (curves 3 to 4), ductility in the temperature range of 293 to 673 K decreases continuously to 8 to 10%. Radiation doses obtained in experiments 2, 3, and 4 did not lead to irreversible changes in the spatial arrangement of polyatomic molecules (Figure 2.7). Therefore, when the testing temperature increases above 673 K (extreme right-hand sections in these curves), ductility recovery is observed; this is indicated by a stable tendency to its increase.

The situation is different when the radiation dose reaches the magnitude of $7.7 \cdot 10^{22}$ n/cm^2 (curve 5). A continuous decrease of ductility is observed here with increase (above 923 K) of the testing temperature. In the range between 1323 to 1523 K, the alloy becomes brittle as the elongation and reduction of area are equal to 1.0 to 0.5% at fracture. The phenomenon of degradation of deformability of austenitic chromium–nickel steels and alloys preexposed to neutron radiation at such high temperatures and irreversibility of the embrittlement process after high-temperature heating is known as "eternal memory of radiation."[3] Some researchers[236] associate this phenomenon with helium embrittlement due to the influence of helium evolving in nuclear reactions (see Figure 5.10 and comments on it). Helium certainly makes a contribution to the embrittlement process, but it is not the principal factor.

The physical essence of this phenomenon is as follows. By changing the shape of the MB factor in directions 14 to 12 to 11 and further on to the left (Figure 1.14), the hard neutron radiation takes a considerable part of the rotoses beyond the cold brittleness temperature T_c (by analogy with position 13). (Its impact on the structure can be formally identified with a super strong localized compression.) As such, the force and thermal effects together transform its shape in the opposite direction (by sequence 11 to 12 to 13 under the influence of force field and 12 to 14 to 15 under the influence of temperature field). This creates a powerful flow of CD phase transformations through T_c temperature (by sequence 4 in Figure 1.15). These processes form the brittle mode of fracture.

5.4.3 Long-Term Strength

Similar to force, thermal, or other fields, radiation fields change the material initial condition according to Equation 3.19. This allows presentation of the equation of durability (4.28) for the case of the joint action of the force, thermal, and radiation fields as

$$\tau = \tau_o \exp \frac{\Omega_o \pm V\sigma \pm \varphi\tau}{kT}, \qquad (5.7)$$

where $\varphi\tau$ n/cm^2 is the radiation dose equal to the product of neutron flow density, σ n/cm$^2 \cdot$ s, and time τ. Comparison of Equations 4.28 and 5.7 shows that the radiation fields should (by analogy with Figure 4.20) change the durability due to equidistant displacement of the graph in the appropriate coordinate system. Experimental data[3] confirm this.

As an example, Figure 5.17 shows the nature of lowering long-time strength of reactor shells made of Cr16Ni13 (16% Cr, 13% Ni) steel. Numbers 1 and 3 indicate the properties of the material before irradiation obtained at the testing temperature of 873 and 973 K, and 2 and 4 after irradiation in the reactor at the same temperatures (doses are plotted on the upper abscissa axis). These and numerous other data lead to an unambiguous conclusion: high-temperature irradiation essentially lowers material durability and Equations 4.28 and 5.7 allow accurate forecasting of the nature of its change.

Comparison of Figures 4.31 and 5.17 reveals that graphs 1 to 4 are end sections of the exponential dependence (4.28) being approximated as straight lines by authors in Troshenko.[3] Equidistant shifting of the graphs is not due to the change of heat content H (although it is also quite important), but to radiation dose $\sigma\tau$.

For some alloys, irradiation consequences can be disastrous. For example, for nimonic alloy KhM77TYR (analogue of Inconel),[3] the fracture stress at long-time strength testing at 1073 K decreases five to six times; the shortening of time to fracture is so great that it is difficult to evaluate. This is due to the durability graph shifting from position A_1 to A_3 (Figure 4.31) and

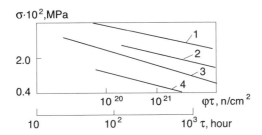

FIGURE 5.17
Lowering of the long-time strength of Cr16Ni13 (16% Cr, 13% Ni) steel at irradiation.

its transformation into a vertical line even at small values of external stresses σ (see Figure 4.20a). The causes for this behavior of the durability graph are discussed in the analyses of these figures. Experiments showed[3] that irradiation directly in the reactor, i.e., in the stressed state, has a greater influence on decreasing durability compared to preliminary radiation and subsequent testing for long-time strength. This is natural because, according to Equation 5.7, in the first case the structure is simultaneously exposed to two destructive factors (σ and $\sigma\tau$) and, in the second, their action is separated in time.

Other kinds of radiation (for instance, ultraviolet, UV) have absolutely similar influence on durability. Regel et al.[10] found that, if polymers are exposed to UV radiation taking place simultaneously with loading, their durability may decrease by many orders of magnitude. If the same radiation dose is applied to a polymer before loading and then it is tested without irradiation, the durability also decreases but by an incomparably smaller value than that when simultaneous action of mechanical stresses and radiation is the case. These authors suggested that a simultaneous occurrence of two parallel processes of destruction proceeded in the loaded body at different rates: mechanical and radiation. However, only the durability equation (5.7) based on the thermodynamic equation of state (2.96) provides a valid physical interpretation of these processes.

Long-term exposure of materials to irradiation gives rise to processes of radiation fatigue.[215] The physical causes for development of this kind of damage and their consequences will become clear after discussion of thermodynamic fundamentals of fatigue in the force fields in subsequent chapters.

5.5 Moisture-Induced Softening of Porous Materials

The equation of interaction (3.19) incorporates environmental parameters (infrared and UV, moisture, atmospheric pressure, etc.) as one of independent variables A_i. They change the technical condition and, therefore, influence the service properties of materials, parts, and structures.

For instance, researchers noticed a long time ago that the change of moisture of porous materials (artificial and natural stones, ceramics, polymers, etc.) largely affects their physicomechanical properties[189,246–250] as well as the load-carrying capacity and deformability of finished structures.[251] Dams[247] presented results of a study to clarify the degree of influence of concrete moisture on its compression strength. Testing specimens in the form of cubes with 20-cm edges made of 15 different concrete compositions showed that moisture variation by 1% on average leads to change of strength by 10%.

Popovich and Nikolko[249] observed a lowering of compression strength from 8 to 17% with variation of moisture from 0.2 to 7% in heavy concretes. Authors of another study[247] reported that a change of water saturation of limestone, marble, shale, and sandstone leads to variation of their tensile and compression strength. The degree of tensile strength reduction is higher than that of compression strength. For instance, for limestone, compression strength decreases by 22 to 48%, while tensile strength decreases by 30 to 55%; for shale it is by 21 to 31% and 28 to 42%, respectively. The authors concluded that the effect of water-induced weakening of rock strength depends on structural features. Eliseev[251] noted that nonuniform moisture content may lead to strength variation of up to 28% within one part. This may result in a decrease of the load-carrying capacity and deformability of reinforced concrete structures by 8 to 10%. A similar effect is produced by moistening polymer materials.[189]

The conclusive experimental data allow (with other conditions equal and by analogy with Equation 3.19) taking into account the influence of moistening on the condition of solids as follows:

$$(\mathbf{P} - W)V = \Omega(T), \tag{5.8}$$

where W is the parameter accounting for the change of moisture content in a porous structure.

Despite the existence of numerous conclusive experimental data obtained by many researchers, the effect of water-induced weakening is still not taken into account when determining the physicomechanical characteristics of porous materials, in design of parts and structures from them, or during their experimental studies. For instance, during the service period of a reinforced concrete structure, the environmental humidity and, therefore, the concrete water saturation, can vary in a broad range. Experience gained in full-scale studies of construction facilities, buildings, and constructions, as well as analysis of emergency situations,[128,129] shows that the number of failures of such structures increases rapidly at considerable moistening of concrete (rain, wet snow, vapor condensation, etc.), even at unchanged applied forces.

In this section the nature of variation of the condition of porous materials at moistening and the degree of its influence on the major parameters of stressed–deformed state under compression and tension (tensile) of heavy (regular and high-strength) and light concrete is clarified. The objects of investigation were specimens — cubes, cylinders, prisms, beams, and structural elements. The number of specimens of each composition was selected so that no fewer than nine cubes were tested at any moisture level. Testing of heavy concrete specimens was performed at the age of 28, 90, and 540 days and that of the light concrete at 28, 60, and 120 days. Water content of each kind of concrete was varied from the dehydrated condition up to complete saturation.

Figure 5.18 illustrates the dynamics of water absorption and dehydration of heavy concrete of design grade 400 at an early age.[252] Buzhevich and Kornev[253] obtained a similar result for keramsit clay–concrete. Moisture equilibrium of heavy concrete of this composition is equal to 3.5% (*aob* horizontal)

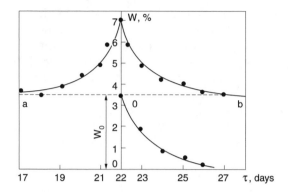

FIGURE 5.18
Dynamics of water absorption and water loss of heavy concrete of design grade 400 at an early age.

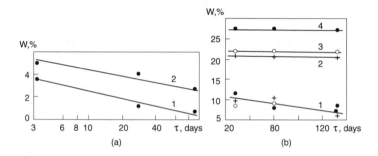

FIGURE 5.19
Time dependence of moisture equilibrium in (a) heavy and (b) light concrete (water–cement ratio of the latter is equal to 1, 0.56; 2, 0.76; 3, 1.03; 4, 1.78).

and corresponds to dynamic moisture equilibrium in the material structure and the environment. Soaking the specimens in an aqueous medium (*ao* section) for 5 days (from days 17 to 22) results in maximum possible water content equal to 7.2% (upper left-hand part of the diagram). At atmospheric temperature–humidity conditions, the equilibrium is restored in 5 days (upper right-hand field of the figure). That is, under regular conditions, the processes of water absorbing *ao* and dehydration *ob* proceed at the same rate.

Heavy concrete at an early age has considerable water retentivity. For instance, in order to lower the moisture from equilibrium to 0.5%, it is necessary to keep the cubic specimen of 20-cm edge size in the heating apparatus at a temperature of 100 to 105°C for 4 days (lower right-hand field).

Moisture equilibrium is a variable value since water consumption of concretes decreases with age. Figure 5.19 presents the time dependency of moisture equilibrium for heavy (a) and light (b) concretes. Irrespective of concrete type, moisture equilibrium decreases with the concrete age (graphs 1 and 2 in Figure 5.19a and 1 in Figure 5.19b). For heavy concretes, the rate of lowering depends on water–cement ratio (graphs 1 and 2 in Figure 5.19a)

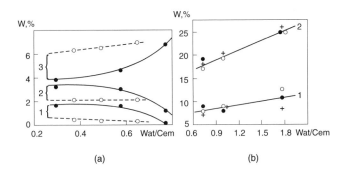

FIGURE 5.20
Dynamics of moisture condition of (a) heavy and (b) light concretes depending on water–cement ratio.

and for light concretes it is approximately the same at any age (all experimental points are grouped around graph 1). In these concretes, the maximal water saturation is the same at any age (graphs 2, 3, and 4 in Figure 5.19b). These features are attributable to the different natures of moisture accumulators. In heavy concretes, it is the porous system of the grout and, in light concretes, the porous and hygroscopic filler. Ramification of the grout porous system is largely dependent on the water–cement ratio. Hydration of the remaining cement grains results in total volume decreasing with time, which leads to lowering moisture equilibrium of heavy concretes.

Figure 5.20 illustrates the dynamics of the moisture condition of heavy (a) and light (b) concretes, depending on the water–cement ratio. Solid lines in Figure 5.20a correspond to the age of 28 days and dashed lines to the age of 540 days. Numbers indicate the following moisture conditions: 1, forced dehydration in desiccators; 2, moisture equilibrium under natural atmospheric conditions; and 3, artificial saturation at atmospheric pressure. Number 1 in Figure 5.20b denotes moisture equilibrium under normal atmospheric conditions, and number 2 shows moisture change at maximal saturation. Points indicate the age of 28 days; circles, 60 days; and crosses, 120 days.

The water–cement ratio influences the change of moisture condition of any kind of concrete. Its nature, however, depends on differing moisture consumption of the structural components. A greater amount of absorbed water increases the distance between the solid components of the concrete mixture; it is fixed as a result of hydration of cement grains and formation of coagulation bonds between the particles of the inert filler. Grout porosity becomes higher. This circumstance plays the decisive role in moisture exchange processes between the environment and the in-depth layers of the material. Presence of a porous hygroscopic filler in the light concrete composition increases the water retentivity of the structure and intensifies the processes of water infiltration from the environment. For this reason, moisture equilibrium of light concrete increases with increase of the water–cement ratio (graph 1 in Figure 5.20b).

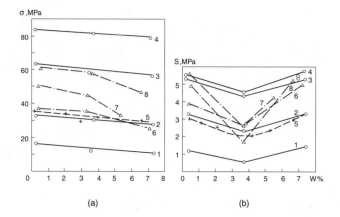

(a)   (b)

FIGURE 5.21
Dependence of the mean strength of heavy concrete in compression (a) and its root-mean-square deviation (b) on moisture.

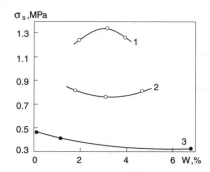

FIGURE 5.22
Moisture dependence of the tensile strength of heavy concrete at the following water–cement ratios: 1, 0.28; 2, 0.57; 3, 0.77.

In heavy concretes, higher porosity intensifies the processes of moisture migration from the material into the environment. As a result, with an increase in the water–cement ratio, moisture equilibrium decreases (curve 2 in Figure 5.20a). Special attention should be paid to the fact that the intensity of moisture exchange processes attenuates with time in heavy concretes (continuous and dashed curves in Figure 5.20a have essentially different curvature), while in light concretes it does not (all experimental points fall near graphs 1 and 2).

Change of moisture has an essential influence on concrete behavior under load. The moisture influence on the major parameters of statistical distribution of compression and tensile strength in heavy concretes is shown in Table 5.2 and in Figures 5.21 and 5.22.[252]

TABLE 5.2

Intensity of the Change of the Strength of Heavy Concrete at the Age of 28 Days at Moisture Variation by 1%

Strength Parameters	Direction of Moisture Change, Compared to Moisture Equilibrium	Parameter Change	Parameter Change (in %), Depending on Average Strength, MPa		
			10–25	30–55	60–80
Average value, σ	From dehydrated state to complete saturation	Decrease	6.0	3.25	2.81
Root-mean-square deviation, S	In drying	Increase	7.58	5.38	2.95
	In moistening	Increase	11.0	6.86	3.51

In Figure 5.21, numbers 1 to 4 (solid lines) designate the compositions of design concrete grades 100, 300, 600, and 800. They were tested at the age of 28 days. Composition 5 (dashed line) of design grade 300 was tested at the age of 90 days, while compositions 6 to 8 (dash-dot line) of design grades 300, 400, and 600 were tested at the age of 540 days. Specimens of composition 5 were tested at moisture variation by 1%, and those of compositions 1 to 4 and 6 to 8 only at limit values.

Decrease in average strength is found for all compositions with the increase in moisture content up to complete saturation. A similar effect was observed for light concretes[253] and polymers.[254] Known for a long time, this is attributable to development of additional tensile strains in microcracks and pores filled with water.[255]

The rate of strength decrease depends on the type of structure. Its arrangement in heavy concretes of low (compositions 1 and 2) and high (compositions 3 and 4) grades is essentially different and, therefore, the decrease in the strength of the latter is more than two times lower (Table 5.2).

Proceeding in the volume of heavy concrete after solidification, structure formation processes not only reduce the total pore volume but also change open porosity to closed porosity. This is indicated by an improvement of concrete water impermeability with age. The intensity of average strength lowering in older concretes (compositions 6 to 8) turns out to be higher than that of young concretes (compositions 1 to 4).

The nature of moisture influence on strength scattering is interesting (Figure 5.21b). Its root mean square value S takes the minimal value at a certain (let us call it equilibrium, W_0) moisture level equal to 3.5% for test conditions (Figure 5.18). The S increase is observed on both sides of W_0. Water absorption is accompanied by a more intensive increment of S than dehydration (Table 5.2). In addition, S depends on concrete structural features to a greater extent than strength (compare compositions 1 to 4 and 6 to 8). This confirms the long-known determinant role of density in concrete durability.[128,129]

At moisture equilibrium, the tensile strength has the maximal value (curve 1 in Figure 5.22). The convexity of the curve becomes more prominent for high-strength concretes. Moisture equilibrium corresponds to the natural atmospheric temperature–moisture conditions. Concrete adjusts somewhat to diverse external conditions; this is natural because any integral system with a certain internal organization, in principle, is capable of self-compensation or self-regulation.[256] A solid, in general, is a set of rotoses in continuous motion that interact with each other (Figure 1.6). For such systems the existence of a feedback is possible.[257]

In light concretes, the change of strength parameters induced by moisture is similar. For almost all the studied compositions, an increase in the moisture level leads to a decrease in the compressive strength, while the tensile strength increases. This tendency is more clearly revealed in younger concretes and more pronounced with smaller water–cement ratios. The nature of this dependence changes essentially when approaching the zone of high water–cement ratios and low strength (less that 10 MPa in compression). The root mean square value S becomes smaller, reaching its minimum at the moisture level of 8 to 12%, and then increases right up to complete saturation. The intensity of the change of average strength and of its scatter depends on the structural features of a specific kind of concrete and its age. Change of light concrete moisture by 1% changes the average value by 1.5% and its scatter, represented by the root mean square value S, by 2.5%.

As shown in References 24, 53, and 54, statistical data on the indirect hardness indicators contain valuable information on the nature of change of the Ω-potential change (Figures 3.15 and 3.27). Therefore, each stage of studs on the influence of moisture was accompanied by the measurement of hardness. Particularly, elastic rebound h of the indenter and plastic imprint made by spherical indenter, d, were measured. Figure 5.23 shows their empirical distributions at the following levels of moisture in heavy concrete: 1, 1.77%; 2, 3.1%; 3, 4.65%.[115] Relative frequencies in percent are designated as m_j on the axis of ordinates.

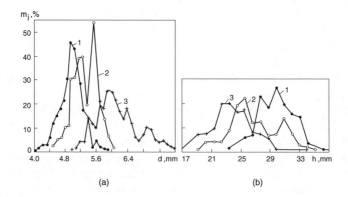

(a)   (b)

FIGURE 5.23
Transformation of hardness indicators: (a) measured as the dent size due to the spherical indenter and (b) measured by the elastic rebound of the indenter at water saturation of heavy concrete with water–cement ratio of 0.57 and the following moisture content: 1, 1.77%; 2, 3.1%; 3, 4.65%.

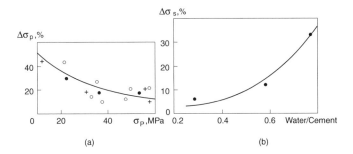

FIGURE 5.24
Range of variations of the compressive (a) and tensile (b) strength of heavy concrete at moisture variation from air-dry condition up to complete saturation.

Concrete aging accompanied by continued hardening processes (see Section 3.8) is equivalent to shifting the G-potential from position 1 to 2 or even 3 (Figure 3.41). The opposite shift of G-potential at hardening and of MB distribution at moistening leads to intensification of destruction processes in force fields. This accounts for a more abrupt lowering of the average strength and an increase of its distribution range (scattering) with concrete aging. (Compare the shape of graphs 1 to 4 and 6 to 8 in Figure 5.21.) Analysis shows that the mechanism of moistening influence on physico-mechanical properties of porous materials is not as simple as is currently believed. It is believed that moistening is due to simple capillary entrapment of moisture by material.[255]

A quantitative measure of concrete susceptibility to moistening can be sensitivity æ and range Δ of strength change. The latter indicates the range of strength variation at the limit moisture values. Figure 5.24 shows the range of change of the compressive strength (a) and that of the tensile strength (b) for heavy concrete at moisture variation from the air–dry condition to complete saturation. With increasing moisture in heavy concrete, the amplitude of σ_p and σ_s oscillations decreases with the reduction in the water–cement ratio. For instance, for σ_p, it decreases from 40 to 45% at the average strength of 20 MPa, to 10 to 15% at the average strength of 60 MPa. Such a nature of resistance change is logical and attributable to an increase of the density of the structure grout part.

The moisture sensitivity æ is understood to be the percentage of strength change due to moisture change by 1%. Figure 5.25 shows moisture sensitivity of the strength at compression and tensile of heavy (a) and light (b) concrete of several compositions. Note the different intensity of moisture transformations in the structures: for heavy concretes, the moisture sensitivity is measured by tens of a fraction of a percent, while for light concrete, it is within several percent. The tensile strength is more sensitive to variation of moisture than σ_p. Function æ = $f(W)$ has the opposite change for different kinds of concretes: for heavy concretes it decreases and for light concretes it increases

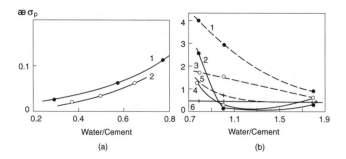

FIGURE 5.25
Moisture sensitivity of (a) heavy (1, age 28 days; 2, 540 days) and (b) light (1 and 2, age 28 days; 3 and 4, 60 and 5 days; and 6, 120 days) concrete. σ_p = solid line, σ_s = dashed line.

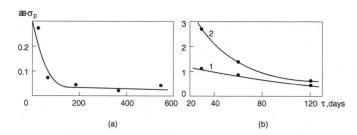

FIGURE 5.26
Time dependence of moisture sensitivity of (a) heavy and (b) light concretes: 1, σ_p; 2, σ_s.

when entering the high-strength zone. Tensile strength σ_s is more sensitive to moisture than compressive strength σ_p (dashed curves in Figure 5.25b located above the solid curves). Sensitivity of σ_p and σ_s to moisture becomes weaker with time (curve for more advanced age 2 in Figure 5.25a falls below curve 1).

Time dependence of moisture sensitivity of σ_p and σ_s for the studied types of concretes is given in Figure 5.26. As seen, the sensitivity of any type of concrete to moisture impact is considerable in the first 3 to 4 months of structure service. Then it becomes weaker and is stabilized for a long time in the vicinity of a constant value equal to 0.05 for heavy concrete and at 0.5 for light concrete.

Moisture sensitivity increases with structure saturation. Figure 5.27 shows the nature of its change for the σ_p parameter of heavy concrete at different degrees of water saturation. Such a shape for this dependence shows that moisture impact has a progressive mode of development: the rate of destruction of the structure increases with the increase of moistening.

At the AM level, moisture softening can be formally regarded as a phenomenon opposite to hydrogen embrittlement (Section 5.3). If the latter is related to the formation of additional molecular orbits of electrons (Figure 5.9b), moistening leads to their reduction, greater distance between the interacting atoms, and lowering of rotos rigidity. It follows from the thermodynamic equation

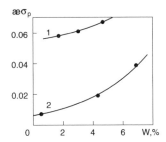

FIGURE 5.27
Dependence of moisture sensitivity of heavy concrete on water saturation (1 — water-to-cement ratio of 0.57; age of 28 days; 2 — water-to-cement ratio of 0.37, age of 90 days).

of state (5.4). Therefore, no change of its AM mechanism or macroscopic consequences, considered in detail for concrete, should be anticipated in considerations of other porous materials (ceramics, composites, polymers, etc.).

It is now easy to explain the processes proceeding in porous structures on moistening. It follows from Figure 5.23 that moistening moves the MB factor as shown by curves 12 to 14 to 15 in Figure 1.14 or in the direction from left to right in Figure 2.24 (which is the same) with a simultaneous increase of the range and flattening. Following the line of reasoning provided in the consideration of Figure 3.34, this should inevitably lead to an appropriate change of parameters of statistical distribution of the strength, which is experimentally confirmed (Figure 5.21). This indicates that the mechanism of moistening influence on the mechanical properties of porous materials is not as simple as is currently believed.[248–251]

5.6 Durability of Unstable Structures

Even in the absence of external fields (force, thermal, radiation, or others), external conditions (composition of the medium, atmospheric pressure and humidity, etc.) change the condition of materials. Also, the condition does not remain stable in time (for instance, metals in metastable condition or concretes at hydration of cement grains which did not react during cement solidification, Figure 3.42). Pettifor[34] has shown that, based on analysis of experimental data, the instability of the structure leads to a change in durability and to a distortion of direct proportional dependence $\log \tau = f(\sigma)$ (see Equation 4.29 and Figure 4.26). Considering concrete and reinforced concrete structures as examples, let us look at possible consequences under actual service conditions.

For this purpose, consider Figure 3.42, which shows the change of concrete condition (more exactly, its resistance to external impacts) with time through

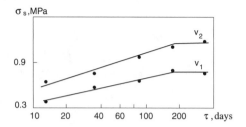

FIGURE 5.28
Time dependence of the tensile strength of concrete at two loading modes ($v_2 > v_1$).

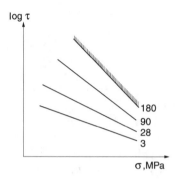

FIGURE 5.29
Durability of the same composition of concrete at structure stabilization at the age of 3, 28, 90, and 180 days.

hydration of excess cement grains. These dependencies were derived by testing specimens of various age, $\tau \neq$ const at a constant loading rate $v =$ const. Up to the moment of testing, the structure-forming processes proceeded in the absence of the external force fields, i.e., $\sigma = 0$. If one of these restrictions is altered, an essential change of the discussed dependence takes place.

Figure 5.28 shows time dependencies of the tensile strength of the same composition of concrete at two loading rates. These dependences were obtained at $\sigma = 0$; $v \neq$ const; $\tau \neq$ const. They should be considered random sections of the graph of durability of a stable structure material (Figure 4.26) at two time levels, τ_1 and τ_2. They enable reconstruction of the durability variation from the results of testing intermediate structures.

Figure 5.29 presents a family of durability curves of the same concrete composition at different ages (3, 28, 90, and 180 days). Similar dependencies can be derived during aging (shifting of positions in Figure 1.14 in the direction 12 to 14 to 15) of metastable metallic structures. These are conditional in a certain sense because they can only be obtained at artificial conservation of the structure at the previously mentioned age by drying it with subsequent hydraulic insulation of the surface or material impregnation with an antihydration composition. They satisfy conditions $\sigma \neq$ const, $v \neq$ const,

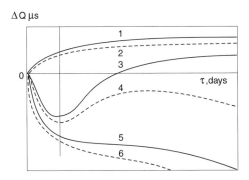

FIGURE 5.30
Changes in concrete structure with time at different levels of compressive stress–strain state.

τ = const, i.e., after the application of a certain stressed–deformed state, the concrete structure does not change any further. In reality, however, this is not the case because it undergoes continuous changes before as well as after loading as a result of concurrent running of the forming and the destruction processes. The resultant effect depends on the kind and level of a current stressed–deformed state as well as on the age and humidity of the material by the moment of loading.

For instance, at the initial stages of compression, the effect of static strengthening of the structure is found (see Figures 3.50 and 3.58). Inducing various levels of compressive stressed state in the concrete, Rozhkov[200] observed the structural transformations for a long time (Figure 5.30). If the acting stresses σ are below the lower boundary of microcracking σ_T (Figure 4.33), ultrasonic measurements show a continuous consolidation of the structure (curve 2). Here $\Delta Q = Q_\sigma - Q_T$, where Q_σ is the time of ultrasound propagation at the last stage of loading. Compression makes the structure-forming processes more active (upper field); at favorable temperature–moisture conditions, it may lead to a considerable increase of durability.[150]

At $\sigma_T \leq \sigma \leq \sigma_v$, where σ_v is the long-time strength,[48] the material undergoes short-time reversible changes (curve 3). These stop with time and the structure recovers (point of intersection with the time axis), then it gradually improves. If $\sigma \geq \sigma_T$, the mechanical destruction processes acquire an irreversible nature. Decompression of the structure (lower field) is a continuous process (curve 5). Progressing propagation of microcracks ultimately brings the material into the limiting state.

The preceding processes change durability graphs in the way shown in Figure 5.31, where the numbers denote concrete ages. They correspond to conditions $\sigma \neq$ const, $v \neq$ const, $\tau \neq$ const. At simultaneous loading of similar type structures of the same concrete composition made at different times, structural stabilization processes will lead to discrepancy in their durability. For instance, at the σ level they take the values of $\tau_3 = \tau_{180}$. As a result, at the initial period of service the number of failures of such structures increases

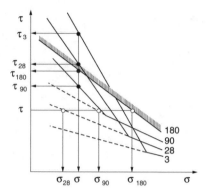

FIGURE 5.31
Durability graphs of structures changing in time.

significantly. Because of its latent nature, this phenomenon does not receive enough attention from engineers.[55,128]

This change in durability is an inevitable consequence of the existing technology of prefabricated reinforced concrete parts production, industrialization, and intensification of the building construction process. To ensure guaranteed durability of the entire construction (for instance, at level τ), it is necessary to differentiate the stressed–deforming state depending on concrete age. Lowering the operating costs is achieved by coordination of the parts fabrication time and sequence of their mounting.

The discussed phenomenon is characteristic only of concrete and reinforced concrete structures. It is versatile and attributable to the mismatch between the methods of design, fabrication, preliminary technological processing, and mounting of the parts and structures and the actual operation of materials in the external fields and active media. It is equally peculiar to the parts made of metallic materials where it is not caused by hydration or moistening but by stress relaxation processes. They bring the material structure from its metastable condition just after fabrication into an equilibrium state corresponding to the service conditions.

The Reference 238 study analyzed the reliability of metal structures operating at static and low-cyclic loading. In the case of static loads, the major part of failures (up to 90% of those failing over 15 years of service) occurs in the first year of operation, which is in agreement with earlier data (see Figure 1.22). For cyclically loaded structures, just 34% of the total recorded failures occurs during the first winter or during testing. Furthermore, service failures do not occur for about 3 years and then their number starts increasing from 4 up to 16% per year. This is the price of inaccurate notions of the actual operation of materials and inability to influence their stressed–deformed state during service.

As shown earlier, the external fields (force, thermal, radiation, and others) and active media (hydrogen-containing, moistened, etc.) have an influence

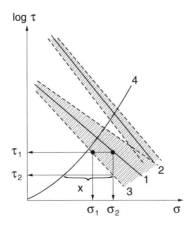

FIGURE 5.32
Moisture influence on concrete durability for two different compositions.

on material condition. The results of theoretical and experimental studies allow forecasting material behavior in time under changing external conditions.

This forecasting procedure is illustrated in Figure 5.32, in which the nature of durability variation is given for two concrete compositions, 1 and 2, differing in their moisture sensitivity.[252]

Let the deviation of the durability graph from the equilibrium condition (solid line) be greater in the first structure than that in the second. It depends on different structural organization of the material and manifests itself in practice as change of the statistical parameters of hardness (Figure 5.23a,b) and strength (Figure 5.21a,b) properties. Continuous decrease of the average value of strength at the change of moisture content from zero to complete saturation points to equidistant shifting of the graph from position 1 into 3. The increase of the distribution spread with moisture content to both sides from the optimal level (solid lines in 1 and 2) is indicative of the difference in the scattering areas around both lines.

Now it becomes possible to explain the known fact: probability of failure of similar building structures of concrete increases with moistening.[2,19,31,250,259] Let durability of a part having equilibrium humidity (corresponding to the minimum in Figure 5.21b) be given by graph 1 in Figure 5.32. By analogy with Figure 4.39, curve 4 in Figure 5.32 is an integral statistical distribution of the actually applied forces. Section x between it and the durability graph (1, 2, and 3) determines the residual resource margin (Equation 4.25). With the increase of moisture level, the durability graph shifts from position 1 into 3. Indeed, as follows from Figure 5.21, moistening of grade 400 concrete leads to a decrease of the average value of strength by 13% with a simultaneous increase of the scattering by 20% on average.

Variation of the moisture content impairs the material's ability to resist dynamic loads. For instance, at moisture equilibrium, concrete stands the peak loads of intensity σ_1 for time τ_1 (Figure 5.32). At maximal moistening (on the level of graph 3), it can resist either a much smaller load σ_2 for time τ_1 or the initial level of σ_1 for shorter time period τ_2. Now, any material transition into the high stress zone inevitably induces the processes of destruction and consumption of the resource (Equation 4.26).

Whether the structure fails or not depends on the duration of hazardous stress action and the ability of the material structure to redistribute these stresses. The described mechanism is valid at one-time moistening and without taking into account force impact intensification. The developing defects will promote a more active progress of the microcracking process. For instance, wind gusts with rain and subsequent icing often cause failure of power transmission poles.[259] Thus, concrete under conditions of alternating moistening and drying accumulates irreversible structural changes that, ultimately, may lead to fatigue failure.[215] In metals a similar effect is produced by multiple variations of the temperature field in the vicinity of Debye temperature θ_s (see Section 2.7 and Figure 1.14). These structural changes are the most critical at an early age of concrete when age shifting of the durability graph C_2 is complemented by moisture-induced shifting C_3 (Figure 5.33).

Clarification of the physical nature of building structure failures in concrete moistening allows one to make practical recommendations on the prevention of their possible development.[252] It is necessary to lower the rate of the equilibrium condition change by elimination or, at least, by decreasing the possibility of water penetration into the material or leaving the material by improving the surface layers or applying sound waterproofing. For concrete, the measures to improve its structure are most effective; they change the nature and also the ramifications of the capillary system as a whole.

In conclusion, we would like to direct attention to the thus far untapped potential for reducing material content in concrete and reinforced concrete

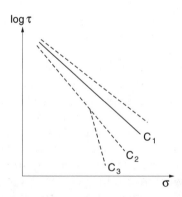

FIGURE 5.33
Combined influence on durability of age C_2 and moisture C_3 of concrete of composition C_1.

structures. Today, statistical characteristics for rating the short-time strength are determined at concrete age of 28 days.[194] This is equivalent to rigid binding of structure durability to the level corresponding to the age of 28 days in Figure 5.31. Such an approach does not allow use of a considerable margin of the short-time strength (a long-known fact) or, what is especially important, potential durability reserves. (Compare the solid and dashed lines of the same graph in Figure 5.31.)

About 30% of consumed cement is conserved by the existing technology in the concrete structure, constituting an unused durability reserve. The need to reduce the material consumption and cost of structures makes it necessary to choose between two approaches: staying within the frames of traditional technology and trying to learn how to make use of the potential of time-dependent structure-forming processes, or cardinally changing fabrication technology with the aim of rational use of deficit materials.

As the internal structure-forming processes become weaker, the physico-chemical processes of aging begin to develop on the surface and in subsurface layers. They become noticeable beginning from the age of 400 to 450 days (see Figure 3.42). In metals, aging is reversible because it is governed by CD phase transitions (Figure 1.15) that involve the entire bulk of the material simultaneously. In concretes this process has the form of irreversible physicochemical transformations. They start from the surface layers and gradually penetrate into the material bulk. Their course and rate depend on the modes of loading, temperature, moisture, pressure, and composition of the medium. The rate of chemical reactions is given by Arrhenius law:[31]

$$B = B_o \exp \frac{U_o}{T_o} \Delta,$$
(5.9)

where B_o is the constant corresponding to mean-statistical service temperature T_o, U_o is the activation energy, and Δ is the actual temperature deviation from the average level, T_o.

Chemical reactions lead to change of the conditions of a material and, as a result, to its embrittlement and deterioration of its ability to resist external impact. Let us assume that the aging processes change the state of a material so that a certain part of its cross section is not involved in the resistance. This results in further decreased in durability by

$$\tau_c = kB,$$
(5.10)

where B is the parameter determined from Equation 5.9, and k is the coefficient accounting for the influence of the surface shape on the intensity of aging processes. When the structure is designed, the durability calculations should account for τ_c.

5.7 Defect Healing and Damaged Structure Restoration

Manufacturing, storage, transportation, installation, and active service of modern engineering objects proceed under the impact of fields of different physical nature: force, thermal, radiation, electrical, chemical, magnetic, climatic, etc. Naturally, they change the technical condition of the materials and, therefore, influence the durability of structural components. As such, several competing physical processes can proceed simultaneously in a material. Some of them improve the material structure by adapting it to the service conditions, while others bring the moment of its destruction closer. Depending on a particular case, improving processes may prevail over those of degradation, or vice versa.

For instance, in the force fields the orientation effect (see Section 3.4) due to compression in the longitudinal direction enhances the resistance in the transverse direction (Figures 3.50 and 3.58). The orientation effect combined with moistening (acting for a long time) leads to a threefold increase of the concrete compressive strength;[150] temperature fields compensate for radiation damage of metals and alloys (Section 5.4.1 and Figure 5.13): deformation and thermal forms of aging can proceed in opposite directions (Section 5.2). Also, other combinations of competing factors can be theoretically substantiated and experimentally confirmed;[12] they will have different signs in the equation of interaction (3.19).

We will refer to the change of the initial structure condition as damage (irrespective of its favorability for the service properties). Distinction should be made between reversible damage, or aging, and irreversible damage, or destruction; the first is restorable and the other is not. Figure 5.34 illustrates the physical essence and fundamental difference between reversible and irreversible damage. The former are accompanied by the transformation of the MB distribution continuously is in the C-phase (for instance, positions 2 and 3, located between vertical lines ar_0 and mr_b) for compresson materials,

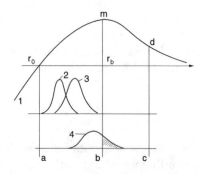

FIGURE 5.34
Recoverable and nonrecoverable phase transformations.

or in the D-zone (between vertical lines *mb* and *dc*) for dilaton materials. When going through the Debye temperature (*mb* vertical line), one part of C-rotoses can safely pass and enter the D-phase (characterized by the unhatched area under the right-hand side of distribution 4 in Figure 5.34), while the other part fails because it has not received the required heat input at the moment of phase transition (hatched area).

Inside the C- and D-phases all the changes are of an elastic or orientation nature. Therefore, the condition defined, for instance, by MB distribution 3 can be brought to the initial position 2 by applying any of the following fields to the material: force, electric, magnetic, thermal, or their combination (see Section 3.3). However, this is not applicable for that part of the rotoses passed through the Debye temperature (distribution 4). In this case, D-rotoses (unhatched region) can be brought into C-phase only by thermal methods. The failed part of the rotoses can no longer be brought into C-phase. (Irreversible reduction of parameter N has taken place in Equation 2.96.)

Correction or improvement and recovery of the structure or relieving its internal stresses are currently performed mainly only in the thermal or mechanical fields (or their combinations) using simultaneous impact of thermal and mechanical stress cycles.[3,228] There are no publications on applying nontraditional methods (for instance, electrical or magnetic) for these purposes, although numerous experiments on magnetization of magnetic materials[29] and polarization of dielectric materials[6,26] point to their effectiveness.

Experiments indicate the possibility of recovery of the technical condition of metals. For instance, Kuznetsov[260] observed the simultaneous processes of structure damage under the impact of mechanical loading and its recovery at the annealing temperature when the mechanical strength was increased compared to the initial one. It is noted that the annealing rate depends on body initial temperature and deformation rate.

The consequences of the destruction processes developing in the material under the impact of external load are not eliminated at subsequent recovery of metals at elevated temperatures. This is indicated by experiments with interrupted loading. For instance, investigation of durability under conditions when the time to rupture was divided into several intervals, between which the specimen was unloaded and was able to recover, demonstrated that the total time is approximately equal to the time to fracture in one cycle.[261]

Studying the strength of solid dielectrics in electric fields, Vorobiov and Boroviov[92] concluded that crystals with lower lattice energy form lattice defects more readily; however, their healing also requires less energy consumption. Therefore, after application of similar external impacts, lattices with lower energy also preserve a lower residual concentration of defects. This conclusion is further confirmed by many kinetic laws of variation of the properties of crystals under simultaneous occurrence of the processes of defect accumulation and their annealing.

Various studies[186,262] contain detailed descriptions of the technical condition of metals and alloys in mechanical and thermal fields. It follows that exposing steels and alloys to different thermal and mechanical impacts allows one to quite markedly change their mechanical properties and recover their initial conditions. It is almost always characterized by the presence of various kinds of residual stresses. The causes for their development in steels are described in Shmidt.[263]

Sammal[94] presented proof of their presence in concretes. He studied the causes for the development of the stressed–deformed state in concretes at their early age using ferromagnetic transducers for measuring normal stresses. These transducers were embedded into concrete prisms so that the occupied area was no more than 10% of the cross section. After concrete consolidation, the prisms were stored under laboratory conditions under no external loads. During this period, the readings of the embedded transducers were taken at regular intervals. The results obtained this way demonstrated the presence of initial compressive stresses of up to 3.4 MPa in the cross-sectional core and of tensile stresses in the region of the prism side faces. Sammal attributed development of the inherent stressed state to a nonuniform drying of concrete at the initial consolidation period.

In concretes, as well as in steels, residual stresses are relieved by special treatments. For instance, Shmidt[263] reports an increase of the short-time compressive strength of concrete up to 27% at low-cycle loading (up to 30 cycles). The objects of study were prisms made of heavy and light concretes of 28.8 and 23.5 MPa strength, respectively. The loading rate was kept constant and equal to 0.5 MPa/sec. Concrete age was in the range of 80 to 90 days. Loading was varied with respect to the working level (selected to be 45% of the fracture stress) within the range from 0 to 73% of the fracture stress.

Regarding steels, an essential influence of the structure elastic prestraining on brittle fracture was reported in Silvestrov and Shagimordinov.[43] If it is performed at temperatures above the cold brittleness temperature T_c (Figure 1.14) when the material can deform in a reversible manner, brittle fracture resistance increases. This means that prestraining changed the Ω-potential by increasing P resistance at the expense of lowering deformability V. Using this way, the state of material located to the left of parameter σ_0 is shifted to the right-hand side (Figure 3.34). This is essentially one of the possible methods of controlling or, more exactly, recovering, the resistance of materials.

Effectiveness of the methods of low-cycle treatment of structures (can be referred to as preaging) to improve their resistance to random peak loads is obvious. They need further study and then extensive practical application. It is quite possible that magnetic or electrical fields, rather than mechanical, will have greater simplicity and effectiveness.[12,122]

A special kind of damaged structure recovery is characteristic for concretes; it is related to the use of the potential ability incorporated in its structure together with excess cement grains. Concrete structure (especially at an early age) is saturated with unreacted cement grains. Their conservation is due to certain specifics of hydration under constrained conditions. During concrete consolidation, colloid particles form on the grain boundaries. They hinder moisture penetration and, as a result, slow the hydration rate. As a result, the strength and deformation properties of cement stone and concrete

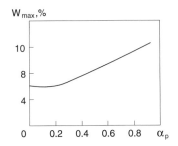

FIGURE 5.35
Change of maximal water saturation at different degrees of concrete deterioration.

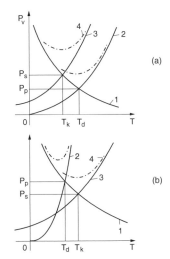

FIGURE 5.36
Recovery of damaged concrete structure in water environment.

are entirely determined by the molecular structure on the colloid particle surface and their adhesion.[246,265]

It is obvious that hydration can become more active under the conditions of unrestrained penetration of moisture to the grains, a condition that can be achieved by inducing mechanical destruction processes accompanied by cracking. Cracking results in formation in the material the chaotically oriented microvoids (see Section 3.9) that facilitate moisture penetration into the material internal layers. Figure 5.35 gives the dynamics of the change of the maximal water saturation, W_{max}, of the same concrete composition taken to different degrees of destruction. In this figure, α_p denotes the level of compression stress. Sannikov[266] observed a similar dependence in fracture of glass-reinforced plastics.

Moistening of loosened material leads to the intensification of hydration of the residual cement grains and, as a result, to complete or partial recovery of the structure. Figure 5.36 shows the nature of recovery of concrete structure of the same composition taken to different degrees of damage by applying compressive load (curves 1 to 3 correspond to compression stress levels $\alpha_{p_1} = 0.4$; $\alpha_{p_2} = 0.6$; $\alpha_{p_3} = 0.85$, respectively). Concrete age at the moment of testing was 60 days. After compression, the specimens were immersed in water of 22°C temperature and periodically subjected to SHF electromagnetic scanning (Figure 3.61a). The intensity of structure recovery was judged by weakening of the propagating SHF signal (Figure 3.61b).

As follows from Figure 5.36, the less the structure is damaged ($\alpha_{p_1} < \alpha_{p_2} < \alpha_{p_3}$), the faster it is restored: curve *I* is the first to cross the level, corresponding to the initial condition (hatched line). Deeper damage is not recovered at all at the expense of internal reserves (curve 3).

In concretes of an older age, potential structural reserves, as a rule, are exhausted. For such concretes, recovery processes can be induced artificially. Because concretes mainly are ion compounds,[267] the charges in such materials are carried by the electrons and the ions in an external electric field.[253,265] The application of the external load excites bound ions and thus "squeezes" nodes forming vacancies out of the crystalline lattice, which leads to the failure of interatomic bonds. Increasing the stressed–deformed level results in an increase of the concentration of such vacancies and free electrons in a unit volume of the solid.

Let a concrete specimen be under a certain stressed state (tension or compression) having intensity σ and corresponding n be the number of free ions. Let us assume that the mechanism of ions going back to the bound state is known. Then, at unchanged external conditions, the stressed level decreases by $d\sigma$ in proportion to the number of annihilated vacancies dn. Writing the differential equation for this process as

$$d\sigma = -dn\frac{\sigma}{n},$$

separating variables, and integrating, we obtain

$$\sigma_1 = \sigma_o \exp\left(-\frac{n_1}{n_o}\right), \tag{5.11}$$

where σ_o and n_o are the initial level of the stressed state and the corresponding number of free ions; σ_1 and n_1 are the final values of these parameters.

The minus sign in Equation 5.11 means a decrease in the number of crystalline lattice defects. Vacancies can be filled by a forced introduction of an ion flow into the defective volume from the outside. In order to facilitate the substitution process, the ions should have lower activation energy. Excess concentration is achieved by material impregnation with ion solutions of rare earth metals. The ion flow is formed using an electric circuit.

Ion substitution current is equal to

$$ezn, \tag{5.12}$$

where e is the electron charge and z is the ion valency. On the other hand, during time τ, the following charge passes through the electric circuit:

$$aI\tau, \tag{5.13}$$

where a is the ionic component of the conduction current and I is the current. Equating Equations 5.12 and 5.13 yields

$$n = \frac{aI\tau}{ez}. \tag{5.14}$$

We can express Equation 5.11, taking into account Equation 5.14, as

$$\sigma_1 = \sigma_0 \exp\left(-\frac{aI\tau}{ezn_o}\right). \tag{5.15}$$

The denominator of the exponential factor,

$$I_o = ezn_o, \tag{5.16}$$

is equal to the magnitude of ionic current corresponding to the initial stressed level σ_0.

Accounting for Equation 5.16, Expression 5.15 becomes

$$\sigma_1 = \sigma_o \exp\left(-\frac{I_1}{I_o}\right).$$

As seen, the process of recovery of damaged concrete structure resource by passing direct electric current through material saturated with ionic solutions has an exponential nature.

Experimental verification of this method was conducted with multiple soaking of the precompressed concrete specimens in weak lime solutions and subsequently passing a direct electric current through them. The objects tested were specimens of three concrete compositions, shaped as cubes and prisms and stored for 3 and 5 years under natural atmospheric moisture conditions. The stressed level, α_p, was varied between 0.1 and 0.9. The current was passed in long-time (for several days) and in pulsed modes and the recovery effect was achieved during the experiments. This method can find an application in renovation of buildings and construction.

During preventive maintenance or scheduled outage, parts or structures of metallic materials should not only rest, but in order to recover their initial condition, they also should be exposed to intensive recovery to compensate for service impacts. As has been noted, traditional (mechanical and thermal) as well as nontraditional (magnetic, electrical, and their combinations) methods can be applied for this purpose. The cost of their development will be generously repaid back through improvement of the failure rate, reliability, and durability. For instance, References 214 through 218 have demonstrated that keeping parts most exposed to tension in service under compression not only results in increased durability, but also compensates for the damaging influence of tension. Thermal fields have proven to have similar effect.[268]

In conclusion, derivation of the equation of state (2.96) and substantiation of the CD essence of solids open up a broad field of activity for development and introduction of effective and, most importantly, inexpensive methods of recovery of damaged structures, from the simplest mechanical to exotic combined methods.

6

Physics of Fracture

6.1 Introduction

This chapter considers physical aspects of the problem of fracture where, as many researchers believe,[5,22] the valuable solutions of many actual practical problems can be found. The main parameter that determines the thermodynamic state of a solid is the MB factor (see Section 3.7). Section 6.2 shows that the MB factor introduces stress concentration into a structure while other factors aggravate it at higher scale levels.[269–276] Failure of interatomic bonds is accompanied with an instantaneous release of internal energy; this leads to mechanical, thermal, ultrasonic, and electron effects (Section 3.11). Their theoretical substantiation and experimental proof are presented and examples of their practical application for prediction of the stress–strain state are given.[12]

Various phenomenological fracture diagrams find application in the known strength theories.[2–4,13,19] Section 6.3 shows that these theories are not in agreement with physical reality. Principles of the rigid-link fracture theory are described. It originates at the depth of the AM structure and propagates to all scale levels. Section 6.4 is dedicated to description of the mechanism of formation of a failure flow in the heterogeneous CD set of the AM bonds. Probability aspects of the deformation diagrams are considered and practical methods for their construction using the results of hardness tests of parts are discussed. Section 6.5 explains the sequence of physical processes that underlies initiation of microdefects and their coarsening as well as the formation of growth of cracks.

Section 6.6 discuses the thermodynamic conditions of growth and arresting of cracks, giving recommendations for arrangement of a structure that would hamper the process of crack propagation as much as possible or make the crack arrest for as long as possible. Obstacles in crack propagation paths can be created by passive methods, such as during structure formation at the manufacture stage, and also by active ones, i.e., through varying the shape of the internal local potential field (Figure 1.8). Section 6.7 shows how the external force and thermal fields can be employed for this purpose.[12]

6.2 Concentration of Stresses as an Inherent Property of Crystalline Structures

To solve problems associated with an increase in strength of parts and structures in operation of engineering objects or a decrease in strength, e.g., in production and processing of mineral resources, one must understand in depth the processes occurring in a structure of solids under the effect of external fields. It is also necessary to study the nature of their microscopic manifestations, followed by synthesis of elementary events of disruption of a bond to describe macroscopic properties. Available engineering experience proves this way to be efficient.[277]

Fracture should be regarded as the final stage of the process of irreversible change in the thermodynamic state of a solid associated with a catastrophic decrease in the number of interatomic bonds in the critical sections of structural members. The elementary event of fracture is none other than transformation of molecular paths of motion of bound atoms from elliptical (Figure 1.7b) to hyperbolic, which makes valent electrons leave for infinity and atoms stop interacting (Figure 1.12). Disintegration of a rotos may take place under the effect of diverse external factors. Whatever the causes of a failure, however, its mechanism remains unchanged. It is based on the relationship of equivalence (2.105) showing that any change in geometrical sizes of a rotos is a thermally dependent process. This leads to thermal dependence of all macroscopic properties (see Section 2.7).

In this connection, one should distinguish three cardinally different mechanisms of fracture. The first is low-temperature, brittle, or compresson fracture. It is realized when energy levels of rotoses located in the left ("cold") tail of the MB distribution pass the brittle fracture temperature T_c (dashed zone in distribution 1 in Figure 3.9). In Figure 1.7a it is identified with point a, i.e., bottom of the potential well, while in Figure 3.9 it is associated with vertical line *ora*. This type of failure is considered in detail in Section 1.10. At low negative temperatures, rotoses of compresson materials (metals and alloys) are prone to this type of failure. The D-phase is located behind the Debye temperature (to the right from vertical line br_b in Figure 3.9) and this makes low-temperature fracture of the CD materials (cast irons, rock, etc.) impossible.

The second mechanism is high-temperature, deformational, ductile, or dilaton fracture. It forms under conditions of excessive thermal energy (to the right from vertical line *bb* in Figure 3.9). Both compresson and dilaton materials may fail by this mechanism; the process is accompanied by high strains.

Finally, the third mechanism is a Debye-type or dislocational fracture, which is most characteristic and decisive for the majority of structural materials (described in Section 1.9). Let us recollect its nature. An external load, nonuniformly distributed over the rigid and cold (compared with dilatons) compresson set, forms a nonuniform temperature gradient (Equation 2.166).

Concentration of stresses grows from the "hot" compresson "tail" to the cold tail (e.g., in distribution 3 from point y to point z in Figure 5.10). The temperature gradient causes formation of local heat flows (Equation 1.71). Dilatons, located on the hot tail of the MB distribution (e.g., on curve 3 to the right from vertical line mH–bH, Figure 5.10), become an ideal source of heat for compressons. The first to supply heat to a compresson is a dilaton located nearby. This forms a dislocation pair (e.g., c_1–d_1 at level ω_1 in Figure 2.59). Compressons and dilatons, while acquiring and losing heat, respectively, start a coordinated opposite movement according to the constitutional diagram cbq_1 (Figure 1.14) toward the Debye temperature v_s (vertical $r_b b$).

During deformation, this process shows up in propagation of dislocations. If the temperature of neighboring dilaton and compresson become simultaneously equal to the Debye temperature, they instantaneously lose the mutual bond, according to the diagram in Figure 1.15, to form a submicroscopic local defect. If heat transfers to a compresson from remote neighbors (compared with sizes of a crystalline lattice), no defect is formed. As loading increases, concentration of defects dissipated in the bulk of a material continuously grows (a process controlled by a space component of the MB factor [Equation 2.2]) until the distance between some of them becomes so small that they start interacting. This results in initiation of a microcrack. The coalescence of such cracks forms an avalanche crack.

Separated by substantial temperature intervals, these mechanisms occur separately, as a rule. Depending on the test temperature, fracture of a material may follow any of these three fracture mechanisms. However, under certain conditions low-temperature and Debye-type, or Debye-type and dilaton, may occur simultaneously. This leads to catastrophic consequences. For example, combination of the first two mechanisms leads to occurrence of a cold-shortness phenomenon and causes numerous failures of steel structures in service;[3] its physical nature is considered in Section 1.8. The second combination may show up at high-temperature production facilities with high-temperature units such as furnaces, turbines, and other similar units.

The Debye-type fracture mechanism has two stages of development. The first is orientational (at the elastic or reversible stage oa of the deformation diagram in Figure 2.10) and the second is destructive or irreversible (region ab). AM peculiarities of these stages are considered in detail in Section 3.4. It is worthwhile to underline here that, during the first stage, the rotoses, whose energy parameters obey the MB factor Equation 2.2, which distributes them in space in a chaotic manner (Figures 2.7, 3.7, and 3.52b), do not change their sizes and shapes. Rather they orient themselves in the potential well (Figure 1.3) in the direction of the external force field (e.g., r in Figure 1.7a). Their own deformation begins only at the second stage. In accordance with Equation 2.105, this deformation can be represented as movement in the potential well upward to the Debye temperature (vertical lines bb in Figure 3.9). Safe transition of any compresson through it to the D-phase is possible only at the presence of a thermal support ΔU. At the absence of such support, disintegration of a rotos into its constitutive atoms occurs (see Section 1.9).

Structure of solids at all scale levels is characterized by heterogeneity. This property is not an abnormality or an exception to the rule, but is a rule itself. Its roots are in the CD ambiguity of rotoses whose energy distribution in a three-dimensional space is assigned by the Maxwell-Boltzmann statistics (Equation 2.2). The plot of its Boltzmann part is shown in Figure 2.1. In accordance with this, at $U(r) = kT$ the value of probability δ becomes equal to 0.37, and at $U(r) = 3kT$ it is 0.05 of the mean value. Although the major parts of rotoses have an energy close to the mean value, at a given time moment, a certain number of them always have energy that is much higher or lower than the mean value.

This means that, in any solid body (even an ideal single crystal), the state of interatomic bonds is always characterized by the energy spectrum: from $U_i = 0$ to $U_i = U_{max}$. As such, the larger the portion of the potential component of energy at a given time moment is, the higher the probability that the rotos is in the compresson phase; predominance of the kinetic component over the potential component is indicative of its being in the dilaton phase.

It follows from Equation 2.2 that the internal potential field of a solid is always heterogeneous, with a complicated spatial configuration of high peaks (one is shown in Figure 1.8) and flat valleys. If an external (e.g., homogeneous force) field having intensity σ is applied to a solid, it first affects all the abnormal regions, for example, first those having configuration $1H$ and then those with configuration 1 (Figure 5.10). That is, the effect of the concentration of internal stresses is not introduced with defects of a structure or with the initial imperfections as has been thought.[3]

Their effect is of secondary importance and is a result of natural heterogeneity of the internal potential field. This effect should be regarded as a fundamental property of crystalline structures. Thus, until now, attempts to obtain "theoretical strength" of even ideal single crystals have failed[10] and will rarely succeed in the future. Strength in tension determined experimentally is three to four orders of magnitude lower than that determined theoretically.[3] From the standpoint of the CD notions, it characterizes no more than the mean concentration of compressons per unit volume of a material at given test conditions.

The submicroscopic structure of a body consists of electromagnetic rotoses (Figure 1.6 and 1.7b), which exist as compressons at some heat content and as dilatons at others (Figure 2.17). The thermodynamic state of their set concentrated in a given volume V is determined by a composition of two distributions: compresson (C) ω_c and dilaton (D) ω_d. In accordance with the Hamilton principle,[31] the MB factor (Equation 2.2) distributes C and D in space so that the internal energy content of each elementary volume is minimum. Thus, the behavior of the CD set of the rotoses under the impact of external effects is determined by a composition of three distributions: ω_c, ω_d, and $V(x, y, z)$ where $V(x, y, z)$ is the spatial layout of the rotoses in volume of a material.

In actual solids, every cubic centimeter comprises about 10^{24} interacting rotoses. In this situation, the existence of interaction forces between the atoms should be considered as well as the potential field that forms as a continuous function in a local region of the space. To describe this function mathematically, we can use Equation 2.176, expanding it in a Taylor series.

Assume that the internal potential field can take an arbitrary shape (e.g., as shown in Figure 1.8) and designate displacement of atoms of a rotos from position of the dynamic equilibrium of a solid affected by external factors as x (in Figure 1.7b it is designated as D). Assuming the displacement to be so insignificant that $U(r) = U(r)_x$ and limiting to three terms of the expansion, we can rewrite Equation 2.176 in the form of linear homogeneous differential Euler equation

$$x^2\ddot{F} + 3x\dot{F} + 6F = 0. \tag{6.1}$$

To obtain the general form of its solution, it is necessary to make complex mathematical operations;[278] however, at $x > 0$ it becomes much simpler. This case is of the highest interest to us because deformation and fracture of C-rotoses at the Debye temperature are accompanied by an increase in their geometrical sizes (Figure 2.8b).

As shown in Kamke,[278] solution of the Euler equation (6.1) is one of the solutions, $y(z)$, where $z = \ln x$, of the following differential equation with constant coefficients:

$$\sum_{v=0}^{n} a_v \frac{d}{dz}\left(\frac{d}{dz} - 1\right)\cdots\left(\frac{d}{dz} - v + 1\right)y = f(e^z).$$

Its solutions

$$F = \begin{cases} c_1 x^{\alpha_1} + c_2 x^{\rho_1} & \text{if } \alpha_1 \neq \rho_1 \\ x^{\alpha_1}(c_1 + c_2 \ln x) & \text{if } \alpha_1 = \rho_1 \\ x^2[c_1 \cos(s \ln x) + c_2 \sin(x \ln x)] & \text{if } \alpha_1 \text{ and } \rho_1 \end{cases}$$

are the complex numbers of the type $(r + is)$. Here c_1 and c_2 are the constants, and α_1 and ρ_1 are the parameters that depend on the value of constant coefficients in the initial equation (6.1). The latter are found from the following conditions:

$$\alpha_1 + \rho_1 = 1 - a \quad \text{(I)}; \quad \alpha_1 \rho_1 = b \quad \text{(II)}. \tag{6.2}$$

Because in our case $a = 3$ and $b = 6$, then $\alpha_1 + \rho_1 = -2$ and $\alpha_1 \rho_1 = 6$.
Obtaining the second condition,

$$\alpha_1 = \frac{b}{\rho_1}, \tag{6.3}$$

and substituting it into the first, we obtain the reduced quadratic equation

$$\rho_1^2 + 2\rho_1 + 6 = 0,$$

which has two roots: $\rho_{1-1} = -1 + 2.2i$ and $\rho_{1-2} = -1 - 2.2i$. Substituting them into Equation 6.3 and multiplying denominators by conjugate complex numbers, we obtain

$$\alpha_{1-1} = -\frac{3}{2}(1 - 2.2i) \quad \text{and} \quad \alpha_{1-2} = -1 + 2.2i.$$

Therefore, the solution of the initial equation (6.1) is

$$F = x^{-1}[c_1 \cos(2.2\ln x) + c_2 \sin(2.2\ln x)].$$

Simplifying and accounting for the initial conditions, we finally find that

$$F = \frac{\sqrt{2}A}{x}\sin\left(2.2\Delta\ln x + \frac{\pi}{2}\right), \tag{6.4}$$

where A is the intensity of the local potential field (Figure 1.8), numerically equal to the maximum force which should be applied to a rotos to change its sizes in the direction of the external field per unit length, and Δ is the bond rigidity coefficient determined by Equation 2.177.

The plot of Function 6.4 at $A = $ const and $\Delta = $ const is shown in Figure 6.1. The stable configuration of a rotos persists for the first quarter of a period

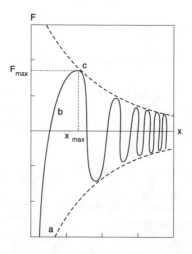

FIGURE 6.1
Dependence of force of atomic interaction on displacement of atoms, which make a rotos.

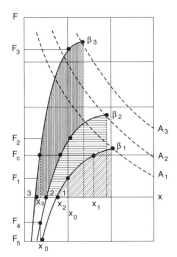

FIGURE 6.2
Variation of resistance in deformation of heterogeneous system of rotoses.

of sinusoidal oscillations (the starting region of curve *abc*). Figure 6.2 shows the ascending regions of the interaction curves at different parameters A ($A_1 < A_2 < A_3$) and Δ ($\Delta_1 < \Delta_2 < \Delta_3$). As seen from Equation 6.4 and Figures 6.1 and 6.2, the resistance of a rotos to external effects is in inverse proportion to its linear sizes. The upper field in Figure 6.2 determines attractive forces F_1, F_2, and F_3, and the lower field determines repulsive forces F_4 and F_5. Equation 6.4 and Figure 6.2 allow description and explanation, respectively, of different situations that actually occur in the CD structure of solid bodies.

In an ideal situation, no internal stresses should be formed. As such, rotoses that have a different spatial configuration (Figure 1.7b), while filling the volume, should be located symmetrically in different azimuthal planes I to V (Figure 2.7). In Figure 6.2 this situation corresponds to points 1, 2, and 3. From the energy standpoint, it is most favorable because it requires a minimum energy; however, in practice it occurs very rarely. The more common situations are those in which the distance between them is the same (e.g., x_0), in which cases noncompensated internal stresses are formed.

It is seen from Figure 6.2 that inequality $F_c < F_4 < F_5$ holds at

$$x_1 = x_2 = x_3 = x_0. \tag{6.5}$$

A repulsive force formed between the rotoses tends to give new sizes (in this case, larger ones) to an elementary volume. Internal stresses may result from technological processing (in metals) or solidification under restrained conditions (for example, in concrete[94]). To avoid them, metals are subjected to additional technological treatments in thermal or force fields.[60] In brittle materials, they are removed by applying a tentative load, which causes disintegration of part of the unfavorable bonds. In this case, a material reacts to it by changing its sizes to the opposite. For concrete, this fact finds an excellent experimental proof[48] (see also Figure 3.11).

The process of disintegration of a set of CD bonds that have different strength and deformability may occur in two cardinally different directions. The first takes place at a uniform distribution of the external load between them (horizontal straight line at level F_c) and a nonuniform distribution of strains (points x_1, x_2, and x_3). The second case occurs at uniform deformation of the bonds with different strength (vertical line x_0 in Figure 6.2), but a nonuniform distribution of an external load (points F_1, F_2, and F_3). Consider both mechanisms in more detail.

As the external load increases (horizontal line F_c rises), fracture begins from the hot rotoses and propagates to the side of the cold rotoses (bonds β_1, β_2, and β_3 are involved in succession). As seen, this mechanism can occur under conditions of free and independent deformation of dissimilar bonds $x_1 > x_2 > x_3$; removing the hot rotoses from the work should be accompanied by a substantial increase in their linear sizes at the initial stages of loading. In a pure form, such conditions may take place only in the case of successive joining of the rotoses of different strength. In actual materials, such a situation is difficult to imagine (Figure 2.7). The existence of strains of opposite signs at the initial stages of loading of brittle materials is reliable proof of this fact.[48] Dashed regions under the curves show a share of the energy released at each event of disruption of interatomic bonds. As seen, the deformation mechanism of fracture should be accompanied by an increase in the energy released at each elementary fracture. Section 3.11 shows that acoustic emission methods do not confirm this fact (Figure 3.71 and comments on it).

The second mechanism is characterized by a nonuniform distribution of the external load over interatomic bonds. Nonuniformity results from a different intensity of local potential fields when distance between the interacting rotoses changes by the same value. The effect of the concentration of internal stresses in interatomic bonds is an objective reality that reflects the energy CD heterogeneity of a structure at the AM level.

Fracture of such materials is based on a principle from a cold and brittle link to a ductile one. Realization of this principle in a pure form takes place in the case of a parallel connection of different-strength bonds. Readily imagined at the AM level (Figure 2.7), such a system consists of two neighboring crystalline surfaces connected to each other by a set of interatomic bonds of the type Δ_1 to Δ_3. At a certain value of the external load, the most rigid bonds (e.g., F_3) are the first to disintegrate. Force F_3 will be redistributed between the remaining bonds (e.g., F_1 and F_2) in accordance with their rigidity parameters Δ_1 and Δ_2. As such, bond F_2 takes up a larger part of the load and F_1 a smaller part. This is accompanied by a jumplike increase in the displacement of a bundle of atoms that have remained in a given place. That is, in restrained deformation (parallel connection of rotoses) the process of failure of the bonds occurs in the direction from the strongest and most rigid link to the weak but ductile one.

This behavior of the system of electromagnetic rotoses (Figure 1.6) under loading is attributable to a fundamental property of the local potential field, which is electric in its physical nature (see Equations 1.15 to 1.26). The work done by force **F** in a given field in any direction does not depend on the shape of the path but is determined only by the coordinates of its beginning and end.

Indeed, rewriting Equation 6.4 in the form of

$$Fx = \sqrt{2}A\sin\left(2.2\ln x + \frac{\pi}{2}\right),\qquad(6.6)$$

we notice that its left part is the Ω-potential of an individual rotos (Equation 1.53), which performs work ω on the separation of constituent atoms of this rotos on distance x in the local potential field (Equation 1.26). As seen, loading the CD AM system by applying the constant external force $F = const$, does not allow a uniform distribution of internal stresses at the submicroscopic level to occur.

Regel and Slutsker's study[62] presented direct proof that a loaded body contains bonds that are substantially overloaded with respect to the mean level of stresses. They observed displacement and distortion of contours of infrared (IR) absorption spectra of high-strength polymers under the effect of tension. As the frequency of thermal oscillations of atoms of a solid lies in the IR region, the external mechanical load, while changing tension of interatomic bonds (Figure 2.8b), causes a distortion of the IR absorption spectrum. A typical character of distortion of the IR spectrum is shown in Figure 6.3, where v is the frequency of the IR spectrum expressed in cm^{-1}, and f is the degree of absorption expressed in percent.

Prior to loading, the absorption band is symmetrical (curve a in Figure 6.3). Under the effect of loading, its contour is deformed toward low frequencies (curve b) and the center of the band shifts to the same side (from positions 1 and 2). It was established that the common areas of the initial a and deformed b bands did not differ from each other. This means that, under the effect of external forces, almost all interatomic bonds are stressed; however, the degree of their excitation is fundamentally different. This shift of the

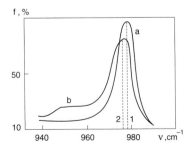

FIGURE 6.3
Variation of the absorption band in the IR spectrum of oriented propylene under the effect of tension: $a - \sigma = 0$; $b - \sigma = 80$ MPa. The curve restores its shape after removal of load.

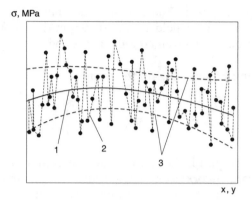

FIGURE 6.4
Stochastic character of distribution of compressive stresses across the section of a concrete specimen: 1, approximated field of stresses; 2, local values; 3, uncertainty boundary.

center of the band indicates that the major part (80 to 90%) is loaded comparatively weakly, whereas the rest take up the major part of the external load. Their thermal oscillations shift further into the region of low frequencies causing deformation of the contour of the absorption band (formation of a low-frequency tail). Although the number of stress concentrators in the total number of the bonds is small, they play a decisive role in formation of fracture centers in the initial period of loading (see comments on Figure 5.34).

Sammal[94] studied transformation of stressed–deformed state of concrete prisms in compression. At some stages of loading, he observed a concentration of stresses inside a section normal to the compressive force that amounted to 150% of the mean value. Based on analysis of numerous measurements, he concluded that the field of stresses in concrete during compression at any time moment is very sophisticated and has a stochastic nature. As an example, Figure 6.4 shows one of the forms of the distribution of stresses in the section of a concrete specimen.[94] Following the Kolmogoroff-Smirnoff test,[279] he has shown that stresses in the section are distributed in accordance with the normal law. Coordinates of the transverse section, x and y, are plotted along the axis of abscissas, and local stresses σ, MPa, are plotted along the axis of ordinates.

6.3   Rigid-Link Nature of Fracture

It follows from the equations of state of a rotos (1.53 and 6.6) that an inverse proportional relationship exists between its resistance **F** and deformability x or r (Figure 2.6). The product of these values expresses the work of deformation in an arbitrary direction r (Figure 1.7). Thus, the Ω-potential in any

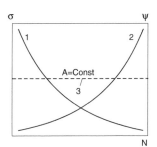

FIGURE 6.5
Relationship between strength (1) and volume
deformability (2) of a set of interatomic bonds.

of the forms of the equation of state of a solid (Chapter 2) is none other than
mechanical work in the volumetric representation performed by consumption of the potential part of energy of a body, i.e., $\Omega(r) = PV$. In accordance
with Equation 1.71, any change in potential energy is replenished through
a decrease in the kinetic, $\Omega(T)$. That is, equations of state of a rotos (6.6) and
body as a whole (Equation 2.96) determine the balance between different
parts of the overall internal energy of the body and thus express the law of
energy conservation $\Omega(r) = \Omega(T)$ (Equation 2.164) in its most general form.
This is the physical nature of the thermodynamic equation of state at all
scale levels.

Suppose that we have managed to sort out the set N of interatomic bonds
by indices of deformability and strength. Arranging them in a decreasing
(increasing) row, we obtain a two-branch diagram (Figure 6.5). As seen, to
meet Condition 2.164, it is required that strength σ (curve 1) and volumetric
deformability ψ (curve 2) of interatomic bonds should be varied in an
inversely proportional manner.

Inverse proportionality between resistance **P** and deformation V of the
system of interatomic bonds can be reliably traced at micro- (Figure 2.6) and
macrolevels (Figure 3.34) and was proven experimentally for metallic[3] and
nonmetallic[48] materials. A good illustration of this is the well-known science
of concrete principle of water–cement ratio[280] (Figure 6.6). An increase in the
amount of water in a concrete mixture leads to a separation of grains in
the cement stone, causing a decrease in the number of coagulation bonds
between the cement grains. Because reaction of concrete to a short-time effect
of load is determined not only by the number but also by the rigidity of such
bonds (Figure 6.2), an increase in the water–cement ratio is accompanied by
a corresponding decrease in strength σ and deformability ψ.

According to the MB factor (Equation 2.2), a given volume always contains
a certain number of bonds with the same resistance and deformability. In
Figure 6.5 they are located at the same energy level (defined by horizontal
lines parallel to line 3). By grouping them using one of the parameters (e.g.,
strength), we obtain the differential distribution law (curve 1 in Figure 6.7).
Curve 2 shows the corresponding change in deformability, and $\delta(\sigma)$ represents probability of the strength.

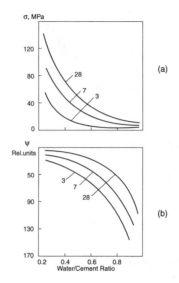

FIGURE 6.6
Dependence of strength and deformability of concrete on the water–cement ratio for concrete aged 3, 5, and 28 days.

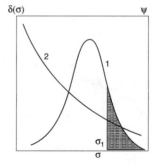

FIGURE 6.7
Rigid-link sequence of fracture of heterogeneous CD system of inter-atomic bonds.

If the external load (e.g., tensile) is applied to such a system, under the effect of stress concentration (Figure 6.2), fracture begins at the base of the right branch of distribution 1, i.e., where the bonds with the highest rigidity are concentrated (dashed zone in Figure 6.7). In Figure 1.10b, they correspond to the crystalline lattice parameter r_{a_1} and level of internal energy U_1. More complying bonds (with parameter r_{a_3} and energy U_3 in the same figure), taking up phonon radiation from the right-side bonds that failed (Equation 3.35), are deformed.

Assume that the impact of certain load σ on the system of N bonds results in failure of those whose strength is higher than σ_1, i.e., $\sigma I > \sigma_1$. In this case, the dashed region under the differential distribution curve (Figure 6.7) determines the probability of accumulation of structural damage. At the macroscopic level, it corresponds to that part of the section that failed.[36,62] The zone to the left from vertical line σ_1 characterizes the area of the section that bears the external load F_1.

It turns out that the right branch of the differential MB distribution 1 (different versions designated by numbers 6 to 14 in Figure 2.47) is a sort of regulator of deformability of the entire AM system. In addition, it controls the process of accumulation of damage. Elementary bonds located on its ascending branch are subjected to deformation during loading and serve as a natural reserve that takes up the load from the failed part of the bonds. This process ends with progressing deformation and fracture at the moment when the areas under curve 1 (to the left and right from moving to the right vertical line σ_1) become equal to each other. At the symmetrical shape of curve 1, this moment coincides with the mathematical expectation of the differential distribution (positions 11 to 14 in Figure 2.47).

This is the physical nature of the mechanism of deformation and fracture of solids. Briefly, it can be formulated as follows: there is no deformation without fracture and fracture is impossible without deformation. Successive connection of new bonds with new bonds is in fact the brake that prevents the AM system from avalanche deformation and fracture. In the equation of state (2.65), its analytical expression is represented by the Ω-potential.

Current engineering practice widely applies the theory of the "weakest link," which looks natural and logical because it postulates that, for microworld, the strength of a complex system is determined by the strength of its weakest component at the macrolevel. This postulate underlies the statistical theory of strength suggested by Weibule,[281] Kontorova and Frenkel,[106] and Kontorova.[107,282] However, because it is mathematically faultless, it takes no account of the effect of interaction of submicroscopic structural elements in actual solid bodies (Equation 2.96). For this reason, the theory of the weakest link is a mathematical abstraction that has little to do with the real picture of fracture of solids. As opposed to this theory, substantiation of the the the "most rigid link" described here follows from peculiarities of AM interaction and, therefore, provides a more exact description of the fracture mechanism formed at the submicrolevel.[283]

A sufficient number of experimental facts have been accumulated to date that provide convincing proof of reliability of the theory of the rigid link at all scale levels. In most cases, mechanical, electric, thermal, and other properties of the same solid body are studied separately and often not correlated to each other, although all of them are determined by the same set of physical parameters: crystalline lattice constant r_a, charge e, mass m, and valence z of atoms that make a rotos (see Equations 1.14 to 1.30 and Figures 1.1 to 1.14). For example, electric as well as mechanical strength of crystals varies proportionally with the energy of the lattice (Equation 2.148.I). Vorobiov and Borobiov[92] studied a simultaneous influence of mechanical load and breakdown voltage. It was established that electric charges in the deformed crystals of rock salt initiate and propagate along the boundaries of mechanically stressed zones, i.e., particularly in locations where the gradient of concentration of internal stresses has the highest value.

Lermit[284] was one of the first to use the acoustic emission method for investigation of the process of fracture of concrete specimens subjected to compression. He noted a peculiarity of acoustic emissions: as the load approaches the fracture limit, the frequency of the acoustic emission signal increases and amplitude decreases. A similar relationship was noted in experiments reported by Nogin et al.[159] An increase in the frequency of acoustic pulses is attributed to passing one of the ends of the MB factor (Figure 3.50 for nonmetals and Figure 5.34 for metals) through the θ-temperature. A decrease in amplitude with loading proves that fracture begins near the base of the differential MB distribution of rigid bonds and propagates along one of its ascending branches (Figure 6.5). The stochastic character of distribution of the dissimilar CD bonds in space (Equation 2.2) leads to successive initiation and development of a large number of submicrocracks dissipated in the bulk of a material.[10]

Figure 6.8 shows variations in amplitude Δu of acoustic signals and their quantity n (pulse/sec) per unit time depending on the level of tensile stresses α_s, in pulling out inserted anchors from concrete (see diagram in Figure 4.5b). All the investigated compositions of concrete exhibit a decrease in amplitude (curve 1) and increase in n (curve 2) with an increase in the level of stressed state, α_s (see Figure 3.58). This is attributable to an increase in the number of elementary events of disruption of the bonds per unit time and serves as convincing proof that fracture of rock materials begins from the most brittle bonds and propagates in a direction of a decrease in their energy consumption (Figure 6.5).

The described manner of variation in the acoustic emission parameters during intensification of loading is universal, found in heavy and light concretes,[159] various metals,[170,171] and natural rocks.[172] For example, in Novikov and Weinberg's study,[171] the test objects were steel 45 and aluminum alloys. The decrease in amplitude of acoustic emission signals at substantial

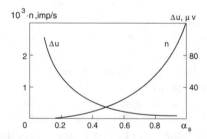

FIGURE 6.8
Variation in amplitude, Δu of acoustic signals and number of their pulses n per unit time, depending on the level of stress–strain state of tension in pulling out inserted anchors from a concrete mass following the diagram in Figure 4.5b.

deformations was observed and attributed to a decrease in the energy of a pulse during significant levels of plastic deformation.

Theoretical notions elaborated in previous chapters as well as obtained experimental results allow reconstruction of the mechanism of fracture of materials at the submicroscopic level. At the initial stage of loading, due to the effect of concentration of stresses, the first to take up the load in metals are the coldest and, therefore, most rigid compresson bonds. They are located on the left end of the MB distribution (positions 1, 3, and 4 in Figure 2.36), i.e., in a dashed zone in Figure 6.7. Loading immediately causes the orientation processes to occur (Section 3.4). This is accompanied by redistribution of the internal energy between its potential and kinetic parts through phonon radiation. The components of the CD set have different sensitivities to it. Dilatons and compressons located near the Debye temperature θ are types of heat traps; they have maximum thermal conductivity and minimum heat resistance (Figure 2.27).

Inflow of kinetic energy to these zones from the closest surroundings leads to differentiation of the compresson part of the volume with respect to deformability. Continuously emitting phonons, one of its parts, i.e., cold, goes to the bottom of the potential well while the other part, absorbing phonons, goes to its surface (Figure 1.7). As the load intensifies, they both inevitably turn out to be in the critical state (position 7 in Figure 2.36): the first acquire the cold-shortness temperature T_c (the left dashed zone under 7) while the second acquire the Debye temperature θ (the right end of distribution 7). When they fail, a fundamentally different share of internal energy is released by thermal and emission radiations: in the first case it is equal to $U(r_a)$ and in the second it is equal to $1/9U(r_a)$ (Equation 1.46). The left cold end of the MB factor approaches the T_c boundary and its maximum is near the Debye temperature. For this reason, an insignificant amount of super-rigid C bonds enter the T_c zone, while a massive failure flow is formed at the Debye boundary.

As a result of these microfractures, the crystalline structure of a material is saturated with local submicrodefects. At the beginning, they are dissipated in microvolumes, being comparatively rarely found in the bulk of material (because both are located in the extreme right-hand region of the statistical distribution shown in Figure 6.7). As the external load increases, compressons and dilatons become involved in the fracture process. Giving up their kinetic energy to deformation of the conjugate compressons and forming the dislocation pairs with them (see Section 2.11), they cool and begin taking up a part of the external load (region *ab* in the diagrams shown in Figures 2.10 and 2.22), not approaching the Debye temperature from the left, as the compressons do, but from the right (Figure 2.36). As the probability of failures of the bonds continuously grows (movement from the right to the left is seen in distribution 1 in Figure 6.7), the concentration of local submicrodefects in volume increases. Further loading leads to coalescence of defects and formation of microcracks.

Therefore, loading a material leads to redistribution of the overall internal energy between its potential and kinetic components. At the AM level this affects the shape of the MB factor, while in macroexamination it impacts the character of distribution of hardness. Investigation into the mechanism of fracture of concretes in tension and compression using methods of a local mechanical impact (rebound of the indenter and plastic strains caused by spherical indenters) proves the validity of the rigid-link fracture concept.[55]

6.4   Probability and Thermodynamic Aspects of the Deformation Diagrams

Depending on the type of a material (metallic or nonmetallic; see Section 2.6) and conditions of heat transfer, interatomic bonds of the same body may be in the compresson or in the dilaton state (see Section 1.8). The force of binding between two neighboring atoms, F_i, in both states (see DC, q_2dc and CD, cbq_1 regions of the constitutional diagram of a rotos in Figure 1.14) may have any value in a range of 0 to F_{max} where F_{max} is found from Equation 1.52 at $r = r_a$ (Equation 1.48.I). For a given material, it is determined by the Debye temperature θ, which, in turn, depends on atomic parameters m, z, and e (Equation 1.87). At $\theta = $ const, F_i is always lower than the F_{max}. This means[279] that F_i can be thought of as a certain function,

$$\delta(F_i < F_{max}) = f(F),$$

while the $f(F)$ function becomes the integral law of distribution of the F_i value. Derivative of the

$$\delta(F) = \frac{df(F)}{dF}$$

function is the density of probabilities or the differential distribution law, which is the MB factor for a solid body (Equation 2.2). It distributes the energy of interatomic bonds following the logarithmic law within a given volume of a solid (positions 8 to 15 in Figure 1.14).

Only super-pure materials consist of similar chemical elements; as a rule, structural materials contain a few of them. In this case condition $\theta = $ const is no longer valid because each of these materials has its own Debye temperature. As such, the mean Debye temperature is calculated as

$$\theta = \sum_{i=1}^{n} \theta_i / n, \tag{6.7}$$

where θ_i is the temperature of a chemical element that is part of the structure of a body and n is their number.

The internal potential field of such materials is not as simple and uniform as shown in Figures 1.7a and 1.8. A schematic of their flat section is shown in Figure 1.10b, where 1, 2, and 3 are the possible realizations of arrangement of the structure consisting of n elements. Therefore, the basic MB distribution is superimposed by another arrangement or structural type, having stochastic character.

This gives rise to a question about how the macroscopic realization, i.e., deformation diagram $\sigma = f(\varepsilon)$ (Figure 2.22) forms out of such multifactorial submicroscopic heterogeneity. To answer this, let us consider a material system consisting of N interatomic bonds, which is subjected to tension under the condition of a dominating Debye fracture mechanism (Section 6.2).

Let dependence of the tension resistance force of ith rotos on strain ε be set as

$$F_i = F_i(\varepsilon) \quad \text{and} \quad 0 \le \varepsilon \le \varepsilon_i, \tag{6.8}$$

where ε_i is the critical value of strain at which $F_i = F_{max}$. It is determined as $\varepsilon = dr/r$, where the absolute values of r are found from Equation 1.14 and their variations are shown in Figure 1.7b. The range of ε in Equation 6.8 corresponds to region $r_a - r_b$ on the axis of abscissas in Figure 1.7a, where point r_a corresponds to point a (the bottom of the well) and r_b corresponds to the inflection point b of the potential curve.

At $\varepsilon = \varepsilon_i$, the transition of the ith compresson to the dilaton phase or its fracture takes place. The former occurs under conditions of thermal support while the latter takes place under conditions of thermal insufficiency (see Equation 1.82 and Figure 1.18). However, in both cases a rotos stops working because in the dilaton phase it cannot bear the tensile load; it just freely deforms. Assume that $F_i(\varepsilon) \ne F_j(\varepsilon)$ at $I \ne j$. This means that, in accordance with the MB distribution (Equation 2.2), the compressons, having the same elastoplastic characteristic (Figure 6.9), may be situated at a different depth in the potential well. (Position of horizontal line U_1 in Figure 1.7a is determined by a local temperature at a given time moment.) Its shape is determined by Equation 1.80 and is identified with region dcb of the ascending branch of the diagram in Figure 1.14.

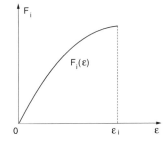

FIGURE 6.9
Elastoplastic characteristic of a compresson–dilaton rotos.

We can arrange values of $F_i(\varepsilon)$ in an increasing order ε_i, i.e., $\varepsilon_1 < \varepsilon_2 < \cdots \varepsilon_N$. In a range of $0 \leq \varepsilon \leq \varepsilon_1$, the value of the external load will be equal to the sum of all resistances of the compresson bonds, i.e.,

$$\sigma_1(\varepsilon) = \sum_{i=1}^{N} F_i(\varepsilon).$$

Because that is in the unloaded state at $\varepsilon = 0 \rightarrow (-F_i) = 0$, the bonds are not stressed and no residual stresses are found between them. If, in transition through point $\varepsilon = \varepsilon_1$, the compresson bond F_i is disintegrated, its share of the external load is redistributed among the closest neighbors. As a result, in a range of $\varepsilon_1 < \varepsilon < \cdots \varepsilon_2$ we have

$$\sigma_2(\varepsilon) = \sum_{i=2}^{N} F_i(\varepsilon),$$

and in a range of $\varepsilon_{j-1} < \varepsilon < \varepsilon_j$ we have

$$\sigma_j(\varepsilon) = \sum_{i=j}^{N} F_j(\varepsilon), \text{ etc.}$$

Consider behavior of the $\sigma(\varepsilon)$ in passing through point ε_j, which is critical for jth compresson. To the left of it, i.e., in a zone of

$$(\varepsilon_j - 0), \qquad \sigma(\varepsilon_j - 0) = \sum_{i=j=1}^{N} F_i(\varepsilon_j - 0),$$

while to the right,

$$\sigma(\varepsilon_j + 0) = \sum_{i-j}^{N} F_i(\varepsilon_j + 0).$$

Note that

$$\sigma(\varepsilon_j - 0) < \sigma(\varepsilon_j) \qquad\qquad\qquad (6.9)$$

and

$$\sigma(\varepsilon_j) < \sigma(\varepsilon_j + 0) \qquad\qquad\qquad (6.10)$$

because elastoplastic characteristics $F_{ij}(\varepsilon)$ are the growing functions (Figure 6.9). It can be concluded from Equations 6.9 and 6.10 that

$$\left.\begin{array}{c} \displaystyle\sum_{i=j}^{N} F_i(\varepsilon_j - 0) = \sum_{i=j}^{N} F_i(\varepsilon_j) - \alpha \\[4mm] \displaystyle\sum_{i=j=1}^{N} F_i(\varepsilon_j) = \sum_{i=j=1}^{N} F_i(\varepsilon_j + 0) - \beta \end{array}\right\} \tag{6.11}$$

where α and β are the numbers larger than zero. It can be written from Equation 6.11 that

$$\sum_{i=j=1}^{N} F_i(\varepsilon_j + 0) = \sum_{i=j=1}^{N} F_i(\varepsilon_j) + \beta = \sum_{i=j}^{N} -F_i(\varepsilon_j) + \beta = \sum_{i=j}^{N} F_i(\varepsilon_j - 0) + \alpha - F_i(\varepsilon_j) + \beta.$$
$$\tag{6.12}$$

With a decrease in an interval surrounding point ε_j,

$$\sum_{i=j=1}^{N} F_i(\varepsilon_j + 0) \rightarrow \sigma(\varepsilon_j + 0), \quad \text{and} \quad \sum_{i=j}^{N} F_i(\varepsilon_j - 0) \rightarrow \sigma(\varepsilon_j - 0).$$

As a result, Relationship 6.12 yields

$$\sigma(\varepsilon_j + 0) = \sigma(\varepsilon_j - 0) + \alpha + \beta - F_i(\varepsilon_j).$$

The last three terms represent a law of redistribution of stresses from the failed rotoses between the remaining ones.

Depending on the sign of trinomial $\gamma = \alpha + \beta - F_i(\varepsilon_j)$, there can be three options: if $\gamma > 0$, resistance increases near point ε_j, at $\gamma < 0$ it decreases, and local maximum or minimum is observed when $\gamma = 0$. Within the frames of the CD concept of a solid, the first case has no physical meaning, the second corresponds to the transition of compresson to the dilaton phase, and the third corresponds to their failure.

Transition from the jth interval $\varepsilon_{j-1} < \varepsilon < \varepsilon_j$ to the $(j + 1)$-th one $\varepsilon_j < \varepsilon < \varepsilon_{j+1}$ occurs without mechanical jumps owing to the smoothness of the process of redistribution of the internal energy and enormous scale-time distances that separate the AM and macroscopic levels, although it is accompanied by thermal, ultrasonic, and electron radiation (see Section 3.11).

According to Equation 6.7, any material consists of a wide variety of elementary bonds. To simplify considerations, we can divide a range of $(0 - \varepsilon_n)$ into k equal intervals of $\varepsilon_k = \varepsilon_n/k$ size so that $(j - 1)\varepsilon_k < \varepsilon < j\varepsilon_k$ holds for the *j*th of them. Designate the number of elementary bonds that fracture in the *j*th interval as m_j. It is apparent that

$$\sum_{j=1}^{k} m_j = N.$$

Then the number of the effective bonds in the *j*th interval is equal to

$$M_j = N - \sum_{i=1}^{j-1} m_i \quad \text{or} \quad M_j = \sum_{i=1}^{k} m_i - \sum_{i=1}^{j=1} m_i = \sum_{i=j}^{k} m_i.$$

Thus, we obtain the polygon of frequencies M_j. Taking into account an enormous scope of sampling and continuity of the AM structure, the integral distribution can be written as

$$M(\varepsilon) = \int_{\varepsilon}^{\varepsilon_n} m(\varepsilon)d\varepsilon$$

where $m(\varepsilon)$ is the frequency of failures of the bonds. It is clear that

$$M(0) = \int_{\varepsilon}^{\varepsilon_n} m(\varepsilon)d\varepsilon = N; \quad M(\varepsilon_n) = 0.$$

Because $M(\varepsilon)$ is a diminishing function (Figure 6.10), then $dM(\varepsilon) \leq 0$.

Dependence of the macroscopic resistance force to tension per unit area of the cross section of a body, σ, on strains, ε, can be plotted in a similar

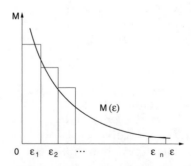

FIGURE 6.10
Rigid-link sequence of failures of CD rotoses.

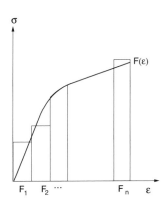

FIGURE 6.11
Variations in microscopic forces of resistance to tension.

way; in fact, M_j forces affect in the *j*th interval. Then its mean value within this interval is

$$\sigma = \sum_{i=j}^{k} \frac{m_i F_i(\varepsilon)}{M_j}.$$

Appropriate statistical processing yields $\sigma = \sigma(\varepsilon)$ (Figure 6.11), where the resistance is

$$\sigma(\varepsilon) = \sum_{i=j}^{k} m_i F_i(\varepsilon) \quad \text{or} \quad \sigma(\varepsilon) = \int_{\varepsilon}^{\varepsilon_n} m(\varepsilon) F(\varepsilon) d\varepsilon.$$

Hence,

$$\sigma(\varepsilon) = F(\varepsilon) \int_{\varepsilon}^{\varepsilon_n} m(\varepsilon) d\varepsilon = F(\varepsilon) M(\varepsilon), \qquad (6.13)$$

where $F(\varepsilon)$ is the statistically processed value

$$F_i = \sum_{i=j}^{k} m_i F_i(\varepsilon) \bigg/ \sum_{i=j}^{k} m_i.$$

To test $\sigma(\varepsilon)$ to extremum, let us take the derivative and equate it to zero:

$$d\sigma(\varepsilon) = dF(\varepsilon) \int_{\varepsilon}^{\varepsilon_n} m(\varepsilon) d\varepsilon - F(\varepsilon) m(\varepsilon) = 0,$$

from where

$$dF(\varepsilon) / F(\varepsilon) = m(\varepsilon) \bigg/ \int_{\varepsilon}^{\varepsilon_n} m(\varepsilon) d\varepsilon.$$

From this equation, expressing

$$F(\varepsilon) = dF(\varepsilon) \int_{\varepsilon}^{\varepsilon_n} m(\varepsilon) d\varepsilon / m(\varepsilon)$$

and substituting it into Equation 6.13 yields

$$\sigma_{\max}(\varepsilon) = \frac{dF(\varepsilon)}{m(\varepsilon)} \left[\int_{\varepsilon}^{\varepsilon_n} m(\varepsilon) d\varepsilon \right]^2. \tag{6.14}$$

Let the cross-sectional area of a specimen containing N interatomic bonds be equal to S_0. Assume that each of such bonds occupies the S_i part of this area. Then, $S_0 = S_i N m(\varepsilon)$ is a distribution of the same elements grouped by deformability property (curve 2 in Figure 6.7) over the plane; in fact, $m(\varepsilon) = N$. Assuming that $S_i = \text{const}$, we can consider that

$$m(\varepsilon) = S_0. \tag{6.15}$$

Loading the specimen to a level of ε yields

$$S_0 = \int_0^\varepsilon m(\varepsilon) d\varepsilon + \int_\varepsilon^{\varepsilon_n} m(\varepsilon) d\varepsilon, \tag{6.16}$$

where

$$S_f = \int_0^\varepsilon m(\varepsilon) d\varepsilon$$

is the fractured part of the section and

$$S_a = \int_\varepsilon^{\varepsilon_n} m(\varepsilon) d\varepsilon$$

is its effective part. We can rewrite Equation 6.14, accounting for the designations of Equations 6.15 and 6.16, as

$$\sigma_{\max}(\varepsilon) = \frac{dF(\varepsilon)}{S_0} S_a^2. \tag{6.17}$$

It can be seen that resistance to tension is directly proportional to the rate of loading (derivative $dF(\varepsilon)$ is given in the numerator), inversely proportional to the cross-sectional area of the specimen (scale factor), and determined by the processes of accumulation of damages in the structure,

$$S_a^2.$$

Expression 4.11 is a thermodynamic analogue of the statistical equation of strength (6.17).

If frequency distribution $m(\varepsilon)$ obeys the normal law, then

$$m(\varepsilon) = \frac{N}{\sqrt{2\pi}S} \exp\left[-\frac{(\varepsilon - \bar{\varepsilon})^2}{2S^2}\right]. \tag{6.18}$$

Integration of Equation 6.18 yields

$$\int_{\varepsilon}^{\varepsilon_n} m(\varepsilon)\,d\varepsilon = \frac{N}{2}\left[1 - \Phi\left(\frac{\varepsilon - \bar{\varepsilon}}{S}\right)\right], \tag{6.19}$$

where

$$\bar{\varepsilon} = \sum_{i=1}^{k} m_i \varepsilon_i \bigg/ \sum_{i=1}^{k} m_i = \sum_{i=1}^{k} \sigma_i = \varepsilon_i; \qquad S^2 = \sum_{i=1}^{k} \sigma_i(\bar{\varepsilon} - \varepsilon_i)^2; \qquad \sigma_i = \frac{m_i}{N};$$

$$\Phi\left(\frac{\varepsilon - \bar{\varepsilon}}{S}\right) = \frac{2}{\sqrt{2\pi}} \int_{0}^{l} \exp\left(-\frac{\tau^2}{2}\right) d\tau$$

is the integral of probabilities in which the upper limit is equal to

$$l = \frac{\varepsilon - \bar{\varepsilon}}{S}.$$

As was assumed,

$$\frac{2}{\sqrt{2\pi}} \int_{0}^{\varepsilon} \frac{\varepsilon - \bar{\varepsilon}}{S} \exp\left(-\frac{\tau^2}{2}\right) d\tau = 1.$$

Substituting Equations 6.18 and 6.19 into Equation 6.14, we have

$$\sigma_{max}(\varepsilon) = \frac{\sqrt{2\pi}}{4} NS(\varepsilon)\exp\frac{(\varepsilon-\bar{\varepsilon})^2}{2S^2(\varepsilon)} dF(\varepsilon)\left[1-\Phi\left(\frac{\varepsilon-\bar{\varepsilon}}{S}\right)\right],$$

which is the maximum of Function 6.13. Writing Equation 6.13 accounting for Equation 6.19 and noting that $\sigma(\varepsilon) = S_0\sigma$, we find an analytical expression for deformation diagram $\sigma = f(\varepsilon)$:

$$\sigma = \frac{1}{2}F(\varepsilon)\rho\left[1-\Phi\left(\frac{\varepsilon-\bar{\varepsilon}}{S}\right)\right] \qquad (6.20)$$

where $\rho = N/S_0$ is the density of a material.

Condition of maximum (6.14) for the normal distribution law is

$$\frac{dF(\varepsilon)}{F(\varepsilon)} = -\frac{2}{\sqrt{2\pi}S}\frac{\exp\left[-\dfrac{(\varepsilon-\bar{\varepsilon})^2}{2S^2}\right]^2}{1-\Phi\left(\dfrac{\varepsilon-\bar{\varepsilon}}{S}\right)}.$$

It follows that the position of maximum of Function 6.20 is determined by a superposition of distributions $F(\varepsilon)$ and $m(\varepsilon)$ (curves 2 and 1 in Figure 6.5). Thus, we have that macroscopic properties of a material, such as strength σ_{max} and deformability ε_{max}, depend not only on the quantitative values of statistical characteristics of submicroscopic bonds $F(\varepsilon)$ and $m(\varepsilon)$ but also on their relative location in the volume. This results from the effect of the MB factor (Equation 2.2) containing a space coordinate on the stressed–deformed state (Equation 2.136). A maximum exists inside an interval of $(0 - \varepsilon_n)$ only in a case where distribution $F(\varepsilon)$ has a local extremum, which should be located in the region of diminishing function $F(\varepsilon)$.

If function $F(\varepsilon)$ continuously grows in a local region (In Figure 1.10b this case corresponds to the movement along the axis of abscissas from the right to the left from parameter r_{a_3} to r_{a_1}), then the extremum is achieved at the end of the interval. This means that the accumulation of damage and progressing development of a microcrack occur here. If this function continuously diminishes (in Figure 1.10b, the movement is from position r_{a_1} to r_{a_3}), the local extremum for $\sigma(\varepsilon)$ shifts to the left from point $\varepsilon = \bar{\varepsilon}$, and a microscopic defect enters a plastic or thermodynamic trap and thus no longer develops.

In general, if a local maximum of $F(\varepsilon)$ shifts to the right (left) from $\bar{\varepsilon}$, the maximum of $\sigma(\varepsilon)$ is achieved at a certain point located more to the right (left) from $\bar{\varepsilon}$. As for a local minimum of $F(\varepsilon)$, which can be associated with point r_{a_3} when the depth of the potential wells increases to the right and to left of it, the moment of achieving the $\sigma(\varepsilon)$ maximum depends on the steepness of fronts of function $F(\varepsilon)$ and its variance value (positions 8 to 15 in Figure 1.14).

Suppose that we have managed to separate a material into compresson–dilaton links, measure their physical–mechanical properties, and count and group them by strength and deformability. Then, using these data, we can figure out the differential or integral distribution (Figures 6.5 and 6.7). It is this sequence of elementary operations that underlies the rigid-link fracture mechanism (Section 6.3). Plot $\sigma = f(\varepsilon)$, which is usually obtained as a result of strength tests of specimens, is one of the natural forms of expression of the integral law of distribution of strength and deformability properties of microvolumes of a material in the force fields.

Objectively, deformation diagrams $\sigma = f(\varepsilon)$ (Figures 2.10, 2.22, 2.50, etc.) are cumulative curves[285] with integral parameters $\sigma = f_1(\sigma_i)$ and $\varepsilon = f_2(\varepsilon_i)$ plotted along the axis of ordinates. Strain ε depends on the quantity of CD bonds failed in a unit volume of the material, and determines the probability of their failure at a given level of external overloading (an accumulated particular feature, according to Dlin[223]).

Many investigators relate strain ε with the integral effect of cracking. For example, while studying the stressed–deformed state of concrete on specimens with grooves of different shapes, Freudental[286] concluded that the redistribution of stresses in the zone of their concentration occurs not because of plastic deformations but because of development of microcracks. He has shown that their propagation can be temporarily arrested by thermodynamic traps (Figure 1.10b). The experimental data presented in this work show that concentration of internal stresses (up to 150%) occurs at the cold dilatons from the very initial stages of loading. These dilations are located on the left end of distribution 8 in Figure 2.47 that proves the validity of the rigid-link diagram of fracture of actual solid bodies (Sections 6.2 and 6.4). Conclusions made in Reference 286 indicate that ε is a statistical value whose integral value is formed as a result of failure of rigid bonds.

Studies by the Swedish Institute for Cement and Concrete[149] further confirm these conclusions. These studies give evidence of existence of the effect of retardation of cracks by the hot dilaton traps (the right-hand end of distributions 1 to 4 in Figure 3.9).

Many different equations for analytical description of deformation diagrams were suggested in the course of developing calculation methods in design. For example, Kroll[287] analyzed 44 publications dedicated to studies of the compression diagram for concrete (Figure 3.11) and presented 13 equations suggested by different authors. In the overwhelming majority of cases, the derivations of such equations were never aimed at revealing the physical nature of the deformation and fracture phenomena. Rather, only the best approximation of the stress–strain diagrams in accordance with experimental conditions was given. To do this, the power, parabolic, hyperbolic, and more complicated functions were utilized.

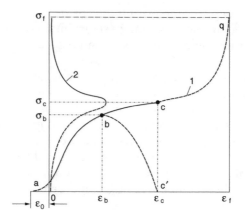

FIGURE 6.12
Schematic of formation of tension (compression) deformation diagrams of compresson–dilaton structures.

Shape of the deformation diagram (Equation 6.20) is determined by three parameters: the most probable value of microscopic strength $F(\varepsilon)$, its scatter S, and density ρ of the material. They characterize peculiarities of the structure arrangement of any material and, thus, are universal. In this connection, the suggested statistical method of description of diagrams $\sigma = f(\varepsilon)$ is applicable to all solids, without exception. Berg[48] indicated the need to use this approach, particularly for the problem of strength and deformability. Function

$$\Phi\left(\frac{\varepsilon - \bar{\varepsilon}}{S}\right)$$

has been tabulated in many handbooks.[279,285]

Formation of the deformation diagram of tension (compression) of CD materials following the law given by Equation 6.20 is schematically shown in Figure 6.12. Considering a specimen as a statistical set of the CD bonds (Figure 2.17) and testing it, we find the integral law of their distribution by deformability. This law is represented by the cumulative curve 1 and can be interpreted on the basis of the rigid-link mechanism of the force destruction processes. The latter are governed by the inherent ability of compressons and dilatons to take up the external loads by direct (compressons) and reverse (dilatons) deformation (see Section 2.11). The first mechanism takes place at absorption of phonons and the second at their radiation.

Graphical differentiation of curve 1 yields the probability density function or the differential distribution law (Curve 2 in Figure 6.12). In accordance with the CD theory of deformation and fracture (see Chapter 1 and Sections 6.2 to 6.4), the upper compresson tail is responsible simultaneously for strength and deformability of the entire stressed system. Performing the function of an energy brake under conditions of excessive thermal energy, the compresson bonds do not allow infinite deformation of the dilatons. At the phonon insufficiency, the role of the brake is played by the dilatons.

The braking effect works until point b, which corresponds to the apex of differential curve 2, is reached. At this moment the areas up and down from ordinate σ_c become equal, i.e., the reserve of resistance of the compressons is exhausted (in Figure 2.47 positions 11 to 14 correspond to this moment). In this situation, the dilaton bonds (dotted region of curve 2) are no longer capable of simultaneously taking up two force flows (in compression), first, due to increasing external load and, second, because of the disintegrated part of the compresson bonds.

Exhaustion of the remaining dilatons in region bc requires substantial deformation of the entire system from ε_b to ε_c. However, it can hardly be utilized because it coincides in time with the process of coalescence of micro-defects and catastrophic growth of an avalanche crack. The known great difficulty of direct measurement of the actual area of the load-bearing section of the specimen at any time moment results not in an ascending bc (in region ab of the true diagram of Figure 2.10) but a descending bc' branch. In Figure 3.11 ε_0 is a reverse deformation formed at the moment of relief of initial volumetric stressed state, which is often observed in materials during recovering processes (see explanations to Figure 3.11).

Diagram $\sigma = f(\varepsilon)$ allows the shape of the microscopic differential distribution law of resistance of microvolumes of a material to be reproduced by graphical differentiation. Figure 6.13 shows the integral curve 1 for limestone at compressive strain rate of $1 \cdot 10^{-4}$ MPa/sec. Experimental data found in Kuntysh and Tedder[193] were used to plot branch oa; region ab was obtained by extrapolation. Differentiating it yields the probability density function 2.

By definition,[285] differential of the function at its certain point graphically is the slope of tangent to the graph of this function at this point. With respect

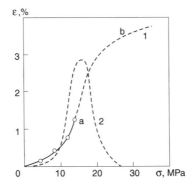

FIGURE 6.13

Integral (macroscopic) (1) and differential (microscopic) (2) function of distribution of resistance of limestone.

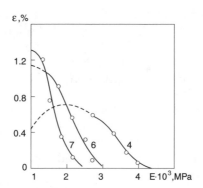

FIGURE 6.14
Reaction of limestone to the loading rate: 1, $1 \cdot 10^{-3}$ s^{-1}; 2, $0.5 \cdot 10^{-4}$s^{-1}; 3, $1 \cdot 10^{-4}$s^{-1}.

to diagram $\sigma = f(\varepsilon)$, the tangent has simple physical meaning that is instantaneous elasticity modulus E. That is, the elasticity modulus introduced into resistance of materials on the basis of phenomenological considerations acquires real physical meaning. It characterizes strength of those CD bonds transformed into the prefailure state (brought up to the Debye temperature θ, Figure 1.7) by a given level of the stressed–deformed state.

Figure 6.14 shows differential (microscopic) distribution curves, which characterize structural changes and sequence of fractures of limestone at different strain rates.[193] An increase in strain rate leads to a change in only one external parameter, i.e., density of the energy inflow to a material from the outside. Distributed at the beginning over the rigid compresson bonds (Figure 6.2), the external load stresses them. Stress concentration of the compressons causes consumption of phonons (Figure 1.18). At a high density of the external energy flow, such consumption is rather high. The thermal energy inflow from outside is determined by thermal conductivity λ and is limited by the heat capacity of the environment.

As such, the hot dilatons (in contrast to compressons) must compensate the shortage of thermal energy from outside by giving up their thermal energy to compressons. While losing heat and cooling down, they are drawn up to the Debye temperature and thus toward compressons to form dislocation pairs with them (Figure 2.59), exerting the same resistance as that of compresson resistance to external loads. For this reason, the elasticity modulus E increases with an increase in strain rate while strains ε decrease (compare positions 1, 2, and 3 in Figure 6.14). This proves that the brittleness of a structure increases under dynamic loading (see Section 4.2 and Figures 4.2 to 4.7).

Macroscopic tests of specimens to obtain the compression (tension) stress–strain diagram allow solution of the direct problem, i.e., using the known integral distribution (Figure 2.11 or 3.11) to obtain the differential distribution law (Figures 3.15 and 3.27, respectively). The latter can be used for estimation of the quality of the microstructure and prediction of its behavior under loading. Statement of the inverse problem is equally allowable: reproduction of the macroscopic integral law from the known microscopic distribution law, i.e., plotting the deformation diagram $\sigma = f(\varepsilon)$ and then, on its basis, estimating the strength and deformability characteristics of a material. This allows formulation of an important practical suggestion, which can provide substantial widening of diagnostic capabilities of the existing nondestructive methods. Its point is that the hardness of materials should be tested and obtained statistical data on its indirect indicators should be used to plot the integral distribution law (Figures 3.15 and 3.27). The law obtained this way serves as the basis for plotting diagram $\sigma = f(\varepsilon)$ without destruction of specimens.

It should be noted in conclusion that understanding the physical nature of deformation diagram $\sigma = f(\varepsilon)$ brings us closer to solving the central problem of materials science: prediction of the properties of a material on the basis of the macroscopic properties of its structure.

6.5 Mechanism of Formation and Development of Cracks

The life of a structural material in external fields (gravitational, force, thermal, radiation, and others) and active environments (see Chapter 5) is a continuous change of thermodynamic states (Equation 2.96) associated with restoration of energy balance between the material and an ever-changing environment. Of the highest interest is the irreversible change of state (in region *ab* of the deformation diagram in Figure 2.10) characterized by decreasing number of AM bonds dN (Equation 2.136) because these bonds determine the processes of initiation, propagation, coalescence, and development of cracks.

In the bulk of a material, we distinguish a certain volume V having the shape of a rectangular parallelepiped with the cross-sectional area S and length L. We designate the number of rotoses per unit volume in the unloaded state as N_0. The orientation processes take place at low intensity of the external force field (in the Hooke's law region and according to the diagram in Figure 2.8a) and at high intensity (in an irreversible region of diagram *ab* in Figure 2.10) rotoses begin to fail (Figure 2.8b). Let their number be equal to $N(F)$ at a certain load **F**. Then the number of those that failed is

$$N_f = N_0 - N(F). \tag{6.21}$$

Designate the number of rotoses that fail with an increase of 1 MPa in the external load as n_1. If the load changes by dF, the number of failed rotoses increases by $n_1 dF$. We can write the equation of balance of the number of interatomic bonds per unit volume as

$$dN = -N n_1 dF. \tag{6.22}$$

Dividing variables and integrating, we obtain

$$N(F) = N_0 \exp(-n_1 F) \tag{6.23}$$

By substituting Equation 6.23 into Equation 6.21, we find the number of the bonds that failed per unit volume in loading to the F level:

$$n_0 = N_0[1 - \exp(-n_1 F)].$$

In the distinguished volume their number will be equal to

$$n = VN_0[1 - \exp(-n_1 F)]. \tag{6.24}$$

What is the physical meaning of n_1?

Designate such an increment of the external load, which causes failure of only one bond as v. Let $dF/d\tau$ be an increment of the load per unit time. Then,

$$\delta = \frac{dF}{d\tau} \Big/ v. \tag{6.25}$$

is the number of the bonds that failed per unit time in a unit volume. On the other hand,

$$\delta = \frac{dF}{d\tau} f(\varepsilon) \tag{6.26}$$

is the probability of their failure per unit time. Here $f(\varepsilon)$ is the statistical distribution of the bonds by strength in a unit volume (see Equation 6.13). Because at any moment,

$$f(\varepsilon) = \frac{dF}{d\tau} \frac{n}{N_0},$$

Equation 6.26 can be expressed as

$$\delta = \frac{dF}{d\tau} \frac{n}{N_0}. \tag{6.27}$$

It follows from comparison of Equations 6.25 and 6.27 that

$$\frac{1}{v} = \frac{n}{N_0}. \tag{6.28}$$

Note that $1/v$ is the number of failures at an increase of 1 MPa in the external load, and

$$\frac{1}{v}dF$$

is the number of the bonds that failed with an increase of dF in the load.
Equation of balance in a unit volume can be written as

$$dN = -N\frac{1}{v}dF. \tag{6.29}$$

Comparison of Equations 6.22 and 6.29 shows that $n_1 = 1/v$ or, accounting for Equation 6.29,

$$n_1 = \frac{n}{N_0} = f(\varepsilon).$$

That is, n_1 is the law of distribution of rotoses by strength. It follows from comparison of Equations 2.2 and 6.29 that

$$dN_1 = d\delta = e^{-\frac{u}{kT}}$$

is the MB distribution of the bonds by energy. The evident equality $V = SL$ allows Equation 6.24 to be rewritten in the following form:

$$n_f = SN_0[1 - \exp(-n_1 F)].$$

This is the law of formation of the flow of failures of bonds between inter-atomic planes located normal to the direction of action of external force **F**.

If N_0 is the number of the bonds per unit volume of the unloaded material, $1/N_0$ determines the mean distance between the atoms. Assuming that failure of one bond results in formation of a defect of a round shape, we can find its area:

$$S_a = \pi \frac{1}{N_0^2} \tag{6.30}$$

The total area of defects in a plane normal to the direction of action of force F is equal to

$$S_f = n_f S_a = \frac{\pi S}{N_0}[1 - \exp(-n_1 F)].$$

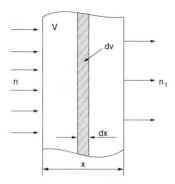

FIGURE 6.15
Model of concentration of defects.

Differentiation of this expression with respect to S yields

$$\frac{dS_f}{dS} = \frac{\pi}{N_0}(1 - e^{-n,F}).$$

Equating its right part to zero at $F \neq 0$ yields $n_1 = 0$. This means that division of a solid into parts at the moment of fracture occurs on that surface that has, at this moment, the minimum total mechanical resistance. This conclusion is in agreement with the fundamental Hamilton principle of the least action, in accordance with which movement of the system of interacting particles occurs on equipotentials characterized by minimum energy[31] (Figures 1.8 and 1.7a).

According to Equation 6.24, a dramatic increase in the concentration of submicrodefects in the volume of a solid takes place at the beginning of region *ab* of diagram $\sigma = f(\varepsilon)$ (Figure 2.10). At subsequent stages it increases slightly. New submicrodefects are formed mostly near those already formed at the previous stage, causing their propagation and transformation into microcracks. Consider the mechanism their formation and development.

In the bulk of a material, we distinguish a certain volume V in a solid that experiences certain stressed–deformed state (Figure 6.15), assuming that at the beginning of observation (under load **F**) the concentration of microdefects was n (the left-hand field of the figure). Under load $F_1 > F$, merging newly formed microdefects with the old ones and coalescence of the latter between themselves result in a decrease in the initial concentration to a certain level of $n(F)$. This phenomenon is similar to the process of attenuation of the flow of particles with density n in the bulk of a material x. In analogy with Equations 6.22 and 6.23, we can write $n(F) = n \exp(-\eta V)$, where η is the number of the new microdefects merging with the old ones and leading to growth of the cracks per unit time through a unit volume. Substituting n from Equation 6.24 into the last expression, we obtain

$$n(F) = N_0 V[1 - \exp(-n_1 F)]\exp(-\eta V). \tag{6.31}$$

This expression describes the concentration of microdefects accounting for their merging.

We can calculate its variation caused by the growth of microcracks accounting for Equations 6.24 and 6.31 as

$$n_c = N_0 V[1 - \exp(-n_1 F)][1 - \exp(\eta V)]. \tag{6.32}$$

Equation 6.32 characterizes intensity of growth of microcracks with an increase in the external load. To find the total area of fractures at a given load, we rewrite Equation 6.32 accounting for Equation 6.30:

$$S_c = \frac{\pi}{\rho}[1 - \exp(-n_1 F)][1 - \exp(-\eta V)]. \tag{6.33}$$

This expression makes it possible to explain many physical–mechanical effects observed in tests of materials in the force fields. For example, it is seen that the intensity of the process of microcracking is inversely proportional to density ρ. Denser materials have better resistance to fracture and, therefore, are characterized by higher strength. The fracture process is greatly affected by parameters of arrangement of a structure, n_1 and η. They account not only for the character of distribution of strength of the bonds by n_1 but also for their distribution manner in the volume of a material, η. In Chapter 3 the first statement was substantiated experimentally in Section 3.8 and the second in Section 3.7.

Experience shows[36] that fracture does not occur because of achievement of some critical value of the concentration of microcracks but because of growth of their relatively small number through the entire cross section of the specimen.

Now we are ready to discuss the mechanism of their formation. Let intensity σ of the external force field Π coincide with the z-axis (Figure 6.16). This

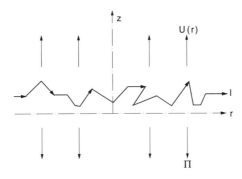

FIGURE 6.16
Propagation of crack in a heterogeneous internal potential field $U(\gamma)$ deformed by external field Π.

external field changes the internal potential field $U(r)$ (Figure 1.8) as shown in Figure 4.25 and thus causes destruction processes to occur. As such, microcracks orient themselves along the axis r normal to the vector of intensity of the external field (Figure 6.16). In general, microcracks move not exactly along axis r but along a complex broken line l. This can be confirmed observing a complex spatial configuration of any fracture surface.[50] Designate the drift path and drift velocity (distance passed by the crack mouth along axis r and its velocity along its axis) as r and v_r, and the true path and true velocity along line l as x and v_x. Then

$$K = \frac{x}{r} = \frac{v_x}{v_r}$$

will be the deceleration coefficient.

To find the relationship between the drift velocity and the intensity of the potential field, i.e., the function

$$v_r = f[U(r)],$$

we should make the following assumptions: (1) movements of a crack at velocities v_r and v_x are independent, (2) all the cracks have the same velocity v_x, (3) events of microbreak follow each other after increments of the load by one and the same value, i.e., condition $v = $ const is met, and (4) all the directions of the true drift of a crack after it was stopped are equally probable. These assumptions allow us to conclude that movement of a crack between stops occurs at the same acceleration, i.e.,

$$r = \frac{1}{2} a \tau^2, \tag{6.34}$$

where a is the acceleration. Using the definition of absolute temperature (Equation 1.24), we can express this acceleration it in terms of the temperature gradient in direction r:

$$a = \frac{K}{m} \frac{dt}{dr}. \tag{6.35}$$

Substituting Equation 6.35 into Equation 6.34 and accounting for the fact that $v_r = r/\tau$, we obtain

$$v_r = \frac{1}{2} \frac{K}{m} \frac{dt}{dr} \tau. \tag{6.36}$$

As $\tau = x/v_x$, then

$$v_r = \frac{1}{2} \frac{K}{m} \frac{dt}{dr} \frac{x}{v_x}. \tag{6.37}$$

Determining v_x from Equation 1.24 and substituting it in Equation 6.37, we write

$$v_r = \frac{\sqrt{K}}{2\sqrt{2}} \frac{x}{\sqrt{mt}} \frac{dt}{dr}. \tag{6.38}$$

Multiplying the numerator and denominator by $\sqrt{t}$ and accounting for Equation 2.145 of the ω-potential near the Debye temperature (where $t \approx \theta$), we find

$$v_r = -\frac{\sqrt{\omega}}{80\sqrt{2}} \frac{x}{\sqrt{mt}} \frac{dt}{dr}. \tag{6.39}$$

It can be seen that velocity of propagation of a crack in direction r (set by a specific stressed–deformed state, i.e., tension, compression, torsion, etc.) depends on the three parameters: ω-potential (Equation 1.53), temperature gradient in the said direction dt/dr, and the ratio

$$\frac{x}{\sqrt{mt}}.$$

The minus sign introduced in Equation 6.39 by the ω-potential indicates that the formed crack propagates in the environment where there are forces that resist its movement.

Figure 1.10 shows flat sections of different spatial shapes (Figure 1.8) of the ω-potential. Steepness of their fronts increases from position r_{a_3} to position r_{a_1}. The steeper the fronts are, the better the conditions for cracking. For example, the activation of destruction processes under dynamic loading conditions is seen in Figure 4.3 while Figure 5.10 shows it in hydrogen-containing environments. In accordance with Equations 1.107, 2.105, and 2.140, temperature gradient dT/dr is the thermodynamic equivalent of mechanical stresses σ. Creating it using external thermal or force fields, it is possible to prevent or, in contrast, favor crack propagation. Section 3.10 considers the effect of the stressed–deformed state on the strength in rupture of nonmetallic materials and Troshenko[3] presents a great number of examples of behavior of metals and alloys subjected to the effect of complex force and thermal fields.

Relationship

$$\frac{x}{\sqrt{mt}}$$

is worth discussing in detail. In analogy with Equation 3.93, its denominator is the square root of rigidity of a local system of rotoses, Δ at local temperature t.

The true length of a crack, x, divided by this value characterizes the size of energy step between the neighboring rotoses (see Figure 3.17). The latter determines the scale of heterogeneity of the internal potential field in a local volume of a material. The degree of the effect of this factor on activity of the destruction process is considered in detail in Section 3.6. Note in conclusion that the destruction process can be controlled, i.e., accelerated or decelerated, by varying parameters of Equation 6.39.

The procedure of the experimental verification of Relationship 6.39 involves thermal measurements that makes it very complicated, but it can be simplified using an indirect way. Its essence is as follows. As shown in Section 3.11, every elementary event of loss of a bond is accompanied by the formation of charged particles (electrons and polarized ions) in the bulk of a material. If an external electric field is applied to the material during the loading process, these particles serve as carriers of electricity. For example, electrical conductivity of dielectrics experiencing certain stressed–deformed states should increase. Let us give quantitative estimation to this phenomenon.

Gradient of concentration of the electric charge carriers,

$$-\frac{dn}{dx}$$

is formed under loading and corresponds to the following pressure gradient:

$$\frac{d\mathbf{P}}{dx} = kT\frac{dn}{dx}.$$

Hence, $d\mathbf{P} = kTdn$. Excessive pressure $d\mathbf{P}$ initiates chaotic displacement of the charges. In the electric field they are affected by force

$$d\mathbf{F}(e) = N(x)e\mathbf{E}(x) = n(x)\mathbf{E}(x)edx,$$

where $N(x) = n(x)dx$ is the number of charges in an elementary layer of a material (Figure 6.15), e is the electron charge, $\mathbf{E}(x)$ is the intensity of the electric field equal to the potential electric drop through the thickness of the specimen. This force causes the directed movement of the charge carriers toward electrodes.

In analogy with $\mathbf{E}(x)$, we can express the intensity of the mechanical force field in the bulk of a material in terms of

$$\Pi(x) = \frac{d\sigma}{dx}.$$

At $\mathbf{E}(x) \approx \Pi(x)$ equality $d\mathbf{P} \approx -d\mathbf{F}(e)$ persists. It characterizes the stationary mode in which all charges are captured by the electric field and transferred to electrodes; this mode takes place under quasi-static loading and has been used for measurements.

This allows the following relationship to be written

$$kT \frac{\partial n}{\partial x} dx = -en(x) \frac{\partial \sigma}{\partial x} dx.$$

Hence,

$$\frac{dn}{dx} = -\frac{ed\sigma}{kT}, \quad \text{or finally} \quad n = n_0 \exp\left(-\frac{e\sigma}{kT}\right),$$

where n_0 is the concentration of the charge carriers at the absence of the external force field. The impact of the electric field leads to the formation of a flow of the charged particles. It provides conductivity of a dielectric:

$$g = (n - n_0)e = n_0 \left[\exp\left(-\frac{e\sigma}{kT}\right) - 1 \right]. \tag{6.40}$$

If the external force and electric fields are not too high (e.g., at the elastic stage of deformation), redistribution of charges takes place not too intensively and the electric potential in volume varies in a quasi-static manner. In this case, the bracket in the last expression can be expanded into a series. By limiting calculations to the first term of expansion, we write

$$g = e^2 n_0 \frac{\sigma}{kT}.$$

As specific electrical resistance R is the value inverse to conductivity g, then

$$R = \frac{1}{g} = B\frac{T}{\sigma}, \tag{6.41}$$

where

$$B = \frac{k}{e^2 n_0}$$

is the constant that depends on the Boltzmann constant k, the charge of electron e, and the initial concentration of the charge carriers n_0. As seen, specific electrical resistance R of a dielectric in a stressed state is directly proportional to temperature T and inversely proportional to the mechanical stress σ.

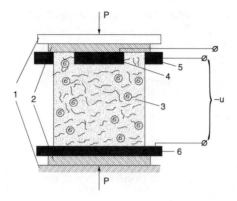

FIGURE 6.17
Schematic of a device for investigating the variation of specific electrical resistance of concrete in the direction of action of compressive load P: 1, supporting plates of press; 2, insulating plates; 3, concrete sample; 4, 5, and 6, central, screen, and receiving electrodes, respectively.

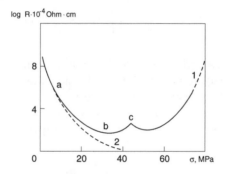

FIGURE 6.18
Variation of the specific electrical resistance of concrete in the direction of action of compressive load: 1, experimental; 2, theoretical.

Figure 6.17 shows a diagram for determination of electrical resistance of a concrete specimen in the direction of action of compressive load. The supplied current flows down from the surface of both electrodes of the probe to receiving electrode 6. A self-compensation device provides equality of potentials of central 4 and screen 5 electrodes. The current cannot flow between them as they close the equipotential surfaces. The current from the central electrode propagates into the bulk of concrete in the direction of compressive force P, i.e., normal to the plane of the receiving electrode.

Figure 6.18 shows the results of measurements of the specific electrical resistance of concrete in the direction of action of compressive forces. Plot 1 represents the experimental data while plot 2 was constructed using calculations (Equation 6.41). Similar results were obtained in investigation of electrical resistance of rock[15,40,288] and concrete.[265] Figure 6.19 shows the variation

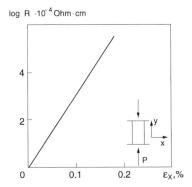

FIGURE 6.19
Variation of the specific electrical resistance of labradorite in the direction normal to action of the compressive load.

of electrical resistance of labradorite in the direction normal to the compressive load.[288]

In accordance with Equations 6.39 and 6.41, temperature anisotropy causes anisotropy of electric properties. It results in cardinally different characters of conductivity along (Figure 6.18) and across (Figure 6.19) the direction of action of the compressive load. The presented experimental data provide sufficient proof to the discussed thermodynamic mechanism of initiation and propagation of cracks (Equation 6.39).

At low values of compressive stresses, a change in the specific electrical resistance is well described by Equation 6.41 and thus curves 1 and 2 to the left of point *a* coincide (Figure 6.18). With further increase in applied stress, differences between theoretical and experimental values become more pronounced (region *ab*). In region *ac* the initiation of charges and their merger occur simultaneously. The first process leads to an increase in conductivity while the second leads to its decrease. In region *ab* the process of accumulation of charges dominates their decrease due to merging and initiation of irreversible microdefects.

An opposite picture is observed in region *bc*. Here the major part of charges in movement to electrodes is captured by growing microcracks. As a result, the electrical resistance increases. Although this process continues until fracture, the mechanism of loss in conductivity changes to the right from point *c* (inflection of curve 1 at point *c*).

The saturation of the volume with microcracks decreases the net section of the specimen. This increases the probability of their meeting with migrating charges and, most important, leads to narrowing of the actual cross-sectional area for the electric current. As a result, the electrical resistance increases. These factors also influence electrical conductivity in the transverse direction. The mechanism described was not accounted for in deriving Equation 6.41 and thus the calculated results are in good agreement with the experimental only at the initial stages of loading. Moreover, intergranular contacts are compacted and total porosity is decreased in the process of deformation of concrete. Although these structural transformations also contribute to the variation of electrical conductivity, they do not exert a decisive effect on the mechanism under consideration.

Substitution of Hooke's law (Equation 2.139) into Equation 6.41 yields

$$R = B\frac{T}{\varepsilon E}. \tag{6.42}$$

It represents the inversely proportional relationship between electrical resistance R and strain ε. As seen from Figure 6.19, this relationship is of the opposite character in the transverse direction. This means that, in the indicated direction, the dominant effect on the electrical resistance is exerted by thermal factors T, rather than mechanical factors ε and E (Equation 6.42). To be more exact, this effect is due to temperature gradient dt/dr (Equation 6.39) created by the external force field.

6.7 Crack Propagation and Restraint

In their physical nature, AM bonds are short range. Thermal radiation from degenerated rotoses (the right-hand part of Equality 3.82) is absorbed by neighbors located at the mean free path of a phonon. Their temperature and, therefore, ductility increase. The probability of their movement to the D-phase is very high; as a result, a disintegrated bond is localized and enters a thermodynamic, dilaton, or plastic trap. (In the energy space, it occupies a position to the right of position r_{a_3} in Figure 1.10). For this reason, as well as because of their stochastic distribution in space (set by the MB factor, Equation 2.2), in the initial loading period the disintegrated bonds turn out to be localized. Materials science has convincing evidence of this fact. As an example, Figure 6.20 shows the variation of the self-correlation function of longitudinal microstrain ε, derived experimentally in tension of an aluminum specimen.[59]

With an increase in the distance between the neighboring grains r, function $f(\varepsilon)$ rapidly decreases. It can be seen that deviation from the mean level of strain of any grain does not depend on how the other grain located at a distance of four to five mean sizes of a structural cell from the first is deformed. Statistical processing of the results of measurements of residual microstrains of grains in the microstructure shows that the distribution law for this parameter is close to normal.

According to the rigid-link theory (see Section 6.3), strains at the macrolevel can increase only due to degeneration of the progressively increasing number of energy-similar rotoses (movement in a direction from right to left along the differential distribution curve in Figure 6.7). The relationship between statistical parameters of deformability at the micro- and macrolevels is in

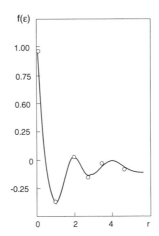

FIGURE 6.20
Autocorrelation function of longitudinal microstrains ε, obtained experimentally in tension.

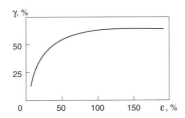

FIGURE 6.21
Dependence of heterogeneity of plastic microstrains γ on the mean macrostrain ε.

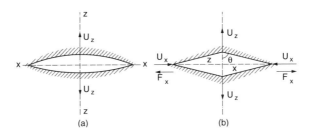

FIGURE 6.22
Model of a microcrack in the linear potential field: (a) actual and (b) idealized.

good agreement with this postulate (Figure 6.21). As deformation ε at the macrolevel increases, the stabilization of the variance γ of microdeformation stabilizes at a certain level. For fine-grained copper, its value is approximately equal to 50%.

Now we try to figure out the necessary shape of the internal potential field surrounding a nucleated microdefect (Figure 1.8) to lock it in a local region of space and to prevent its further development. To answer this question, distinguish an elementary volume in a solid body experiencing a linear stressed–deformed state. Assume that a defect of a lens shape (Figure 6.22a) is nucleated in it as a result of disruption of an interatomic bond under the effect of the external force potential field U_z.

Assume that the dimensions of the cavity are so small that they are close to interatomic distances. To simplify, we consider a defect of diamond shape (Figure 6.22b). The linear stressed–deformed state is characterized by axial symmetry of the field potential about the z-axis, i.e., at $x = const$;

$$\frac{\partial U_z}{\partial \varphi} = 0, \tag{6.43}$$

where φ is the azimuthal angle. This symmetry is a result of the uniaxial stressed–deformed state induced by a tensile load. In the Cartesian coordinate system, Condition 6.43 can be rewritten as

$$U(x) = -U(-x), \tag{6.44}$$

i.e., substitution of x for $(-x)$ does not change the field, or, in other words, the field is certainly determined to the left and right of the cavity.

To prevent further propagation of the considered defect and lock it, the field should have the radial component U_x. As the external load tending to change configuration of the local potential field (Figure 1.8), it generates force F_x which tends to cause a "collapse" of the interatomic cavity. Therefore, the main condition of localization of a microdefect is the presence of the radial component of force F_x. As such, in the development of the defect (if the cavity does begin increasing in size), force F_x, while preventing its growth, must change with an increase in x obeying a certain law. This law can be found from geometrical or energy relationships following from Figure 6.22b:

$$tg\theta = \frac{x}{z} = \frac{U_x}{U_z}. \tag{6.45}$$

Multiply the numerator and denominator of the left-hand part of Equality 6.45 by mass m of structure-forming particles. It is known from mechanics of interacting particles[25] that impact of mU_x is equal to a change in the pulse of force Fdt, i.e., $mU_x = Fdt$. Substitution of this equality into Equation 6.45 yields

$$F_x = mv_z \frac{x}{z}.$$

As seen, the force preventing the development of a local defect is directly proportional to its size x; and this force increases with an increase in the rate of variation of the axial component of the potential field

$$v_z = \frac{dU_z}{dt},$$

and depends on the physical nature of a material: mass m of particles and interatomic distance z.

Now we can find the path of development of the microcrack opening in the axisymmetric potential field. It is characterized by two components of the resistance forces: longitudinal

$$F_z = \frac{\partial U(x,z)}{\partial z}$$

and radial

$$F_x = \frac{\partial U(x,z)}{\partial x}.$$

At any point of a body both components depend on the two coordinates $U = U(x, z)$. Assuming that distribution of potential in a space of the axisymmetric field has the form of a power series, we can write

$$U(x,z) = U_0(z) + U_1(z)x + U_2(z)x^2 + U_3(z)x^3 + \cdots$$

Condition 6.44 is met due to axial symmetry and thus only the terms that contain even powers x remain

$$U(x,z) = U_0(z) + U_2(z)x^2 + U_4(z)x^4 + \cdots \qquad (6.46)$$

The Laplace equation for the axisymmetric case is

$$\frac{\partial^2 U}{\partial z^2} + \frac{1}{x}\frac{\partial U}{\partial x} + \frac{\partial^2 U}{\partial x^2} = 0. \qquad (6.47)$$

Substituting Equation 6.46 into Equation 6.47, we obtain

$$U_0'' + U_2''x^2 + U_4''x^4 + \cdots + 2U_2 + 12x^2U_4 + \cdots + 2U_2 + 4x^2U_4 + \cdots = 0, \quad (6.48)$$

where U'' is the differential with respect to the z coordinate. This expression is equal to zero when and only when the sum of the coefficients at equal powers x is equal to zero, i.e., when

$$U_0'' + 4U = 0, \qquad U_2'' + 16U_4 = 0, \qquad \text{etc.}$$

Therefore, we can find values of all U_i by expressing them in terms of U_0, namely:

$$U_2 = -\frac{1}{4}U_0''; \qquad U_4 = -\frac{1}{16}U_2'' = \frac{1}{4\cdot16}U_0''; \qquad \text{etc.}$$

Accounting for all these relationships, Series 6.48 can be written as

$$U(x,z) = U_0(z) - \frac{1}{4}U_0'x^2 + \frac{1}{4\cdot16}U_0''x^4 - \cdots$$

Considering that $U_0 = U(0, z)$ is the function of distribution of the field along the axial symmetry axis that passes through the center of a local defect, and taking $U_0 = U$, we can rewrite Equation 6.48 in the following form:

$$U(x,z) = U - \frac{1}{4}U''x^2 + \frac{1}{4\cdot16}U''x^4 - \cdots \qquad (6.49)$$

That is, to find potential at any point of the axially symmetrical field, it is necessary to know its axial component and solve Equation 6.49. While considering microdefects in which x has a value of interatomic distances (when $x \to 0$), all the terms in Expression 6.49 whose power x is higher than 2 can be ignored, i.e.,

$$U(x,z) \approx U - \frac{1}{4}U''x^2. \qquad (6.50)$$

The latter expression characterizes the distribution of potential of the axially symmetrical field. Differentiating Equation 6.50 with respect to x and z, we find the radial and axial components of the field:

$$F_x = -\frac{x}{2}U'' \quad \text{and} \quad F_z = U'.$$

The radial component determines the intensity of development of microdefects.
    We can determine the path of the crack opening in the axially symmetrical field. In accordance with the number of components of resistance force **F**, we write two equations of movement of the microcrack,

$$m\ddot{x} = -\frac{1}{2}xU''; \qquad m\ddot{z} = U', \qquad (6.51)$$

where the coordinate derivative is designated by primes and the time coordinate is designated by dots. In addition, consider that, during the process of propagation of a microcrack along the z-axis, the energy consumption is

$$\frac{m\dot{z}^2}{2} = U. \qquad (6.52)$$

We can transform time derivatives into coordinate ones as follows. Let

$$x' = \frac{dx}{d\tau},$$

then

$$x' = \frac{dx}{d\tau} = \frac{dx}{d\tau}\frac{d\tau}{dz} = \frac{\dot{x}}{\dot{z}}.$$

Therefore,

$$x'' = \frac{d}{dz}(x') = \frac{d}{d\tau}(x')\frac{d\tau}{dz} = \frac{\ddot{x}\dot{z} - \ddot{z}\dot{x}}{\dot{z}^2}\frac{1}{\dot{z}}$$

or, simplifying, we obtain

$$x'' = \frac{\ddot{x}}{\dot{z}^2} - \frac{\ddot{z}}{\dot{z}^2}z' = \frac{1}{\dot{z}^2}(\ddot{x} - \ddot{z}x'). \tag{6.53}$$

Substitution of Equations 6.51 and 6.52 into Equation 6.53 yields the principal law of propagation of microcracks in the axisymmetric potential field:

$$x'' + \frac{U'}{2U}x' + \frac{U''}{4U}x = 0. \tag{6.54}$$

Solving it, we find the path of movement of microcracks.

As seen, the intensity of the development of microfractures depends on the configuration of the internal potential field at successive moments of time along the line of action of the external force field U. To be more exact, it does not depend on the magnitude of U determined by the structure of a particular solid body, but rather on its gradients

$$F_z = U' \quad \text{and} \quad v_z = \frac{dF}{dz} = U''.$$

This explains the known dependence of strength and deformability of materials on the loading rate (see Section 4.2).

Equation 6.54 does not include any parameters that characterize structure of a particular material, so it is valid for a wide range of structural materials. Because it is linear with respect to x, a deviation of several times in the linear dimensions of an investigational object, the length of a defect changes by the same factor. This is another reminder that the scale effect plays a very serious role in the problem of fracture (see Section 3.6). Equation 6.54 raises the possibility of predicting the path of propagation of a microcrack according to a given distribution of potential of the internal field, as well as allowing distribution of the potential to be reproduced by the known traces of fracture.

The development of a microcrack in the axisymmetric field depends on the configuration of equipotentials (terms U' and U'') at the microcrack

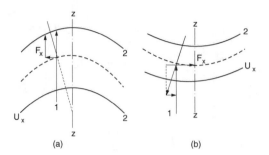

FIGURE 6.23
Effect of (a) convex and (b) concave shapes of the internal potential field on the process of microcracking.

opening. A convex (with respect to direction of its movement) configuration (Figure 6.23a) favors opening of a microcrack (force F_x tends to widen its leading front), and a concave configuration (Figure 6.23b) creates conditions for "collapse" of a microcrack, thus preventing further crack propagation. In Figure 6.23, 1 designates the direction of movement of the microcrack opening, and 2 the direction of equipotentials.

In accordance with the law of conservation of internal energy of a rotos (Equation 3.82), there is an increase in local temperature in a region with decreased intensity of the potential field (zone B). Dilatons are concentrated here and serve as the thermal, dilaton, or plastic traps for microcracks. For a crack to propagate further on, dilatons should lose part of their kinetic energy while approaching the Debye temperature. Compressons are located in the zones with an increased intensity of the potential field, A. While transferring into a critical state at the θ-temperature, they predetermine the direction of propagation of a microcrack in a local volume. The crack propagates until it again enters a dilaton trap with its tips.

A great number of objective proofs of existence of such traps can be found in metal[30,59,90] and concrete[48,289–291] studies. Their existence proves the validity of the CD concept of fracture and indicates ways to control the fracture process. For example, Bunin, Grushko, and Ilyin[289] presented experimental data which indicate that regardless of the particular type of stressed–deformed state (tension, compression, torsion, etc.) formed in concrete, a homogeneous field of local stresses and strains, with multiple concentrations, forms in this concrete (Figure 6.24). The concentration level is highest in the contact zones, decreasing with the distance from these zones. This means that, in concrete, the dilaton traps are located in the integranular crystalline aggregate of cement stone. Table 3.6 gives direct experimental proof of this fact.

Further experimental proof of the existence of plastic dilaton traps can be found in studies by Finkel[36] and Chogovadze.[292] For example, Finkel presented results of a study on the process of fracture of lithium fluoride crystals using a polarization microscope, noting that the intensity of glow of crack opening depends on its mobility. Glow of a moving crack is not intensive. When the

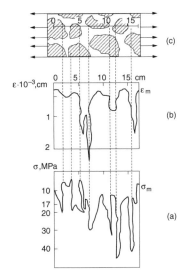

FIGURE 6.24
Variation of local microstresses (a) and microstrains (b) in a concrete macroelement (c) subjected to tension.

external load is slowly increased, a gradual increase in brightness of the "torch" is observed in the crack opening until the crack stops this. This is attributable to an increase in the concentration of stresses accompanied by progressing phonon radiation (since $d\sigma/dr = dT/dr$, according to Equation 2.105). A jump then follows and stresses begin to concentrate at a new location of the stopped crack. Similar phenomena were noted by Chogovadze in radiographic investigation of the processes of cracking in concrete under tension.

Interesting results of x-ray analysis of development of the microcracking process in concrete are given in a study by Hsu et al.[173] At the earlier stages of loading concrete of a comparatively low strength (about 20 MPa), nucleation of microcracks occurs at the interfaces between the cement stone and the filler. Microcracks propagation was observed in the cement stone between particles of the coarse filler starts only at subsequent stages. The authors paid attention to the fact that concrete initially contained cracks at the mentioned interfaces, but these cracks did not move until the external load had reached a certain level. Formation of new cracks occurs independently of them. Intensification of loading causes the development of mostly new cracks, whereas the old ones serve as a passive reserve increasing in size of new cracks during coalescence.

Formation of microscopic cracks in many locations in the material during loading is followed by a qualitatively new stage of fracture, i.e., formation of a macroscopic crack. If the level of the mean stress does not decrease, the smallest cracks start interacting with each other because of thermodynamic instability. Propagating from point to point at a velocity of v_r (Equation 6.39), the majority of them coalesce to cause formation of macrocracks. The process

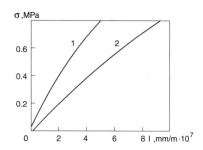

FIGURE 6.25
Deformation diagram of expanded-clay light (1) and heavy (2) concretes in tension.

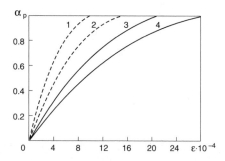

FIGURE 6.26
Dependence of longitudinal strain of heavy (1 and 3) and fine-grained (2 and 4) concretes on the level of compressive stressed state.

of formation of the latter sometimes takes more than half of the time required for fracture of a specimen.[36]

It is likely that the larger the number of rigid stress concentrators and the smaller the number of plastic localizers per unit volume of a material, the more favorable the conditions are for development of the cracking process and, hence, the higher (other conditions being equal) the macroscopic strains. With a naturally substantially different ratio of localizers and concentrators, heavy and light concretes, for example, must differ in their deformability. Figure 6.25 shows deformation characteristics of light (1) and heavy (2) concretes in tension;[253] they confirm the validity of this postulate.

The same ratio is valid for different types of heavy concretes. Figure 6.26 shows the relationship between the relative level of compressive stressed state α_{pp} (see Figures 3.50 and 3.58) and longitudinal strains for heavy (1 and 3) and fine-grained (2 and 4) concretes.[293] At the same level of stress, strains of low-strength ($\sigma = 10.8$ MPa) concrete 1 are lower compared to increased-strength (48 MPa) concrete 3. This is attributable to the fact that the structure of the first concrete is saturated with a large number of plastic traps at which cracks are restrained and thus stop propagating. Therefore, low-strength concretes feature a lower concentration of cracks per unit volume.

The differential microdistribution is responsible for development of the destruction process: it has a gently sloped shape (position 1 in Figure 3.27) in concretes of type 1, whereas the third composition is characterized by steep fronts (positions 3 or 4). Similar relationships are valid for fine-grain concretes 2 and 4 (whose strength is 10.5 and 44 MPa, respectively), as well as between coarse-grain (1 and 3) and fine-grain (2 and 4) concretes. Therefore, deformability of a material can be evaluated from the shape of differential microdistribution (obtained by testing materials to hardness) and the level of strain indicates the degree of structure degradation.

The degree of localization of microfractures depends not only on the CD heterogeneity of structure of a material but also on the kind of stressed–deformed state; this has highest importance in compression, lower in tension, and the lowest in torsion and, especially, in bending. For this reason, the consequences of deformation of concretes by compression, tension, and bending are different. Experiments conducted by Bunin, Grushko and Ilyin[289] showed that, under a load of 85% of the fracture load, residual plastic strains in compression amount to 30 to 50%, and those in tension and bending are equal to 8 to 10%. Therefore, critical cracks are formed more quickly in bending and more slowly in compression.

A formed crack first propagates at a low, almost constant velocity, and then the intensity of its movement increases in an avalanche manner. At present, the best-studied fracture mechanism is that due to development of one critical crack.[22,36] However, this character of fracture is observed in a limited range of materials (glass, some types of ceramics, some plastics, etc.). The prevailing number of materials fracture due to spontaneous initiation of several cracks in different locations and their subsequent spatial interaction. Because of extreme complexity, the mechanism of fracture of such materials has been insufficiently studied to date.

In our opinion, the fracture process is affected fundamentally by the CD arrangement of a structure. Changing this arrangement purposefully, we can accelerate, retain, or eliminate the destruction processes.

6.7   Retardation of Cracks

It is rather difficult to obtain the general solution of the equation of cracking (6.54). However, it can provide helpful information even without being solved. For example, assuming that

$$x'' = 0 \quad \text{and} \quad x' = 0,$$

we find a condition for nonpropagation of cracks:

$$\frac{U''}{4U} x = 0. \tag{6.55}$$

At x = const, it takes the form of $U'' = 0$. Equality of the second derivative of potential U (for one rotos it is equivalent to the ω-potential) in direction x to zero means constancy of the constitutional diagram of a rotos (Figure 1.14) and, therefore, of the field itself (Figure 1.8) in this direction, i.e.,

$$U' = F = \frac{dU}{dx} = \text{const.} \tag{6.56}$$

Noncontrolled variation in the potential of the field in the x-direction under the effect of service factors can be eliminated only by its synchronous correction in the y- and z-directions, i.e., due to a deliberately induced heterogeneous stressed–deformed state.

Experience shows that a heterogeneous stressed–deformed state does increase resistance of materials under static and dynamic loading conditions. For example, the results of investigation on the extent of the impact of compressive state of stress on the tensile strength of concrete in quasi-static tension at pulling out are given in Section 3.10 (see Figure 3.58). Analysis of the effect of a heterogeneous stressed–deformed state on fatigue limit[3] showed that this limit increases in the presence of the mentioned stress state as compared with the homogeneous state, finding this to be the case for linear (tension, bending) and flat (torsion) kinds of stressed state.

To illustrate, Figure 6.27, where σ_T is the yield strength in uniaxial tension and σ_a is the cyclic stress due to tension–compression, gives experimental data obtained by several investigators and tabulated by Ekobori.[44] As seen, the presence of a static tensile stressed–deformed state ($-\sigma_m$) decreases; compression ($+\sigma_m$) increases the fatigue limit σ_R of different grades of materials. This study also presented confirmation that cyclic durability in torsion, bending, and other kinds of stressed state increase in the presence of compressive stresses. Further, results of analysis[3] of experimental data on the impact of the mean cyclic stress and heterogeneous state of stress on the fatigue limit of steels and alloys (based on tests of aluminum, magnesium, nickel, and copper) showed that the presence of compressive stresses substantially

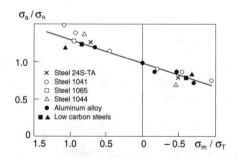

FIGURE 6.27
Effect of the mean tensile and compressive stresses on cyclic durability of various steels and alloys.

increases the fatigue limit in tension, bending, and torsion. Similar results can be found in Indenbom and Orlov.[294]

Consider the physical nature of one of the effective methods for delay of fracture employed in fabrication of composite materials. Using technological methods, the maximum possible heterogeneity of the G-potential (Equation 2.35) is formed in the structure during manufacturing. It widens the MB distribution (Figure 2.2). Rewriting Equation 1.39 and accounting for Equations 1.37.I and 1.48, we find dependence of the shape of the ω-potential on the crystalline lattice parameter r_a

$$\omega = f(r_a) = -\frac{1}{r_a}. \tag{6.57}$$

During the process of formation of a composite material, its atoms m_i form stable AM bonds at different distances from the center of the local potential field (point O in Figure 1.7) because, in accordance with Equation 1.48.I,

$$r_a = \frac{a}{m_i z_i^{1/3}},$$

where

$$a = \frac{\beta_2 \hbar^2}{4\beta_1 q^2}$$

characterizes the quality of packing of the structure. Expression 6.57 is an inversely proportional function, the plot of which is shown in Figure 1.10b.

As seen, the depth of the potential well and its shape are entirely determined by the atomic parameters (m_i and z_i) and the quality of packing of a structure, a. The value of the former is assigned by the number of chemical elements in the periodic Mendeleev system, and the quality of packing depends on the efficiency of technological operations. Combinations of fundamentally different chemical elements allow one to obtain drastically heterogeneous potential fields with centers r_{a_1}, r_{a_2}, and r_{a_3} to be generated (Figure 1.10b). Depending on particular manufacturing technology, this series can take positions C_1, C_2, and C_3 in Figure 5.33.

Such structures feature high resistance to external force and thermal fields. For example, let a composite material after solidification have structure of the type of C_2. Both C and D rotoses can coexist in this structure (Figure 2.17). For example, a dilaton may be in the well with center r_{a_1} and a compresson may be in the well with center r_{a_3}. Failure of the former occurs with the loss of heat and that of the latter with it absorption. In both cases, a disintegrated bond always has the opposite nearby, which would not allow a submicrodefect to develop further on.

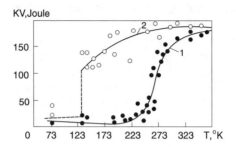

FIGURE 6.28
Temperature dependence of fracture energy (Charpy) for mild steel (1) and laminated material (2).

These conclusions find confirmation in practice. Figure 6.28 shows a temperature dependence of impact toughness (Charpy) of homogeneous mild steel (curve 1) and steel having a layer of solder on its surface (curve 2), which prevents crack propagation. When a crack reaches this layer, substantially higher energy is needed to propagate in it. The presence of such a layer leads to a substantial decrease in critical brittleness temperature (from 223 K in the original material to 123 K in the laminated material).

It is not difficult to follow the reaction of a composite material to temperature differences. Supposedly, temperature of the environment decreased dramatically. The overall energy changes so that the level of U_1 goes down and, at temperature T_2 ($T_2 < T_1$), it takes position of U_2 (Figure 1.10b). In this case, the AM bonds, whose energy state is described by the extreme right-hand potential curves, fail (e.g., point U_3 goes beyond the limits of the level of U_2). However, the left-hand bonds, which are in position r_{a_1}, do not lose their capacity. In addition, rotoses, which are in the extreme right-hand positions, have greater kinetic energy compared with the potential energy. For this reason they possess higher kinetic inertia, which prevents lowering of the level of U_1 and embrittlement of a structure.

Because of the nonhydrostatic nature of the stressed state (see Section 3.10), the potential methods for retaining of fracture (Equation 6.56) are characterized by certain inertia. Also, they require a special mechanical system to be implemented.[12,24] Thermal methods have no such drawbacks. What is their physical nature? Assuming, as above, that $x = $ const, we can rewrite Equation 6.55 accounting for Equations 3.82 and 1.71 as

$$U''x = F'x = \pm \frac{F'dU}{F} = \pm F'dU\frac{1}{g}, \qquad (6.58)$$

where

$$g = C_v \frac{dT}{dx}$$

is the heat flow in the direction of crack propagation. This condition implies that crack development ceases if heat flows are created in the most probable direction of its propagation due to heat removal from D materials (minus sign) or, on the contrary, introducing it from the outside into C materials (plus sign).

In addition to the force and thermal fields, electric, magnetic, and other fields, or their combinations, can also find application to control fracture.[12]

7

Fatigue: Physical Nature, Prediction, Elimination, and Relief

7.1 Introduction

Cyclic loading is one of the varieties of the force field. The differential form of the thermodynamic equation of state (Section 2.4) makes it possible to derive an equation for thermodynamic fatigue (Section 7.2) that relates the number of loading cycles (known also as cyclic fatigue life) to working parameters of the alternating force and thermal fields. It also characterizes mechanical fatigue at constant temperature and thermal fatigue in constant force fields. Comparison of theory and experiment yields good coincidence. Section 7.3 shows that the physical nature of fatigue is in the active occurrence of compresson–dilaton phase transitions (following the diagram in Figures 1.15) under thermal insufficiency at alternating deformation. This stimulates formation of the flow of failures of interatomic bonds at the Debye temperature.

Sometimes fatigue fracture develops at the presence of x-ray radiation. Section 7.4 gives expressions for thermal-radiation, mechanical-radiation, and thermal–mechanical–radiation fatigue as well as their comparison with experiment. The relationships derived allow calculation of fatigue life and also indicate methods for fatigue relief and elimination (Section 7.5). In many respects, they are similar to those described in Section 5.8 and work efficiently at the reversible stage of occurrence of deformation processes.

7.2 Equation of Thermomechanical Fatigue

Fatigue implies a process of gradual accumulation of damages in the initial structure of a material under the effect of dynamic variations in the parameters of one or several external fields (force, thermal, radiation, chemical, etc.). This leads to initiation of a fatigue crack and its propagation, and ends with sudden fracture.[295] Fatigue is considered to be mechanical, radiation, thermal and thermomechanical, mechanical–radiation, thermal-mechanical radiation, etc. In turn, in the force fields, fatigue is subdivided into low cycle and high cycle.[3] The former takes place when stresses are equal to or higher than the yield strength of the material; thus fatigue fracture is accompanied by changes in shape of the specimen.

The latter occurs at stresses much lower than the yield strength and at a large number of loading cycles to fracture, at which geometrical dimensions of a section hardly change. In other words, low-cycle fatigue is characterized by massive transitions of rotoses at the hot end of the MB distribution into the D-phase (positions 11 to 15 in Figure 1.14) accompanied by CD phase transformations at the Debye temperature. High-cycle fatigue processes are extended in time. Fatigue in the force fields has been a subject of extensive studies, while that in the thermal and radiation fields has been studied to a lesser degree. Only fragmentary data are available on other types of fatigue.

It is thought that the fatigue phenomenon is characterized by extreme complexity and a variety of processes occurring in the structure of materials under the effect of alternating external fields and by high sensitivity of these materials to impact of various technological, design, and service factors.[3,19] All this hampers investigations of fatigue fracture and thus prevents development of a comprehensive theory of fatigue that could serve as a reliable basis for ensuring the strength of machine parts under diverse external conditions. According to Ekobori,[44] existing theories and suggested fatigue mechanisms just describe the known experimental facts in analytical form. His conclusions are pessimistic in that he indicates that neither now nor in the near future might we expect a detailed and correct explanation of all aspects of fatigue using the same (e.g., atomistic or phenomenological) notions. Reasons for such pessimism persist because, despite very important results of experimental studies in this area, the majority of failures in practice are associated particularly with fatigue.

Discovery of the two-phase CD nature of solids[12,24] and derivation of the thermodynamic equation of their state[24,296] can be considered the first step toward development of a generalized and physically sound theory of fatigue. Next, fatigue is explained by the CD heterogeneity of the structure and appears to be its definite manifestation.[215]

Ekobori[44] thought that such a theory should successively reflect atomistic, microstructural, and continual aspects of fatigue and meet the following requirements:

- The major assumptions should not contradict reality.
- Any experimental fact should have explanation within the frame of this theory.
- Each elementary mechanism should be as simple as possible and, at the same time, preserve features of the phenomenon as a whole.

- The number of parameters describing the fatigue process should be as small as possible and each parameter should be a subject of experimental determination.

Also, Ekobori considered that a fatigue theory claiming to correctly explain to real facts should describe the dependence of limiting stresses or strains upon the number of cycles or time to fracture.[44] We leave it to readers to judge to what extent the information given below meets these requirements.

We can write the thermodynamic equation of state (2.96) at $\Delta V = 0$ (see Section 3.6) in parametric form,

$$\mathbf{P}V = kTN(\mathbf{P})_T, \tag{7.1}$$

where $N(\mathbf{P})_T$ is the number of interatomic bonds that depend on internal pressure $\mathbf{P}$ at constant temperature T. The linear differential with respect to both parts of Equation 7.1 has the form

$$\mathbf{P}V\left(\frac{\dot{V}}{V} + \frac{\dot{\mathbf{P}}}{\mathbf{P}}\right) = kTN(\mathbf{P})_T\left[\frac{\dot{N}(\mathbf{P})_T}{N(\mathbf{P})_T} + \frac{\dot{T}}{T}\right]. \tag{7.2}$$

Equation 2.13 suggests that Equation 7.2 can be rewritten as

$$\mathbf{P}V\left(\frac{\dot{V}}{V} + \frac{\dot{\mathbf{P}}}{\mathbf{P}}\right) = kTN(\mathbf{P})_T\left[\frac{\dot{N}(\mathbf{P})_T}{N(\mathbf{P})_T} + \dot{s}(T)_P\right].$$

Accounting for the expression of entropy for quasi-static processes (Equation 2.17), $\dot{s}(T)_P = k\dot{N}(T)_P$ $\dot{s}(T)_P = k\dot{N}(T)_P$, we find that

$$\mathbf{P}V\left(\frac{\dot{V}}{V} + \frac{\dot{\mathbf{P}}}{\mathbf{P}}\right) = TN(\mathbf{P})_T\left[k\frac{\dot{N}(\mathbf{P})_T}{N(\mathbf{P})_T} + \dot{N}(T)_P\right]$$

or

$$\frac{\mathbf{P}V}{TN(\mathbf{P})_T}\left(\frac{\dot{V}}{V} + \frac{\dot{\mathbf{P}}}{\mathbf{P}}\right) - \dot{N}(\mathbf{P})_P = k\frac{\dot{N}(\mathbf{P})_T}{N(\mathbf{P})_T}. \tag{7.3}$$

This equation describes thermomechanical fatigue, i.e., the process of fracture of materials in alternating force and thermal fields. Corresponding equations for particular cases of mechanical and thermal fatigue follow from it.

Indeed, in a constant thermal field where $\dot{T} = 0$, no variation occurs in the number of interatomic bonds, due to formation of thermal stresses, i.e., $\dot{N}(T) = 0$. Then,

$$n_M = \frac{N(P)}{\dot{N}(P)} = \frac{kTN(P)}{(\mathbf{P} \pm \sigma)V\dot{s}}, \tag{7.4}$$

where n_M is the number of cycles equal to the ratio of the total number of interatomic bonds, $N(P)$, to their decrease per cycle

$$\dot{N}(P), \quad \dot{s} = \frac{\dot{V}}{V} + \frac{\dot{\mathbf{P}}}{\mathbf{P}},$$

or, in analogy with Equation 4.15 and accounting for Equations 1.69, 2.13, and 2.105: $\mathbf{s} = s_p + s_V$ is the structural dynamic coefficient determining the ability of the AM system to change entropy (or, in accordance with Equation 4.26, residual life) in a reversible (Figure 2.8a) or irreversible (Figure 2.8b) manner in constant or slowly varying thermal and force fields. The latter expression characterizes fatigue of materials in alternating mechanical fields, where the effect of the thermal field can be neglected.

As seen, in addition to the level of cyclic mechanical stresses ($\pm\sigma$), fatigue life (number of cycles to fracture, n_M) depends on many other factors: temperature T, packing density of structure $N(P)$, scale effect V, loading conditions $\dot{V}/V$, and kind of stressed–deformed state $\dot{\mathbf{P}}/\mathbf{P}$. Practical experience proves these points.[3]

Dependence $n_M = f(\sigma)$ is adopted as the major correlation in the existing theory and practice of fatigue.[19,297,298] It follows from Equation 7.4 that it is inversely proportional; the rest of the parameters just determine position of the curve of the cyclic fatigue life in the coordinate system. Consider the limits to which this dependence tends.

As shown in Section 2.3, under quasi-static loading conditions, it can be approximately considered that $\dot{s}_p \approx \dot{s}_V$. Using this assumption, we can rewrite Equation 7.4 as

$$n_M = \frac{kTN(P)}{2(\mathbf{P} \pm \sigma)\dot{V}}.$$

Under quasi-static (i.e., almost equilibrium) conditions, the equation of state (7.2) holds, i.e., $2P\dot{V} \approx kTN(P)$. It follows then that, with a slow increase in stresses (where $P \to \sigma$) $n_M \to 1$. Also, it is apparent from Equation 7.4 that, with a decrease in cyclic stresses (where $P \to 0$) $n_M \to \infty$ (at $T = $ const and $N(P) = $ const). The tendency of the number of cycles n_M to infinity also takes place when $\dot{V}$ and $\dot{\mathbf{P}}$ become incomparably small values. This case corresponds to superhigh frequencies of loading.

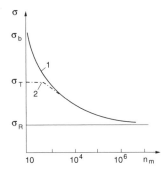

FIGURE 7.1

Typical fatigue curves.

   In the last case, infinity has a physical limit that can be estimated on the basis of the following considerations. It is known[30] that one cubic centimeter of a solid material contains about 10^{24} atoms. The number of interatomic bonds in this volume is of the same order of magnitude; their number in a cross-sectional area of 1 cm^2 is equal to 10^{16}. This is the ultimate cyclic fatigue life of a material, provided that no more than one interatomic bond fails per cycle. Therefore, at the number of cycles to fracture equal to $n = 10^8$, about 10^8 bonds fail in each of these cycles.
   Results of fatigue tests in force fields are usually shown as curves on coordinates "stress σ — number of cycles to fracture n" ("soft" loading conditions). Tests under rigid conditions are conducted at constant values of strains, $\Delta\varepsilon$ = const. Figure 7.1 shows a typical fatigue curve reproduced from GOST 25.504-82, "Strength design and tests. Methods for calculating fatigue resistance characteristics." Curve 1 corresponds to the total range of variations in the loading frequencies: from quasi-static conditions where fracture takes place at σ_b to superhigh frequencies. Curve 2 represents conditions where the cyclic stresses hardly differ from yield strength σ_T, i.e., where the processes of elastoplastic deformation occur. As seen, Equation 7.4 provides a correct description of these major fatigue characteristics.
   Moreover, Equation 7.4 predicts the fatigue limit. The latter characterizes the ability of a material to resist fracture under conditions of high-cycle loading and is equal to the maximum stresses at which fracture will never occur, however large the number of the loading cycles (horizontal line σ_R). We can rewrite Equation 7.4 in the following form:

$$n_M = \frac{kT\rho}{(\mathbf{P}\pm\sigma)\dot{s}}\,, \qquad (7.5)$$

where $\rho = N(P)/V$ is the density of a material.
   It is seen that the position of the inversely proportional dependence $n = f(\sigma)$ in the coordinate system is determined by the set of parameters

$$\frac{kT\rho}{\dot{s}}\,. \qquad (7.6)$$

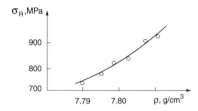

FIGURE 7.2
Dependence of fatigue limit on the density of a chromium-bearing steel (15% Cr) made using different casting methods.

In particular, the position of asymptote σ_R is directly dependent on the density of a material, ρ. As follows from Figure 7.2, which shows the dependence of the fatigue limit in circular bending on the density of a chromium-bearing steel (15% Cr) made using different methods of casting,[299] this is fully confirmed experimentally. It cannot be otherwise because, in accordance with Equation 2.175, the resistance of a body to any external field is directly dependent on its density (see Section 3.8).

Accounting for different types of fracture and pursuing practical purposes, the mechanical fatigue curve (Figure 7.1) is usually divided into three regions.[3] In region I, where stresses are close to the tensile strength, fracture occurs as a result of directed plastic deformation to the limiting value (characteristic to the material under investigation) at a monotonic increase in load. Fracture in this case does not differ from that under static loading. This type of failure is called quasi-static.

In regions II and III, fracture occurs as a result of initiation and propagation of a fatigue crack. Two regions are readily distinguishable on fractured surface: one of has a fine-fiber structure representing the region of propagation of the fatigue crack, and the other has a more or less coarse-grained structure characteristic of a brittle fracture, which is a region of the final stage of fracture.

Regions II and III differ in the number of the loading cycles and the character of fracture. In region II ($n < 2 \cdot 10^5$), the propagation of the fatigue crack occurs under a high level of the effective stress. Therefore, it is accompanied by considerable plastic strains, which often lead to marked changes in the shapes and sizes of the specimen or part being investigated. In region III, fracture occurs after a high ($n > 2 \cdot 10^5$) number of loading cycles at stresses much lower than the yield strength and elasticity limit of the material; therefore, no visible traces of residual strains are found after fracture. Given these points, region II is referred to as low-cycle fatigue and region III as high-cycle fatigue.

Dependence of the cyclic fatigue life on the temperature is much more complicated. In fact, representing Equation 7.3 in the following form,

$$n_M = \frac{kTN(P)}{(\mathbf{P} \pm \sigma)V\dot{s} - TN(P)\dot{N}(T)},$$
(7.7)

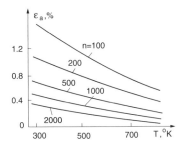

FIGURE 7.3
Dependence of amplitude of elastoplastic strains ε_a on the number of cycles at different temperatures for steel Kh18N10T (0.05% C, 18% Cr, 10% Ni, 1% Ti).

we note that temperature affects the number of cycles directly (through parameter T in the numerator) and also indirectly through the number of interatomic bonds $N(P)$ and rate of their variations $\dot{N}(T)$. This may lead to a decrease in strength and ductility of materials and to their increase, which has a different effect on transformations of the fatigue diagrams. The AM causes of this phenomenon are considered later. Troshenko's study[3] presented a number of examples which show that, in general, a change in temperature leads to an increase in the contrast of cyclic properties of structural steels attributable to a different intensity of processes of structural transformation ṡ following diagrams shown in Figures 1.15 and 2.8.

The general trend on heating metals is characterized by a decrease in their cyclic strength. This is graphically shown in Figure 7.3 for steel Kh18N10T (0.05% C, 18% Cr, 10% Ni, 1% Ti)[300] for a temperature range of 293 to 823 K and fatigue life of 100 to 2000 cycles to fracture. Other steels are also characterized by a similar variation in fatigue life with temperature increase.[3]

As follows from Equation 7.6, this dependence is possible at a simultaneous increase in the numerator and denominator only when the denominator changes more than the numerator. This may occur only at a substantial decrease in term $\dot{N}(T)$ and takes place because an increase in the distance between the atoms of a rotos (Figure 5.9) is accompanied by a decrease in the electron system they share caused by disengagement of the first, second, and then next levels. This results in a decrease in the total number of interatomic bonds that depend on the temperature $N(T)$, at a simultaneous decrease in their rigidity and strain rate $\dot{N}(T)$.

When temperature decreases, one might expect the reverse course of the temperature dependence of the cyclic fatigue life. This is what happens in reality: a decrease in temperature leads to an increase in static and cyclic strength (based on the same number of cycles). Troshenko[3] presented a great amount of supporting evidence. For example, at 293 K, the fatigue limit of steels and alloys 12Kh18N10T (0.12% C, 18% Cr, 10% Ni, 1% Ti), 000Kh2N16ATG (0.001% C, 2% Cr, 16% Ni) and high-strength aluminum alloy D20 is 295, 265, and 210 MPa, respectively, while at 77 K it is equal to

410, 655, and 275 MPa, respectively. It is noted that the major difficulty in realization of the effect of an increase in fatigue life with a decrease in temperature is in the simultaneous occurrence of a decrease in the resistance to brittle fracture and transition from the fatigue to brittle mechanism of fracture. Next, causes of this behavior of materials are explained on the basis of the CD notions of strength.

Depending on the application of machines and mechanisms, their parts operate at different kinds of stressed–deformed state and are subjected to impact of different kinds of alternating loads having different intensities, frequencies, and cycle amplitudes. In Equations 7.5 and 7.7, these peculiarities are accounted for by the dynamic parameter $\dot{s}$. As follows from Relationship 7.6, along with temperature and density, this parameter determines the position of the fatigue diagram in the coordinate system and, therefore, the fatigue limit.

In addition to the strain characteristic $\frac{\dot{V}}{V}$, it includes the vector value, i.e., relative variation in the volume resistance of a material, $\frac{\mathbf{P}}{\mathbf{P}}$. Depending on the direction of an external force field, it may have either positive (in tension) or negative (in compression) value. Compression increases $\dot{s}$, which is adequate to an increase in the fatigue limit whereas tension, on the contrary, leads to its decrease. Experiments such as those shown in Figures 3.50 and 3.68 fully confirm these conclusions. Extra evidence in favor of these conclusions can be found in Troshenko.[3] Studies[214–218] have shown that decrease in the strain rate $\frac{\dot{V}}{V}$ and, moreover, in the time of holding under tension, leads to a decrease in the number of cycles to fracture, i.e., the longer the time of holding, the more intensive the decrease.

Finally, Equation 7.6 includes volume of the material, V. Its position in the denominator makes the impact of the volume on cyclic fatigue life more pronounced. Based on analysis of many experiments, Troshenko[3] concludes that fatigue limit increases with a decrease in volume of a material located in the zone affected by external stresses. In addition, it was found that an increase in size of specimens is accompanied by an increase in their sensitivity to stress concentration. In turn, this increases the probability of dangerous manifestations of the scale effect. Explanation of the objective causes of this phenomenon is given in Section 3.6 and various studies.[23,96,97]

7.3 Compresson–Dilaton Nature of Fatigue

Fatigue as a phenomenon is definitely determined by the CD nature of solids (see Section 2.6) where the differences in the fracture mechanism of compresson–dilaton or brittle (cast iron, ceramics, natural, and synthetic stones, etc.) and compresson (metals and alloys) materials are discussed. CD materials imply primarily a large number of D-rotoses chaotically arranged in the bulk of the material (Figure 7.4). Each of them (e.g., 1 in Figure 7.4a) consists of two

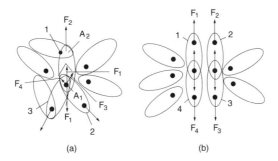

FIGURE 7.4
Orientation nature of fracture of compresson–dilaton or brittle materials.

atoms, A_1 and A_2, surrounded by common molecular electron shells (shown by ellipses).

The nature of the D-rotoses is such (see Section 1.8 and Figure 2.7) that both atoms under the effect of repulsive forces F_1 and F_2 continuously tend to move apart from each other. In turn, each is bound to the other and also to its nearest neighbors to form, for example, rotoses 2 and 3. In interactions with these rotoses, the repulsive forces (F_2 and F_3) form resultant force F_4, which compensates for force F_1. This resultant force causes transformation of atom A_1 into the state of stable dynamic equilibrium. Each local micro-volume, as well as the CD material as a whole, is in this state. Therefore, the AM systems can be in the state of dynamic equilibrium under the effect of the mutual attractive forces (which is logically faultless, as their rotoses are grouped in the zone of thermal barrier *cbq* [see Figure 1.14], and is universally recognized now) and under the effect of the submicroscopic repulsive forces (formed in the acceleration well $q_2dc^{12,18,24}$). In this case, the imperative condition of dynamic equilibrium is a chaotic arrangement of D-rotoses in space.

Transformation of any local part of such AM systems into the ordered state (e.g., by applying an external load) leads to its natural or spontaneous disintegration (rotoses 1, 2, 3, and 4 in Figure 7.4b) under the effect of the internal dilaton pressure (repulsive forces F_1, F_2, F_3, and F_4). The role of the external force field in fracture of CD materials is reduced to no more than performing the AM orientation processes. For this reason, such materials fracture by the brittle mechanism on completion of the orientation process at all cross sections. In principle, they cannot fracture by the plastic mechanism; plastic deformation requires transfer of rotoses from the D to C state. However, such transformations are practically impossible in the CD materials because temperature of their phase transition is in a long-range negative region (see Sections 2.7 to 2.9).

The AM system of compresson materials behaves differently in external (force) fields. In contrast to the dilaton ones, C-rotoses, also chaotically arranged in the bulk of the material, are characterized by attractive forces

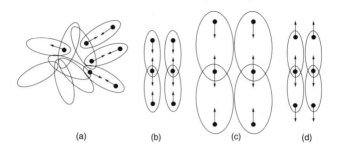

FIGURE 7.5
Types of failures of compresson–dilaton interatomic bonds.

acting between atoms of which they are composed (Figure 7.5, position a). They are located (in energy domain) in the C-region (see Figure 1.14) and temperature of their CD-phase transition is close to room temperature (Figure 1.21). Therefore, they can easily reach it during the deformation process and perform (or not perform, in the case of thermal insufficiency) CD phase transitions.

Assume that, under the effect of an external force field (static, quasi-static, or cyclic — it does not matter), the C-rotoses, which are at the hot end of the MB distribution, turn out to be in the D-phase (see Figure 5.34). Let the local energy allow part of them to complete transition successfully into the D-phase (nonhatched region), whereas the other part, having no such possibility, turn out to be in the D-phase in the C-metastable state (hatched region). Orientation effects during deformation cause the former to move to position d and the latter to position b from the initial position a (Figure 7.5).

Multiple repetition of such a situation (e.g., under cyclic loading) or intensification of deformation (under quasi-static conditions) leads to fracture of the former like conventional dilatons and to formation of a local defect. The latter, bound by mutual attraction forces, deform (illustrated in position c in Figure 7.5 and in Figure 2.8b). Their deformation continues until external molecular orbits (4 and 3 in Figure 5.9) are transformed from ellipses into hyperbolas. This moment corresponds to disintegration of the metastable C-rotoses.

Therefore, if the fracture mechanism of CD materials consists of one stage and is comparatively simple, it is sufficient to orient D-rotoses in the direction of the external field due to deformation. After this is accomplished, they fracture under the effect of natural repulsive forces. The mechanism taking place in the compresson materials is more complicated. Several mutually related processes occur simultaneously in their structure: first, CD transition (following the diagram in Figure 1.15), second, failure of the D-rotoses, and third, deformation and subsequent fracture of the metastable C-rotoses. It is these peculiarities that determine cyclic fatigue of both materials; they have found numerous experimental confirmations.

Direct and indirect evidence in favor of occurrence of the CD transitions exists. For example, Ekobori[44] established that fatigue is accompanied by changes in a microstructure (term $\dot{\theta}/\theta$ in Equation 4.22 is responsible for this, according to Equation 1.87) and presented data that prove that precipitation of cementite takes place at stresses a bit higher than the fatigue limit of steel SAE 1020. It was established at the early stage of investigations into the nature of fatigue fracture[295,301] that such fracture is accompanied by the formation of plastic deformation traces on individual grains of polycrystals.

Ekobori[44] showed that plastic deformation is accompanied by processes of restructuring and loss in strength of a local microstructure (following the diagram in Figure 1.15), which lead to a substantial concentration of strains in isolated zones. A crack is formed in one of these zones particularly, propagating from grain to grain and leading to fracture. Measurements showed that the length and width of the plastic zone at the crack tip is proportional to a current value of the crack length. Formation of plastic shears in some grains of polycrystalline metals at stresses lower than the yield strength or even the elasticity limit is evident if we take into account that deformation is impossible without heat replenishment of C-rotoses from their nearest neighbors, their transition to the D-phase, and subsequent orientation in the direction of action of the external force, accounting for the Coriolis effect (see Section 1.11).

It was established[3] that plastic strains in some microvolumes of a material are dramatically heterogeneous. This is natural, taking into account the energy and the space distribution of the MB factor (Equation 2.2). As a result, even at low strains in a specimen as a whole, residual strains in some of its microvolumes may amount to substantial values. This means that CD transitions did take place here, ending with irreversible changes of the microstructure. They are schematically shown in positions c and d in Figure 7.5.

The presence of the plastic deformation zones (i.e., CD transitions and associated irreversible structural changes in local volumes of the material under cyclic loading) was established by direct observations using optical and electron microscopes. They are confirmed by studies of the hysteresis loop (see Section 2.10), attenuation of free oscillations during heating of specimens, and other manifestations of these previously discussed processes.

Establishment of the fact that the fatigue crack initiates at locations of plastic shears (i.e., in the locations of the CD transitions) gave impetus to further studies. It was proved that the crack initiates in the most plastically deformed volumes of a material, i.e., on the surface, as a rule. Let us consider some well-established facts and comment on them from the standpoint of the CD notions. First, grain boundaries in polycrystalline specimens restrain plastic strain in the bulk of a grain: the fatigue crack develops inside the grain and does not propagate along its boundaries; upon crossing the grain boundaries, it moves at a slower speed. This is attributable to the fact that the cold part of the compresson set (left-hand end of distribution 4 in Figure 5.34) is concentrated at the grain boundary, whereas the hot part, i.e., that which is the first to approach the boundary of the CD phase transitions (which crossed vertical line *mb* in Figure 5.34), is concentrated in the bulk of grain. This fact is proved by direct microhardness measurements in metals[302] and nonmetals (see Table 3.6). Naturally, plastic strain starts developing in the bulk of grain because it has been retarded at its boundary.

At stresses higher than the fatigue limit, the cyclic load causes loosening bands in metal grains. At the same time, there is an increase in hardness of those grains or that part of a grain where there are no bands. This should mean that loosening bands are locations of failure of the generating D-rotoses (as shown in position d in Figure 7.5); the law of conservation of energy dictates an increase in hardness. Transition of C-rotoses into the D-phase occurs by consuming the kinetic energy (see Equations 1.81 to 1.83 and Figures 1.17 and 1.18) supplied to them by their nearest neighbors which, while doing this, move toward the cold tail of the MB factor and are embrittled.

Plastic deformation under cyclic loading is characterized by a high degree of localization. Troshenko[3] gives examples that a static load above the yield strength causes substantial deformation of grains of polycrystalline copper. Here, the dominating case is the mechanism of orientation and subsequent deformation following the diagram shown in positions b and c in Figure 7.5, whereas under rapid alternating loading, the plastic strain is concentrated in local volumes (locations of CD transitions followed by orientation and disintegration of the D bonds according to position d). In both cases, disorientation of the initial arrangement of grains and distortion of crystalline lattice occur. This proves a postulate that the orientation processes play a significant role in deformation and fracture (see Section 3.4).

Almost all modern theories of fatigue fracture of metals relate initiation of a fatigue crack to localized plastic strains, but provide different explanations for the mechanism of crack initiation and propagation. The CD theory gives these processes a reliable physical sense. Moreover, it points to the effective methods for crack retardation to explain causes of their initiation (see Section 6.7 and Reference 12).

It has been noted[3] that strengthening takes place at stresses below the fatigue limit, which leads to a decrease in the amplitude of cyclic deformation in individual grains and causes no distortions of the atomic lattice. This means that the processes of energy exchange between atoms involve no changes in configuration of rotoses (following the diagram in Figure 2.8a). As such, the MB distribution, while transforming, does not go outside the C-phase (distributions 2 and 3 in Figure 5.34). At stresses above the fatigue limit, structural loosening starts playing the prevailing role, attributable to the fact that the hot tail of the MB factor entering the D-phase (distribution 4 in Figures 5.34) and thus to the beginning of CD transitions (following the diagram in Figure 1.15) and to their consequences (diagram shown in Figure 7.5). Eventually, they cause nucleation, growth, and propagation of the fatigue crack.

Extrusion (pressing out of thin lobes of metal along the slip lines) and intrusion (formation of respective grooves) have been detected in some fatigue studies.[3,44] These phenomena serve as a reliable proof of the collective character of orientation processes occurring in the D-system of rotoses precipitated from the C-matrix (Figure 2.17) and leading to brittle decoupling of atoms following diagram d (Figure 7.5).

Results of investigations of the structural changes in metals suggest that fatigue fracture can be subdivided into three stages:[3,44,295]

> The first stage is preparatory or incubation and is characterized by the transformation of the MB distribution inside the C-phase (distributions 2 and 3 in Figure 5.34) and its gradual shift toward the Debye temperature (vertical line *mb*). It is associated with a change in the energy state of rotoses under the effect of the energy input into a material from the outside.

> The second stage begins when hot C-rotoses cross the Debye temperature. Beginning of the CD transitions, development of the orientation processes under cyclic deformation, and failure of the D-rotoses following the diagram in Figure 7.5 characterize it. Because of a random character of distribution of failures in the volume of the material (assigned by the space component of the MB factor, Equation 2.2), this stage is called loosening. It leads to saturation of the volume of the material with a chaotically arranged system of cracks.[10]

> The third stage is final. It is characterized by the development of the formed system of cracks, formation of a macrocrack, and its growth to fracture.

It has been shown that, for some materials, the number of cycles to fracture since the moment of formation of microcracks in the first grain is about 90% of the total life of the specimen. The duration of the development of a visible crack in a smooth specimen is 10 to 15%, and that in a notched one 20 to 50%, of their total life.[3]

If the fatigue fracture of metals is really based on the dilaton mechanism whose diagram is shown in Figures 5.34 and 7.5, it should be brittle. A direct proof of this can be found in Ekobori.[44] Unlike the avalanche brittle fracture, which takes place under quasi-static loading of the CD materials (e.g., cast iron or concrete), a specific feature of the macrostructural fatigue fracture of a metal is its configuration in the form of rings arranged concentrically about the point of initiation of a microcrack. As shown by electron microscopy, each ring consists of several fine contours with width of about a few microns. These rings correspond to the number of C-rotoses transferred into the D-phase and their orientation in the direction of action of the external load and failure, according to diagram d (Figure 7.5), per cycle. As the MB factor moves to the D-zone (distribution 4 in Figure 5.34), their number continuously increases. Therefore, the distance (according to Ekobori[44]) between these contours increases during the process of growth and propagation of a crack. As might be expected, it also depends on the level of external stresses.

The common AM mechanism of fracture of metals and nonmetals under cyclic loading gives rise to similarity, while the existence of preliminary transition of rotoses from the C- to D-phase in the former characterizes a difference in their fatigue behavior. Many investigators[3,44,295–301] have pointed out this fact. The achieved level of knowledge allowed them to relate fatigue processes in metals to variations in their dislocation structure and those in brittle materials (e.g., polymers) to inter- and intramolecular disruptions.[44,303] A convincing proof of similarity, as well as difference, of these mechanisms is heat release, which is characteristic of both brittle materials[188,303] and metals.[44,35] In addition, it is more substantial in the former than in the latter.

As shown in Sections 1.6 to 1.7, heat release is in fact a direct proof of failures of interatomic bonds. It is clear that the flow of failures of single-phase per cycle is greater in naturally brittle materials than in two-phase metallic materials. This explains the experimentally observed difference in the intensity of heat release in testing the above materials. Independent of the physical nature of a material, an increase in effective stresses causes an increase of the flow of failures and, hence, an increase in heat release. Figure 7.6 shows the variations in temperature of polymetalene oxide with variations in the amplitude of cyclic loading (numbers of the curves stand for its values).[303] Variations in heat release in concrete under high-rate loading by compression and tension are shown in Figure 4.13.

These differences also affect the major fatigue characteristic $\sigma = f(n)$. Figure 7.7 shows fatigue curves of two types of polymeric concrete in bending;[3]

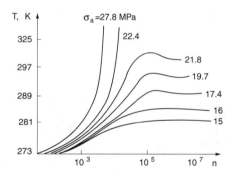

FIGURE 7.6
Variations in temperature of polymetalene oxide at different amplitudes of cyclic loading.

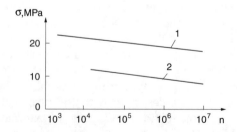

FIGURE 7.7
Fatigue curves of polymeric concrete of two compositions in bending.

the fatigue curves have a very small inclination for such materials. This means that even an insignificant increase in the effective stress causes a substantial increase in density of the flow of failures of D-rotoses in each cycle and, therefore, a decrease in durability. In metals, this development of events cannot take place because CD phase transitions control the intensity of this flow by slowing it. For this reason, their fatigue curves have a steeper front. This can be easily seen in known studies.[3,19,295,301]

7.4 Prediction of Thermal–Mechanical–Radiation Fatigue

Many modern parts and structures operate under conditions of alternating force or thermal and also radiation, electric, or other fields, as well as their combinations. In this connection, it is interesting to consider thermal and radiation fatigue in addition to mechanical fatigue, as well as their most characteristic combinations.

Assuming in Equation 7.3 that $\dot{N}(P) = 0$, we obtain the equation of thermal fatigue,

$$n_T = \frac{N(P)}{\dot{N}(T)} = \frac{PV\dot{s}}{T},$$ (7.8)

where, in analogy with Equation 7.4, the number of cycles of thermal loading is designated as

$$n_T = \frac{N(P)}{\dot{N}(T)}.$$ (7.9)

Comparing cyclic fatigue life in force (Equation 7.4) and thermal (Equation 7.8) fields, we see that

$$\frac{n_T}{n_M} = \frac{\dot{N}(P)}{\dot{N}(T)}.$$ (7.10)

As seen, this relationship is determined by the rates of variations of the number of interatomic bonds

$$\dot{N} = \frac{dN}{d\tau}$$

with variations of the intensities of force **P** and thermal T fields.

Section 2.2 showed that the degree of the effect of these fields on the AM set is fundamentally different. The force field causes only flattening of the MB distribution (changes 2, 3, and 4 in Figure 5.34), gradually bringing its hot end closer to the phase transition temperature θ (shown in Figures 5.34 by vertical line *mb*) and cold end closer to the brittle fracture temperature T_c (vertical line $r_0 a$); however, the thermal field simultaneously affects the entire AM system, not only flattening it but also shifting its major part to the D-phase (Figure 2.2 and position 4 in Figure 5.34).

This means that $\dot{N} \gg \dot{N}(P)$. Therefore, it might be expected that thermal fatigue life should be much shorter than the mechanical one, i.e., $n_T \ll n_M$. This expectation has an excellent practical proof. According to Ekobori,[44] fracture as a result of thermal fatigue occurs only within a few thermal cycles, whereas fatigue life under mechanical loading amounts to 10^5 or more cycles, as a rule.

If in the equation of thermomechanical fatigue (7.3) we take parameter $N(P)$ out of brackets, then, accounting for Equation 7.8, we can write that

$$n_T - \frac{1}{n_T} = k \frac{N(P)}{n_M}.$$

Bringing the left-hand part to the common denominator and assuming, to the first approximation, that

$$n_T^2 > 1,$$

we obtain that

$$n_{TM} = n_T n_M = k N(P), \tag{7.11}$$

where n_{TM} is the cyclic fatigue life in alternating thermomechanical fields. Taking into account the orders of magnitude of terms in Equation 7.11 ($k \approx 10^{-16}$; $N(P) \approx 10^{24}$), we find that $n_{TM} \approx 10^8$. This is the maximum possible thermomechanical fatigue life under noncontrolled conditions. Practical experience confirms this estimate.[3] It can be increased to any magnitude by controlling one of the fields and letting the other change arbitrarily. Physical principles of the control methods are described in Komarovsky[12] and in chapters that follow.

Comparison of Equations 7.9 and 7.8 makes it possible to create controlled service conditions. As seen, mechanical and thermal types of fatigue are the inversely dependent values. Thus, the controlled antiphase conditions of variations of the force and thermal fields prevent fatigue fracture. This issue is considered in detail in Section 2.7 for quasi-static loading conditions, while an indirect proof of the possibility of controlling fatigue fracture can be found in Troshenko.[3] Based on the results of analysis of several publications dedicated to investigation of thermomechanical fatigue, it is stated that durability under co-phase loading decreases compared with antiphase loading, by more than an order of magnitude. We can note that the impact of the antiphase conditions is attributable to the fact that mechanical effects tend to push out the hot tail of the MB factor to the dilaton phase (Figure 5.34), whereas the thermal effects hamper this process, preventing fracture.

Consider the effect of the radiation fields on fatigue life. Differentiation of the equation of state in such fields (Equation 5.7) yields

$$PV\dot{s} + V\varphi\tau\dot{s}(\varphi\tau) = TN(P)\left[k\frac{\dot{N}(P)}{N(P)} + \dot{N}(T)\right], \tag{7.12}$$

where $\dot{s}$ is determined from Equation 7.4 and

$$\dot{s}(\varphi\tau) = \left(\frac{\dot{V}}{V} + \frac{\dot{\varphi}\tau}{\varphi\tau}\right)$$

is the dynamic parameter of the radiation field. If the thermal field does not change, $\dot{N}(T) = 0$, then, in analogy with Equation 7.4, we obtain the equation of fatigue in alternating force and radiation fields (mechanical–radiation fatigue):

$$n_{Mr} = \frac{kTN(P)}{PV\dot{s} + V\varphi\tau\dot{s}(\varphi\tau)}. \tag{7.13}$$

Accounting for Equation 7.4, we can write

$$n_{Mr} = \frac{1}{n_M^{-1} + n_r^{-1}}, \tag{7.14}$$

where

$$n_r = \frac{kTN(P)}{V\varphi\tau\dot{s}(\varphi\tau)} \tag{7.15}$$

is radiation fatigue.

By rewriting Equation 7.14 as

$$n_{Mr} = \frac{n_M n_r}{n_M + n_r}, \tag{7.16}$$

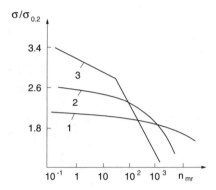

FIGURE 7.8
Low-cycle fatigue of steel 10KhSND at the following radiation doses: 1, initial state; 2, dose
$3 \cdot 10^{19}$ n/cm²; 3, $1 \cdot 10^{20}$ n/cm².

we note that mechanical–radiation fatigue life is much lower than that in
the force fields. From an elementary estimation assuming to the very first
approximation that $n_t \approx n_M$, we obtain that

$$n_{Mr} \approx \frac{n_M}{2}.$$

A very limited number of investigations on fatigue under radiation effect
is available. Few tests have been conducted because of the great complexity
and cost involved. However, even in available studies, we can find confirmation of the obtained results.[3] Figure 7.8 shows results of low-cycle fatigue
tests of steel 10KhSND (a low-carbon, low-alloy steel), depending upon the
dose of preliminary radiation. As seen from this figure, irradiated specimens
exhibit a dramatic decrease in fatigue life with an increase in the radiation
dose. It has been noted[3] that, within a fatigue life range from 0.5 to 10^3 cycles,
embrittlement takes place, so, at $n_M > 10^3$ cycles, reduction in area is close
to zero. It is worthwhile to note great similarity between radiation embrittlement and that taking place at low temperatures. This proves theoretical
predictions that, at the AM level, the embrittlement mechanism in force (see
Section 4.2.2 and Figure 4.3), thermal (Section 2.7), and radiation (Section
5.4.2) fields is the same.

Under actual service conditions, parts and structures are often subjected
to simultaneous action of force, thermal, and radiation fields. Hereafter
fatigue under such conditions is referred to as thermal–mechanical–radiation
fatigue. We can derive the equation of durability for this kind of fatigue by
rewriting Equation 7.12 in the form

$$\frac{\dot{N}(P)\dot{N}(T)}{N(P)} = \frac{PV\dot{s} + V\varphi\tau\dot{s}(\varphi\tau)}{TN(P)\left[\dfrac{k}{\dot{N}(T)} + \dfrac{N(P)}{\dot{N}(P)}\right]}.$$

The use of Equations 7.4, 7.8, and 7.15 makes it possible to write this expression as

$$n_{TMr} = m_r \left(\frac{n_T}{N(P)} + \frac{n_M}{k} \right),$$

(7.17)

where

$$n_{TMr} = \frac{N(P)}{\dot{N}(P)\dot{N}(T)}$$

is the thermal–mechanical–radiation fatigue life.

It is seen that n_{TMr} is always higher than n_r. The value of the integrated fatigue life n_{TMr} depends on the time coincidence or noncoincidence of phases of cyclic changes in thermal, n_T, and mechanical, n_M, fields. Under the antiphase conditions one might expect a substantial increase in n_{TMr}; this conclusion is confirmed by experimental data.

Sosnovsky et al.[304] presented the results of integrated experimental investigations into properties of stainless steel 08Kh18N12T (0.08% C, 18% Cr, 12% Ni, 1% Ti) in the initial state and after service for about 100,000 h under conditions of the first loop of the main circulation piping at the Novovoronezhskaya nuclear power station (Russia). As an example, Figure 7.9 shows the effect of radiation field.

In addition to high-frequency tests, low-frequency tests were conducted in symmetrical and asymmetrical loading cycles. The results of the integrated

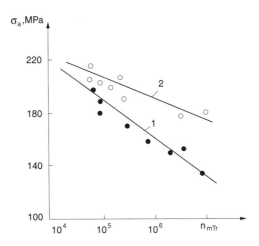

FIGURE 7.9
Effect of service duration on durability of steel 08Kh18N12T (0.08% C, 18% Cr, 12% Ni, 1% Ti) under high-frequency loading (1, initial state; 2, after service).

tests suggest the following general conclusion: service led to growth of fatigue resistance and increase in fatigue limit at any test bases and durability at a preset level of stresses. Qualitatively, this conclusion does not change for any type of test, low-frequency, high-frequency, or two-frequency, and symmetrical and asymmetrical loading cycles, although, quantitatively, there are certain differences. For example, in the high-frequency tests by the symmetrical cycle, the fatigue limit after service increased by more than a third. This led to a high increase in cyclic fatigue life that is beyond estimation. When tested using asymmetrical cycles, the difference between the amplitude values of the fatigue limits of the specimens before and after service was 1.6 times. It was found that the higher the level of stresses, the smaller the difference.

A change in the mechanism of fatigue fracture after service was found: the ability of the steel to delay the processes of accumulation of fatigue damages increased after service, which is associated with radiation strengthening of the structure. However, embrittlement led to a decrease in its ability to resist development of the already formed crack. This allows the following conclusion: steel 08Kh18N12T (0.08% C, 18% Cr, 12% Ni, 1% Ti) can be utilized under these conditions for a much longer time, although it is extremely important that no cracks or cracklike defects are formed on its surface.

In conclusion, the equations of fatigue life in single or combined fields, derived here, bring the problem of fatigue from its current phenomenological stage to the scientifically predictable one. These equations are based on the thermodynamic equation of state of a solid and the conclusion drawn from it, which can be found in Chapter 2. This opens the way for decreasing materials consumption of parts and structures and increasing their reliability and durability.

7.5 Prevention and Relief of Fatigue

The derived equations relate cyclic fatigue life n in single (Equations 7.4, 7.8, and 7.15) and combined (Equations 7.11, 7.14, and 7.17) fields to parameters of these fields and physical–mechanical characteristics of a material (contained in the Debye temperature, Equation 1.87).

Figure 7.10 shows historical development of methods for ensuring reliability and durability.[21] Probability of failure-proof operation of materials, parts, and structures in percent is plotted along the axis of ordinates. Number 1 designates the qualitative transition from classical mechanistic approaches to the methods of material reservation, 2 from the methods of material reservation to statistical theory of reliability,[3,59] and 3 from statistical theory of reliability to the physical principles of reliability.[21,207,208] Although the latter are highly promising, they are at earlier stages of the development in theoretical[12,16,18,21,24] and practical (Figure 2.1) aspects. Horizontal line BD shows the level of technogenic safety achieved by now. Separating the distance x from the 100% level can be judged from the data given in Klyuev.[8]

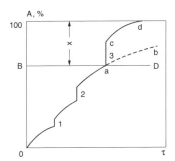

FIGURE 7.10
Development of methods for ensuring reliability and durability.

What should be done if the required strength, σ, and durability, τ, exceed those predicted by Equations 7.4 to 7.17? Staying within the frames of traditional methods, the problem can be solved following the evolutionary path *ab*. First, it is necessary to transform AM bonds of a suitable material into a metastable state by technological processing, i.e., give the MB factor shape 11 instead of natural positions 12 or 14 (Figure 1.14). This moves its end from critical temperatures T_c and θ_p and thus increases strength and extends durability.[12,24] If this approach fails to give the desirable result, then, varying function $\theta_p = f(m, z, e)$ within the periodic system of chemical elements, it is necessary to create a new material superior to the known ones.[296]

At present, these approaches are the best developed[41,60,61,228] and, therefore, are most extensively applied.[3] By utilizing the reserve embedded in a material structure at the stage of manufacture, the known design methods try to ensure reliability and durability of parts and structures. Although the only possibility so far, this way is not free from important drawbacks. Among these are high inertia, unpredictable probability of failures, noncontrollable stressed–deformed state, and absence of active control of the deformation and fracture processes. Due to these drawbacks, there are substantial limitations on the capabilities of the evolution way *ab*. In this connection, we often must agree with reduced durability or with increased risk of failures.

It follows from Equations 4.27 and 7.4 to 7.17, that durability in alternating and quasi-static fields of any physical nature is determined by variations in entropy, $\dot{s}$, i.e., by formation in the structure of a material of the processes followed (diagrams shown in Figures 1.15, 2.8, and 7.5). Notions elaborated in this book allow complementing the existing arsenal with the conceptually new methods for ensuring reliability and durability (path *acd* in Figure 7.10). They function as follows.

The controlling effect (**E, H, T**, or other fields), competing with (having different signs in the left-hand part of Equation 3.19) the service fields (most often a force field) should ensure that parameter $\dot{s}$ in Equations 7.4 to 7.17 always equals 0. This restrains the MB factor in the preset region of the energy domain and does not allow its end to approach critical temperatures

T_c and θ_p (Figure 1.14). If this happens anyway, it is necessary to assure its immediate return to the initial state to prevent development of destruction processes.

The principles of this method were suggested for the first time in Komarovsky.[12] Many indirect data prove the feasibility and effectiveness of the proposed method. For example, Ekobori[44] presented data showing that annealing increases durability when performed at an early stage of cyclic deformation, when maximum strain hardening has not yet been achieved. This fact proves the possibility of returning the MB function from position 3 to initial state 2 (Figure 5.34) or, in other words, relief of early fatigue by annealing.

In addition, it is well known[3,44] that durability can be increased by meeting at least one of the following conditions: increasing duration of periodic unloadings or their frequency and increasing the test temperature up to a certain limit; that is, letting a material structure return on its own to the initial state (or close to it) under the effect of internal (formation of the metastable D-state in the AM set) or external (temperature) factors.

A powerful measure to affect fatigue is to create such stressed–deformed state in the volume of a material that it would compete with the service effects (see Section 6.7). This follows from theoretical considerations (Chapters 2 and 3) and is confirmed experimentally (Chapters 4 and 5). Based on analysis of numerous experiments, Ekobori[44] concluded that superposition of static tension with the alternating force field decreases cyclic strength, whereas that of compression increases it. The combined effect of the superposed static tension and bending leads to a marked decrease in cyclic strength in torsion, while static compression increases it. This is attributable to the compression increasing steepness of the cold and hot fronts of the MB factor (Figure 5.34) and moving its hot end farther from the Debye temperature (vertical line *mb*). Tension, on the other hand, by flattening it, promotes transfer of hot compressons into the D-phase with catastrophic consequences for integrity of some of them (see Chapter 1).

A study[3] in Troshenko presented experimental results on the degree of the effect of bi- and triaxial state of stress on ductility and cyclic fatigue life of metals. It is definitely stated that the triaxial state of stress formed by the effect of hydrostatic pressure favors an increase in cyclic strength, ductility, and durability in the low-cycle fatigue tests. Another study[12] showed that the required stressed state can be created not only in the force fields but also in the electric and magnetic fields. This increases efficiency and decreases inertia of the methods for prevention and relief of fatigue.

As an example, consider the physical mechanism of prevention and relief of thermomechanical fatigue. To do this, rewrite Equation 7.2, accounting for the definition of parameter δ (Equation 2.141),

$$\dot{s} = \frac{\dot{P}}{P} + \frac{\dot{V}}{V} = \frac{kTN(P)\delta}{PV}\left(\frac{\dot{N}(P)}{N(P)} + 4\frac{\dot{T}}{T} - 3\frac{\dot{\theta}}{\theta}\right). \tag{7.18}$$

Fatigue fracture would not occur if deformation $\dot{V} = 0$ and failure of inter-atomic bonds $\dot{N} = 0$ are prevented. Under these conditions, Equation 7.18 has the form

$$\dot{P} = -3\frac{G\delta}{V}\frac{\dot{\theta}}{\theta}. \tag{7.19}$$

Relationship of equivalence of thermal and mechanical microstructural manifestations (Equation 2.105) allows writing the following equality:

$$\frac{1}{V} = \frac{1}{T}\frac{dT}{dx}.$$

Substituting it into Equation 7.19, we can write

$$\dot{P} = -3G\delta\frac{1}{T}\frac{\dot{\theta}}{\theta}\frac{dT}{dx}. \tag{7.20}$$

It can be seen that dynamic loading in direction $x(\dot{P})$ is compensated for without consequences for a structure by the reverse thermal pulse

$$\left(-\frac{dT}{dx}\right).$$

As such, the material reacts to the alternating force field not by deformation, $\dot{V} = 0$, or fracture, $\dot{N} = 0$, but by adequate change in the phase transition temperature $\theta \neq 0$ (vertical line *mb* in Figure 5.34 that makes cyclic oscillations about its equilibrium position, and rotoses [Figure 1.7b] of polyatomic molecules [Figure 1.4] make reversible CD phase transitions following the diagram in Figure 1.15).

It can be shown (see Section 3.3) that using electric or magnetic fields for this purpose leads to a similar effect. The only difference lies in the fact that these fields reliably perform at the elastic stage of deformation, *oa*, whereas the thermal field covers also the elastoplastic stage, *ab* in Figure 2.10.

Consider some more examples showing the possibility of restoration of the technical state of materials by using different fields. The fact of simultaneous occurrence of the processes of damage of a structure under the effect of mechanical loading, and its restoration due to heating this structure at the annealing temperature, was described long ago.[260] Also, mechanical strength may sometimes even increase, compared with the initial strength. Note that the rate of annealing depends upon the initial temperature of a body and the deformation rate.

Manifestations of fatigue in metals and nonmetals are also relieved by aging in the force fields. It is reported in Babich et al.[264] that the compressive strength of concrete increases by 27% after low-cycle (up to 30 cycles) loading; Silvestrov and Shagimordinov[43] conclude that preliminary elastic deformation of structures has a substantial effect on brittle fracture of steel. Data on the efficiency of methods of low-cycle aging of structures aimed at increasing their resistance to fatigue fracture can be found also in other literature sources.

Investigations into durability under conditions where the time to fracture of specimens was broken into several intervals between which a specimen was unloaded and received a restoration relaxation showed that the total time of action of stresses remains constant and is approximately equal to the time of fracture in one stage.[10] This indicates that consequences of the destruction processes cannot be liquidated by any subsequent manipulations. They must be prevented, or it is necessary to create conditions that would make development of a fatigue crack impossible (see Sections 6.5 to 6.7).

However paradoxical it may seem, structural members of supercritical engineering objects, which do not operate under controlled conditions, should not only relax during maintenance operations or scheduled outages but should also be subjected to intensive effects competing with the service effects in order to accelerate the process of restoration of the initial state. This can be done using mechanical, thermal, electric, and magnetic methods or their combinations. Costs of their development will be more than repaid due to extension of failure-free operation and the increase in reliability and durability.

8

Diagnostics of Technical State and Prediction of Service Life

8.1 Introduction

The equation of state of a solid (2.96) makes it possible to substantiate the thermodynamic theory of strength (Chapters 1 and 2) and to explain all physical–mechanical effects that accompany deformation and fracture (Chapters 3 to 7). This equation also makes it possible to suggest a solution for direct and inverse problems to ensure technogenic safety of engineering objects (Chapters 8 and 9).

In melting (Figure 2.2), the MB factor successively takes positions 11 to 12 to 14 to 15 (Figure 1.14) in the energy domain, changing the Cartesian, N, and energy, δ_s, parameters of entropy (Equation 2.97). The real conditions, under which melting, solidification (formation of the reverse sequence 15 to 14 to 12 to 11, Figure 1.14), or fracture (Figures 1.15, 2.8, and 7.5) occur, inevitably affect the structure of a material. Entropy serves as a carrier of this information (Equation 2.113). In other words, entropy contains information on structural arrangement at levels of local microvolumes and the whole volume of a part. This knowledge is a reliable basis for understanding how the material behaves under given external conditions and is the essence of the so-called inverse or prediction problem.[18]

The essence of the direct problem is in the purposeful arrangement of structure of a part during solidification and subsequent technological processing in order to meet scheduled service conditions. Solutions of the direct and inverse problems are interrelated; it would be naive to think that all aspects of these problems have already been addressed. However, we can foresee that the future of science and technology is associated with the ability to control entropy or life (Equation 4.26) at all the scale levels for the entire lifetime of a part.

Next we consider conceptual issues of diagnostics of the stress state of materials and structures and their service recourse.

8.2 Diagnostics of Stress–Strain State

More than a hundred physical methods and thousands of types of instruments are used now for diagnostics of structural materials.[8] All use empirical relationships between a studied property and its indirect manifestations. It seems that no generalized theory of diagnostics of materials and prediction of their properties is available so far. [53,54]

Such a theory can be built around the thermodynamic equation of state of a solid (2.96), which establishes equality between the potential (left part) and kinetic (right part) components of the overall internal energy of a material. Deviation of the state of a material from the equilibrium state is accompanied by mechanical, thermal, ultrasonic, magnetic, electric, and electron effects (Section 3.11) and thus their parameters contain information on a current state of this material.

The differential (Equation 2.140) with respect to both parts of Equation 2.96

$$\mathbf{P}V\dot{s} = kTN\delta\left[\frac{dN}{n} + \frac{dT}{T} + 3\left(\frac{dt_i}{t_i} - \frac{d\theta}{\theta}\right)\right] \tag{8.1}$$

written accounting for Equations 2.141 and 7.4, enables understanding of the physical processes which occur in the structure of materials in exhaustion of their resource, $\dot{s}$. As seen, resistance $d\mathbf{P}/\mathbf{P}$ to deformation dV/V is accompanied by a qualitative (change of positions of the MB factor in a direction of 12 to 14 to 15 [Figure 1.14] transfers an increasing number of C bonds into the D-phase) and quantitative (movement of its ends beyond the T_c and θ limits) changes in interatomic bonds dN/N, as well as displacement of the center of the MB factor (sequence of positions 12 to 14 to 15 in Figure 1.14) in the energy domain dT/T with an increase in amplitude (expression in parentheses) and probable changes in physical–mechanical properties of the structure, $d\theta/\theta$.

We can rewrite Equation 8.1 at $\Delta V = 0$ (see Section 3.6) and $T =$ const as

$$d\mathbf{P} = kTN\delta\left[\frac{dN}{n} + 3\left(\frac{dt_i}{t_i} - \frac{d\theta}{\theta}\right)\right] \tag{8.2}$$

where $\rho = N/V$ is the density of a material. It is seen that an increment in resistance $d\mathbf{P}$ can occur only due to an increase or decrease in the energy content of a unit volume of a material (expression in brackets). Its averaged measure is density ρ and a local measure is hardness. This is the physical meaning (not any other, as in Akchurin and Regel[305]) of the term "hardness." [53]

Figure 3.47 shows empirical distributions of hardness indicators obtained by the method of plastic deformations, d, and elastic rebound, h, during the process of solidification of concrete cubic samples with an edge size of 20 cm (age of concrete in days is shown in the plots). As seen, consolidation of the material is equivalent to transformation of the MB factor in the energy domain (Figure 1.14) in a direction of 15 to 14 to 12 to 11. Similar results for metals can be found in studies.[3,302] So, the methods for measurement of hardness allow determination of energy capacity of a unit volume of a material and shape of its MB distribution.

At $\mathbf{P}$ = const, T = const, N = const, and ΔV = 0, but at $dV \neq 0$ and $d\delta \neq 0$, Expression 8.1 assumes the following form:

$$\mathbf{P}dV = kTNd\delta. \tag{8.3}$$

This corresponds to the elastic stage of deformation (*oa* in Figures 2.10 and 2.22) in testing standard specimens in the constant thermal field. Substituting Equation 2.137 into Equation 8.3 and omitting η, we can write

$$\mathbf{P}\varepsilon = kT\rho d\delta \tag{8.4}$$

Comparison of Equation 8.4 with Hooke's law (Equation 2.139) makes it possible to give parameter P the conventional physical meaning (in a linear approximation, equivalent to elasticity modulus E) and write that

$$\sigma = kT\rho d\delta. \tag{8.5}$$

As seen from Equation 8.4, the stressed–deformed state can be estimated by strains ε. This approach is the simplest and, therefore, is widely applied. It underlies diverse linear displacement meters, strain gauges, moiré methods, coordinate grids, etc.[42,95] It follows from Equations 8.5 and 8.2 that the stressed–deformed level can be estimated from variations in hardness indicator parameters $d\delta$.

Figure 3.50 shows the variations in elastic rebound of the indenter (mean values h, curve 1; standard deviations S, curve 2; and asymmetry A, curve 3) in tension (due to off-center compression) (a) and central compression (b) of concrete. Parameter P in Equation 8.4 is vectorial and thus external force fields oriented in different directions can decrease it (in tension) or increase it (in compression) by differently transforming the MB distribution. These data allow determination of the mirror reflection of the energy state of a material in space of indirect hardness indicators[129] in tension (a) and compression (b) (Figure 3.51).

In tension (Figure 3.50a), in full compliance with Equation 8.5, the center of the MB distribution (curve 1) remains almost fixed up to fracture: only its amplitude 2 and asymmetry 3 change. For metals, this process is identified with positions 11 to 12 to 13 in Figure 1.14 and with position 1 in Figure 3.51a for nonmetals. Metals consist mostly of C bonds, and nonmetals of D bonds. The first may perform CD phase transitions on passing through the

θ-temperature (determined by vertical lines br_b in Figure 1.7, and mn in Figure 3.51a) and deform, whereas D bonds entering this boundary ends with brittle fracture (Section 2.6). Increment in function h by value ε_b (in Figure 3.50a designated as x) indicates an insignificant shift of the distribution center 1 (vertical line ab) toward the θ-temperature.

This turns out to be enough for its left end to be in the C-phase (dashed region) that initiates the brittle fracture mechanism and is confirmed by a drastic decrease in parameters S and A after point b in Figure 3.50a. In metals, the fracture process develops simultaneously from the "cold" T_c and "hot" θ ends of the MB factor (position 13 in Figure 1.14). Unlike brittle materials, in metals it is preceded by a substantial deformation (change of positions in a direction of 11 to 12 to 13). Compressive state of stress is formed following a more complicated law (see comments to Figure 3.51).

In the equation of state (2.96), all parameters (except for θ) are of universal character. Only the θ-temperature determines the individual peculiarities of a specific material (Equation 1.87). Depending on its physical–chemical composition, this temperature differs from T_c by no more than 25 to 30% (Section 1.10). Both temperatures can be expressed in terms of the velocity of ultrasound v in a given material (Equation 3.90). At $r_a = $ const, it follows from Equation 8.5 that $\sigma = f(v)$. This serves as the physical substantiation for the ultrasound methods of technical diagnostics of materials that find wide practical application.[145,159,160]

Physical substantiation of the acoustic emission method of diagnostics can be found in Section 3.11 and is also described in a great number of literature sources.[171–181] Our purpose was simply to show that the acoustic emission pulse is emitted when a bond reaches the Debye temperature.

Despite its attractiveness and objective character, the acoustic emission method of diagnostics has two important drawbacks. First, it works only when the hot end of the MB distribution crosses the θ-temperature (right dashed zone in position 13 in Figure 1.14) and does not show anything at the early stage of energy restructuring (transition from position 11 to position 12, or vice versa). Second, it is absolutely insensitive to failure of bonds at the T_c temperature (left end of position 13). Indeed, according to Equation 3.94, $\Delta = 0$ at point r_a. Hence, $g_v = 0$, and this method can only be used to study the development of the Debye fracture mechanism (see Section 1.10). Thermal methods have no such drawbacks.

Using transformations of trigonometric functions and assuming that variations of these functions as well as their arguments are extremely small at the AM level, we can rewrite Equation 6.6 in the following form:

$$\omega = -3A\Delta \ln x. \tag{8.6}$$

Comparison of Equations 1.53, 2.64, and 8.6 shows that the last formula expresses the law of variation of the state of an interatomic bond in linear representation. It becomes clear that the kinetic part of the function of state $\Omega(T)$ (right-hand parts of Equations 2.64, 2.153, 2.174, etc.) defines a spatial configuration of the system of the AM bonds and its dynamic equilibrium provided by a given absolute temperature of a body, T. The relationship of equivalence 2.105 allows us to write Equation 8.6 in the form conventional for the thermodynamic equation of state (2.96):

$$\omega = -3A\Delta \ln\left(t\,\frac{dx}{dt}\right). \tag{8.7}$$

As seen from Equations 8.6 and 8.7, the mechanical effects (parameters of resistance P, F, and deformation V, x included into the ω- or Ω-potential) and the thermal ones (terms on the right-hand side) are interrelated.

Taking logarithms of Equation 2.2 at $\Gamma = N$, we can write

$$N = \frac{U}{kT}. \tag{8.8}$$

In the equilibrium state, the potential and kinetic parts of the internal energy are equal to each other. Otherwise, a body exchanging thermal energy with the environment would arbitrarily change its sizes and shape. This does not happen, so it should be assumed in Equation 2.29 that $H = -\Omega$. Then $U = 2H$. In this case Equation 8.8 takes the form of $N = 2H/kT$. Substituting the expression for the kinetic energy (Equation 2.145) in this equation, we obtain

$$N = \frac{1}{2k}\,\frac{12\pi^4}{5}\,kN\left(\frac{t}{\theta}\right)^3.$$

Taking into account that the factor standing behind coefficient $1/2k$ is none other than heat capacity (Equation 2.143), we can write

$$N = \frac{1}{2k}\,C_V. \tag{8.9}$$

This expression is of high practical importance. It may be used for substantiation of new thermodynamic methods of technical diagnostics. It allows tracking the process of exhaustion of resource (service life) of a material (the number of effective interatomic bonds in a given volume at a given time moment) using variations in heat capacity.

The acoustic[37] and electron[38] emission methods allow detecting irreversible CD phase transitions such as disintegration of interatomic bonds. These methods make it possible to determine the rate of exhaustion of a resource and the most probable time of fracture. They do not, however, characterize the dynamics of the stressed–deformed state; these methods require direct contact with the object under study.

Derivation of the thermodynamic equation of state (2.96), proof of the thermal nature of internal mechanical stresses (Equation 8.5) and strains (Equation 2.105) reveals the role and importance of reliability and durability of the noncontact thermal nondestructive test methods in fracture studies. These methods form the fundamental basis for reliable estimation of the effective level of the stressed–deformed state of materials, parts, and structures, and open up wide prospects for prediction of its trend under different service conditions.

We can write Equation 8.5 in the following form:

$$\sigma = k\frac{T}{V}(N_c + N_d)d\delta. \tag{8.10}$$

It follows from this equation that CD phase transitions $d\delta$ cause changes in temperature conditions of a unit volume of a material, T/V, which is confirmed by numerous experiments.[3,92] Expression 8.10, while directly relating thermal effects to mechanical stresses σ, underlies thermodynamic methods for diagnostics and prediction of the stressed–deformed state of materials. One of its analogues, realized in the light spectrum of electromagnetic radiation, is the polarization-optical method.[306] In contrast to the latter, the thermodynamic method does not require making any models of a polarization-active material to determine and visualize the actual stressed–deformed state.

Thermal radiation of actual structural members is analyzed using thermographs (Figure 8.1).[53] For this, they are equipped with different readout devices like recorders or infrared (IR) imagers. Equations 8.5 and 8.10 allow their calibration in mechanical stresses (with a correction for the natural thermal background). The thermodynamic method enables not only the diagnostic of the actual stressed–deformed state, but also identification of the zones of critical temperatures T_c and θ, i.e., the locations of zones in a workpiece where cracks are nucleate and propagate at a given time moment or where these cracks might be expected in the future.[53]

FIGURE 8.1
Thermodynamic analyzer of mechanical stresses.

FIGURE 8.2
Thermography pattern of a fragment of wall of the five-floor concrete-panel building obtained during decoding of the stressed state.

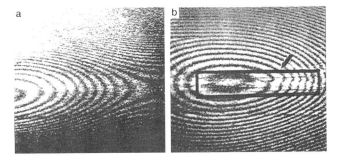

FIGURE 8.3
Interference holograms of a weld in aluminum alloy in the as-welded state (a) and at the moment of initiation of a local defect (b).

These zones show up as a bright glow that forms in disintegration of a bond following the law given in Equation 3.35.

As examples of the use of the thermodynamic method, Figure 8.2 shows results of thermography of a fragment of the wall of a five-floor concrete-panel building and Figure 8.3 shows interference holographic patterns of a weld in aluminum alloy in the as-welded state (a) and after initiation of a local defect (b).[1] From the brightness of the glow and the character of variations in the intensity of thermal radiation (other conditions being equal), we can judge the level of the stressed–deformed state. For example, Figure 8.2 shows the variation from the first to fifth floor. Color pictures of thermography, combined with automatic recording of quantitative indicators of thermal background after decoding, yield convincing proof of the efficiency of the thermodynamic investigation methods. Control duplication using traditional methods proves its objective character.

The equation of state (2.96) enables consideration of magnetic, electric, and all other diagnostic methods from similar standpoints.

8.3   Determination of Strength of Materials Using Elastoplastic Hardness Indicators

It follows from Equations 8.4 and 8.5 that at $T = const$, the strength σ depends on the rate of consumption of specific resource, $S_V = k\rho d\delta$, and is determined by elastic, P, and plastic, ε, parameters of a material. This section presents theoretical substantiation to the integrated sclerometric nondestructive method, which allows determination of the strength from indirect elastoplastic characteristics.

When an absolutely rigid sphere hits the surface of an elastoplastic material, elastic strains developed in the contact zone can be estimated from the rate restoration coefficient[307] as

$$K = \left\{ 0.4r^{1/2} \left[\frac{5(\pi\sigma)^5(1-\mu^2)^4}{E^4 W^6} \left(W + \frac{0.015(\pi\sigma)^5 r^3(1-\mu^2)^4}{E^4} \right)^5 \right]^{1/6} \right\}^{1/2}, \quad (8.11)$$

while the maximum value of plastic strains is determined from expression

$$X_{max} = \left(\frac{W}{\pi\sigma r} + \frac{0.015(\pi\sigma)^4 r^2(1-\mu^2)^4}{E^4} \right)^{1/2}, \quad (8.12)$$

where r is the radius of the sphere, W is the kinetic energy of the impact, and σ is the dynamic tensile strength of a material having elasticity modulus E and Poisson's ratio μ.

Numerical values K and X_{max} are obtained using instruments based on the methods of elastic rebound of the indenter and plastic indentation.[308] For example, these methods and instruments are standardized for concrete.[197]

Elastic strains that can be developed by very ductile materials are small. Therefore, Equation 8.11 can be simplified as

$$K = \left[0.4r^{1/2} \left(\frac{5(\pi\sigma)^5(1-\mu^2)^4}{WE^4} \right)^{1/6} \right]^{1/2}. \quad (8.13)$$

Equations 8.11 to 8.13 establish the relationship between dynamic tensile strength of a material, σ, and hardness indicators K and X_{max}.

As seen, the accuracy of determination of the strength of elastoplastic materials by hardness measurement methods (indentation-hardness and dynamic-hardness measurements) is affected by two independent groups of parameters. One of them that includes r and W is determined by design peculiarities of testing tools and should be assigned to achieve maximum sensitivity of the measurement method. The other group,

$$\left(\frac{1-\mu^2}{E}\right)^4,$$

includes physical–mechanical constants of a specific type of a material. The second group has the highest effect because E and $(1 - \mu^2)$ are in the fourth power.

Maintaining the radius of a steel ball at a constant level (for example, in the Brinell hardness test its diameter is 10 mm[308]) during indentation-hardness measurements is not difficult, and fluctuations of the impact energy of the indented in the dynamic-hardness measurements are comparatively easily accounted for by periodical calibration of the tester. However, variability of the second group of parameters cannot be accounted for by the procedural or maintenance methods because it reflects the objective reality (Equations 8.4 to 8.5) described by Equation 2.96. The latter greatly decreases accuracy and limits the area of application of one-parameter nondestructive test methods based separately on elastic or plastic properties of materials.

Accuracy of nondestructive tests can be increased by using an integrated two-parameter nondestructive method, which was developed as a synthesis of the methods of elastic rebound of the indented and plastic indentation of the surface of a material. The integrated method was implemented in practice using a standard KM tester having a ball indenter (Figure 8.4).

FIGURE 8.4
Determination of strength across the section of a steel billet by the elastoplastic method.

The idea behind the proposed method can be explained as follows. If we write Equation 8.12 in the form of

$$\left(\frac{1-\mu^2}{E}\right)^4 = 63\frac{\pi\sigma r X_{max}^2 - W}{(\pi\sigma)^5 r^3} \tag{8.14}$$

and substitute it into Equation 8.13, the measurement error formed due to uncertainty of elasticity modulus and Poisson's ratio can be avoided. After transformation, we obtain

$$\sigma = \frac{4(0.76K^{12} + 1)rW}{\pi d^4}, \tag{8.15}$$

where d is the diameter of the plastic indentation, which can be found from simple geometrical considerations. This diameter is related to current values of local strains through the following expression:

$$d = (2rX)^{1/2}. \tag{8.16}$$

Because Equation 8.13 was derived under the assumption that plastic strains X_d exceed the elastic ones X_h, then $K < 1$. Moreover, $0.76\,K^{12} \ll 1$. The strength of elastoplastic materials in this case is determined primarily by the size of the plastic indentation. Therefore, indentation-hardness methods are more efficient for strength tests of ductile materials.

In some materials (e.g., synthetic and natural stones, cast iron, etc.) elastic strains become commensurable with the plastic ones or often exceed the latter significantly. In this case, substitution of Equation 8.14 into 8.11 yields

$$K = \left[1.3(\pi\sigma r)^5 \chi_{max}^2 (\pi\sigma r \chi_{max}^2 - W)\right]^{1/12} W^{-\frac{1}{2}}. \tag{8.17}$$

Assuming that $\chi_d \approx \chi_h$, we obtain an approximate relationship for the kinetic energy of the impact

$$W \approx \frac{4}{5}\pi\sigma r \chi_{max}^2. \tag{8.18}$$

Substitution of Equation 8.18 into Equation 8.17 and transformation yield

$$\sigma \approx \frac{5}{4}\frac{W}{\pi r} - \left(\frac{K}{\chi_{max}}\right)^2.$$

It follows from Equation 8.16 that

$$\chi_{max} = \frac{d^2}{2r};$$

then,

$$\sigma \approx \frac{5}{\pi} rW \frac{K^2}{d^4}.$$ (8.19)

The rate restoration coefficient can be expressed in terms of the ratio of the energy before, W_0, and after, W, the impact:

$$K = \frac{v}{v_0} = \left(\frac{W}{W_0}\right)^{1/2}.$$ (8.20)

The potential energy of elastic strains W_d transfers into the kinetic energy of the reverse movement of the indenter. If no energy losses are accounted, its value calculates as

$$W_d = W_h = mah.$$ (8.21)

where h is the elastic rebound, m is the mass of the indenter, and a is the acceleration of damping that depends upon the operation principle of the tester and, in the simplest case (when vertical impact is the case), equal to the acceleration of free fall.

Substituting Equations 8.20 and 8.21 into Equation 8.19 and taking into account that it is convenient to measure plastic strains using the diameter of the indentation d, we write

$$\sigma = \frac{80rmah}{\pi d^4}.$$ (8.22)

Measurement error is found from Equation 8.22 as

$$\frac{\Delta\sigma}{\sigma} = \frac{\Delta r}{r} + \frac{\Delta m}{m} + \frac{\Delta a}{a} + \frac{\Delta h}{h} + \frac{4\Delta d}{d}.$$ (8.23)

Contribution of the first three terms to the total error is not high, as the mass of the indenter does not change during the measurement process and it is not difficult to ensure constancy of the radius of the spherical indenter and acceleration of damping.

Therefore, the error of measuring the strength of elastoplastic materials using the combined sclerometric instrument does not depend on fluctuations of kinetic energy of the impact (as Formula 8.22 does not contain term W) or on physical–mechanical properties of a test object (E and μ are excluded). It is determined only by the accuracy of measurements of the elastic rebound, h, and diameter of the plastic indentation, d.

Analysis of Equation 8.22 allows two important conclusions:

To increase accuracy of estimation of strength of elastoplastic materials by the local dynamic effect methods it is appropriate to use
ratio h/d as an indirect indicator.

The integrated sclerometry method is characterized by versatility and,
therefore, is more suitable for general conditions than both its
components.

Theoretical relationships 8.12, 8.13, and 8.22 were verified on concretes
(Figure 3.19c) and metals (Figure 8.4). Figures 8.5, 8.6, and 8.7 show correlation point fields for the dynamic-hardness measurement method, indentation-hardness measurement method, and the integrated elastoplastic
method, obtained on different types of concretes.

These data are a generalization of strength tests of concretes conducted
for a number of years using the KM instrument with a ball indenter. The
investigation objects were reference cubic specimens with an edge size of 20
cm, made from concretes of cardinally different compositions (from light to
heavy high-grade concretes), the mean compressive strength of which was
varied from 5 to 115 MPa. Specimens of natural consolidation and steam
curing, from 1 to 180 days of age, were utilized. Conditions of manufacturing
of concretes varied from laboratory to production.

Each point in Figures 8.5, 8.6, and 8.7 corresponds to a compression test of
3 to 21 cubes; the total number of specimens tested was over 1500. The accuracy of each test method can be judged from the spread of experimental points.

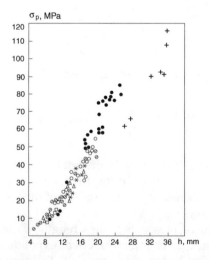

FIGURE 8.5
Correlation point field for the dynamic-hardness measurement method for concrete. Differences
in the composition of concretes and the technology of their fabrication are marked by symbols.

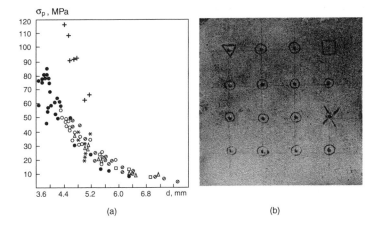

(a)   (b)

FIGURE 8.6
Correlation point field for the indentation-hardness measurement method (a) and diagram of location of single measurements at one of the faces of a cubic specimen (b).

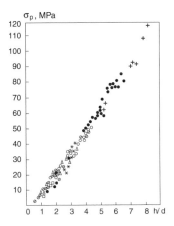

FIGURE 8.7
Calibration for the integrated elastoplastic method.

As seen, the accuracy of the proposed combined method is much higher than that of standard methods, containing almost no systematic errors associated with structural peculiarities of concretes and different manufacturing conditions.

Correlation coefficients of 0.816, 0.791, and 0.924 were calculated from the data in Figures 8.5, 8.6, and 8.7, respectively. Therefore, the error of determination of the strength of concrete using the integrated elastoplastic indicator decreases by 17 to 20%, compared with each of its one-parameter components.

Verification of the integrated method on other materials in different situations, as well as its practical utilization for research and applied purposes, confirms its high efficiency.

8.4   Prediction of Residual Resource and Durability

To raise the level of technogenic safety of engineering objects and to decrease economic, moral, and social losses that accompany accidents and catastrophes,[8] it is extremely important to be able to predict residual resource, $s(\tau)$, and durability τ (Equation 4.27). The thermodynamic equation of state (2.96) creates all required prerequisites for it.

Differentiation of Equation 2.97

$$ds_1 = k(dN\delta + Nd\delta) \tag{8.24}$$

shows that exhaustion of resource occurs in two stages. An experimental evidence of this is a true deformation diagram (Figure 2.10). CD phase transitions and, therefore, plastic deformation of brittle materials are impossible (see Sections 2.4 and 2.9). Therefore, on completion of the orientation process, when $T_i = \theta_f$, equality

$$ds_f = k\delta_f dN \tag{8.25}$$

assigns a brittle character of fracture.

In the region of plastic deformation

$$ds_s = 3kN \frac{dT_i}{T_i}\left(\frac{3}{8}\frac{\theta_s}{T_i} + 1\right) \tag{8.26}$$

(the expression in brackets was derived by differentiation of Equation 2.94.II at θ_s = const), the redistribution of the kinetic part of the internal energy from the dilaton part of set N (proved by the presence of term dT_i/T_i) into the compresson part takes place. This leads to a change in the initial shape of the MB factor (positions 11 to 15 in Figure 1.14), formation of dislocations,[47,308–314] and disintegration of the N_i part of set N. So, orientation, CD phase, dislocation and destruction restructuring, which involve consumption of resource (Equation 2.112), occur in the external fields.

Using Hooke's law (Equation 2.139), we can rewrite Equation 4.27 as

$$\tau = \tau_0 \frac{S_0}{S} \exp\frac{1}{Q}\left(\frac{P_0 \pm G}{kT} - \rho(\tau)\delta\right), \tag{8.27}$$

where P_0 is the maximum possible resistance that a specific structure can develop in the direction of action of the force field σ at temperature T

(in Figure 2.22, identified with point *b*), G is the current value of resistance (located in any region of the deformation diagram, corresponding to the achieved level of stressed–deformed state), and $\rho(\tau) = N/V$ is the density of a deformed and partially degraded structure. The local measure of density is hardness; it has been shown in studies[53,54] that the indirect hardness indicators can adequately represent any kind of stressed–deformed state.

A plot of Function 8.27 is shown in Figure 4.31, where the exponent is expressed as *A* and inequality $A_1 > A_2 > A_3$ corresponds to different levels of stresses $\sigma_1 > \sigma_2 > \sigma_3$. Position of the plot in the coordinate system is determined by the external force field σ and also by the heat content of a material, *H* (Equation 4.45). As it increases (e.g., at a dramatic decrease in temperature of the environment), the durability curve moves in a direction from A_1 to A_2 due to a change in positions of the MB factor following diagrams 15 to 14 to 12 to 11 in Figure 1.14. In this situation, even at a constant level of effective stresses, σ_0, and relatively high resistance of the crystalline lattice *P*, a dramatic decrease in durability occurs, i.e., from τ_1 to τ_3, due to the transfer of an increasing number of members of set *N* through the θ_s-temperature (in analogy with position 13 in Figure 1.14). As a result, the probability of failure of structures increases.

In fact, this character of variation of durability is observed in practice. (See Figure 1.22 and comments to it.) This behavior of materials and structures can readily be explained within the frames of the developed notions of the nature of deformation and fracture. In a new structure, the location of the MB factor (positions 11 to 15 in Figure 1.14) within the range of local temperatures $T_c - \theta_s$ is random. It may not correspond to force and thermal fields, i.e., the structure is in a metastable state with respect to them. (Identify it, for example, with position 11.) Within a short period of time service, the service conditions cause the MB factor to move to position 12, drawing its ends closer to critical temperatures T_c and θ_s. A dramatic decrease in temperature causes its major part to move beyond vertical line br_b, leading to a brittle disintegration of the structure according to diagram 4 (Figure 1.15).

Not only the MB factor but also the diagram of state of a rotos (curve $fdcbq_1$ in Figure 1.14) change in the transition to the equilibrium state. Radiation of the kinetic part of the overall internal energy into the environment causes lowering of its level in the potential well (Equation 1.26.II). As a result, the T_c temperature describes a certain curve in the temperature field.[24] Its macroscopic analogue is shown in the lower field of Figure 1.22.

While T_c varies under the effect of external conditions, the θ_s temperature is a constant value for a given structure (see Section 2.5 and Komarovsky[24]). It is worthwhile to emphasize here the decisive role of these temperatures in the destruction processes. It follows from comparison of Equations 2.65 and 4.44 that resource directly depends on the thermal expansion coefficient

$$s = \frac{\alpha}{\alpha + v}. \tag{8.28}$$

This means that frequent and sudden temperature gradients lead to intensification of the cracking processes and, hence, to a decrease in life.

Materials whose θ_s-temperature is close to the service temperature are especially sensitive to CD phase cracking. Under such conditions, parts of these materials have extremely low durability. Phase cracking of steels and alloys at θ_s by its end result is similar to loosening of porous materials (e.g., concrete and ceramics) at fluctuations of the ambient temperature at about 0°C. However, while stringent requirements for low-temperature resistance are imposed on such material,[42] specification of the θ_s-temperature of metals is neither available nor taken into account in assignment of the service temperature.

Using Representation 1.24.II, apparent equality $F = \partial T_i / \partial r$ following from Equation 1.51, and definition of the heat flow,[12] we can rewrite CD Inequality 1.70 as

$$\left(\frac{\partial g}{\partial r}\right)_c > \lambda \ \text{(I)}; \quad \left(\frac{\partial g}{\partial r}\right)_d < \lambda \ \text{(II)}, \tag{8.29}$$

where λ is the thermal conductivity. It becomes clear that CD phase transitions (any change in positions in the lower field of Figure 1.14 under the effect of parameter δ) change not only mechanical properties of solids (their quantitative measure in Equation 2.97 is parameter N) but also the thermal–physical ones. Expression 8.5, written accounting for Equation 3.51 at $T = $ const,

$$\sigma = kT\rho(\tau)\delta\frac{dN}{N}, \tag{8.30}$$

shows that consumption of resource at uniaxial state of stress occurs due to the formation of the flow of failures dN/N of the AM bonds that decreases density ρ and is accompanied by a variation of thermal radiation δT. These facts have numerous experimental evidence.[10,12,35,36,53]

Methods of thermography,[54] optical holography,[1] and determination of thermal–physical properties using noncontact scanning[315] show high potential for diagnostics of the stressed–deformed state, residual resource, and durability. As an example, Figure 8.3 shows holograms of regions of a weld in quasi-static consumption of life (a) and at the moment of initiation of a local defect (b). Equation 8.30 explains the causes of formation of interference patterns (Figure 8.3) or thermography images (Figure 8.2) and also allows their calibration in units of mechanical stresses, creating prerequisites for quantitative estimation of the stressed–deformed level. Along with hardness measurement methods,[54,59,308] they make it possible to achieve reliable control of the residual resource.

Equation 8.27 creates a scientific basis for the methods of prediction of durability of structural materials. Parameters S_0 and S are found by comparative nondestructive tests of parts and assemblies in loaded and free states or by direct stereology methods.[120] Information on a particular value of θ is contained in heat[35] or acoustic emission flows.[37] Values of P and G are determined using the results of standard tests of specimens of materials, i.e., technical deformation diagram (points b_1, b_2, G_1, G_2, and G_3 in Figure 2.22, respectively). As such, a working point may be located in plastic zone G_1 (where the life is consumed due to CD phase transitions and destruction processes), at elastic stage G_2 (here the reversible orientation of the AM bonds occurs at a constant life), or in compression region G_3, where the life is restored due to inverse CD transitions (that part which has not degraded into microcracks at the moment of observation). Naturally, the highest efficiency would be for the diagnostics systems that comprise a developed network of transducers placed into the structure and computer processing of their readings according to the program that follows from Equation 8.27.

Emergence of materials and structures adaptable to service conditions (see Chapter 9) will involve increasing requirements for the efficiency of diagnostics and prediction systems.[12]

9

Physical Principles of Adaptation of Materials and Structures to Service Conditions

9.1 Introduction

Analysis of the reference and periodical literature[38] shows that standard methods of regulating properties of structural materials gradually exhaust their capabilities, and disadvantages of traditional technologies become more and more evident. This is a manifestation of the distinctive "advance paradox."[56] Its point is simple: technical progress makes us use different materials and technologies long before we begin to understand their physical nature. This paradox is most pronounced in industries where practice develops faster than the corresponding science. Such science grows on the basis of advanced technologies and does not lead the development serving as a descriptor of the empirical results. Only a conceptually new nontraditional approach to the problem of reliability and durability allows change in this disproportionality in development.

Historically, the known theories of strength, reliability, and durability, with rather rare exceptions, are not based on physics.[2-4] Using a phenomenological approach, they often diligently elaborate logically faultless and mathematically elegant problems. Thus, not many specialists pay attention to how schematic and groundless their initial physical prerequisites are. As a result, these theories and methods make a sort of legalization of the deep-rooted notions of a fatal inevitability of fracture as a no-option outcome of structural processes that inevitably develop in a material under the effect of various external fields and aggressive environments. However, this is not the case in reality. For this reason, they cannot in principle ensure a qualitatively new breakthrough in the field in order to increase the technogenic safety of engineering objects. Ungrounded belief in the omnipotence of modern methods of fracture mechanics[1] and vain expectations associated with its quantomechanical approach[3,4] will hardly help here.

Materials science and different industries have begun lately to pay great attention to behavior of materials in various external fields: force,[316-318] electromagnetic (in thermal[319-321] and light[322] ranges), magnetic,[323-325] electric,[326] and others. This is not incidental; successful attempts are made to use the external fields for delaying fatigue fracture,[316–321] improvement of physical–mechanical properties of materials,[320,321,324–326] production of high-quality alloys,[323] heat treatment, and welding,[322,325] and for other purposes. Unfortunately, no unambiguous theory of these processes is available now because no theory of adaptation of materials and structures to diverse service conditions exists as yet.[1–5]

Let us define certain terminology to be used below. Traditional resistance of materials to deformation and fracture[3] suggests putting the maximum possible resource into a preset volume of a material at the stage of solidification,[41] its corrections and adjustments by technology methods,[61,62] and passive spending of this resource during service[1–3] Adaptation implies the possibility of ensuring a 100% guarantee of failure-free performance of engineering objects under diversity of external effects during the entire service time due to input of certain types of energy into the most stressed regions of structures and control of its stressed–deformed state.[12,18,21,24,53]

The idea of controlling the stressed–deformed state of a material using various external effects of a different physical nature is not new. The possibility of its realization follows from the equation of state (2.96). Even in the absence of a developed theory of adaptation, this idea makes its way empirically. Figure 9.1 shows results of comparison of the efficiency of different methods for retardation of fatigue cracks from the commonly accepted and simplest methods (welding up of cracked regions) to affecting a material by fields of different physical nature.[1] It is evident that methods of fracture prevention rather than those controlling its consequences have higher efficiency.

This chapter aims to develop theoretical substantiations and to present experimental verifications of methods for controlling the processes of structurization, deformation, and fracture (Section 9.2).[21,24] This can be accomplished only

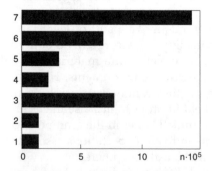

FIGURE 9.1
Cyclic fatigue life *n* (cycles) of steel specimens after application of different methods for retardation of fatigue cracks: 1, initial state; 2, drilling holes in crack tips; 3, drilling holes in crack tips and installation of high-strength bolts in these holes; 4, static overloading 1.5 times; 5, local explosive treatment; 6, local heating; 7, welding of cracks.

through utilizing the thermodynamic equation of state (Chapters 2 and 3). As follows from our analysis of this equation, strength, reliability, and durability of parts can be ensured in uncontrolled or passive conditions and in controlled or active conditions (Section 9.3). The former are described in Sections 9.4 to 9.7 and the latter are covered in Sections 9.9 and 9.10.

Properties can be controlled from the moment of design of a material for performing application-specific functions (Section 9.4), at the stage of manufacturing (Sections 9.5 to 9.7), and during service of parts and structures (Sections 9.8 to 9.10). The possibility of the transition from descriptive to predictive materials science is discussed in Section 9.4, which shows that such transition is possible due to the periodical law of variation in state of solids (Section 2.8). It is an objective outcome of Mendeleev's system of chemical elements and the equation of state (Section 2.4).

The essence of passive methods is in ensuring maximum possible energy resource in a material at the stage of design (Section 9.4) and manufacture (Section 9.5 to 9.7), the adjustments of this resource to the expected service conditions (Section 9.6), and its rational consumption in a subsequent service period. Section 9.5 considers methods for the development of an artificial anisotropy to increase resistance to deformation and fracture in certain directions. If anisotropy cannot be arranged during solidification, the initial state is adjusted using technological methods (Sections 9.6 to 9.7). The principle of variatropic technology is also suggested; this allows arrangement of a structure having variable resistance across the section of a structural member.

Active methods (Section 9.9) ensure a 100% guarantee of failure-free performance under any external effects during the entire service life by providing energy support to a material in mechanical (Section 9.9.1), electric (Section 9.9.2), magnetic (Section 9.9.3), and thermal (Section 9.9.4) fields. Such support aims to prevent changes in its initial state.

Section 9.10 suggests a new methodology for resolving topical scientific and technical problems: construction of engineering objects with artificial intelligence (Section 9.10.1), ensuring of thermomechanical strength of an independent super-deep exploration geoprobe (Section 9.10.2), as well as technology for decreasing energy consumption and improving workability of metals in cutting (Section 9.10.3).

9.2   Control of Physical–Mechanical Properties

Modern engineering objects operate in diverse (homogeneous and combined) external fields (force, thermal, radiation, electromagnetic, etc.), aggressive environments, and vacuums. In fact, the selection of structural materials to work in any given environment is empirically based. No generalized theory for controlling physical–mechanical properties of materials is available as yet[3,6,7] and neither are physical principles for controlling the stressed–deformed state or a sound theory of adaptation of materials and structures to service conditions.[1-5]

The development of such principles and theory can be based on the thermodynamic equation of state of a solid (2.65). Along with the equation of interaction (3.19), it suggests a whole set of possible methods that can be used for regulation of the process of structurization during solidification, adjustment of resource by technological methods, or by arrangement of different structures and, eventually, prevention of fracture by controlling the stressed–deformed state. Some of the suggested methods are comprehensively studied and have been in use for a long time. Another group are methods at the development and application stage, while the third group can be considered prospective for the future.

In Equation 2.65, the θ-temperature contains AM peculiarities of a specific structure. This temperature, expressed in terms of atomic parameters (Equation 1.87), assigns specific points on the constitutional diagram of a rotos (Figure 1.14), determines electrodynamics of phase transitions (Figure 1.15), and transforms Equation 2.65 into the periodic law (Section 2.8). This is confirmed by a periodic character of variations in the atomization energy,[92] Debye temperature (Figure 1.21),[270] rigidity and heat of evaporation of chemical elements,[3] density and thermal expansion coefficient,[65] melting point (Figure 1.20), elasticity modulus, density, and tensile strength of simple solids (Figures 2.45 and 2.46).[3]

Using the periodic Mendeleev system as the basis and varying function $\theta = f(m, z, e)$, we can control structurization processes to make materials with desired properties. This is exactly how it is done intuitively in materials science.[60,61] Properties of materials are varied within wide ranges by fixing the MB factor of metals and alloys in positions 11 to 12 and that of nonmetallic materials in positions 8 to 9 (Figure 1.14), through varying alloying elements in the first case or quality and content of initial components in the second. Equation 2.65 makes this procedure predictable.

It follows from Equation 2.65 that mechanical properties are determined by the thermodynamic potential Ω (Equation 1.111) rather than by **P** or V considered separately. The value of this potential is equivalent to the stored kinetic part of the overall internal energy sT. It is seen that temperature T is a powerful regulator of the state of any physical body, changing the Ω-potential and entropy **s**. Heating increases not only T, but also t_i. Displacement of energy levels of atoms toward the surface of the potential well causes transformation of the MB factor in a direction of 12 to 14 to 15 (Figure 1.14) and is accompanied by progressive transition of bonds from the C- to D-phase. While the center of distribution is inside the C-phase (up to position 12), only dimensions of a body, dV_s, change. When it comes to the D-phase (position 15), the body is no longer capable of preserving its shape dV_m.

Cooling at $dV_m = 0$ causes the MB factor to move in a reverse direction 15 to 14 to 12. As such, dramatic overcooling may change its initial shape from 12 to 11. Saturation of volume with the C-phase leads to the strengthening of a body and an increase in its hardness. This is the physical nature of all (without any exception) heat treatment methods of metals and alloys.[327] Thermal restructuring in some is characterized by the following relationships:

$$dV_m \gg dV_s \quad \text{at} \quad \mathbf{P} = \text{const.} \tag{9.1}$$

In this case, we deal with materials that have shape memory, which find application in adaptive structural systems.[328]

For materials in stationary force and thermal fields, Equation 3.19 assumes the following form:

$$(\mathbf{P} \pm \varepsilon\sigma)V = kN(\sigma)\delta(\sigma)T. \tag{9.2}$$

A single impact by the force field changes the initial orientation of the AM set, $N \neq N(\sigma)$, and CD phase ratio $\delta \neq \delta(\sigma)$ so that the material becomes anisotropic. This is the physical nature of deformation strengthening methods.[3] Indeed, at $T = \text{const}$, the compressive force field, coinciding with direction r, causes the MB distribution to move from position 12 to position 11 (Figure 1.14). This is indicative of the adequacy of structural transformations occurring in heat and deformation treatment of materials.[3]

Composite materials are produced by filling volume V with components whose MB factors are located in opposite regions of the constitutional diagram (positions 8 to 15 in Figure 1.14).[3] For example, reinforced concrete, i.e., a composite material capable of resisting compressive and tensile loads equally, is produced by reinforcing the concrete matrix (position 9), having a high compressive strength, with a steel frame (position 11; see Section 4.2.4). The same approach is used for other composite materials.[3]

Inactive modern technologies (in terms of their influence on the structurization process) are not capable of obtaining values **s** ordered in direction and homogeneous in volume. Therefore, the kinetic part of the overall internal energy, T, nonuniformly distributed in imperfect field **s** in accordance with Equation 2.65, causes high residual stresses **P** in volume V.[3] Only structures homogeneous in value of **s** and ordered in direction of **s** are characterized by uniform internal thermal field T, in which no overstressing at micro-, meso-, and macroscopic scale levels occurs (Equation 2.113). Promising new technologies should use one of the remarkable peculiarities of molten metals (position 15 in Figure 1.14), i.e., the ease with which they form ordered structures of the anisotropic type (not only configuration $\mathbf{s}_c$ but also phase $\mathbf{s}_f$ anisotropy) under the effect of external control fields (Equation 3.19).

In practice, this means that the structure of a given material should be given particularly that form of ordering which would correspond as closely as possible to future service conditions. The higher the suitability of this structure to service conditions, the higher the initial level of technogenic safety of a finished product. Conflicts and paradoxes of traditional technologies are often caused by lack of knowledge or by ignoring those remarkable capabilities that can be introduced into technology by controlling the structurization process. There is nothing extraordinary in this suggestion, as there is no alternative method (except for those indicated by Equation 3.19) for arranging atoms in a part in the desired direction over a sufficiently large (on macroscopic scales) length.

It is likely that, by certain technological methods applied to a structurization process or to a solidified structure, it is possible to produce the desired values of entropy s in the first case or make its adjustment in the second. It allows us to make a very important conclusion: if a material at the stage of solidification is given certain structural ordering by fixing paths of movement of the maximum possible number of members of set N in direction r (positions 1 or 2 in Figure 3.7), and phase heterogeneity by forming the MB factor so that its ends are located as far as possible from critical temperatures T_c and θ (position 11 in Figure 1.14), resistance of this material to external factors can be fundamentally increased.

In production of materials for various application, some methods of controlling entropy have been intuitively used for a long time[3,41,329,330] and others are now at the stage of experimental verification;[1,12,318–325] some methods have emerged very recently.[12–16,323–326] For example, solidification of building materials,[330] as well as steels and alloys,[41] takes place under conditions of static (applying a deadweight or spinning) and dynamic (compression, impact, vibration) loads. Compared with consolidation using vibration, the application of the external load during solidification increases compressive strength of concrete by 30 to 40%.[330] According to the Chalmers data,[41] structurization of steel in spinning is accompanied by the effect of modification of structure due to decrease in size of grains and their orientation in the direction of the external force field. Ultrasonic peening of welds creates favorable compressive stresses in surface layers of metals.[1,316,317]

The phase component of entropy, s_f, can be radically affected only through the θ-temperature. Of all other possibilities, this seems to be not the best, but so far the most elaborated,[60,61] method for regulation of entropy.

After solidification (when $\theta = $ const and after the formation of plots 1 to 8 in Figure 1.13), a part or a structure may be placed in differing external fields (force, thermal, radiation, electromagnetic, and others), aggressive environments, or a vacuum. The effect of some shows up in service, while others can be used for adjustment of the stressed–deformed state of the structure. As seen from Equation 3.19, the technical state of a material can be controlled before (by varying the θ-temperature) and also during the process of solidification and, especially important, after this process is complete,[12,18,316–318] including the period of active service.

The methods of active fracture prevention allow practical realization of the idea of creation of structural systems, which can adapt themselves to service conditions. Such system withstand any type of loading without deformation and fracture and, when used together with the technical diagnostics methods (Chapter 8), provide a 100% guarantee of failure-free performance within service period.

9.3 Controllable and Noncontrollable Modes of Ensuring Strength, Reliability, and Durability

Behavior of a solid in any external field and aggressive environment is entirely determined by its thermodynamic potentials and their derivatives (see Chapters 3 to 5). Three of them, Ω, G, and F, are included in the equations of state (Equations 1.111 and 2.65). Methods for their calculation are well elaborated in statistical physics[20,49,50] and thermodynamics.[11,51,52] Supported by mechanical– and thermal–physical measurements, these parameters provide reliable data on the actual performance of structural materials under diverse external conditions (see Section 3.11). They create a fundamental basis for the design and manufacturing of engineering objects with guaranteed reliability and durability and also indicate efficient methods for controlling the processes of deformation and fracture during service.

We can rewrite expressions for thermodynamic potentials Ω, Equations 1.111 and 2.45, F, Equations 2.39 and 2.63, and G, Equations 2.35 and 2.63, accounting for the periodic law of variation in state given by Equation 2.174, the right-hand part of which is determined by two kinetic parameters: absolute temperature T and Debye temperature θ:

$$\Omega = -PV \qquad \text{(I)}$$

$$F = U - Ts = -kT \ln Z(T;\theta) \qquad \text{(II)} \qquad (9.3)$$

$$G = F + PV = -kT \ln Z(T;\theta;P) \qquad \text{(III)}$$

As seen, all the potentials that form the state of a solid depend on potential, P and V, and kinetic, T and θ, parameters. It should be especially noted that absolute temperature T is a vital characteristic of the environment in which a solid is situated (including the natural thermal background) and of the solid.

The Debye temperature θ is the most important structurizing parameter of the solid. In accordance with Equation 1.87 and Figure 1.21, it is determined by the atomic nature of chemical elements constituting the solid. At $\theta = $ const, its resistance is found from one of the following formulae:[20]

$$P = kT\left(\frac{\partial \ln Z}{\partial V}\right)_T \quad \text{(I);} \quad P = -\left(\frac{\partial F}{\partial V}\right)_T \quad \text{(II);} \quad P = T\frac{\partial P}{\partial T} - \frac{\partial U}{\partial V} \quad \text{(III),} \quad (9.4)$$

and the required volume V from the following equation:

$$V = \left(\frac{\partial G}{\partial P}\right)_T = -kT\frac{\partial \ln Z(P;T)}{\partial P}. \qquad (9.5)$$

As seen from Equations 9.3 to 9.5, all the thermodynamic potentials and parameters that determine the state (except for thermal T and θ) are expressed in terms of function Z (Equation 2.2).

Thermodynamic potentials (Equation 9.3) allow the equation of state (1.111) to be given the form

$$\exp \frac{PV}{kT} = \frac{Z(\mathbf{P}, V, \theta)_T}{Z(T, \theta)_P} .$$ (9.6)

Substituting Equations 9.4.I and 9.5 into 9.6, we obtain

$$\exp kT \left[\frac{\partial \ln Z(\mathbf{P}, V, \theta)}{\partial V \partial \mathbf{P}} \right]_T = \frac{Z(\mathbf{P}, V, \theta)_T}{Z(T, \theta)_P} .$$ (9.7)

In analogy with Equation 2.119, we can represent the numerator of the right-hand parts of Equations 9.6 and 9.7 as the product of independent terms

$$Z(\mathbf{P}, V, \theta) = Z(\mathbf{P}, \theta)_T Z(T, \theta)_P .$$ (9.8)

For a specific material (when θ = const), substitution of Equation 9.8 into Equations 9.6 and 9.7 yields

$$\exp \frac{PV}{kT} = \exp kT \left[\frac{\partial \ln Z(\mathbf{P}, V)}{\partial V \partial \mathbf{P}} \right]_T = Z(\mathbf{P})_T .$$ (9.9)

This representation of the equation of state clearly shows the causes for any change in the state of a structure during deformation of a solid under isothermal conditions (T = const). As seen, a change in the MB distribution (the part of the exponent in brackets) occurs at the unchanged ω-potential (Equation 1.53). This means that the constitutional diagram of a rotos in Figure 1.14 retains its shape. The MB distribution, tending to ensure constancy of the right-hand part of the equality, changes its shape depending on rate ∂V and direction $\partial \mathbf{P}$ (compression, tension, bending, torsion, etc.) of deformation (see the lower field in Figure 2.47). Whether the deformation is accompanied by fracture of the structure or not depends on the location of the MB factor relative to the θ-temperature (vertical line m_1b). For example, transformation of the distribution from position 7 to position 6 does not lead to failure of the part of interatomic bonds that passed the θ-temperature (dashed zone in the right end of curve 7).

Value of function Z depends on the peculiarities of variations in the energy state with variations in temperature, external pressure, or volume. That is, the stressed–deformed state of a solid can be determined theoretically providing that energy levels of its AM bonds and statistical weights of the respective states are known. However, these parameters are almost always hard to determine theoretically, so an experiment helps in this case. It reveals physical meaning of the laws that govern the thermodynamic behavior of solids under diverse external conditions, introducing an insignificant change in the most general laws set by Equations 9.3 through 9.9.

The general methodology of estimation of durability in resource consumption (Equation 2.112) is given in Section 4.3.6. Variations in entropy given by Equation 4.48 may occur reversibly, when the MB factor, while transforming, all the time remains only in the compresson ($r_a - r_b$) or dilaton ($r_b - \alpha$) phase (Figure 2.47) or, irreversibly, when one of its ends goes beyond the θ-temperature (positions 7 and 8 in the same figure). In the first case the structure retains its integrity whereas, in the second, under conditions of thermal insufficiency (Equation 9.9), the destruction processes occur in this structure. The presence of reversible changes of the state in external fields yields a 100% guarantee of reliability and durability of materials, parts, and structures. When irreversible processes take place, it is necessary to conduct durability tests.

The way for identification of the type of structural processes that accompany force and thermal deformation can be described as follows. We can write the equation of interaction (3.19) in the following form:

$$\frac{\mathbf{P} \pm \sigma}{kT} = 20\rho\left(\frac{t}{\theta}\right)^3. \tag{9.10}$$

The plus sign of σ corresponds to tension and minus means compression. The range of variation of the left-hand part of the equality is calculated by determining resistance $\mathbf{P}$ of a solid from Equation 4.44 and substituting the expected range of variations in the intensities of the external force, σ, and thermal, T, fields. The right-hand part is found from the known density ρ and Debye temperature θ of a structural material selected for the application. If, at any combination of parameters of the force and thermal fields, the left-hand part is less than the right-hand part, the deformation processes would occur in a reversible manner. If the inverse inequality is obtained, then the destruction processes occurring in the material lead to the accumulation of damages and, hence, increase in the probability of unexpected fracture.

If the desirable result cannot be obtained by selecting a suitable material (i.e., by varying parameters ρ and θ), there can be three ways out from this situation. The first way consists of the design of an absolutely new material, which would meet all the imposed requirements. (Problem-solving method is described below.) The second way is to accept the fact of occurrence of the destructive processes and thus to limit durability of the designed object to the safe level. The third way is to make and equip the object designed with an individual system of prevention of deformation and fracture (see Sections 3.10, 5.7, 6.7, and 7.5).[12,18] The decision is to be made accounting for cost and functional importance of the object, as well as material, financial, labor, and other expenses for realization of this variant or the other.

The second way is most elaborated thus far and, therefore, most common in world practice. The known design methods try to ensure reliability and durability of products due to gradual consumption of the energy resource s (Equation 2.112) set into a material at the stage of solidification. As the only possibility so far, this way has important drawbacks. The major drawbacks of this method include a high degree of inertia, unpredictable probability of failures, and noncontrolled stressed–deformed state.

The previously discussed thermodynamic relationships play fundamental roles in the resistance of materials to various external effects. Determining durability of materials, these relationships fill the known statistical strength theories with a clear physical content and serve as a reliable ground for the physical principle of reliability.[55] Let us show this.

Multiplying both parts of Equation 2.95 by the Boltzmann constant k and taking into account the main thermodynamic relationship (2.17), we can write

$$kN^2 = -k \ln Z(\mathbf{P},T) = -s(\mathbf{P};t). \tag{9.11}$$

This allows Equation 4.48 to be expressed as

$$N_f = N_b - N_l = \alpha V \Delta \sigma, \tag{9.12}$$

where N_b, N_l, and N_f are the initial, maximum permissible, and fractured number of interatomic bonds at a given time moment, respectively. Failure of dN bonds occurs during time $d\tau$

$$d\tau = \frac{dS}{\alpha V}, \tag{9.13}$$

where $dS = dN/d\sigma$ is the crack resistance of a material, which is numerically equal to an increment in the fracture surface, ds, caused by an increase in external stresses by $d\sigma$. The time required for fracture of Nf bonds is

$$\tau = \frac{S_c}{\alpha V} = \frac{S_l}{\alpha} = \frac{N_f}{dN}, \tag{9.14}$$

where $S_1 = S_c/V$ is the maximum possible concentration of cracks per unit volume. A working volume is determined by assigning this value, based on the consideration of inadmissibility of failure of a part. It follows from Equation 9.14 that durability is inversely proportional to the volume. This apparent paradox is attributable to the scale effect; its physical nature was revealed in Section 3.6, which shows that crack resistance increases with a decrease in the working volume.

Therefore, to estimate performance of materials in the force and thermal fields reliably, it is necessary first to figure out and accept as given the service range of temperatures T and mechanical stresses σ. Then the following procedure should be followed:

Determine heat content in the range (Equation 4.45).

Measure thermal expansion coefficient α and thermal pressure coefficient v (Equation 4.43).

Calculate thermodynamic potentials (Equation 4.44).

Select the type of structural material using parameters ρ and θ, keeping in mind the required durability (Equation 9.14) and then calculate its working volume (Equation 4.48).

The preceding calculation procedure is based on fundamental physical laws, which makes it universal. It allows general principles of deformation and fracture of materials in alternating force and thermal fields (and, naturally, in each of them separately) to be described from unified positions.[24] It indicates the main direction for improving methods of calculating the strength and reducing the scope of experimental studies, including bench tests, for substantiation of reliability and life of machines and mechanisms. However, compared with traditional approaches, it requires extra costs associated with an increase in the scope of source information and upgrading of regulatory procedures and documentation. This is the cost of accurate prediction of behavior of materials, parts, and structures during operation, and trustworthy conclusions on their reliability and durability.

Study of the physical nature of deformation and fracture (see Chapters 1 and 2) allows us to make an important comment concerning a decisive role of the Debye temperature in structural processes. As seen from Equation 9.14, durability is inversely dependent on the thermal expansion coefficient α. This means that frequent and dramatic gradients of temperature lead to intensification of cracking processes and, therefore, to a reduction of resource (see Section 7.4). Materials with Debye temperature close to the service temperature are especially sensitive to CD phase cracking.

Durability of parts made of such materials under the above conditions is extremely low. Phase cracking of steels and alloys is similar in the end result to loosening of porous materials (e.g., concrete and ceramics) at seasonal fluctuations of the ambient temperature at about 0°C. However, whereas stringent requirements on the low temperature strength are imposed on such materials,[42] no determination of the Debye temperature for these materials has been made yet. Moreover, this temperature is not even considered in the selection of materials for parts and structures to be used at low temperatures.[2-4]

The situation becomes especially dangerous when the Debye mechanism of fracture coincides with one of the temperature mechanisms (see Section 6.2). This takes place in the case of a dramatic decrease or a considerable increase in the ambient temperature. Both cases are almost always associated with catastrophic failures of structural members.[3] The following sections present theoretical substantiation of energy methods for control of the stressed–deformed state to prevent sudden failures. Such methods make it possible to produce structures that can adapt to variable external conditions; engineering objects with artificial intelligence can be built on this basis.[12,18,21,24]

The possibility of controlling the stressed–deformed state (movement along path *acd* in Figure 7.10) was discussed in previous chapters. As seen from Equation 9.14, $\tau \to \infty$ if $dNf \to 0$. This is equivalent to meeting the condition of $ds(\sigma,T) = 0$. If we substitute two-parameter entropy $s(\sigma,T)$

expressed in terms of a product of its components that depend only on mechanical stresses $s(\sigma)_T$ and temperature $s(T)_\sigma$, i.e.,

$$s(\sigma,T)=s(\sigma)_T s(T)_\sigma,$$

and, differentiating, yield

$$ds(\sigma,T)=s(\sigma)ds(T)+s(T)ds(\sigma).$$

At $ds(\sigma,T) = kdN = 0$, we find that

$$\frac{ds(\sigma)}{s(\sigma)}=\frac{ds(T)}{s(T)}. \qquad (9.15)$$

It follows from this equation that the force and thermal fields provide a competing character of variation in entropy. As a result, the numerator in the exponent in Equation 4.28 becomes constant and equal to Ω_0. This means that the control of one of them can compensate for the negative effect of the other. Condition 9.15 fixes the MB factor in a preset range of the energy spectrum and does not allow its ends to come close to critical temperatures T_c and θ_s (Figure 1.14). Even if this does happen, it immediately returns it to the initial state to prevent any development of the destruction processes according to the law set by Equation 8.25.

Figure 9.1, position 6, gives an idea of the efficiency of the simplest method of thermal compensation for the fracture process. If both the external fields are actual and their intensities vary in a random manner, electric, magnetic, and other fields or their combinations may be applied as compensating fields for fracture prevention.[12,21] If the use of the compensating fields is not feasible or is economically unjustifiable, fracture of a structure is unavoidable. Damage that will inevitably lead to fracture accumulates with time following the law given by Equation 8.26.

Because of the lack of knowledge on physical–mechanical processes in a structure of solids at all the scale levels in diverse external fields, the known approaches, based so far on their external phenomenological manifestations, consider deformation and fracture phenomenologically using mathematics as a descriptor of the available experimental results. This imposes considerable limitations on accuracy, reliability, and durability of traditional design and calculation methods. As a rule, they are based on probability calculations, so a certain (even very small) risk of failure is permitted. The new thermodynamic concept of strength, which has been described in previous chapters, makes it possible to substantiate a purely physical approach to the problem of strength of materials.

The system of ensuring reliability and durability common in practice now implies consumption of the energy resource set in the structure of a material during solidification. As a result, it is passive, inertial, and noncontrollable. It does not provide for adjustment of the service state aimed at prevention (or, at least, retardation) of destruction processes and prompt replenishment of the spent part of the resource. An inevitable consequence of this is fracture of materials and failure of parts and structures, which often leads to accidents and catastrophes. In contrast to this system, the subsequent sections of this chapter substantiate a new concept of strength of materials aimed at conservation of the resource during the scheduled service life by controlling the stressed–deformed state of the structure. The controlling regime allows deformation and fracture to be facilitated (if necessary) or prevented. The control action is in providing energy support to the material in extreme situations, i.e., when it is needed.

Structural members and parts by which the stressed–deformed state is controlled should, in principle, differ from those we use now. Such design components are equipped with special systems that provide a material with ability to adapt to variable service conditions. Structures containing such systems have the features of artificial intelligence.[331] On the other hand, the new concept allows identification of the most efficient ways of decreasing energy consumption of technological processes associated with production and processing natural resources or operations associated with shaping of structural materials (cutting, forming, drawing, etc.).[16]

At the modern level of development of strength of materials, the dominant opinion is that the statistical approach is the most common and promising area of handling the problem of reliability and durability. Based on this opinion, many statistical strength and reliability theories have been developed on the basis of a number of assumptions and simplifications. Some have been recognized by specialists and have turned out to be helpful for practical application.

Troshenko[3] gives a general classification of such theories as well as analysis, main advantages, and drawbacks. Using probability theory methods, these methods suggest procedures for prediction of mechanical properties of a material and degree of damage of its structure in an assigned stress field based on a certain, particular model (specific to each approach) of the deformation and fracture process. Thus, the efficiency of any theory depends in full on how accurately the accepted model corresponds to the actual fracture mechanism. Inadequacy of the model to the physical nature of the process leads to a serious decrease in value of statistical conclusions and forecasts. It is likely that their credibility would increase significantly if the fundamental physical law (Equation 4.44) is set based on the discussed statistical approaches.

Accounting for Equation 9.11, Equation 9.14 can be expressed as

$$\tau = \frac{1}{\alpha V} \int_{\sigma_{min}}^{\sigma_{max}} \frac{1}{Z(\sigma, T)} d\sigma \qquad (9.16)$$

assigns statistical nature to the mechanical characteristics of resistance of materials.

In this form, however, it is based on the generalized physical parameter characterizing the heterogeneity of the energy state of the AM set, i.e., MB statistics $Z(\sigma,T)$ rather than on formal outcomes of macroscopic experiments or phenomenological model. MB heterogeneity is a natural feature, characteristic of all the AM systems without any exception. It shows up in different loading systems, in different materials, and under different test conditions.[2-4] Formed in the depth of the AM structure, it propagates to macroscopic levels, passing vast scale-time distances. Despite the fact that in this way secondary effects superimpose it, it still remains most important and decisive. Therefore, the statistical theories of strength and reliability complemented by a physical content (Equation 9.16) constitute the solid basis for reliable prediction of the limiting state of parts and structures under noncontrollable conditions.

The use of the refined AM mechanism of fracture presented in Chapter 6 makes statistical models more objective and engineering calculations utilizing the criterion of operational reliability more accurate.[332] Upgraded statistical theories of strength will make it possible to account for not only technical, but also economic factors. Construction of highly critical engineering objects with a requirement of a 100% guarantee of their failure-free performance within the entire service life (i.e., where economic factors play secondary roles and possible social, moral, and human losses become of the paramount significance) is impossible without development and application of the stressed–deformed state control systems.[333]

9.4 Principles of the Theory of Design of Materials Properties

It follows from Equations 4.43, 9.11, 9.14, and 9.16 that strength and durability are determined by the resource or the energy content of a unit volume of a material. Setting maximum possible resource s (Equation 2.112) into the volume of a material at the stage of design of its chemical composition increases reliability and durability of parts and structures under noncontrolled service conditions.

It follows from the laws of atomic mechanics (Chapter 1) that energy of a rotos is fully determined by the crystalline lattice parameters r_a and U_a (Equations 1.37.1, 1.39, 1.48, and 1.63), which it acquires at cold-shortness temperature T_c (Equations 1.95 and 1.100 and Figure 1.14). Expression 1.63 and Figure 1.10 show that, on solidification of the AM system, they fix it in the Cartesian and energy spaces to form the MB statistics (Equation 2.2). Moreover, it becomes clear from Section 2.9 that, in change of the aggregate state of a solid in thermal fields, all the characteristic phase transition (solidification T_c, melting T_m, and evaporation T_l) temperatures are expressed in fractions of these parameters (see Table 2.1).

Regularities shown in Table 2.1 are of a general character and valid for all solids without exception. Peculiarities of each of them depend on particular values of parameters r_a and U_a. In turn, the latter are determined by the AM structure of a specific chemical element (Equation 1.46) and density of packing of atoms per unit volume of a material (Equation 1.63). It follows from Equations 1.61 and 1.62 that r_a means the minimum distance to which neutral atoms of a given chemical element should be drawn together at temperature T_c to form first a rotos (Figures 1.5 and 1.6), and then a solid. Equation 1.63 estimates this distance with respect to the third quantum level (counting from the nucleus). As seen, parameter r_a is fully determined by the valence of a given chemical element.

By the moment of solidification, the internal energy determined for each rotos by U_a (Equation 1.48.II; in Figure 1.7a, designated by vertical line ar_a) is stored in the crystalline structure. It is numerically equal to the work of the external source consumed for the earlier discussed drawing together. In this case, energy U_a and deformation r_a parameters are functions of three arguments: mass m of a chemical element, its valence z, and common charge e (Equation 1.48). All the known chemical elements are grouped by these indicators in the periodic Mendeleev's system. It follows from the periodic law of variation of state (see Section 2.8) that the laws of formation and subsequent variation of physical–mechanical properties of solids are associated with it.

Indeed, the energy of an interatomic bond (depth of the potential well in Figure 1.7a) is proportional to the common charge of an atom, e, in the fourth power. If the concepts developed in Chapter 1 of the mechanism of its formation are true, then, with an increase in the total number of electron layers, the increasing number of them (especially starting from the third from the nucleus) would participate in the formation of molecular orbits (see sequence of positions a to b to e to d in Figure 1.3). Therefore, as the number of a chemical element in the periodic system increases, its AM bonds tend to increase in rigidity and strength. Valence z of chemical elements has a substantial effect on value of parameter U_a (Equation 1.48.II). In Mendeleev's system, it varies following the periodic law. This means that, with a general trend to increase in strength and rigidity of the AM bonds, periodic variations in these properties should clearly show up inside the periodic table.

Equations 1.48.I and 1.48.II allow reproduction of the variations in functions $r_a = f(m, z, e)$ and $U_a = f(m, z, e)$ within the table of chemical elements (Figure 9.2). In each period (I to VI), number 1 designates curves, which have a minimum, and number 2, those having a maximum. The former determine the dimensional parameter of a rotos r_a while the latter determine the energy parameter U_a. In odd rows (5 to 9), they are indicated by numbers with a prime. Zero values of the functions, lying on the axis of abscissas, correspond to inert gases. They are incapable of creating AM bonds, as all their electron shells are fully filled. Numbers 2, 3, 4, 5, and 6 on the axis of ordinates designate minima of function r_a (left-hand side) and maxima of U_a (right-hand) in each period (II to VI).

As follows from Table 2.1, parameters r_a and U_a are fundamentals for the solid state. As seen from Figure 9.2, they vary periodically, depending on

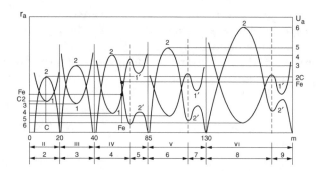

FIGURE 9.2
Variations of structurizing parameters r_a and U_a within the ranges of periodic system of chemical elements.

the location of a chemical element in Mendeleev's table. All physical–mechanical properties of solids are derivatives of these parameters. As with the chemical properties of solids, they are periodic functions of the number of an element (see Figures 1.20, 1.21, 2.45, 2.46, and 3.69). As shown in Chapter 1, this is associated with the mechanism of formation of interatomic bonds and accounts for the relationship between physical–mechanical properties of solids, with the most important thermodynamic potentials determining the state of a material at any moment of time (Equations 1.111 and 2.65).

The existence of the same fundamental principle underlying physical, mechanical, and chemical properties of a solid explains causes of chemical transformations in mechanical deformation[3,44,45] and creates prerequisites for the control of some of them (e.g., mechanical) through the others (e.g., physical). This possibility is used further for the substantiation of methods of in-process control of the stressed–deformed state of materials and parts during operation. Chapters 1 and 2 consider in detail the physical nature of elementary processes that determine these properties, which makes solution of the discussed control problem much easier.

It can be seen from Table 2.1 that all physical–mechanical properties of simple solids depend on r_a and U_a and, therefore, on atomic parameters m, z, and e. Periodic character of variation in the melting point is shown in Figure 1.20, that of the Debye temperature in Figure 1.21, elasticity modulus and tensile strength in Figure 2.45, density and thermal expansion coefficient in Figure 2.46, and evaporation heat and coefficient of rigidity of interatomic bonds in Figure 3.69. Depending on the location of a chemical element in Mendeleev's table, these vary following the periodic law. Confirmation that the energy of atomization of all chemical elements obeys this law can be found in Bogoroditsky et al.[26]

Analysis of these relationships shows that an increase in mass of an atom, m, and its charge e (the number of an element) is accompanied by a monotonic increase in each subsequent maximum. (Compare, for example, ordinates of

TABLE 9.1

Chemical Elements with a Maximum Energy Capacity of Interatomic Bonds

1	Series number of element equal to a total number of electrons		6	14	24	33	42	51	74	83	106
2	Element designation		C	Si	Cr	As	Mo	Sb	W	Bl	-
3	Atomic mass		12.0	28.0	51.9	74.9	95.9	121.7	183.8	208.9	-
4	Group		IV	IV	VI	V	VI	V	VI	V	VI
5	Location of electrons at energy levels (numerals designate their quantity)		2(2)	3(3)	4(4)	4(5)	5(6)	5(7)	6(8)	6(9)	7(10)
6	Location of electrons in energy levels (numerals stand for their quantity)	Q									2
		P							2	5	12
		O					1	6	12	18	32
		N			1	5	13	18	32	32	32
		M		4	13	18	18	18	18	18	18
		L	4	8	8	8	8	8	8	8	8
		K	2	2	2	2	2	2	2	2	2

the plot in Figure 1.20 at points 14, 24, 42, and 74.) The similar variations in valence of elements inside each period assign similar variations in rigidity and energy capacity of interatomic bonds (Figure 3.69), which in turn is indicative of the existence of extrema of parameters r_a and U_a inside each period (Figure 9.2). This highlights a new physical–mechanical aspect of the periodic Mendeleev law.

Table 9.1 shows groups of elements whose interatomic bonds have a maximum energy capacity. Elements with series numbers 6, 14, 24, 42, 74, and 106 form main maxima, and 34, 52, and 84 form auxiliary maxima.

The first are located in even rows and the second in odd rows. One electron level participates in formation of the AM bond in elements 6, 14, 34, 52, and 84; two electron levels participate in that in the rest of the elements. For example, in formation of a bond in the sixth element, the atoms draw together to the first or Kth level to form a common L-level that contains eight electrons (four from each atom). For this reason, high heat resistance, hardness, and strength are characteristic for carbon compounds. In formation of silicon bonds, the distance between atoms is counted from level L. Now the interacting atoms share levels M, which again contain eight electrons. As r_a of silicon becomes higher than that of carbon, its interatomic bonds have lower energy capacity. This principle persists in movement from element to element along line 6 in Table 9.1.

Two valent energy levels participate in the process of formation of solids of elements 24, 42, 74, and 106. Despite an increase in the total charge, each of them contains 14 valent electrons. In bond formation, their quantity becomes equal to 28: 18 fill up the lower valent level of one of the atoms, and the remaining 10 form two common energy levels that simultaneously service two interacting atoms. The latter contain two and eight electrons,

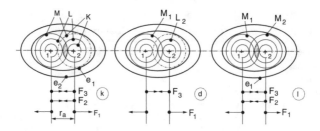

FIGURE 9.3
Schematic of formation of interatomic bond in chromium molecules.

respectively, which rotate about both polarized ions along the elliptical paths
(Figure 1.12). Figure 9.3 shows how this occurs, for example, in chromium.

Let two atoms be at a distance of r_0 from each other and be designated by
numbers 1 and 2 at position k. Each has two filled levels: K and L. The first
level has two electrons, and the second has eight electrons. Unfilled levels
M and N contain 13 and 1 electrons, respectively. For level M to become
neutral, it must contain 18 electrons. For example, in drawing together, atom
1 receives five electrons, lacking for its filling up, from atom 2. Valent levels
M and N of this atom become free (conditionally shown in position k by a
dashed line). The other 10, which became free, form two new common energy
levels, e_1 and e_2. The first has two electrons and the second has eight. Because
they must service both atoms to form the AM bond, these electrons perform
movement along the molecular paths. Atoms 1 and 2 become polarized ions
located at the focuses of the elliptical path (Figure 1.12).

Vorobiov and Boroviov's study[92] presented a proof of splitting circular
energy levels into the elliptical levels during formation of the NaCl crystal. It
is shown that, as free ions Na^+ and Cl^- approach each other and the Coulomb
interaction energy increases, the shape of their own circular levels becomes
increasingly deformed. When the distance between them becomes equal to a
lattice period r_a, wave functions of both ions are overlapped to transform paths
of movement of valent electrons and ions into ellipses (Figure 1.12).

According to Condition 1.63, parameter r_a is smaller than a geometric sum
of radii of levels M and L. In this connection, their electron shells are
deformed (in position k a dashed zone between level M of the first atom and
level L of the second atom). Dynamic equilibrium of such a structure depends
on the relationship of three kinds of forces generated by three charges: two
positive charges concentrated in massive nuclei and, therefore, inertial, and
one easily transformable negative mobile charge formed by the electron
system surrounding nuclei 1 and 2. Consider how these forces vary with an
increase in distance between the nuclei.

First, dispersion of valent electrons in molecular orbits e_1 and e_2 leads to
positive charges 1 and 2 becoming noncompensated. This generates repulsive
forces $\mathbf{F}_1$ (position k in Figure 9.3). Second, deformation of natural circular
orbits at level M (atom 1) and L (atom 2) leads to the formation of a forbidden

zone (dashed region), where movement of electrons is impossible. At these levels, they perform a circular movement, omitting this zone. These levels are transformed from a regular sphere into something close to a hemispherical solenoid. While moving along it, electrons create a magnetic field that pushes positive charges to each other. This results in attractive force F_2.

Electrons rotating on molecular orbits e_1 and e_2 are equivalent to a circular electric current that exists around both nuclei. It is not difficult to see that this current creates a magnetic field, which generates attractive forces F_3 between two positive charges. Inequality

$$F_3 + F_2 > F_1 \qquad (9.17)$$

provides an internal reduction between atoms. In Chapter 1, this state of an interatomic bond is called compresson or C-state (position k in Figure 9.3). As the distance between atoms increases, it persists up to the Debye temperature θ, where $r_b = 1.5r_a$ (Figure 1.7a).

At this moment the situation radically changes. Individual electron shells M_1 and L_2 no longer exert any effect on one another (position d in Figure 9.3). As a result, force F_2 disappears and Relationship 9.17 is transformed into inequality $F_1 > F_3$. A relative repulsion becomes dominant between atoms 1 and 2 and the distance between them starts growing. The state of a rotos, whose characteristic feature is relative repulsion of constituent atoms, is called dilaton, or D-state, And persists until $r = r_a = 2r_0$ (see column 8 in Table 2.1).

Here a material is transformed from solid into liquid state. Size of a molecular path e_1 near focus 2 becomes commensurable with an individual circular orbit of level M_2 and electrons transfer from it to more profitable level M_2. (In position l in Figure 9.3, this situation is conditionally shown by replacement of the dashed contour by a solid one.) In turn, electrons from orbit e_2 transfer to e_1 and thus orbit e_2 ceases to exist. In the liquid state (position l in Figure 9.3), natural orbits M_1 and M_2 become deformed, and Inequality 9.17 is restored. However, structure of the interatomic bond becomes fundamentally different compared with position k. In this connection, the liquid state can be considered the C-state of the second kind.

A distinctive feature of the most stable AM bonds of simple bodies (elements 6, 14, 24, 42, and 74) is filling of all electron levels, including natural circular and molecular elliptical levels. A similar situation takes place in the odd rows: in row 5 in arsenic (element 33), in row 7 in antimony (element 51), and in row 9 in bismuth (element 83); each has five valent electrons. While joining together to form the AM bond, they fully fill up two molecular levels, e_1 and e_2. Two electrons function at the first and eight electrons function at the second. As a result, the AM structure formed is similar to that shown in Figure 9.3. However, having large relative sizes (their fourth and fifth natural energy levels are filled), they are characterized by a lower binding energy (see Figure 1.20).

It is not difficult to show that the rest of the elements have AM bonds of a different structure. For example, palladium (its number is 46) does not change valent electrons. Therefore, it has no molecular levels; its interatomic bond is maintained only due to existence of repulsive, F_1, and attractive, F_3, forces.

A generalized rule follows from analysis of the mechanism of formation of the AM bond of different elements of Mendeleev's periodic table: energy of the crystalline lattice of simple solids depends on their ability to create common molecular levels and the completeness of their filling. This ability is determined by valence; within one period, the energy consumption of the bond decreases with an increase in radius of the last complete circular orbit. Other conditions being equal, the higher the charge of the nucleus, the higher the orbit. Structures with a minimum geometrical parameter r_a have the highest energy consumption. The higher the binding energy, the higher the stability of a structure with respect to external thermal, force, and other fields.

Analysis of Figures 1.20, 1.21, 2.45, 2.46, 3.69, and 9.2 allows a conclusion that all physical–mechanical properties of solids are determined by a generalized thermodynamic equation of state (2.96). The same conclusion follows from Table 2.1, and is directly or indirectly confirmed by many experiments.[22,45,65,92,114,263,264,334–338]

It can be seen from Figure 9.2 that creating different combinations of chemical elements, it is possible to obtain a wide range of parameters r_a and U_a and, therefore, fundamentally change the physical–mechanical properties of solids. For example, in metallurgy, properties of steels can be varied over wide ranges by varying combinations of alloying elements and their content. As an example, Figure 9.4 shows the CD mechanism of formation of different properties of carbon steels and Figure 9.5 shows their variations depending on the carbon content.[327] Particular values of parameters r_a and U_a for carbon and iron are found from Figure 9.2, where they are designated on the axis of ordinates by chemical symbols Fe and C.

Referring to Figure 9.4, let the form of Function 1.26.II in the Cartesian space V in direction r for iron (Fe) be set by curve 1, and that for carbon (C) by curve 2. This arrangement of curves 1 and 2 is not incidental. It is assigned by a generalized constitutional diagram of a rotos (Figure 1.14), in accordance with which the minimum of the first coincides with θ_s, and that of the second with the θ_f-temperature. Plots 3 and 4 determine the character of variation in the Lorentz forces (Equation 1.16) generated by their EM dipoles (Figure 1.7b). Vertical lines passing through points F_e and θ_s delineate the compresson phase of iron designated as F_k, where the MB distribution of this element is located (curve 5). The MB factor of carbon occupies interval C_d formed by vertical lines outgoing from points θ_f and n, and it is designated by numbers 6, 7, and 8, depending on the concentration per unit volume.

Formation of iron–carbon molecules (Figure 2.7b) is characterized by summing up of energy (1 and 2) and force (3 and 4) functions. Their geometrical sums are shown by plots 9 and 10. Formed by vertical lines F_e and n, interval F_c determines the CD boundary of iron–carbon alloy. Location of the MB factor of the alloy inside it depends on the carbon content (distributions 6 to 8); it forms as the superposition of plot 5 and one of 6 to 8 distributions.

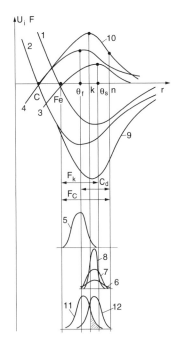

FIGURE 9.4
Compresson–dilaton mechanism of formation of properties of carbon steels.

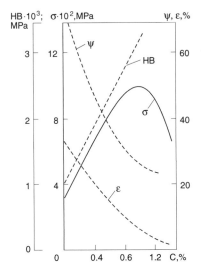

FIGURE 9.5
Variations in properties of steel, depending on the carbon content.

At a low carbon content (curve 6), the MB distribution of the alloy is almost completely located in the C-phase of the matrix (interval F_k). Its maximum is also located there (curve 11). As the concentration of carbon increases (distribution 6 shifts to 7 and then to 8), the MB factor of the alloy 11, while changing in shape, gradually shifts from the C- to D-phase (interval k to n). Its maximum also moves there (distribution 12). In this case, the volume of the D-phase in the alloy (dashed region in distribution 11) continuously grows, and that of the compresson phase decreases. As a result, the alloy transforms from the C-state (steel) to the D-state (cast iron). This transformation occurs when the maximum of the MB distribution of the alloy coincides with the CD boundary (vertical line passing through point k).

The described CD phase processes explain the variations in mechanical properties of steels with variations in carbon content (Figure 9.5). As seen, an increase in carbon leads to a nonmonotonic variation in the tensile strength σ, decrease in ductility ε, ψ, and an increase in hardness HB. At a low carbon content (below 1%), σ increases due to an increase in the compresson pressure in excess of the dilaton pressure. Then an increase in volume of the D-phase changes this ratio to the opposite: a decrease in the C-pressure is accompanied by a decrease in tensile strength (descending region of curve σ), increase in compressive strength, and transition of the alloy into the brittle D-state. It is the embrittlement process that accounts for a continuous decrease in ductility (plots ε and ψ). An increase in the volume of the D-phase shows up as an increase in resistance to the penetration of indenters of different sizes and shapes into a material. This causes a continuous growth of hardness of the alloy (plot HB).

The effect on properties of pure iron by phosforus, sulfur, nitrogen, and other similar elements can be considered using a similar approach. The presence of these elements leads to a dramatic decrease in ductility and toughness. They take the 7th, 15th, and 17th places in the periodic table of chemical elements, occupying the end regions of the descending branches of periodic functions r_a and U_a (Figures 1.20 and 9.2). Compared to iron, they have a much higher parameter r_a and a much lower energy U_a. Their U and F curves are a smeared shape and located not to the left from point F_e, as in the case of carbon (Figure 9.4), but to the right from this point. From practice, this results in well-known deterioration in mechanical properties of steel[3,327] associated with the presence of these elements in its composition.

Manganese, silicon, chromium, molybdenum, and tungsten occupy positions 12, 14, 24, 42, and 74, respectively, in Mendeleev's table. It can be seen (Figures 1.20 and 9.2) that they are located on the ascending branches of the preceding functions or coincide with their maximum. Being on coordinates $U = f(r)$ and $F = f(r)$ to the left from the matrix element (point F_e in Figure 9.4), they improve the properties of steel.

It should be noted in conclusion that the periodic law of variations in structurizing parameters r_a and U_a of the known chemical elements (Figure 9.2) and the generalized equation of state (2.96) open up wide prospects for targeted control of physical–mechanical properties of solid bodies. In analogy with the periodic Mendeleev law, which can be formally regarded as a one-plus-one combination of 108 elements, they make it possible to compile periodic systems for combinations of 2, 3, 4, etc. chemical elements. Also, they can be used to predict properties of the corresponding compounds (by preliminarily performing the required calculations and analyzing dependencies of the type shown in Figure 9.4), i.e., design materials with preset properties.

Moreover, filling volume V with initial components θ_1, the MB factors of which are located in the opposite regions of the constitutional diagrams (positions 8 to 10 and 11 to 15 in Figure 1.14) due to formation of different types of rotoses near θ_f and θ_s (Figure 2.7a and b), we can produce various composite materials. In particular, reinforced concrete, a composite material capable of resisting compressive and tensile loads equally,[330] is fabricated by reinforcing a concrete matrix (position 8) having a high compressive strength with a steel grid (position 11 or 12). Different types of composite materials are fabricated similarly, thus making different combinations of materials having a fundamentally different δ_f and δ_s (Equation 2.94). Technologies of their fabrication are covered in voluminous special[3] and periodical[328] literature.

9.5 Formation of Anisotropic Structures

If it is impossible to meet Condition 9.10 by selecting parameters ρ and θ, another possibility to produce the required mechanical properties should be considered. This includes locating the MB factor in the safe zone during manufacture of a material (e.g., in position 11 or 12 in Figure 1.14) and can be achieved by controlling the structurization processes.

Absolute size and shape of elliptical paths of atoms depend on the value of t_i (Figure 1.7a) while orientation of orbital planes A with respect to direction r is determined by the vector of mechanical moment M (angle α in Figure 2.9). Members of set N can form different angle α with r (Figure 3.7) and thus the depth $r_a a$ and shape of the potential wells (Figure 1.7a) become dependent on the choice of r. This gives parameters $\mathbf{P}$ and $\mathbf{s}$ in Equation 2.65 vector character and makes physical–mechanical properties of materials dependent on the shape and location of the MB factor (positions 11 to 15 in Figure 1.14) in the energy, δ_s, domain and Cartesian, V, space and also on the orientation of the AM bonds.

Because they are electrically charged, structurizing particles have electric $M(\mathbf{E})$ and magnetic $M(\mathbf{H})$ moments in addition to mechanical moment M (see Section 3.3). No inherent submicroscopic anisotropy (Figure 2.7) is detected in practice (as a rule) because of the averaging of properties in transition from a microworld to macroscopic scales. However, it immediately manifests itself in the external fields through the orientation effects (OE). In magnetic fields, it is called magnetization,[30] in electric fields, polarization,[6] and in mechanical and thermal fields, elastic deformation.[2,3] In all fields, the elastic stage of deformation changes the configuration part of entropy s_c, while the inelastic stage that follows changes the destruction part s_d of entropy (Equation 2.113).

Equation 3.15 allows the suggestion of active methods for the control of the structurization processes and stressed–deformed state after solidification.[12,18,21] Because it is a mean-static value of AM set N, the internal pressure **P** is regarded at the macrolevel as an isotropic parameter, whereas any of the external fields **E**, **H**, σ, and T is characterized by a clearly pronounced anisotropy. The kinetic part of the overall internal energy sT can be redistributed in the volume of a material in the direction of the corresponding field by affecting the macroscopic parameter **s** through submicroscopic moments $M(E)$, $M(H)$, $M(\sigma)$, and $M(T)$, thus creating artificial anisotropy.

In a direction of the small axis of the ellipse (position 1 in Figure 3.7), the interatomic bond exerts a 1.8 times higher resistance to external effects than in the perpendicular direction (position 2; see Sections 1.6 to 1.8). Orienting the bonds in external fields in the desired direction and fixing them in the solidification process, resistance **P** can be almost doubled due only to a decrease in s_c.

In fact, large, a, and small, b, semi-axes of elliptical orbits of the constituent atoms of a rotos at the Debye temperature θ (in Figure 1.7a, corresponding to point b, which is the inflection point of curve r_1ab) can be found from Equation 1.38, accounting for Equation 1.45 as

$$a_\theta = \frac{r_a}{1-g^2} = 1.125 r_a; \quad b_\theta = \frac{r_a}{\sqrt{1-g^2}} = 1.07 r_a \qquad (9.18)$$

In the compresson phase, the potential energy of a rotos decreases from U_a to $0.88\,U_a$, i.e., by 12% (see columns 4 and 6 of Table 2.1) with an increase in radius vector r from r_a to $r_b = 1.5 r_a$. Therefore, a deviation of 0.1% in r_a causes a change of 2.4% in the binding energy. This means that the movement of an atom from the center of the rotos to a distance equal to the large axis of the ellipse decreases the binding energy by $1.125 \times 2.4\% = 3\%$, and that to a distance corresponding to the small axis by $1.07 \times 2.4\% = 1.68\%$.

It turns out that, during the process of rotation of atoms on the elliptical orbits about the center of the rotos (points 0 in Figure 1.7b), its binding energy changes from 1.68 to 3% to both sides from U_a. If we take into account that the entire range of its variation in the compresson phase amounts only to 12%, it should be admitted that this variation of energy is substantial. That is, in the direction of the small axis of the ellipse (position 1 in Figure 3.7), the rotos provides resistance

$$\frac{3}{1.68} = 1.785 \qquad (9.19)$$

times higher than that in the perpendicular direction.

It follows from Equations 1.26 and 1.31 that energy of an individual rotos depends on mechanical moment M (Equation 1.27). This means that the internal energy of a body can be redistributed in volume V in the direction of the maximum projection of moment M, thus creating artificial anisotropy. Its effect on mechanical properties can be estimated from Equation 9.19.

Sometimes the preferred orientation of electric, magnetic, or mechanical moments can be formed naturally during the process of solidification of bodies. In the first case, such bodies are called segnetoelectrics;[26] in the second, they are called ferromagnetics[30] and, in the third case, anisotropic materials.[336] The latter will be our focus in further considerations because mechanical orientation allows optimal structures to be created for given working conditions. Controlling the solidification process can solve the problem of optimal packing of a structure. Anisotropic structures intended for resistance to specific types of the stressed state in given external fields can be formed this way.

Due to chaotic arrangement of rotoses in volume of a material, internal resistance $\mathbf{P}$ (Equation 9.10) is an isotropic value, whereas internal stress σ, according to Equation 3.73, is anisotropic. Formation of artificial anisotropy causes an increase in $\mathbf{P}$ in one of the directions and its decrease in the other direction. This may affect the left-hand part of Equation 9.10 so that phase rearrangement of the structure in the working direction will occur in a reversible manner. For example, in one of the directions, the formed MB distribution can be such that it will be fully located in the C-phase, whereas in the other (nonworking direction) its major part will be in the D-phase (positions 6 and 8, respectively, in Figure 2.47). This causes a decrease in energy consumption of the shape transformation processes and improvement in quality of treatment.[16]

The orientation effect can appear under a long-time action of external factors or can be deliberately formed at any stage of the service life of a part, starting from the moment of structurization of a material to the final stage of service. In this connection, it is necessary to distinguish the following types of orientation effect: crystallization, technological, operational, and restoration. In turn, the technological type of orientation effect includes deformation and composite subtypes. The crystallization type is arranged in the process of structurization and fixed in a structure during solidification.

Technological anisotropy is caused by altering the texture of metals and alloys using appropriate technological treatments of parts and semifinished products. Composite anisotropy is deliberately arranged in a structure combining matrix materials and fillers. Operational orientation effect is formed in the most stressed zone of a part only for the time of action of the external factor. Anisotropy achieved at the manufacturing stage of a material gradually attenuates under the effect of varying service conditions and can be restored by repair, relaxation, or maintenance. This is the nature of the restored orientation effect.

As to duration, orientation effects can be further divided into long- and short-time. The former includes crystallization and technological anisotropy, while the latter includes operational anisotropy. As to the degree of the effect of a human factor, the orientation effect can be referred to as natural or deliberately formed. With respect to external fields, orientation effect can be noncontrollable or passive and controllable or active. Finally, as to the physical nature of the generating field, the orientation effect can be mechanical, thermal, electric, magnetic, or combined.

All the previously mentioned types of orientation effect are well known in the respective branches of knowledge, although the extents to which they are studied and, moreover, the extents of their readiness for practical application in materials science and strength are different. The efficiency and importance of some of them have not yet been fully understood. In this section, we consider the basic types of the orientation control of resistance of materials to deformation and fracture. Their efficiency will be illustrated by examples, where possible; some will be considered for the first time.

The essence of the crystallization method is that an anisotropic structure is produced during solidification of a material. External force, thermal, electric, and magnetic fields, or their combinations, generate orientation effect. A structure formed under conventional conditions due to chaotic orientation of rotoses is isotropic at the macroscale. Crystallization in the force fields leads to their preferred orientation in the direction or counterdirection of the field. Mechanism of such an orientation is described in detail in Section 2.6. Mechanical effects on solidifying materials can be exerted by applying static and dynamic loads (pressing and impact), gravitational fields (application of dead loads), or centrifuging. They find application in production of steels and alloys[41,323,329] and in building materials such as concrete.[330] According to experimental data collected by producers of prefabricated reinforced concrete, centrifuging leads to a 30 to 40% increase in the compressive strength of concrete compared with preliminary consolidation by vibration.[330] Solidification of concrete under an extra load of 100 g/cm^2 results in a 3 to 5% increase in density and 15 to 25% in the compressive strength.[188]

The influence of the external fields on rotoses (Figure 1.7b) is described in Section 3.3 and in Komarovsky,[24] as well as illustrated in Figure 2.9. Any external field is characterized by a clearly defined anisotropy; therefore, it not only transforms isotropic polyatomic molecules (Figure 2.7a and b) according to the diagram shown in Figure 2.8a, but also orients them in the direction of the field, thus creating artificial anisotropy. The problem of formation of anisotropic structures intended for resistance to specific kinds of stress state can be solved through controlling the solidification process. The theory of control can be based on the control of interactions (Equation 3.19).

Figure 9.6 shows temperature dependence of the compressive strength of concrete related to setting conditions: 1, solidification of cast concrete under natural conditions; and 2, setting of concrete of the same composition while applying impacts (concrete spraying).[188] As seen, if the temperature dependence retains its general form, the impact effect leads to almost doubled strength over the entire temperature range. The situation is the same with ultrasonic impact treatment (peening), which creates favorable compressive stresses in the sublayers of metal during solidification of the weld and increases reliability and durability of a weldment.[1]

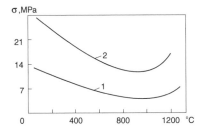

FIGURE 9.6
Temperature dependence of the strength of concrete under different setting conditions.

Chalmers[41] presented the results of comparative investigations of the process of structurization of steel in the fields of gravity and centrifugal forces. The effect of modification of structure is observed in both cases. It shows up in a decrease in size of grains and their orientation in the direction of the centrifugal force. It was found that centrifuging exerts a more intensive effect on a structure and can be used in practice. Chalmers also observed an orientation effect when vibrations accompanied solidification. It was found that the preferred orientation of grains increases with an increase in the vibration amplitude.

According to Equation 8.29, heat flows have a strong orientation effect.[24] Major factors that determine the process of solidification of metals and alloys poured into a mold are the heat removal rate and the nucleating ability.[41,329] It was noticed that when cooling rate is relatively low and the volume of material is relatively large, the crystals have preferred orientation in the direction normal to the mold wall, so their size in this direction is many times larger than that in other directions. This is explained by the fact that rotoses, while transforming from the liquid to solid state (first in dilaton and then in compresson phase; Figure 2.47) and losing excessive kinetic energy, are oriented in the direction of the heat flow, i.e., normal to the wall. Also, an increase in the cooling rate causes activation of the orientation process and an increase in the number of crystallization centers.

Figure 9.7 shows the intensity of the orientation effect of the thermal field (curve 1) and variation in the size of grain (curve 2) depending on the distance from the mold wall L in solidification of steel.[41] The angle between the direction of the preferred orientation of grains and the wall surface is designated as β and the quantity of grains in 1 mm^2 of the section normal to this direction is designated as n. As seen, a decrease in thickness of the wall of a casting is inevitably accompanied by an increase in the orientation effect and in the number of grains per unit volume of the material.

Figure 9.8 shows the manifestation of the thermal orientation effect as the dependence of the tensile strength of cast iron on the wall thickness of the casting.[337] As seen, the strength significantly increases with a decrease in wall thickness. Because the compressive strength of this grade of cast iron is 4.5-fold higher than the tensile strength, one should expect a corresponding

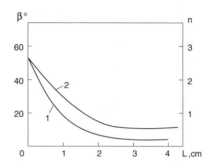

FIGURE 9.7
Orientation effect of thermal field (1) and variation in size of grains (2) on solidification of steel.

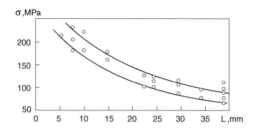

FIGURE 9.8
Dependence of the tensile strength of gray cast iron on the wall thickness of the casting. The upper curve corresponds to cooling in the chamber and the lower curve to that in open air.

increase in the compressive strength as result of the thermal orientation effect. It follows from analysis of experimental data presented in Reference 337 that other grades of cast iron exhibit a similar character of manifestation of the thermal orientation effect. We can conclude that the efficiency of the orientation effect in thermal fields can be controlled by cooling (rate and direction).

The orientation effect is especially pronounced in single crystals (Figure 9.9) and eutectic alloys (Figure 9.10);[191,338] its manifestation varies from crystal to crystal. Figure 9.9 shows curves "stress σ to strain ε" in shear for single crystals of zinc of different orientations at a temperature of 294 K.[191] Significant dependence of strain characteristics on the orientation of a crystal is clearly seen.

The orientation effect increases resistance of materials under static and under cyclic loading conditions.[339] The extent of its influence on the tensile strength of a eutectic alloy can be evaluated using the data shown in Figure 9.10.[340] Numbers 1, 2, and 3 designate temperature dependencies of the tensile strength of eutectic alloy (Co, Cr) to $(Cr, Co)_1C_3$ in the longitudinal and transverse directions and at an angle of 45°, respectively, and, for comparison number 4 designates similar dependence for alloy MAKM-302, close in composition to the eutectic one. As seen, in accordance with theoretical predictions

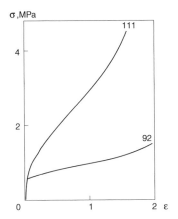

FIGURE 9.9

Deformation curves of single crystals of zinc with a different orientation of crystallographic planes at a temperature of 294 K.

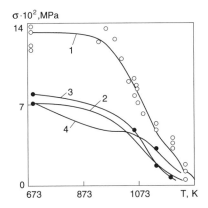

FIGURE 9.10

Effect of the degree of anisotropy on tensile strength of eutectic alloy.

(Equations 9.10 and 9.19), the tensile strength can be almost doubled over the entire temperature range due to directional crystallization. It has been shown[340] that such materials have increased characteristics of creep and corrosion resistance at high temperatures.

If heat flows are not arranged in a special way, the orientation effect attenuates with distance from the cooling surface, as can be seen from Figure 9.7. This results in a decrease in its efficiency.

Electric and magnetic fields have the highest efficiency in terms of controlling structurization processes. They allow anisotropic structures to be created for specific kinds of stressed state: tension, compression, torsion, bending, or their combinations.[341]

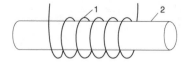

FIGURE 9.11
Schematic of formation of the orientation effect in a magnetic field.

The physical nature of the orientation effect in electric (polarization) and magnetic (magnetization) fields is considered in Section 3.3. Their efficiency depends on the mobility (polarizability) of rotoses; they acquire the highest mobility in a liquid state (see Figure 2.47). Consumption of energy for anisotropy in this state is minimal and its fixation in solidification is the best.

Figure 9.11 shows a schematic of formation of the orientation effect in a magnetic field. Structurization processes occur in the magnetic field generated by coil 1, and billet 2 is placed inside. Combination of solidification and magnetization makes it possible to produce anisotropy in the axial direction. Anisotropy of dielectrics in the electric field can be formed in a similar way.

Gerasimov et al.[30] have shown that polarization in the electric field leads to an increase in the binding energy in the direction set by the field due to an increase in van der Waals attractive forces. This increase varies from several percent for light elements to several tens of percent for the heavy ones. Besides, the higher the polarizability of both, the larger the increase.

Magnetically controlled melting of titanium[323] enables an increase in the density of ingots and the degree of dispersion of their structure, a decrease in impurities, an increase in workability under various treatment conditions, and production of different shapes of ingots. Electromagnetic control of the weld pool[325] provides the effect of improvement of the weld shape and increase in deposition efficiency.

Because modern technologies are not intended to intervene in the structurization process, they cannot provide directionally structured and volume-homogeneous entropy s. Therefore, the kinetic part of the overall internal energy T, while distributing nonuniformly according to Equation 2.65, causes high residual stresses P in volume V.[3] On the contrary, homogeneous and ordered systems s are characterized by a uniform internal thermal field T, in which no overstresses are formed at micro-, meso- and macroscopic levels. New advanced technologies should make use of one of the remarkable peculiarities of molten metals, i.e., the readiness with which they form ordered structures under the effect of external controlling fields (Equation 9.15), thus forming optimal values of the configuration, s_c, and phase, s_f, components of entropy.

In practice this means that the structure of a given material should be given particularly that form of ordering which would meet future service conditions as much as possible. The lower its adaptation to these conditions, the lower the initial level of technogenic safety of a finished product. Conflicts and paradoxes of traditional technologies are often caused by lack of knowledge or by ignorance of those remarkable capabilities that are introduced into technology by controlling the structurization process. There is nothing extraordinary in this, as no alternative method (except for those indicated by Equation 3.19) is available for packing atoms in the volume so that a required resource s is provided.

9.6 Correction of Resource after Solidification

After solidification, when θ becomes a constant value, a material starts its life following the law prescribed by Equation 3.19. Its subsequent behavior, up to exhaustion of the resource, can be explained within the frames of this law. It is likely that entropy can be corrected by varying other parameters of the state (see Section 2.5) in a special enhancing process. It is seen from Equations 5.3 and 9.10 that, after solidification, the mechanical properties can be changed due to redistribution of the internal energy in a given direction in the force, σ, thermal, T, or other fields, and by varying CD ratio

$$\delta = e^{\frac{t_i}{\theta}}.$$

Graphical interpretation of Equation 5.3 is shown in Figure 9.12. Resistance σ and mechanical moment M (Equation 1.27) change, depending on the CD ratio δ following a graph represented by a cubic parabola. Its slope is set by temperature T (see Figures 2.23 and 2.24). The origin of coordinates, where $\delta = 1$, $\sigma = 0$, and $M = 0$, coincides with the Debye temperature θ (Figure 1.14). As shown in Section 1.7, the AM bonds cannot exist here because they dissociate into individual atoms due to thermal deficiency or immediately move to the dilaton or compresson phase by absorbing or radiating a phonon

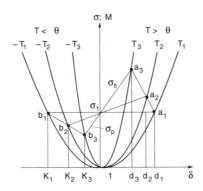

FIGURE 9.12
Dependence of the resistance ability and deformability of the structure on anisotropy of the CD ratio δ.

because in the D-phase $T > \theta$, $\delta > 1$ (right half-plane). In the C-phase, $T < \theta$ and thus $\delta < 1$ (left-hand part of the plot). For the majority of solid bodies, the Debye temperature is close to room temperature (Figure 1.21). Therefore, it can be considered that the $T_3 < T_2 < T_1 < \theta$ inequality is met in the D-region, and $T_3 < T_2 < T_1 < \theta$ is met in the compresson one.

Consider first how the technical state of a material is changed in different types of heat treatment (including cryogenic), and the effect of the CD ambiguity of the structure of this material. As an example, consider a material having no internal stresses in its structure. Continuing to refer to Figure 9.12, let its state be determined by horizontal line a_1b_1 at level σ_1, and the MB distribution at temperature $T = T_1$ have spread k_1d_1. The absence of internal stresses is possible only under conditions of thermodynamic equilibrium, where the concentration of dilatons and compressons is the same in any microvolume. If this is the case, the areas of figures limited by parabolas T_1 and $(-T_1)$ in the dilaton, $b_1k_1 1b_1$, and compresson, $a_1d_1 1a_1$, phases should be equal to each other. Equality of the C and D pressures implies that a given material has the same resistance to external compressive and tensile loads, i.e., $\sigma_1 = \sigma_p = \sigma_s$. Assume that this material is steel.

Let this material be subjected to heat treatment by heating it to a certain temperature (e.g., to T_3) at different tempering temperatures. An increase in temperature from T_1 to T_3 is accompanied by an increase in the concentration of dilatons (determined by ordinates of points a_1 and a_3) from d_1a_1 to d_3a_3, due to the CD transitions, and decrease in the compresson concentration from k_1b_1 to k_3b_3. Point a_1 located on parabola T_1 in the D-phase moves to position a_3 on temperature curve T_3 along path $a_1a_2a_3$. Point b_1 in the C-phase takes position b_3 by describing path $b_1b_2b_3$ in the phase space. As a result, initially horizontal line a_1b_1 transforms into inclined line a_3b_3. Position of the latter sets the difference between compresson and dilaton pressures (area of figure $a_3d_3 1a_3 >$ area $b_3 1k_3b_3$).

Assume that tempering (i.e., fixing a new CD ratio in the structure) is performed first at temperature T_3 determined by point a_3, then at temperature T_2 determined by point a_2, and finally at temperature T_1 determined by point a_1. In the compresson phase this corresponds to points b_3, b_2, and b_1 on curve $b_1b_2b_3$. After completion of the restructuring process, inclined lines a_2b_2 and a_3b_3 characterize the CD ratio fixed by a given tempering temperature. In this state, the structure is metastable and characterized by the following peculiarities. First, its compressive and tensile strength are not equal, i.e., $\sigma_s \neq \sigma_p$. Second, it contains the internal stresses (tensile or compressive $\Delta\sigma_s = \sigma_s - \sigma_p$, depending on the type of heat treatment). Third, it is unstable. Returning to the initial state sometimes requires a very long time; as proven by experience, it may take tens of years (see Section 5.2.1 and Table 5.1). Experiments fully confirm theoretical predictions.

Shmidt's study[263] presented classification of residual stresses induced by different technologic processes. As stated, the absence of such stresses is a rare exception rather than a rule. This is confirmed by the fact that horizontal line a_1b_1 is one particular case in a wide variety of inclined positions: a_3b_3,

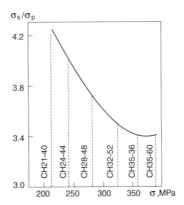

FIGURE 9.13
Relationship between tensile and compressive strength values of gray cast irons.

a_2b_2, etc. To meet certain service requirements, the structure of a material is modified by changing its state of thermodynamic equilibrium into another state using certain technological methods. As a result, most materials have different compressive and tensile strength. As such, the higher the σ_s, the lower the σ_p, and vice versa (see Figure 9.12). Experience shows[3] that this relationship always holds. As an example, Figure 9.13 shows the relationship between tensile, σ_s, and compressive, σ_p, strength values for gray cast irons with flake graphite.[337]

Phase trajectories $a_1a_2a_3$ and $b_1b_2b_3$ are formed in the dilaton and compresson zones when one changes the state of metals and alloys using heat treatments. In fact, these trajectories characterize a change in the physical–mechanical properties, depending on the tempering temperature. In addition, the C-phase determines their resistance and the dilaton phase determines deformation. As seen from Figure 9.12, the former decreases and the latter increases with an increase in the tempering temperature. Experience proves theoretical expectations: Figure 9.14 shows the influence of the tempering temperature on mechanical properties of quenched steel with 0.44% carbon.[327] The following designations are used in this figure: σ_b and σ_T are the yield and tensile strengths, HB the hardness, ε and ψ the longitudinal and volumetric strains.

It is well known (see Troshenko,[3] for example) that mechanical properties of steels can be varied within a wide range by selecting the tempering temperature. Thus, the following general principle is maintained: when the tempering temperature increases, a decrease in strength and hardness takes place with an increase in ductility and impact toughness. We would fully agree with this statement if only the tensile strength is meant. When a material works under alternating loads, it is necessary to take into account the increase in compressive strength that takes place in this case (see Figure 9.12).

As was noted in Section 2.7, temperature T is a powerful regulator of the state of any physical body. Playing a leading role in redistribution of the

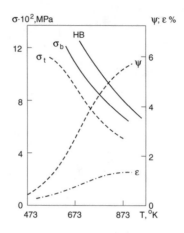

FIGURE 9.14
Effect of tempering temperature on the mechanical properties of quenched steel having 0.44% carbon.

kinetic part of the overall internal energy (Equation 5.3), it changes the Ω-potential and entropy **s**. Heating, like drastic cooling, changes not only the mean, T, parameters but also local, t_i, parameters initiating the CD phase transitions (Figure 1.15). Without exception, they underlie all methods of heat treatment of metals and alloys at high[3,60,61] and cryogenic[223] temperatures. Thermal restructuring in some of them (Figure 2.24) is characterized by the following relationship: $dV_m \gg dV_s$ at $P = \text{const}$ (Equation 2.134), where dV_m is the change in shape (according to the diagram in Figure 2.7) and dV_s is the change in size (according to the diagram in Figure 2.8) of a body. In this case we deal with materials that have a shape memory; they find application for the manufacture of adaptive structural systems.[328]

Cooling forms a CD flow directed into the C-phase, which increases the concentration of compressons due to a decrease in the dilaton density. As such, the MB distribution transforms in a direction of 8 to 7 to 6 (Figure 2.47), decreasing its spread and becoming sharply pointed. As a result, phase lines a_1b_1, a_2b_2, a_3b_3, etc. in Figure 9.12 change their inclination to the opposite. A decrease in the dilaton pressure (recall that it is directed outward from volume) and an increase in the compresson pressure (tending to compress a material) lead to an increase in σ_p, whereas σ_s decreases. This trend is equal to a decrease in deformability of materials.

Dlin's study[223] describes the efficiency of treatment of steels and aluminum- and magnesium-based alloys by cold. Note, in particular, that cold treatment combined with subsequent heating has a substantial effect on stabilization of aluminum alloy D16. One cycle of cooling to a temperature of –70°C and heating to 180°C is more efficient than aging at a temperature of 190°C for 5 h. The most common conditions of treatment of aluminum and magnesium alloys are cooling to temperature of –50 to –100°C and heating to 100°C,

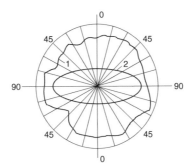

FIGURE 9.15
Manifestation of structural anisotropy in hardness tests of wrought materials.

subsequently bringing it to a temperature of conventional annealing. The higher the initial internal stresses, the more complex the shape of a part, and the lower the cooling temperature, the higher the efficiency of cold treatment. Cooling to –70°C leads to a decrease of 20 to 40% in internal stresses. This effect is explained by phase lines a_3b_3 and a_2b_2 moving closer to horizontal line a_1b_2 (Figure 9.12). No structural changes occur in alloys; cold treatment of steel parts is required only when they are made from quenched steels containing residual austenite in their structure.

Technological deforming is an equally common method for adjustment of the structure of metals and alloys. It is based on the orientation effect illustrated in Figure 2.8 and described in detail in Reference 24 and in Chapter 3. Indeed, the system of chaotically oriented rotoses (Figure 2.7) in hardness tests (Figures 3.15 and 3.27) reproduces a relatively symmetrical line 1 in the polar coordinate system (Figure 9.15) showing variation of the indirect indicators of hardness. Curve 2 shows a flat section of anisotropic system after plastic deformation by tension in a direction of vertical line 00.

To illustrate, Figure 9.16 shows the variations of the tensile, σ_b (curve 1), and yield, σ_T (curve 2), strength of steel EI 811 depending on angle γ between the direction of deformation (90 to 90 in Figure 9.15) toward transverse section (0 to 0 in Figure 9.15) in cold rolling. They were plotted using the experimental data of Berner and Kronmuller's study.[338] Deformation by tension usually creates a "direct" anisotropy, i.e., values of σ_T in the longitudinal direction are maximum. Naturally, cold rolling often creates "reverse" anisotropy, where σ_T (and σ_b, of course) of longitudinal direction is lower than that of transverse.

Combined (i.e., thermomechanical) methods which include intensive strain hardening at high temperatures and structural transformations accompanied by a rapid cooling process, find an increasingly wide application for technological improvement of steels.[223,327] As shown in Section 5.2, these methods are based on fundamentally different mechanisms taking place independently of one another: in thermal fields, following diagram 11 to 12

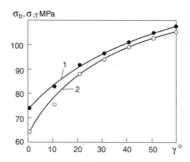

FIGURE 9.16
Variations of the tensile, σ_b (curve 1), and yield, σ_T (curve 2), strength of steel EI 811 depending on angle γ between the direction of deformation (90 to 90 in Figure 9.15) toward transverse section (0 to 0 in Figure 9.15) in cold rolling.

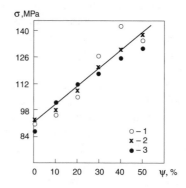

FIGURE 9.17
Yield strength σ of hot rolled low-carbon steel preliminarily deformed at different temperatures—1, –79°C; 2, 27°C; and 3, 100°C—at a temperature of –150°C.

to 14 to 15, and in force fields, following diagram 11 to 12 to 13 (Figure 1.14). Experimental proof of this fact can be found in Ekobori's study.[44]

Figure 9.17 shows dependence of the tensile strength of hot rolled low-carbon steel preliminarily deformed at different temperatures—1, –79°C; 2, 27°C; and 3, 100°C—at a temperature of –150°C. As seen, the fracture stress due to cleavage, σ, strongly depends on preliminary deformation $\psi\%$ and hardly depends on the temperature at which this deformation was performed.

There are two types of thermomechanical treatment: high-temperature thermomechanical processes, where strain hardening is performed at a temperature above the recrystallization threshold, and low-temperature thermomechanical processes, where deformation occurs in a temperature range below the recrystallization threshold. The effects of high- and low-temperature thermomechanical processes on the strength and deformabilty of steels are discusses in detail in Ivanova and Gordienko.[329]

It should be noted in conclusion that the equation of state (5.3) gives the correct description of processes that occur in solids under thermal and force effects. It is likely that such fields (or their combinations) lead to a short-time (only for a time of action of dangerous factors) increase (or decrease) in deformation and resistance of materials during operation, which results in delaying or fully eliminating the process of their fracture. Such methods are described in the next sections.

Voluminous literature is dedicated to composite anisotropy.[3] The existence and wide application of this type of anisotropy confirm validity of the notions of the deformation and fracture processes elaborated in this book. They allow composite anisotropy to be referred to as one of the types of orientation control of properties of materials, probably not the simplest and most efficient.

9.7 Technologies for Formation of Variatropic Structures

If, at the design stage (Section 9.4), it is impossible to meet the requirement for reversibility of the deformation processes (Equation 9.10) and adjustment of properties after solidification is unfeasible or inefficient (Section 9.6), one must accept that part of a working volume might fracture during operation (Section 4.3). Fracture is caused by simultaneous occurrence of two interrelated processes: formation of microcracks (Section 6.3) and their growth (Section 6.5). The first is regulated by the submicroscopic level of a structure and the second by peculiarities of its micro- and macroscopic state.

Understanding the laws of initiation and propagation of cracks (Sections 1.7, 1.9, 3.6, 5.7, and 6.7) creates a realistic prospect for solving the problem of retardation of destructive processes (Section 6.7). It can be realized at the manufacturing stage by formation of a type of structure that would prevent development and propagation of cracks. In this connection, let us formulate the following the idea behind our considerations: if we bring the structure of a material in correspondence with the service conditions of the part, or (more efficiently) if we make this structure particularly so that it matches these conditions, the cracking process can be limited in time and space or fully eliminated.

To prove the feasibility of this idea, we should conduct experiments using a material having a structure that would allow manipulation of its components at several scale levels. Concrete is the most suitable material for this purpose. Its structure is made up of three different components: binder (cement), fine (sand), and coarse (crushed stone) fillers. They allow variation in their size parameters and also in structurizing activity (cement). Arranging concretes in rows in increasing order of their short-time compressive strength, it is worthwhile to note the shift of characteristic phases of the mechanodestructive process, depending on their location in this row.

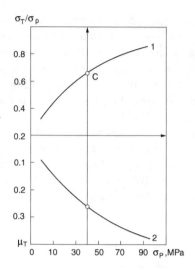

FIGURE 9.18
Transition of the microscopic phase of fracture into a macroscopic one (1) and mechanical–destructive sensitivity (2) of different compositions of concrete.

The dependence of the critical concentration of microdefects per unit volume of the stressed material on the value of strength σ_p is shown in the upper field of Figure 9.18. Curve 1 divides the cracking process into two phases: incubation (lower zone) and macroscopic (upper zone). The transition moment between these two zones coincides with the end of a linear section of the differential curve of accumulation of damages (point a in the deformation diagrams of Figure 3.11). Here the quantity (increase in the concentration of local microtears) transforms into quality, i.e., a new phase of the mechanodestructive process begins (coalescence of microdefects and formation of microcracks). It corresponds to the moment of emergence of residual microstrains (see Section 3.9) and is referred to as the lower boundary of cracking, designated as σ_T (Figure 4.33).[48]

Variations in mechano-destructive sensitivity μ_T for different compositions of concretes is shown in the lower field of Figure 9.18 (plot 2). It characterizes dynamics of the growth of cracks and shows the ratio of relative change in mean length of cracks per unit volume of the stressed material to relative change in value of external load.

As mentioned, curve 1 divides the cracking process into two phases: incubation (lower zone) and macroscopic (upper zone). In low-strength concretes (to the left from point C), cracks are initiated at an early stage and exhibit passive behavior (low sensitivity μ_T). Although the incubation period of crack formation does not last long (downward from curve 1 along the arrow) in these concretes, they possess considerable resistance potential (upward from curve 1 along the arrow). For this reason, they have little sensitivity to random overloads.

High-strength concretes, on the other hand, are characterized by a long incubation phase (up to 0.6 to 0.8 of fracture stress) and catastrophic final phase. Peculiarities of the fracture mechanism of both give ideas for the rational fields of their application. For conventional concretes, it is reasonable to use them at low levels of the stressed–deformed state and employ them under dynamic loading conditions. Efficiency of employment of high-strength concretes rises at high levels of stresses and under static or quasi-static loading conditions.

In high-strength concretes, the cracking process lasts until high levels of the stressed–deformed state are achieved (upper field in Figure 9.18), while conventional concretes have well-developed immunity to fine cracks (lower field). Ignoring these peculiarities in extreme situations leads to catastrophic consequences. (The author drew public attention to this many times in the mass media while discussing the problem of "shelter" at the Chernobyl Nuclear Power Station.[14,17,18,342–344])

The differences described can find practical application, for example, in fabrication of bent reinforced concrete structures of the combined type. A gradient of strength and toughness is formed in such structures, propagating from the stretched zone to the compressed zone and caused by layer-by-layer arrangement of different compositions of a concrete mix. In this case, the most effectively employed properties of a structural material will be the deformative and strength properties.

One of the ways to raise the technical level in building manufacturing is mastering production and application of high-strength concretes of grades 800 and higher. This enables material consumption to be decreased to 40%, consumption of steel to be decreased up to 15%, and their costs to be cut by 8 to 15%.[345] Choice of the rational field of application of such concretes provides extra advantages associated with ensuring the assigned level of reliability and durability and reduction of operational costs. We should emphasize here that a trend to achieve extra homogeneity of structural materials makes no sense because the long-term arrest of growth is possible exclusively due to the presence of thermodynamic heterogeneities in the structure of these materials (see Section 6.6).

Pinus[119] also noticed the dependence of fracture and deformation on the presence of such heterogeneities. The possibility of existence of blocking traps in the deformation of solids was theoretically predicted by Shilkrut.[346] The trend to homogeneity of a structure leads to growth of the flow of failures of elementary bonds and, thus, to a dramatic decrease in durability (see Sections 6.4 and 4.3.4).

Is it possible to exercise meaningful control over the structurization process? If so, how efficient is this process? To answer these questions, we studied the effect of quantity and quality of the binder, dissolved part, and content of crushed stone per unit volume of concrete on the structure formation processes. For this purpose, three groups of compositions were used. The first group was characterized by constant water–cement ($W = 0.27$) and cement–sand ($n = 1{:}0.75$) ratios and by a variable volume of crushed stone per cubic meter

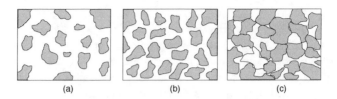

FIGURE 9.19
Basalt (a), threshold (b), and contact (c) types of concrete structure.

of the material. In the second group, the cement–sand ratio was varied at constant values of $W = 0.42$ and content of crushed stone ($K = 1.25$). In the third group, the water–cement ratio was varied from 0.27 to 0.60 at constant values of cement–sand ratio and content of crushed stone per unit volume of a material ($K = 1.25$).

This enabled three types of structure to be produced (Figure 9.19). According to classification suggested in Grushko et al.,[188] they are called basalt (a), porous (b), and contact (c). In the first group, the major portion of the volume is the dissolved part of cement adhesive and fine filler (nondashed zone). Crushed stone (dashed regions) is seen periodically. In the second group, volume fractions are approximately equal and, in the third group, crushed stone forms a rigid frame joined together by the cement adhesive in contact regions.

The specimens (9 to 12 samples from each composition) were made in rigid metal molds with polished surfaces. At an age of 28 days, the hardness indicators of concrete were measured on cubic specimens by the methods of plastic deformation and elastic rebound and then compressive strength was evaluated (Figure 3.19c). Variations in the integrated macroscopic and local microscopic statistical parameters of strength of concrete with a variation in main structurizing factors are shown in Figure 9.20. The short-time compressive strength under standard loading conditions was used as an integrated macroproperty.[196]

The varied factors affect different scale levels of the structure. At $n = \text{const}$, $K = \text{const}$, and $W \neq \text{const}$ (Figure 9.20a), the deep submicroscopic and microscopic elements of the structure–new crystalline formations, i.e., the environment where interatomic bonds initiate, interact, and fail at the θ-temperature (see Section 1.9), undergo changes. Variations in the cement–sand ratio at $\eta \neq \text{const}$, $W = \text{const}$, and $K = \text{const}$ affect the matrix of the dissolved part, i.e., change its block arrangement (Figure 9.20b). Part of the volume of new crystalline formations is substituted by particles of the fine filler connected to the solidified cement adhesive by contact layers. Systems of individual microcracks are initiated and developed here during the mechanodestruction process. Variations in the quantity and relative arrangement of the coarse filler and dissolved part per unit volume induce changes in the upper order of the structure ($K \neq \text{const}$, $W = \text{const}$, and $n = \text{const}$, Figure 9.20c).

The total length and configuration of brittle spatial surfaces edging the particles of the coarse filler change in volume in this case (see Table 3.6).

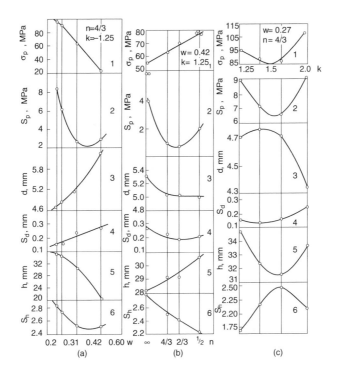

FIGURE 9.20
Variations in mean statistical values of σ_p and standard deviation of empirical distributions of short-time compressive strength S_p (1 and 2), ductility d (3 and 4), and elasticity h (5 and 6) indicators of hardness, depending on the content of solution part K per unit volume of concrete (c) and variations in water–cement W (a) and cement–sand n (b) ratios.

These surfaces are separated by the solution interlayers, where the dilaton traps that retard the fracture process are concentrated (see Section 6.6). The boundary surfaces are the potential reserve that leads to a sudden growth of macrocracks, while the contact zones serve as a sort of bridge that facilitates their coalescence.

Analysis of Figure 9.20 shows that heterogeneity of a structure can be regulated within wide ranges and, therefore, macroscopic properties of concrete can be controlled by varying the structure-forming factors. As these factors affect different structural layers, the degree of their effect on these layers is different. Radical changes in macroproperties take place with variation in W, i.e., those components that have a direct impact on the processes of formation of the MB factor and its subsequent dissociation in force fields (see Section 3.7).

With W varied from 0.6 (composition 9) to 0.27 (composition 1 in Figure 9.20a), the mean value of the short-time strength grows from 20 to 96 MPa (upper field in Figure 9.20a) and the heterogeneity value (standard deviation S_p, MPa) increases from 1.5 to 9 MPa (field 2 in Figure 9.20a). A shift of the center of a macroscopic distribution to the zone of high-strength values results from the saturation of a unit volume with rigid dilaton bonds (change of position 9 to position 8 in Figure 2.47). This is evidenced by a continuous increase in height of the elastic rebound of indenter h (field 5 in Figure 9.20a).

In addition, heterogeneity of such bonds increases with a decrease in W (field 6 in Figure 9.20a). This is accompanied by a continuous decrease in the concentration of plastic zones, which can be seen from the reduction of mean sizes of plastic indentations in field 3 of Figure 9.20a. This leads to degradation in the blocking ability of plastic traps (homogeneity of local plastic strains increases—field 4 in Figure 9.20a). Therefore, transition of concrete to the high-strength region is characterized by an increase in the concentration of the rigid dilaton part of elementary bonds, which leads to the degradation of the blocking ability of the structure due to a decrease in its plastic phase. In other words, the number of centers of initiation of microcracks increases and the speed of their nonstop propagation grows (see Section 6.6)—a restructuring responsible for brittle-explosive fracture of concrete.

Structure of the dissolved part was altered by varying the cement–sand ratio, n. Micro- and macroscopic structural levels in this case remained unchanged (W = const, K = const). An increase in the sand content of the solution part (variation of n from ∞ to $1/2$) causes an increment in the mean compressive strength from 56 to 76 MPa (field 1 in Figure 9.20b). This is attributable to an increase in the volume concentration of rigid bonds edging the grains of the fine filler, evidenced by the effect of growth of height of the elastic rebound of the indenter (field 5 in Figure 9.20b) and an increase in the homogeneity of this hardness indicator (field 6 in Figure 9.20b). Variations in the standard deviation of the strength are characterized by a curve with a minimum (field 2 in Figure 9.20b); this is formed due to a change in the percentage of dilaton and compresson phases in the microstructure. With continuous increase in the concentration of rigid bonds and improvement in their homogeneity (fields 5 and 6), the quantitative (field 3) and qualitative (field 4) indicators of the dilaton traps hardly change. The oversaturation of the structure with brittle bonds leads to higher heterogeneity of the strength (left- and right-hand branches of curve $S_p = f(n)$).

Variation in the amount of the coarse filler in volume of concrete leads to radical change in the macrostructure (Figure 9.20c). Statistical characteristics of elastic (fields 5 and 6) and plastic (fields 3 and 4) phases undergo ambiguous changes. With a certain space relationship between the coarse filler and the dissolved part (from 2 to 3 in plot 1 of Figure 9.20c), the direction of change of these parameters changes into the opposite. The minimum on plots $\sigma_p = f(K)$ and $S_p = f(K)$ is observed at the moment of this change, which indicates the existence of the optimum volume arrangement of concrete at the macroscopic level.

As seen from Figure 9.20, a purposeful variation in the arrangement factors means that the center of the macroscopic distribution of the strength may not only move to the zone of a higher or lower resistance (e.g., at $W \neq$ const or $n \neq$ const) but also have a minimum ($K \neq$ const). Macrohomogeneity has a minimum in any case (field 2 in Figure 9.20a, b, c). This fact indicates the possibility of a flexible regulation of the processes of structurization and production of concrete accounting for given service requirements.

The level of structure-organizing possibilities of a technological process should meet the requirements imposed on reliability and durability of parts. A product with improved parameters cannot be manufactured using old methods. Therefore, high reliability and durability at minimum consumption of materials can be provided only by appropriate technological support. Potentialities of concrete can be realized to their full extent by developing a flexible and controllable process of concreting, which requires a radical revision of the basics of the established notions of concreting technology.

In traditional technology, the concrete mix, i.e., intermediate product in our considerations, is the object or starting point. Its peculiarities determine the consumption of materials and quality of structures. Its "exclusion" from the technological process leads to a fundamental decrease in the power, materials, and labor consumption of a unit product at the stage of fabrication of structures and at the subsequent stages of life of a construction object. As proven by experience,[330] improvement of individual technological operations does not lead to qualitative growth of reliability and durability, or economical indicators as a whole, because it does not affect the essence, origin, and traditions of the process.

An advanced technology should provide for combination of such operations as transportation of source constituents, preparation of the concrete mix, placement, consolidation, and possibility of flexible and separate regulation of feed of the source components (cement, water, sand, crushed stone, and additives) with simultaneous intensification of the process of hydration and solidification in hot water or vapor. This technology can be called variatropic; a schematic of its realization is shown in Figure 9.21.

The major features of variatropic technology are as follows. Using air jets, source components 1 to 4 are fed separately into mixing chamber 6 (a closed ventilated volume). The reinforcement frame is installed preliminarily or fed simultaneously with the layer-by-layer placement of cement, sand, and crushed stone. The speed of feeding of each component is adjusted separately.

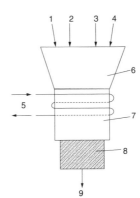

FIGURE 9.21
Schematic diagram of the variotropic technology: 1 to 4, separate feed of concrete components; 5, heater; 6, mixing chamber; 7, matrix; 8, finished structure; 9, structure yield direction.

Water temperature may vary from 0 to 100°C, depending on the selected solidification conditions. Cement is fed in parallel or opposing flows. Components of concrete are placed in layers by reciprocating movement of the nozzles. The priority and sequence of their feed are determined depending on the expected state of stress of the structure and are established experimentally. The degree of consolidation is set by varying pressure and rate of flows 1 to 4. The layer-by-layer placement is performed in a vapor–air environment. This leads to intensification of the processes of hydration and solidification of cement.

External heating 5 of matrix 7 serves the same purpose. Its technical realization involves no difficulties. Finished structure 9 is extruded through the matrix by excessive pressure or using a mechanical device. This technology enables manufacture of thin-wall flat or extended linear members, as well as structures of any configuration, thus making it possible to rationally utilize cement and fabricate structures with desired strength, reliability, and durability.

The suggested technology allows regulation of density and homogeneity of concrete across the section of parts and production of the latter with smooth or, if necessary, sculptured surfaces. It provides radical improvement in the quality of products, makes the production cycle shorter, and enables a several-fold reduction in materials, power, and labor consumption.

Achievement of this purpose should be preceded by:

- Selection of efficient methods for separate pneumatic transportation of concrete components
- Investigation of the completeness of the processes of hydration of cement in vapor–air flows, the rate of solidification of the cement–sand mixture in layer-by-layer placement, and variation of temperature of the water–air environment
- Analysis of the rational sequence of deposition of layers of the cement slurry, sand, and crushed stone
- Study of the interaction of parallel and opposing flows with each other and with the reinforcement frame
- Development, manufacture, and study of mixing chambers and punches for different types of structures and different types of their extrusion
- Identification of the necessity and substantiation of methods and conditions of heating of the mixing chamber and punch
- Investigation of physical–mechanical properties of samples and structures with strength and homogeneity of concrete varied across the section

That is, it will be necessary to solve a package of theoretical and applied problems, the outcome of which should be the introduction of variatropic technology into the building industry.

9.8 Control of the Stress–Strain State

The technical state of a material can be controlled prior to solidification, by varying the θ-temperature value (Section 9.4), in the process of solidification (Section 9.5), and, especially important, after solidification (Section 9.6), including the period of active service.[12] The control effect is achieved varying the shape and location of the MB factor (positions 11 to 15 in Figure 1.14) in the energy domain under the impact of external factors a_i, A_i, and T.

Substituting the general expression for the MB factor (Equation 2.2),

$$Z(\mathbf{P}, T) = \Gamma \exp\left(-\frac{U}{kT}\right),$$

at $\Gamma = 1$ into Equation 9.11, we can write

$$N^2 = \frac{U}{kT}. \tag{9.20}$$

This equality is of fundamental importance for the resistance of materials to external fields and for solid-state physics. It shows that, in the formation of a solid, the interatomic bonds are formed only by the compresson–dilaton (Section 1.8) or dislocation (Section 2.12) pairs (the square of the number of interatomic bonds, N, is located in the left-hand part). This is explained by peculiarities of atomic mechanics (Chapter 1), which are caused by the motion of structurizing particles according to the law of conservation of mechanical moment (see Equations 1.31 and 1.27 and Figure 1.7). It allows them to move only along the flat paths. Because they are located during solidification in the azimuthal planes in a chaotic manner (in analogy with Figure 1.4), they form the solid aggregate state (see Section 2.9).

Interatomic bonds are energy formations. Their number N depends on the overall internal energy of a solid, U, and its kinetic part T (to be more exact, on their relationship). Their number does not remain constant in various external fields and active environments (see Chapter 5). It continuously changes quantitatively (see Chapter 6) and qualitatively (compresson–dilaton transformations of structure are considered in Sections 1.8, 2.6, 2.9, and 2.11).

The initial arrangement of a structure depends on the degree of perfection of the technology and is determined by its stability (Sections 9.4 and 9.5). It continuously varies under the effect of different factors that result from technological treatment (Section 9.6), storage, transportation, assembly, and, finally, service (Chapters 6 and 7). This structure can be improved (if it is related to the given external conditions) through structurization or restoration processes (Section 5.7) or deteriorated from aging, force destruction, corrosion, etc. (Chapter 5).

The known statistical theories of strength and reliability[2-4] are based on incomplete knowledge of the internal structure of materials and do not account for probable consequences of these processes. They enable only the probability prediction of strength and reliability, rather than a credible estimation of their variations with time under given external conditions. Understanding the physical mechanism of deformation (Sections 3.4 to 3.5) and fracture (Sections 1.6 to 1.10 and Chapter 6), combined with the possibility of determining the technical state of a material (Chapter 8), leads to a substantial decrease in the role of chance in prediction of behavior of parts and structures and creates conditions for the development of the theory of design and reliability on a basis different from probability (Sections 4.3.6, 9.2 to 9.4).[12,18,21,207-209,215,283,331,333,347]

Equation 9.20 allows formulation of the physical principle of reliability of structural materials (see Section 4.3):[18] reliability of a material (or a part made from it) in randomly varying external fields is determined only by the value of the exhausted resource s (part of the interatomic bonds that failed and changed the energy potential following the diagram in Figure 2.8) and does not depend on how it has been exhausted. This principle was first suggested for complicated technical systems by Sedyakin[208] and Sedyakin and Kabanov.[347]

The thermodynamic equation of state (1.111) and Equation 9.20 show that resistance **P** and deformation V can be controlled by varying the energy potential of a body (parameters U, T, G, F), not allowing for variations in entropy s and, moreover, for fracture of interatomic bonds, i.e., at $N = \text{const}$. Consider the physical principles of the energy control methods.

In analogy with Equation 2.164, the equation of state (5.3) can be given a symmetrical form

$$PV = G(T)V_e,\qquad(9.21)$$

where $G(T) = kNT$ is the Gibbs potential (Equation 2.35) that depends only on the temperature, and

$$V_e = \delta^3 = \left(e\frac{t_i}{\theta}\right)^3$$

is some volume in the energy space whose linear component is equal to δ (Equation 2.92).

Let us analyze the function of state given by Equation 9.21. Strength, **P**, and deformative, V, properties of solids are determined by the left-hand side of the equation. If the Ω-potential monotonically grows, then

$$\dot{\Omega} > 0.\qquad(9.22)$$

Therefore,

$$\dot{\Omega}(T) > 0.\qquad(9.23)$$

Thus we can write Equation 9.22 in the form of an inequality

$$PV\left(\frac{dV}{V} + \frac{d\mathbf{P}}{\mathbf{P}}\right) > 0. \tag{9.24}$$

It is met when:

$$(a) \ PV > 0; \ \left(\frac{dV}{V} + \frac{d\mathbf{P}}{\mathbf{P}}\right) > 0$$

or

$$(b) \ PV < 0; \ \left(\frac{dV}{V} + \frac{d\mathbf{P}}{\mathbf{P}}\right) < 0.$$

As the volume cannot be negative, the first pair of inequalities has physical sense when the following inequalities are met:

$$(d) \ \mathbf{P} > 0, \ V > 0 \text{ and } \left(\frac{dV}{V} + \frac{d\mathbf{P}}{\mathbf{P}}\right) > 0,$$

and the second pair at

$$(c) \ \mathbf{P} < 0, \ V > 0 \text{ and } \left(\frac{dV}{V} + \frac{d\mathbf{P}}{\mathbf{P}}\right) < 0.$$

Expressing the volume in terms of a product of the cross section area S and the length of a working part of a body L, i.e., $V = SL$, and the pressure in terms of the ratio of force $\mathbf{F}$ acting on area S, i.e., $\mathbf{P} = \mathbf{F}/S$, we find that

$$dV = V\left(\frac{dS}{S} + \frac{dL}{L}\right)$$

and thus

$$d\mathbf{P} = \mathbf{P}\left(\frac{d\mathbf{F}}{\mathbf{F}} - \frac{dS}{S}\right). \tag{9.25}$$

Accounting for this equation, inequalities (d) and (c) assume the following forms:

$$\left.\begin{array}{l} (d) \ \ \mathbf{P} > 0; \ V > 0; \ \left(\dfrac{dL}{L} + \dfrac{d\mathbf{F}}{\mathbf{F}}\right) > 0 \\[3mm] (c) \ \ \mathbf{P} < 0; \ V > 0; \ \left(\dfrac{dL}{L} + \dfrac{d\mathbf{F}}{\mathbf{F}}\right) < 0 \end{array}\right\}. \tag{9.26}$$

As D-rotoses (Figure 2.17) develop a positive pressure (directed outward from the volume), the inequalities (*d*) characterize behavior of the D-materials in force fields, and the inequalities (*c*) characterize that behavior in the compresson fields (see Section 2.6). Reducing the third inequality in both systems to a common denominator and noting that $A = Fl$ is the deformation work, we can write

$$(d) \ \frac{FdL + dFL}{A} > 0; \quad (c) \ \frac{FdL + dFL}{A} < 0. \qquad (9.27)$$

These inequalities have physical sense when

$$\begin{aligned}
(d_s) \ & \mathbf{F} > 0; L > 0; (FdL + d\mathbf{FL}) > 0; \ d_s)\mathbf{F} < 0; L > 0; (FdL + d\mathbf{FL}) < 0 \\
(c_p) \ & \mathbf{F} > 0; L > 0; (FdL + d\mathbf{FL}) < 0; \ c_s)\mathbf{F} < 0; L > 0; (FdL + d\mathbf{FL}) > 0
\end{aligned} \right\} \qquad (9.28)$$

from where it follows that

$$\begin{aligned}
(d_s) \ & d\mathbf{FL} > -FdL; \ d_p)d\mathbf{FL} < -FdL \\
(c_s) \ & d\mathbf{FL} < -FdL; \ c_p)d\mathbf{FL} > -FdL
\end{aligned} \right\} \qquad (9.29)$$

The latter inequalities explain a known fact that is perceived as natural, but until now has had no satisfactory physical interpretation, namely, radically different reactions of brittle (dilaton) and ductile (compresson) materials (Section 2.6) to tensile and compressive loads (Figures 2.10 and 2.22). In compression, resistance (dFl) of the D-materials (cast iron, concrete, ceramic, and other brittle materials) exceeds deformability ($-Fdl$) (Inequality d_s, 9.29), whereas, in tension, deformability dominates resistance (Inequality d_p, 9.29). For this reason, D-materials have good compression and poor tensile resistance. Characterized by the opposite tendency, the C-materials have high tensile resistance.

External manifestations of deformation and fracture are attributable to the right-hand part of the equation of state (9.23), which can be represented, accounting for Equation 9.24, as

$$G(T)V_e\left(\frac{dV_e}{V_e} + \frac{dG}{G}\right) > 0.$$

Two pairs of inequalities follow from it:

$$GV_e > 0; \left(\frac{dV_e}{V_e} + \frac{dG}{G}\right) > 0, \quad GV_e < 0; \left(\frac{dV_e}{V_e} + \frac{dG}{G}\right) < 0$$

Because temperature T and quantity of interatomic bonds N are positive values, these inequalities have physical sense if

$$(d) \quad G > 0; V_e > 0; \left(\frac{dV_e}{V_e} + \frac{dG}{G} \right) > 0$$

$$(c) \quad G > 0; V_e < 0; \left(\frac{dV_e}{V_e} + \frac{dG}{G} \right) < 0$$

Using definitions of potentials G (Equation 2.35) and V_e (Equation 9.21), we can express these conditions as

$$(d) \quad T > \frac{1}{e}\theta; N > 0 \left(4\frac{dT}{T} + \frac{dN}{N} - 3\frac{d\theta}{\theta} \right) > 0;$$

$$(c) \quad T < \frac{1}{e}\theta; N > 0 \left(4\frac{dT}{T} + \frac{dN}{N} - 3\frac{d\theta}{\theta} \right) < 0, \qquad (9.30)$$

where

$$\frac{dG}{G} = \frac{dT}{T} + \frac{dN}{N}; \quad \frac{dV_e}{V_e} = 3\left(\frac{dT}{T} - \frac{d\theta}{\theta} \right) \qquad (9.31)$$

The system of inequalities (9.30*d*) determines the behavior of the dilaton materials and (9.30*c*) that of the compresson materials. In the first, the rotoses are located to the right (in a range of r_b to ∞ in Figure 1.7a) and in the second to the left from the phase transition temperature θ (in region r_b to r_a). The external effects that cause deformation and fracture (system of Inequalities 9.26) are accompanied by three interrelated processes. First, the kinetic energy of rotoses, which determines the order in the AM system (see equation of entropy 2.13), changes $-dT/T$. Second, the Debye temperature deviates to one side or the other from the equilibrium value $-d\theta/\theta$ (which changes configuration of the local potential field in Figure 1.8, following the diagram shown in Figure 1.10, and changes chemical composition of a material within the fracture zone[3]). Third, the quantity of rotoses decreases due to their failure, $-dN/N$. The first two processes reveal the physical nature of the elastic or reversible stage of deformation, *oa*, and the third process reveals that of the next irreversible or destructive stage, *ab* (Figure 2.10).

As the last stage is of highest practical interest, we separate it from the third inequalities of system (9.30):

$$(d) \quad \frac{dN}{N} > \left(3\frac{d\theta}{\theta} - 4\frac{dT}{T} \right), \quad (c) \quad \frac{dN}{N} < \left(3\frac{d\theta}{\theta} - 4\frac{dT}{T} \right). \qquad (9.32)$$

Under the same conditions, fracture occurring in the D-materials (Inequality 9.32*d*) leaves behind the processes of phase restructuring (right-hand parts of both inequalities), whereas in the C-materials they precede fracture (9.32*c*). Therefore, it is not by chance that brittle fracture takes place with D-materials (almost without plastic deformation, curve 1 in Figure 2.22) and ductile fracture is the case with C-materials (curve 2).

Fracture of a structure does not take place if $dN = 0$. In this case Inequalities 9.32 have the form

$$(d) \quad ds > \frac{3}{4} \frac{d\theta}{\theta}, \quad (c): ds < \frac{3}{4} \frac{d\theta}{\theta}, \tag{9.33}$$

where, in accordance with Equation 2.13, $ds = dT/T$ is the entropy differential. Integration yields

$$(d) \quad \exp(s - s_0) > \left(\frac{\theta}{\theta_0}\right)^{3/4} \quad \text{and} \quad (c) \quad \exp(s - s_0) > \left(\frac{\theta}{\theta_0}\right)^{3/4},$$

where s_0 and θ_0 are the entropy and temperature, respectively, of phase transition of the initial structure. Expressing the ratio θ/θ_0 in terms of power function $\theta/\theta = e^n$, we can write

$$(d) \quad s - s_0 > \frac{3}{4} n \quad \text{and} \quad (c) \quad s - s_0 < \frac{3}{4} n,$$

where number n sets the safe range of variations in entropy (in the elastic region of the deformation diagram in Figure 2.22). If the last inequalities are met, under any external effect, no fracture of both types of materials will take place.

The necessary indicator of the extremum of the function of state (9.21) is equality to zero of its first derivative with respect to time

$$\dot{\Omega} = \dot{\Omega}(T) = 0.$$

The use of the definition of the $\Omega(T)$-potential (see comments to Equation 9.21) yields the equation of the stressed–deformed state:

$$\dot{\Omega}(T) = GV_e\left(\frac{\dot{G}}{G} + \frac{\dot{V}_e}{V_e}\right). \tag{9.34}$$

Accounting for the structure of the potential and kinetic parts of the Ω-potentials, Equation 9.34 can be broken into two equations:

deformability

$$\frac{\dot{V}}{V} = \frac{GV_e}{PV}\left(\frac{\dot{G}}{G} + \frac{\dot{V}_e}{V_E}\right) - \frac{\dot{P}}{P} \tag{9.35}$$

and strength

$$\frac{\dot{P}}{P} = \frac{GV_e}{PV}\left(\frac{\dot{G}}{G} + \frac{\dot{V}_e}{V_e}\right) - \frac{\dot{V}}{V} \tag{9.36}$$

Expressing Equation 9.34 as

$$\tau = \frac{\dot{G}}{G} = \frac{PV}{GV_e}\left(\frac{\dot{V}}{V} - \frac{\dot{P}}{P}\right) - \frac{\dot{V}_e}{V_e}, \tag{9.37}$$

we obtain a generalized equation of durability.

This gives rise to a logical question: if the structure of materials is so sensitive to the stressed–deformed state, is it possible to exclude any consumption of resource under various service conditions by its targeted control? Analysis of Equations 4.27, 4.28, and 7.3 from the standpoints formulated in Chapters 1 and 2, as well as in many outcomes of the indirect (for example, References 2 to 4, 10, 19, and 22) and direct (References 1, and 316 to 326) experiments make it possible to give the positive answer to this equation. Equality of the differential with respect to both parts of Equation 3.19 to zero provides such conditions:

$$d\left(\Omega_0 \pm V\sum_{i=1}^{n} a_i A_i\right)T = \left(\Omega_0 \pm V\sum_{i=1}^{n} a_i A_i\right)dT \tag{9.38}$$

It follows from this equation that the integrity of the resource can be ensured in a constant or variable thermal field. In the first case,

$$d\Omega_0 = \pm d\left(V\sum_{i=1}^{n} a_i A_i\right) \tag{9.39.I}$$

and, in the second case,

$$\frac{d\Omega}{\Omega} = \frac{dT}{T}, \tag{9.39.II}$$

where

$$\Omega = \Omega_0 \pm V \sum_{i=1}^{n} a_i A_i .$$

Hooke's law (Equation 2.139) and the classical definition of entropy (Equation 2.13) allow us to rewrite Equation 9.39 as

$$d\mathbf{P}_0 = dE_1 - dE_2 \ \text{(I)}; \quad d\mathbf{s}(\sigma) = C_v d\mathbf{s}(T) \ \text{(II)}, \tag{9.40}$$

where E_1 and E_2 are the elasticity moduli in the mutually perpendicular directions, $d\mathbf{s}(\sigma)$ and $d\mathbf{s}(T)$, as in Equation 9.15, are the variations in entropy in external force and thermal fields, and C_v is the heat capacity of a material. Now it becomes clear that the development of the destruction processes, following the law presented in Equation 8.25 or Equation 8.26, can be avoided using a controlling effect (force, thermal, electric, magnetic, or other), competing with the service effect (confirmed by experiment; see Figure 9.1), thus providing integrity of the resource (Equations 2.112 and 2.113) for the assigned service life.

The foregoing analysis provides a solid basis on which to regard the existing system of ensuring strength, reliability, and durability of engineering objects as extremely conservative and inertial. The statistical theory of reliability, considered to be promising by many investigators,[1-4] fails to provide an unambiguous estimate of the outcome of processes that occur in materials under the effect of various external factors. It merely allows a probability prediction of them, thus accepting the probability of fracture (Figure 7.10). The application of materials and structures having self-adaptation means to service conditions[12,18,21,24] is the only proper way to go.

9.9 Prevention of Deformation and Fracture in Competing Fields

Depending on a particular type of field generating compensating effects, the method of control of the stressed–deformed state is subdivided into mechanical, electric, magnetic, and thermal. Condition 9.39.I can be met using the first three of these, and Condition 9.39.II using the last. These conditions allow the following formulation: no fracture of the structure of a material occurs if an increase in external loads is accompanied by the opposite change in the longitudinal and transverse strains. In other words, for example, no increase in transverse strains should be permitted in compression; they should be decreased by any available method, thus preventing development of orientation processes according to the diagram in Figure 2.7 or 2.8. In tension,

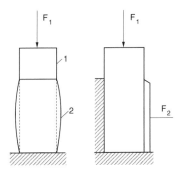

FIGURE 9.22
Schematic diagram of delay of fracture in vertical compression: $\mathbf{F}_1$ and $\mathbf{F}_2$, working and compensation loads.

their decrease is inadmissible. This means that all methods for retardation of fracture are based on an immediate restoration of initial entropy due to creation of a complex stressed–deformed state.

9.9.1 Mechanical Methods

Let the MB factor of a material in specimen 1 subjected to compressive stressed–deformed state (Figure 9.22) first be located in position 12 (Figure 1.14). Under noncontrollable conditions, an increase in loading leads to the transformation of the shape of this distribution in the longitudinal direction according to a diagram of 12 to 11, and in a transverse direction according to a diagram of 12 to 13. Escape of the ends outside the ranges of critical temperatures T_c and θ causes fracture of brittle materials and plastic deformation (position 2 in Figure 9.22) of metals and alloys, which is also interpreted as failure.[3]

If the process of vertical loading by useful load $\mathbf{F}_1$ is accompanied by the horizontal reduction by controlling force $\mathbf{F}_2$, deformation and fracture can be avoided (at least, they will begin at much higher loads; see explanations to Figure 3.59). This is attributable to the fact that flattening of the MB distribution in the transverse direction (according to a diagram of 12 to 13 in Figure 1.14) under the effect of working load $\mathbf{F}_1$ is compensated for by an increase in steepness of its fronts in a longitudinal direction (position 11) under the effect of correcting force $\mathbf{F}_2$. This brings destruction processes into the zone of high stresses.

The proof that compressive stresses cause slow-down or braking fracture, i.e., prevent an increase in the s_d component of entropy, can be found in many literature sources (for example, References 1 to 4, 10, 36, 44, and 330). This is natural because tension (change in positions in a direction of 11 to 12 to 13) causes the "hot" end of the MB factor to move closer to the θ-temperature (vertical line br_b in Figure 1.14), and the "cold" end closer to T_c

(vertical line r_0c), thus inevitably damaging the structure of a material by microdefects distributed in volume V.[10,21]

In contrast, compression leads its ends out of the dangerous zones while increasing steepness of both fronts (reverse change of positions 13 to 12 to 11). Periodic changes in the state compression–tension raise the probability of escape of the end regions of the MB factor outside the ranges of critical temperatures T_c and θ, thus activating the process of accumulation of damages (dashed regions in position 13) and inducing fatigue of a material.[215] Compressive stresses may be induced not only by force,[1,12,318,330] but also by electric,[12,326] magnetic,[12,320,323–325] and thermal[12,320–322] fields.

The deformation diagram (Figure 2.22) of tension (upper field) and compression (lower field) can be used to interpret mechanical methods for retardation of fracture. Let a material be deformed only at the elastic stage (in region $0G_2$ in tension and $0G_3$ in compression). The equilibrium state is set by point 0 and, under uniaxial loading, the resistance formed in tension is equal to $(-\sigma_2)$ and that formed in compression is equal to $(+\sigma_3)$. Assume that prior to tension the material was reduced to a stressed–deformed state determined by point G_3. If so, tension begins not from the coordinate origin 0, but from point G_3, where molecules are transformed from the isotropic state shown in Figure 2.7a or b into the anisotropic state shown in Figure 2.8a. If compressive stresses are not relieved during the entire tension cycle, tensile forces must overcome the natural resistance of the material and also the external compressive field applied to it.

Therefore, to transform a body from the compressive state G_3 into the tensile state G_2, it is necessary to pass path G_30G_2 in the diagram (Figure 2.22), change strain from ε_3 to ε_2, and overcome resistance $\sigma = \sigma_2 + \sigma_3$. Therefore, tension under the compressive stressed–deformed state causes the destruction phase of deformation, ab (Figure 2.10), to move into the zone of higher resistance levels. This is the physical essence of mechanical methods for prevention of fracture.

Expressing differential of resistance from Equation 9.36 in the explicit form

$$\dot{P} = P\left[\frac{GV_e}{PV}\left(\frac{\dot{G}}{G} + \frac{V_e}{V_e}\right) - \frac{\dot{V}}{V}\right]$$

and noting that it takes the maximum value at

$$\frac{\dot{V}}{V} = 0,$$

that is, when

$$\frac{ds}{S} = -\frac{dL}{L},$$

we find that

$$\dot{P}_{max} = \frac{GV_e}{V}\left(\frac{\dot{G}}{G} + \frac{\dot{V}_e}{V_e}\right)$$

or, accounting for Transformation 8.36, we can write

$$\dot{P}_{max} = \frac{GV_e}{V}\left(\frac{\dot{N}}{N} + 4\frac{\dot{T}}{T} - 3\frac{\dot{\theta}}{\theta}\right). \qquad (9.41)$$

At the stage of elastic deformation,

$$\dot{N} = 0;$$

therefore,

$$\dot{P}_{max} = \frac{GV_e}{V}\left(4\frac{\dot{T}}{T} - 3\frac{\dot{\theta}}{\theta}\right).$$

As seen, the slope of the elastic region of diagram $\sigma = F(\varepsilon)$, which is the differential of $\dot{P}_{max}$, is fully determined by the redistribution of the kinetic part of the internal energy. Thus, the maximum resistance is found from condition

$$4\frac{\dot{T}}{T} = 3\frac{\dot{\theta}}{\theta}.$$

After integration, we find another value of the most favorable service temperature $T_n = \sigma^{3/4}$ (see Equation 2.172).

Multiple evidence and experimental results prove that the described mechanism takes place in practice in different situations. For example, its static analogue has long been used in the construction industry to increase crack resistance of bent reinforced concrete structures due to preliminary compressive stressing the stretched zone of concrete.[330]

Under quasi-static loading, its implementation increases the rupture strength of concrete by 35 to 40% (see Section 3.10). Presenting the results of Bridgman's experiments[46] on investigation of the effect of biaxial loading on the deformation and fracture processes, Ekobori[44] noted that an induced anisotropy of mechanical properties, as well as an associated marked increase in toughness and fracture stress of low-carbon steel, have been observed under these conditions.

Vorobiov and Borobiov[92] reported that, in three-dimensional compression created by directed explosion, the resistance of rock to subsequent uniaxial compression increases from 5 to 20 times. Figure 9.23 shows the diagram of

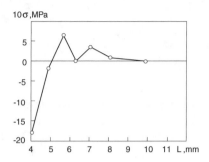

FIGURE 9.23
Diagram of residual stresses induced in steel specimens by local explosive treatment.

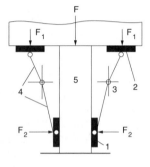

FIGURE 9.24
Schematic of realization of fracture prevention method in centrally compressed members by separation of working load **F** into longitudinal, F_1, and transverse, F_2, components (1, fragment of a structure; 2, mobile support plates; 3, axis of rotation of the lever system; 4, lever system; 5, fragment of a building structure).

distribution of residual compressive stresses induced in steel specimens ahead of the tip of a fatigue crack (*L* is the distance from the tip) by local explosive treatment.[318] An idea of the efficiency of this method for crack retardation can be seen from position 5 in Figure 9.1.

A study in Troshenko[3] presented the results of analysis of experiments on the effect of the mean cyclic stress and heterogeneous stressed–deformed state on the value of fatigue limit of steels as well as aluminum-, magnesium-, nickel-, and copper-based alloys. It was concluded that the fatigue limit of all investigated materials in tension, bending, and torsion fundamentally grows at the presence of mean compressive stresses as well as under conditions of heterogeneous stressed–deformed state. We mention these data not only to confirm the possibility of retardation of fracture by inducing heterogeneous stressed–deformed states, but also to illustrate the efficiency of mechanical methods for controlling reliability and durability.

As an example, Figure 9.24 shows a possible realization of mechanical fracture retardation methods under static or quasi-static conditions of loading of structural members working on compression. They involve a system of levers (positions 1, 2, 3, and 4), which splits working load **F** into two components: vertical F_1 and horizontal F_2. When **F** is increased by *d*F, the centrally compressed element 5 is deformed in a vertical direction by *d*ε. Upper support plates 2 are lowered by the same value, causing the corresponding

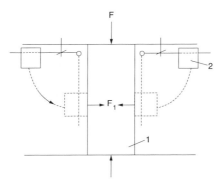

FIGURE 9.25
Schematic showing realization of fracture prevention by applying a short-time antidestructive effect (1, structural member; 2, massive dead weights supplying a transverse pulse of mechanical energy).

displacement of vertical plates 1 in a horizontal direction. In accordance with the explanations to Figure 3.59, the latter are placed in the most stressed regions of structures, whose transverse reduction prevents horizontal deformation and, therefore, eliminates the major cause of fracture in compression.

The effect of dynamic strengthening (Figure 9.25) (see Section 4.2) can also find application in retardation of fracture. The moment of application of antidestruction disturbances is timed with the periodicity of occurrence of extreme situations. Every event of the dynamic effect leads to substantial changes in the structure (Figure 4.3). The character of transformation of micro- (Figure 4.14) and macrodistributions (Figures 4.8 to 4.9) of elastoplastic properties of a material indicates the direction of such changes. The concentration of compresson bonds dramatically increases at this moment. Their short-time saturation in the volume of a material leads to an increase in its resistance in the vertical direction.

If, despite everything, a microcrack does appear, many possible centers of prefracture will be situated on its propagation path (Figure 6.32); they activate the process of branching the crack and loss of its energy due to unloading of the crack mouth. Finkel[36] indicated the possibility of practical application of this method, which is particularly efficient for medium-strength concretes where dynamic strengthening has the highest value (Figures 4.11 to 4.12). It can be employed to raise the efficiency of operation of centrally compressed reinforced concrete members by applying dynamic pulses of transverse compression F_1 when the working loads F increase to dangerous levels (Figure 9.25).

9.9.2 Electric Methods

The stressed–deformed state required for the prevention of deformation and fracture can be created by mechanical and also by electric and magnetic fields. All physical–mechanical properties of solids are initiated at the AM level;

therefore, they are interrelated (see Sections 1.2 to 1.4). They are various manifestations of the same laws associated with movement, interaction, and dynamic equilibrium of charged particles: polarized ions and valent electrons (Section 1.4). For this reason, variations of some of them inevitably affect others. As shown in Sections 3.3 and 3.11, consideration of deformative and strength properties in close relation with their thermal, electric, and magnetic manifestations results in promising prospects for control of reliability and durability of materials, parts, and structures placed in these fields. What is the physical mechanism of this control in electric fields (E-fields)?

An E-field is generated by fixed electric charges. Moving charges are surrounded by an electromagnetic (EM) field (Equation 1.32) that exerts an effect on other charges located in it. Electric charges in dielectrics (this type of solid is sensitive to the E-field) are related to each other and joined in rotoses (Figure 1.6). In the theory of dielectrics,[26] the static model of a rotos is called a dipole. The electric part of the EM field has a force effect on charged particles and thus can perform certain work (Equation 1.16). This work forms the Ω-potential of an individual rotos (Equation 1.53) and a solid as a whole (Equation 1.111). In practice, it shows up always and everywhere in the thermal range of EM radiation (right-hand, kinetic part of the equation of state, 2.65).

This creates a reliable foundation for utilization of the E-fields for the prevention of deformation and fracture. In practice, loading rates fall on a time range from 10^{-6} (liquid and gas storage tanks in filling) to 10^7 s^{-1} (explosive loading). As seen, both electrostatic and variable E-fields can find application for the preceding purposes.

The variations in the state of dielectric materials in the electrostatic field is considered in detail in Section 3.2 (see Equations 3.7 to 3.18 and Figures 3.2 and 3.3).

In analogy with Equation 9.21, Equation 3.17 can be rewritten as $PV = GV_e - pVE$. After differentiation at

$$\dot{E} = 0 \quad \text{and} \quad \dot{V} = 0$$

(which corresponds to deformation of a material at the elastic stage at the presence of the constant E-field), we have that

$$\dot{P} = \frac{GV_e}{V}\left(\frac{\dot{G}}{G} + \frac{\dot{V}_e}{V_e}\right) - p\mathbf{E} \tag{9.42}$$

Comparison of Equations 9.42 and 9.36 shows that a change in sign ahead of the second term caused by a change in direction of the E-field leads to a change in differential $\dot{P}$ and, therefore, in the resistance of a material to external mechanical effects. The extent of this effect can be seen from Figure 3.3.

Unfortunately, experimental data on the combined effect on the stressed–deformed state of materials by mechanical and electric fields are scanty.

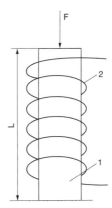

FIGURE 9.26
Magnetic (1) in force, **F**, and magnetic, **H**, fields.

Vorobiov and Borobiov[92] note that, when the intensities of mechanical and E-fields are perpendicular to each other, the total pressure increases (i.e., compressive strength grows), whereas when they are parallel, the total pressure decreases (i.e., tensile strength decreases). This situation is illustrated in Figure 3.2.

Desnenko and Peckarskaya's study[326] considered the effect of the E-field with an intensity of up to 20 kV/cm on plastic deformation of polycrystals of high-purity copper and aluminum at temperatures of 77 and 300 K. It was established that, in the E-field, the deformation stress decreases and the process is characterized by higher strain hardening. An increase in microhardness of the surface layers of metals was also detected, i.e., direct evidence exists of a change in state predicted by the theory (Equations 3.17 and 3.19).

9.9.3 Magnetic Methods

Some structural materials are magnetics (for example, iron- and nickel-based alloys); they change their state in the magnetic field.[30] Some have a clearly defined magnetostriction (i.e., they are capable of changing their dimensions on magnetization) while in others magnetic polarization involves no visible deformations. Magnetostriction is a convincing proof of the analogy between microstructural processes (Figures 2.7 to 2.10) occurring in force and magnetic fields (as well as in electric fields). They are described by similar laws,[12,30,283] which allows us to use results discussed in the previous section.

Referring to Figure 9.26, we consider magnetic 1 placed in magnetic field **H** of coil 2 and subjected to external stresses σ under the effect of force **F**. Like the electric field, the magnetic field affects all the parameters of state of a material; in analogy with Equation 3.17, its equation of interaction can be written as

$$(\mathbf{P} \pm \mathbf{B}\mathbf{H})V = \mathbf{F} - \mathbf{G}, \qquad (9.43)$$

where **B** is the magnetic induction. The deformation effect exerted by the magnetic field is estimated by magnetostriction

$$\left(\frac{\partial V}{\partial \mathbf{H}}\right)_{\mathbf{P}},$$

while a violation of magnetic equilibrium in the external force field is estimated by the piezo-magnetic effect,

$$\left(\frac{\partial \mathbf{B}}{\partial \mathbf{P}}\right)_{H}.$$

It can be seen from the condition of existence of differential

$$\left(\frac{\partial V}{\partial \mathbf{H}}\right)_{P;T} = -\left(\frac{\partial \mathbf{B}}{\partial \mathbf{P}}\right)_{H;T}, \tag{9.44}$$

which follows from the first law of thermodynamics, written in analogy with Equation 3.14 for a body located in magnetic field:

$$Td\mathbf{s} = dU + PdV - \mathbf{H}d\mathbf{B}, \tag{9.45}$$

that these effects have competing characters; an increase in the effect of one causes a decrease in that of the other, and vice versa. Relationship 9.44 is valid for isothermal magnetization, which takes place during the elastic stage of deformation. It can be interpreted as follows: if external load **F** changes the magnetic state of a material induced by the magnetic field generated by a coil, it can be readily restored to its initial state by varying the electric current in the electric circuit of this coil. Returning to the state of magnetic equilibrium causes, in turn, damping of the external load, which saves the structure from undesirable deformation and fracture.

To calculate magnetistriction, we can write down Equation 9.44, taking into account that $\mathbf{B} = \eta \mathbf{H}V$,[52] as

$$\left(\frac{\partial V}{\partial \mathbf{H}}\right)_{\mathbf{P}} = -\left(V\frac{\partial \eta}{\partial \mathbf{P}} + \eta\frac{\partial V}{\partial \mathbf{P}}\right)\mathbf{H},$$

where η is the magnetic susceptibility. Using analogy with the thermal expansion coefficient (Equation 2.152) and thermal pressure coefficient (Equation 4.43), we can introduce the isothermal compressibility

$$\beta = -\frac{1}{V}\left(\frac{\partial V}{\partial \mathbf{P}}\right)_{T}.$$

Integrating and taking into account equality $\ln(1 + \psi) \approx \psi$, which is true at $\psi = dV/V \ll 1$, we obtain

$$\psi = \frac{H^2}{2}\left(\beta\eta - \frac{\partial\eta}{\partial P}\right). \tag{9.46}$$

At $d\eta = 0$, Equation 9.46 transforms into

$$\psi = \frac{1}{2}\beta\eta H^2. \tag{9.47}$$

If the magnetic field is generated by a coil with the winding number n, then $H = 0.4\pi nI$, where I is the electric field current.[29] Then,

$$\psi = 0.8\beta\eta n^2 I^2. \tag{9.48}$$

As seen, deformation can be counteracted in a given direction and thus the structure can be saved from fracture by selecting physical–mechanical characteristics of the material (β and η), varying the number of windings n, and by immediately reacting by current I.

To calculate the thermal effect of magnetization, we can rewrite the formula of free energy[52] F as

$$F(\mathbf{B};T) = F_0 9t0 + \frac{\mathbf{B}^2}{8\pi\eta}V,$$

where the first term $F_0(T) = U - T_s$ is the free energy at the absence of the field and the second term indicates its dependence on magnetization $\mathbf{B}$. Entropy of a body is calculated as

$$\mathbf{s} = -\frac{\partial F}{\partial T} = \mathbf{s}_0 - \frac{V\mathbf{B}^2}{8\pi}\frac{1}{\partial\eta\partial T},$$

where

$$\mathbf{s}_0 = -\frac{\partial F_0}{\partial T}$$

is the entropy of the body, which is not subjected to any external field. Differentiation of this expression yields

$$d\mathbf{s} = \left(\frac{\partial \mathbf{s}_0}{\partial T} - \frac{V\mathbf{B}^2}{8\pi}\frac{1}{\partial^2\eta\partial^2 T}\right)dT - \frac{V\mathbf{B}}{4\pi}\frac{1}{\partial\eta\partial T}d\mathbf{B}.$$

As

$$dQ = C_v dT = \frac{d\mathbf{s}_0}{\partial T} T dT,$$

then the first term in brackets can be written as

$$\frac{\partial \mathbf{s}_0}{\partial T} = \frac{C_v}{T},$$

where C_v is the heat capacity of a nonmagnetized body. The second term expresses dependence of the heat capacity on magnetization. When **B** is low, the term containing $\mathbf{B}^2$ can be ignored. Because reversible adiabatic processes,[24] at which condition $ds = 0$ (Equation 2.134) is met, occur at the elastic stage of deformation, we find that

$$dT = \frac{TV\mathbf{B}}{4\pi C_v} \frac{d\mathbf{B}}{\partial \eta \partial T} \tag{9.49}$$

It follows from this equation that magnetization and demagnetization of a body are accompanied by a change in its temperature. The magnetic field may cause either heating or cooling. Results of experiments on the behavior of structural materials at low temperatures in magnetic fields confirm these conclusions.[320]

Confirmation of validity and efficiency of magnetic methods for controlling the stressed–deformed state can also be found in the literature. Thus, it is reported in Rzhevsky and Novak's study[40] that an increase in uniaxial pressure is normally accompanied by a decrease in magnetic permeability of ferromagnetic rock in the direction of compression, while, in the transverse direction, it first rapidly grows and then, on reaching saturation, no longer changes. Figure 9.27 shows variations in magnetic properties of sheets made of steel E-42 in compression (a) and tension (b).[29]

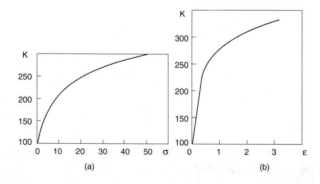

FIGURE 9.27
Variations in magnetic properties of electrotechnical sheet steel E-42 in compression (a) and tension (b).

In the first case compressive stresses σ, measured in MPa, are plotted on the axis of abscissas; in the second case, the strains at fracture, ε, are plotted on this axis. In both plots, variations in magnetic properties are shown as relative variation (expressed in percent) of magnetic permeability $K = \psi_1/\psi_2\%$, where ψ_1 is the initial value of magnetic permeability and ψ_2 is its value under the stressed–deformed state. As seen, magnetic properties are strongly changed at the elastic stage of development of the stressed–deformed state. Therefore, during this period of loading, the stressed–deformed state can be fundamentally lowered and fracture can be prevented by increasing intensity of the magnetic field.

9.9.4 Thermal Methods

Based on the orientation effect, mechanical, magnetic, and electric methods work reliably only at the elastic stage of deformation. In addition, the last two, featuring a fast response and comparatively simple principle of operation, are characterized by selectivity with respect to materials differing in physical nature; magnetic field fails to affect dielectrics and electric field fails to affect magnetics. The thermal method has no such drawbacks. Retaining positive qualities of the rest of the methods, it covers the entire variety of structural materials and affects the entire range of deformation.

The relationship of equivalence,

$$d\mathbf{s}(T) = d\mathbf{s}(V), \tag{9.50}$$

formulated in Section 2.5 and written in relative form (Equation 9.15), follows from the EM nature of forces of interaction between atoms (Equation 1.16). It establishes adequacy between the force and thermal deformation. Substitution of Equations 2.137 and 2.13 into Equation 9.50 allows us to derive the following expression:

$$d\mathbf{s}(T;V) = \varepsilon + \eta, \tag{9.51}$$

which is extremely important in the theory of adaptation. This expression relates variation in entropy $d\mathbf{s}$ to parameters measured in practice: transverse η and longitudinal ε strains.

Expressing Equation 9.50, accounting for Equation 2.152 as

$$\frac{1}{T} = \frac{1}{V}\left(\frac{\partial V}{\partial T}\right)_P = \alpha \ \text{(I)}; \quad \frac{1}{V} = \frac{dT}{dV}\alpha \ \text{(II)}, \tag{9.52}$$

and substituting Equation 9.52.II into the equation of state (2.65), we obtain

$$\mathbf{P} = \alpha \mathbf{s} T \frac{dT}{dV}. \tag{9.53}$$

Comparing Equation 9.53 with the definition of the heat flow,[32]

$$\mathbf{q} = \lambda \frac{dT}{dx},$$

in direction x shows that $\mathbf{P}$ is its mechanical equivalent. Flow $\mathbf{q}$ is initiated by external mechanical loads and causes redistribution of the kinetic part of the overall internal energy to restore dynamic equilibrium in the AM system in the loaded state. Equality $\lambda = \alpha sT$ means that the redistribution processes are accompanied by radiation or absorption of heat.

The foregoing analysis yields a fundamental conclusion that the notions "concentration of mechanical stresses" and "temperature gradient" are adequate for each other. On applying an external load, a heat flow is formed in the bulk of the material in the direction of load application. Therefore, the residual stresses can be relieved and the stressed–deformed state can be regulated (preventing deformation and fracture), by controlling the internal heat flows. Position 6 in Figure 9.1 provides evidence of the efficiency of these methods.

Analysis and comparison of mechanical and thermal properties of various classes of materials are presented in Ashby,[65] where numerical estimations of thermal equivalent of mechanical stresses are considered. For many analyzed materials, an increment in stress σ with a temperature variation of 1°C is 1 MPa. This means that a dramatic change in temperature may induce high mechanical stresses and thus shows up as a thermal shock.[3] We suggest that it could be used to prevent deformation and fracture.

The thermal field changes the energy potential of the AM bonds (causing the direct, 12 to 14 to 15, or reverse, 15 to 14 to 12, change in positions of the MB factor in Figure 1.14). Also, it has an orientation effect (Equation 9.53), i.e., it simultaneously affects the $\mathbf{s}_c$ and $\mathbf{s}_f$ parts of entropy (Equation 2.113). Moreover, substantial differences in intensities of the internal T and external T_b fields induce thermal fatigue,[3,44,215,321] i.e., increase component $\mathbf{s}_d$ in Equation 2.113.

Differentiation of Equation 2.148.II at $C_v = $ const yields

$$dA = \mathbf{P}dV = \frac{1}{12}C_v dT - Vd\mathbf{P}. \tag{9.54}$$

This is just the first approximation. To conduct more stringent analysis, it is necessary to account also for nonconstancy of heat capacity C_v, as deformation changes temperature conditions of a solid (Figure 2.29). As seen from Equation 9.54, a solid may perform useful work (left-hand part of equality) due to variations in temperature or pressure, or both simultaneously. In the first case, it expands on heating and, counteracting boundary conditions, performs the work to transform thermal energy into mechanical. In the second case, taking up the external load, it changes its linear dimensions to transform mechanical energy into thermal. In both cases, the useful work has an opposite variation.

It follows from Equation 9.54 that dV is the differential of parameters $\mathbf{P}$ and T, i.e.,

$$dV = \left(\frac{\partial V}{\partial \mathbf{P}}\right)d\mathbf{P} + \left(\frac{\partial V}{\partial T}\right)dT. \qquad (9.55)$$

Let us introduce the following designations:

$$\alpha = \frac{1}{V}\left(\frac{\partial V}{\partial T}\right)(\text{K}); \quad E = -V\left(\frac{\partial \mathbf{P}}{\partial V}\right)(\text{MN}/\text{cm}^2), \qquad (9.56)$$

where α is the coefficient of volume thermal expansion (Equation 9.52) and E is the volume elasticity modulus. Substituting Equation 9.55 into the left-hand part of Equation 9.54 and accounting for the adopted designations, we can write

$$da = \alpha P dT - \frac{1}{E}\mathbf{P}\, d\mathbf{P}, \qquad (9.57)$$

where $da = \frac{dA}{V}$ is the specific work.

Changing the variables in differential Equation 9.57 as

$$T = \ln \mathbf{P}; \quad dT = \frac{d\mathbf{P}}{\mathbf{P}}; \quad d\mathbf{P} = \mathbf{P}dT, \qquad (9.58)$$

we find the work of resistance of a body to external force fields

$$da = \alpha d\mathbf{P} - \frac{1}{E}\mathbf{P}\, d\mathbf{P}.$$

After integration, we can write that

$$A(\mathbf{P}) = -\frac{1}{2E}\mathbf{P}^2 + \alpha\mathbf{P}. \qquad (9.59)$$

As an incomplete square trinomial, this formula determines the value of the work depending on the intensity of external force fields (direct work). Its variation is set by parameters α and E. Points of intersection of the parabola with the axis of abscissas are found from conditions

$$A(\mathbf{P}) = 0; \quad \mathbf{P}_1 = 0; \quad \mathbf{P}_2 = 2\alpha E,$$

and coordinates of the apex are determined by equality

$$\frac{dA(\mathbf{P})}{d\mathbf{P}} = 0.$$

They are equal to

$$\mathbf{P}_{max} = \frac{1}{2}\mathbf{P}_2 = \alpha E \quad \text{and} \quad A(\mathbf{P})_{max} = \frac{1}{2}\alpha^2 E.$$

To calculate the work done by a solid over external objects during the heating process, introduce a new variable:

$$\mathbf{P}^2 = T; \quad dT = 2\mathbf{P}d\mathbf{P}; \quad \mathbf{P}d\mathbf{P} = \frac{dT}{2}. \tag{9.60}$$

As such, Equation 9.57 assumes the form

$$da = \alpha T^{1/2}dT - \frac{1}{2E}dT. \tag{9.61}$$

Integration yields

$$A(T) = \frac{2}{3}\alpha T^{3/2} - \frac{1}{2E}T. \tag{9.62}$$

Equating this expression to zero, we find the coordinates of points of intersection of the plot with the axis of abscissas

$$T_1 = 0; \quad T_2 = \frac{9}{4}\frac{1}{(2\alpha E)^2}. \tag{9.63}$$

Using the condition of equality of differential (9.61) to zero, we can determine coordinates of its maximum

$$T_{max} = \frac{1}{(2\alpha E)^2}; \quad A(T)_{max} = -\frac{1}{6}\frac{1}{E(2\alpha E)^2} = -\frac{1}{6}\frac{T_{max}}{E}. \tag{9.64}$$

Using Substitution 9.60, move the term $A(T)$ (the work done by a solid in external thermal fields, inverse work; Equation 9.62) and its extremes (Equations 9.63 and 9.64) into the coordinate system containing the direct work (Equation 9.53):

$$A(T) = \frac{2}{3}\alpha\mathbf{P}^3 - \frac{1}{2E}\mathbf{P}^2 \text{ (I)}; \quad \mathbf{P}_1 = 0; \quad \mathbf{P}_2 = \frac{3}{2}\frac{1}{\alpha E} \text{ (II)}$$

$$\mathbf{P}(T)_{max} = \frac{1}{2\alpha E}; \quad A(T)_{max} = -\frac{1}{24\alpha^2 E^3} = -\frac{1}{48E^2 A(\mathbf{P})_{max}} \text{ (III)}. \tag{9.65}$$

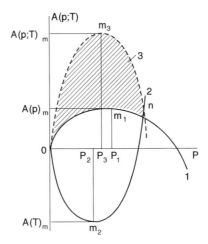

FIGURE 9.28
Work done by a body in the force and thermal fields.

Plots of Functions 9.59 (curve 1) and 9.65.I (curve 2) are shown in Figure 9.28; curve 3 represents their superposition. Curve 1 designates the work done by a material during deformation in a constant force field. As shown by experience (see Figures 2.10, 2.22, and 3.11), the form of a practical deformation diagram coincides with that of the theoretical (except for peculiarities near point a [Figure 2.22] introduced by CD transitions). The area between superposition curve 3 and curve 1 (dashed zone) forms a kinetic reserve of strength $\Delta\sigma$ and can only be produced under controlled work conditions, i.e., where loading matches a corresponding thermal support. It is this zone that contains the whole temperature "fan" of the deformation diagrams produced at different test temperatures (Figure 2.39).

Point n (designated so in Figure 3.11, and corresponding to point $b_{1,2}$ in Figure 2.22) characterizes the moment of fracture. Its abscissa is found by combined solution of Equations 9.59 and 9.65.I. It is equal to constant value $P_n = 3/2$, which numerically coincides with the crystalline lattice parameter at the Debye temperature (Equation 1.87) at $r_a = 1$. This means that the kinetic reserve has been exhausted and thus redistribution of load between interatomic bonds, that have failed and those that continue working, is no longer impossible (see Section 6.3).

It should be emphasized once more that there are differences between thermal and mechanical effects (see Section 3.11), which accompany deformation (Equation 9.59) and heating (Equation 9.65.I). This difference is in the number of interatomic bonds that participate in them: only the bonds oriented under the effect of external stresses σ participate in deformation, while heating involves the entire isotropic set N. Therefore, there are different relationships between parameters $\mathbf{P}$ and T in both cases: logarithmic (9.58) in the first case and quadratic (9.60) in the second.

When deformation is forbidden (requirements that are almost always imposed on structural members of engineering objects), i.e., when $dV = 0$, the differential of the Ω-potential has the following form:[23]

$$d\Omega = -sdT - Nd\omega. \tag{9.66}$$

Substituting the expression for ω-potential near the Debye temperature (where $T \approx \theta$) in Equation 9.23, we find that

$$\dot{P} = -\frac{s+kN}{V}\dot{T}. \tag{9.67}$$

It can be seen from Equations 9.15, 9.39, and 9.67 that any predictable or random attempt to change the ω-potential can be immediately compensated for by the opposite change in temperature T. As such, the material would not deform (as parameters of state, s and V, remain constant) and will not fracture (N = const). Note that no limitations (in terms of isothermicity or reversibility of processes) exist here. The foregoing analysis allows us to formulate the major principle of adaptation of materials: if a mechanical impact $d\sigma$ (or any other effect) is accompanied by the opposing impact dT, coordinated with the former in time and capacity (or any other compensating factor), deformation and fracture of the material would not occur at all (within the limits of a dashed zone in Figure 9.28). Therefore, the methods of control of the stressed–deformed state based on this principle can be called energy methods.

Figure 9.29 presents the AM mechanism of this method for C-materials (MB distributions 2, 3, and 4 are located on resistance curve 1 between characteristic temperatures T_c and θ, which are determined by vertical lines passing through points r_0 and r_b.) Let us assume that, at the initial state of a body, the MB factor occupies position 2. When the body is affected, for example, by a compressive impact, it moves while transforming to the left to position 3.

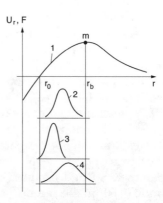

FIGURE 9.29
Combined effect of competing mechanical and thermal impacts on the AM structure.

If the mechanical impact is accompanied by a specially arranged and controlled thermal impact (in this case, dramatic heating), this causes its opposite change (in accordance with Figure 2.2, flattening and shifting to position 4). The combined effect of both competing processes is such that the MB distribution will not change but will remain as shown in position 2.

We can write Equation 9.53 accounting for Designations 9.56 in the following form:

$$\varepsilon E = \alpha s T \frac{dT}{dV}. \tag{9.68}$$

This equation explains the macroscopic mechanism of the thermal control method. The schematic diagram of the thermal method for controlling the stressed–deformed state becomes clear from Figure 3.2. Let position 2 designate actuators of the control system having heat supply system 3 and controlling the state of structural component 1 that is working in irregularly changing force and thermal fields. Their possible changes with time are shown in Figure 9.30.

Loading can be nonisothermal and isothermal.[3,348] If the level of effective mechanical stresses in a material does not depend on the character and interval of temperature cycles, loading is considered to be nonisothermal. When they are related in time and space, loading is said to be isothermal. Antiphase conditions of deformation are characterized by a combination of stresses and temperatures in which compression takes place in a heating half-cycle and tension in a cooling half-cycle. As such, the extreme values of temperature and load (deformation) coincide in time. All other combinations of temperature and force effects correspond to co-phase loading conditions. Positions 1 and 2 in Figure 9.30 characterize antiphase isothermal conditions, whereas co-phase nonisothermal conditions last for the rest of time.[3] Control systems should provide the antiphase conditions for purposes of decreasing the level of the stressed–deformed state. It has been proven experimentally[348] that durability under such conditions is higher by an order of magnitude than in the co-phase conditions.

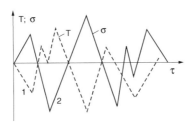

FIGURE 9.30
Temperature-force loading.

This experimental fact can be easily explained if we return to the equation of thermomechanical fatigue (7.4). It follows from this equation that, under antiphase loading conditions, the stressed–deformed state due to mechanical load, σ, is compensated for by a synchronous change in temperature T.

Analysis of the relationship between adiabatic

$$\beta_s = -\frac{1}{V}\left(\frac{\partial V}{\partial P}\right)_s$$

(Equation 9.46) and isothermal,

$$\beta_T = -\frac{1}{V}\left(\frac{\partial V}{\partial P}\right)_T, \tag{9.69}$$

deformability yields the same conclusion:[20]

$$\frac{\beta_S}{\beta_T} = \frac{C_V}{C_P}.$$

In accordance with Equation 2.143, equality $C_p \approx C_V$ is always met for solid bodies. This allows us to write that $\beta_s = \beta_T$ or

$$\left(\frac{\partial V}{\partial P}\right)_s = \left(\frac{\partial V}{\partial P}\right)_T.$$

A fundamental conclusion follows from this: deformation of a solid at a constant temperature does not change its entropy s, i.e., eliminate the initial cause that favors development of the deformation and fracture process (see Equation 9.51). In this case, if deformation does occur accidentally, it will always be at the elastic stage. In other words, if any deformation is accompanied by a corresponding thermal support, fracture will not take place. So what resistance can a body develop in this case?

To answer this question, let us differentiate the equation of state (2.96) with respect to T at $N = $ const and $\theta = $ const that yields

$$\frac{d(PV)}{dT} = \left(\frac{\partial P}{\partial T}\right)_V V + \left(\frac{\partial V}{\partial T}\right)_P P = C_V.$$

Using the definition of thermal expansion coefficient α (Equation 2.152) and thermal pressure coefficient v (Equation 4.43), and taking into account Equation 2.143, we can write

$$\Omega = PV = \frac{C_V}{\alpha + v}.$$

Assuming α and v for a given material to be constants, we can differentiate the left- and right-hand parts with respect to pressure at $T = \text{const}$ that yields

$$\mathbf{P}\left(\frac{\partial V}{\partial \mathbf{P}}\right)_T + V = \left(\frac{\partial C_V}{\partial \mathbf{P}}\right)_T \frac{1}{\alpha + v}.$$

Recalling the definition of isothermal deformability (Equation 9.69), we find that

$$\mathbf{V}(1 - \mathbf{P}\beta_T) = \frac{1}{\alpha + v}\left(\frac{\partial C_V}{\partial \mathbf{P}}\right)_T.$$

Substituting the known relationship between derivatives of thermodynamic values[23]

$$\left(\frac{\partial C_P}{\partial \mathbf{P}}\right)_T = -T\left(\frac{\partial^2 V}{\partial T^2}\right)_P,$$

and accounting for Equation 2.143, we obtain

$$1 - \mathbf{P}\beta_T = -\frac{T}{(\alpha + v)V}\left(\frac{\partial^2 V}{\partial T^2}\right)_P. \tag{9.70}$$

Expressing the coefficient of thermal expansion (Equation 9.52) as

$$\alpha V = \left(\frac{\partial V}{\partial T}\right)_P$$

and differentiating both parts with respect for T at $P = \text{const}$, we write:

$$V\left[\left(\frac{\partial \alpha}{\partial T}\right)_P + \alpha^2\right] = \left(\frac{\partial^2 V}{\partial T^2}\right)_P.$$

Substitution of this expression into Equation 9.70 yields

$$\mathbf{P}\beta_T - 1 = \frac{T}{\alpha + v}\left[\left(\frac{\partial \alpha}{\partial T}\right)_P + \alpha^2\right].$$

Ignoring the second-kind effects (i.e., assuming that $\partial \alpha = 0$), and assuming to the first approximation that $\mathbf{P}_T \gg 1$, we can write that

$$\mathbf{P}_T = \frac{\alpha^2 T}{\beta_T(\alpha + v)}.$$

It follows that the resistance of materials under isothermal conditions is directly proportional to temperature T. Under nonisothermal conditions, this dependence is inversely proportional (Equation 2.170). Equations 2.105 and 9.51 show that there are equivalent reactions of structure of materials to mechanical and thermal effects at all scale levels. This suggests that, for solids, $\alpha \approx v$. Then,

$$\mathbf{P}_T = \frac{\alpha T}{2\beta_T}. \tag{9.71}$$

Experimental confirmation of the validity of these laws can be found in studies.[349,350]

As follows from Equations 9.68 and 9.70, the efficiency of thermal regulation of the stressed–deformed state directly depends on the value of the coefficient of thermal expansion α. It is much higher in metals and alloys than in stone materials. For this reason, such methods will have the highest efficiency for metals. Using combined thermomechanical methods can increase their effect on stone materials.

The operating principle of thermomechanical methods can be demonstrated using examples shown in Figure 9.31 in which the stressed–deformed state of bent reinforced concrete members is controlled. Consider Figure 9.31a. Hollow metal bars 2 and 3 (made of a suitable metal having high coefficient of thermal expansion) are used as reinforcing bars and as stressing elements

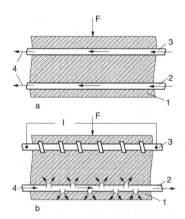

FIGURE 9.31
Schematics of control of the stressed–deformed state of bent reinforced concrete members by thermomechanical methods: (a) regulated prestressing (1, fragment of the bent reinforced concrete member; 2, 3, tubular reinforcement bars of the stretched and compressed zones, respectively; 4, feeding of heated or cooled liquid). (b), variation in concentration of thermodynamic traps (1, fragment of a structural member; 2, perforated tubular reinforcement bars; 3, reinforcement bars with electric regulation of temperature; 4, pressure feeding of hot water or vapor).

and placed into the stretched and compressed zones. Their temperature may be varied by different methods: by supplying electric current, by feeding a liquid having a required temperature, etc. Variation of temperature of the reinforcing bars leads to a double effect: first, it creates a required temperature gradient in concrete, and second, it leads to formation of a mechanical prestress in its structure.

The temperature gradient and level of prestress are regulated depending on the activity of external force effects F, for example, by feeding a cooling liquid into the hollow reinforcing bars in stretched zone 2 or a heated liquid to those in the compressed zone 3. Variation of temperature will cause compensation of the dynamic component of the external load without fracture of the structure. Temperature elongation of hollow reinforcing bars in the compressed zone and shortening in the stretched zone induce additional stresses of the opposite sign in concrete, which inhibit the cracking process (following the diagram in Figure 9.22). They serve as reliable barriers in the way of propagation of random cracks (see Section 6.6).

The process is triggered, i.e., the liquid is fed, by a signal of the servo system whose sensing elements react to any variation in external loads, strains, or intensity of the acoustic emission signals (see Section 3.11). Such structures possess self-adjusting properties and can be used in highly critical constructions under dynamic loading conditions.

Sections 5.2 and 5.5 have shown that temperature and humidity change the microstructure of concrete, exerting a substantial effect on its technical state. This can find application in systems for retardation of the cracking process. The essence of the proposal consists of an artificial formation of the humidity or temperature gradient (or both). An increase in temperature extends the crack incubation period to high stress levels, while water saturation, causing multiplication of plastic traps, localizes the formed microcracks and prevents their further development and coalescence (see Section 6.6).

The diagram for realization of this proposal is shown in Figure 9.31b. Water saturation of the structure 1 in the stretched zone can be realized by pressure feeding of water 4 into perforated tubular bars 2, while temperature of concrete in the compressed zone can be increased by supplying electric current I through bars 3.

The passive analogue of a system that uses the effect of static strengthening of concrete in compression (see Section 3.10 and Figure 3.58) has long been widely applied in the construction industry for increasing crack resistance of prestressed reinforced concrete structures.[330] Prestressing is made at the manufacturing stage and kept unchanged for the entire service life of a construction. Such a system is in a state of passive waiting with respect to external effects. It has substantial drawbacks: high metal consumption, unavoidable losses of prestresses, and lack of flexible control of the stressed–deformed state. The systems discussed earlier (Figure 9.31a,b) have high efficiency and are free from these drawbacks.

Existing design methods[2–4] are intended to ensure failure-free performance of engineering objects and their structural members on the basis of probability approximation, using excessive amount of materials (safety due to redundancy) to "cover" emergency situations. They operate with volume, mass, or structural parameters of a material only at the stage of manufacture. Therefore, they are passive or inertial. The degree of their reliability can be evaluated from news that appears from time to time about accidents and catastrophes of engineering objects, especially in connection with earthquakes and technogenic phenomena.

Mechanical, thermal, thermomechanical, and other adaptation methods provide reliability and durability not by varying quantity or quality of materials but by a conceptually new approach, i.e., affecting them by certain kinds of energy. Unlike traditional methods, they allow an immediate reaction to variations in service conditions, increasing resistance of a material through short-time consumption of this or that kind of energy. These methods were first described by the author.[12]

9.10 Promising Technologies and Ingenious Design Solutions

It is now thought that deformation and fracture appear fatally inevitable, inherent in all engineering objects.[1-4] This is not the case, however, as shown by the concepts of physics of strength and mechanism of fracture of solids elaborated in this book and confirmed by numerous experiments on resistance of materials.[3] It turns out that these processes can be accelerated,[16] decelerated,[215] or eliminated.[122] Moreover, by controlling them, it is possible to achieve a 100% guarantee of failure-free performance of machines and mechanisms as well as to fundamentally decrease power consumption of many technological processes.

They make it possible to solve technical problems considered unsolvable within the frames of existing technologies and provide the following economic, social, and moral effects:

- Materials consumption of machines and mechanisms can be decreased by increasing efficiency of utilization of materials.
- Capabilities of design can be widened to a substantial degree by using nontraditional design solutions.
- New advanced technologies and materials can be developed.
- Material, social, and moral losses, which are inevitable consequences of accidents and catastrophes, can be avoided.

Next we present some original solutions of conceptually different technical problems: possibility of building engineering objects with artificial intelligence (Section 9.10.1), ensuring thermomechanical strength of independent super-deep exploration geoprobe (Section 9.10.2), and technology for improvement of workability of metals by cutting (Section 9.10.3).

9.10.1 Structural Systems Adaptable to Service Conditions (ASS)

The energy methods for controlling the stressed–deformed state allow fabrication of engineering objects adaptable to service conditions. It is impossible to imagine them within the frames of existing technologies. Such objects should meet the following requirements:

- They should have preset consumer properties (reliability, safety, durability, capacity, comfort, etc.).
- They should have a 100% guarantee of retaining these properties for the entire service life, allowing no emergency situation under any circumstances.
- They should consume minimum materials, power, and labor during manufacture, construction, and functioning.

First, we mean expensive and highly critical objects, such as spacecraft, aircraft, sea ships, heat and nuclear power stations, bridges, tall buildings, entertainment facilities, etc., whose failure involves high material, social, and moral losses. From the formal standpoint, each can be represented as a pyramid structure (Figure 9.32) made up of a set of subsystems that perform specific consumer functions. The location of any in the pyramidal structure characterizes the extent of its responsibility for reliability and durability of an object as a whole. Service subsystems 1 are located at its top: heating, ventilation, lighting, water supply, etc. They perform auxiliary functions and their failure will cause deterioration of consumer properties of the object, exerting almost no effect on its reliability and durability.

Responsibility of life support systems 2 (hydraulic, power supply, automation, transport service lines, etc.) is a bit higher. Failure of any of these does not lead to catastrophic consequences but has a marked effect on reliability of the object. It is only the failure of structural members (pyramid foundation 3) that is inevitably accompanied by the collapse of the entire pyramid system. Even proper functioning of all other subsystems of the higher levels becomes

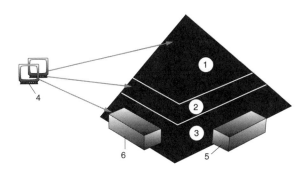

FIGURE 9.32
Subsystems of engineering objects shown as a pyramidal structure.

useless in this situation. Thus, elements of the lower, i.e., structural, level such as bodies, fuselages, beams, columns, etc. have the highest extent of responsibility.

As proven by experience, parameters of most subsystems can be readily controlled. This allows making automated systems 4, which maintain the optimal level of their operation. Structural subsystems are characterized by the highest conservatism in this respect. As has been noted many times, modern design methods are capable of making probability predictions for them,[1-4] leaving substation margins for randomness and providing no adjustments of resource during operation.

The energy methods of controlling the stressed–deformed state (Section 9.9) radically change the situation, allowing building of structural members with self-regulating mechanical properties. Such members take up dynamic, seismic, or random technological loads without fracture and will find application in construction of objects in seismic regions,[351,352] as well as under conditions of periodic or random peak loads, in regions with dramatic gradients of low or high temperatures,[353] with a high level of radiation, etc.

The adaptation effect is provided by solving the following problems (Figure 9.33):

- Inclusion into load-carrying structures 1 special devices 2 and 3, through which one of the kinds of energy described in Section 9.9 is introduced into a material; installing on them the acoustic emission, stress, strain, temperature, etc. transducers 4 required to solve prediction problems (Chapter 8)

- Equipping the objects with control system 5, which generates the specified kind of the energy and directs it to structures when needed

- Equipping the objects or their complexes with servo system 6, which takes up external loads 7 or predicts their variations and sends signal 8 to trigger control system 5

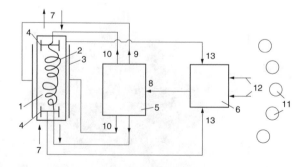

FIGURE 9.33
Principle of operation of structural systems adaptable to service conditions.

The final control system should service the most responsible structural members 1. Most of the time it remains in a waiting state; during dangerous activity of service factors, it is triggered by system 6 (signal 8). As a result, control system 6 generates a required kind of the energy and sends it, 9 and 10, to the most stressed zones of load-carrying members, making the structure ready to take up varying loading conditions.

Sensitive elements of the servo system are a branched network of stress and strain sensors 4 installed in different regions 11 of structure 1. In addition, the servo system can be switched on manually (12) by an operator.

Active protection of the construction from fracture is realized as follows. Servo system 6 is always in the active waiting mode. Triggered by one of the previously discussed methods, it sends signal 8 to actuate final control system 5. The latter generates and sends the appropriate kind of energy 9 or 10 to the most stressed zones of structural members 1. Energy replenishment causes a short-time increase of strength and improves crack resistance of the material by one of the methods described in Section 9.9. The energy flow to each zone is regulated by stress and strain sensors.

On completion of external activity, both systems and all structural members return to the initial state. Such objects have all properties of adaptable systems. No specific problem could prevent realization of such a system; servo systems of the required type are well known. Electric circuits, simple hydraulic devices, and other similar units that exist in any object can be used.

Not only new objects but also those under repair or reconstruction can be equipped with systems for active protection from sudden loading. The possibility of building adaptable engineering objects was reported in the special literature[18,21] and in the mass media.[17,354]

9.10.2 Ensuring Thermomechanical Strength of Independent Super-Deep Exploration Geoprobe

Earthquakes are the most destructive catastrophes for humans. They annually take tens and hundreds of human lives.[355] Their initial cause always lies in the bowels of the Earth at a depth from several kilometers to several hundreds of kilometers.

Experimental studies of the centers of earthquakes are still conducted using instruments located near the Earth's surface. So far no equipment can bring measuring instruments to a large depth directly to the earthquake centers. Therefore, studies of the true nature of super-deep earthquake centers take place in the absence of reliable data on the chemical composition and physical state of matter at the super-deep horizons of the Earth's crust and mantle.

Vashchenko[355] suggested a concept, considered theoretical, and engineering aspects of making of an independent super-deep geoprobe. A conceptually new super-deep technique, economically viable and having no analogues in world practice is discussed. It can be used to obtain cardinally new information about the depths of the Earth and also for commercial prospecting and geotechnical production of deep-lying energy raw materials and mineral resources, as well as in many other research, development, and technological applications.

To penetrate deep into the Earth, such a probe should be capable of operating under a pressure of 7000 bars and at temperatures of up to 1600°C under high radiation levels. As such, the major physical–technical and engineering problem appears to be heat resistance of structural materials. It is understood that strength characteristics of casing parts of the geoprobe should be maintained.

Multifunctional materials are required to design the main subsystems of the probe: heat generator, heat-releasing elements, body, shell, and bit. They should combine high strength, chemical resistance, thermal conductivity, heat and radiation resistance, preset level of electrical conductivity, and a number of other properties. Making such materials and, moreover, structural systems is hardly possible without the use of theoretical developments (Chapters 1 and 2) and original ideas (Chapters 6 through 9).

Such super stable devices will be needed not only for penetration deep into the Earth. They also will be required for space engineering, nuclear reactors, metallurgy, geotechnical industries, geothermal systems, and other scientific and technical applications.

9.10.3 Technology for Improvement of Workability of Metals by Cutting

After more than a century-long discussion, specialists at last came to an obvious conclusion that the process of metal cutting is a purposeful destruction or fracture of the work material, and that all its aspects should be treated particularly from this standpoint.[356] Figure 9.34 shows a model of a cutting process comprising tool 1, billet (workpiece) 2, and chip 3. The main task of cutting is to separate chip 3 from base metal 2 with minimum power consumption and thus create surface A of the best possible integrity.

This task is handled more or less successfully by varying only the external factors—first of all, geometry of cutting tool 1 (and thus creating different stressed–deformed states in the cutting zone, which results in different

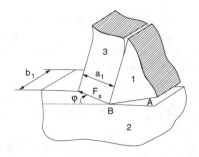

FIGURE 9.34
Model of orthogonal metal cutting.

angles φ), parameters of the chip (sizes a_1 and b_1), cutting speed, and cutting feed. Astakhov[356] was probably the first to state directly that plastic deformation in cutting has an absolutely harmful effect. It causes a substantial increase in temperature within the machining zone and, thus, wear of the tool, residual stresses, and strains (cold working of the machined layers) in the workpiece. This is the difference between the process of metal cutting and all other forming processes, such as forging, stamping, rolling, drawing, etc., in which the major process objective is achieved through plastic deformation. Unfortunately, this important aspect is not yet clearly understood by many other specialists in the field (see, for example, References 357 through 359).

Despite certain success achieved in the theory of cutting, the following issues remain open:

- How to minimize plastic deformation, especially in machining of tough metals and alloys
- How to decrease resistance of metal to fracture temporarily (for the time period of machining)
- How to transform ductile stainless steel, e.g., 303, into a structure close to gray cast iron, which is known for its excellent machinability

These problems cannot be solved without a change in the technical state of the workpiece material.

Depending on the AM composition of metal (Section 9.4), solidification conditions (Section 9.5), and technological treatments (Section 9.6), the MB factor (Equation 2.2) arranges the AM bonds so that they are in the compresson (positions 1 and 3 in Figure 5.7) or dilaton (position 4) phases. As such, if most of N bonds are in the C-phase (positions 1 and 3 in the same figure), a material deforms following the elastic mechanism and fractures by the brittle mechanism (plot 1 in Figure 2.22). If dilaton bonds are in its volume (position 4), depending on their quantity, it will have ductile or even tough, highly plastic behavior (curve 2 in Figure 2.22).

This figure shows a combined deformation diagram, which characterizes structure in the stress state of tension in longitudinal (upper field) and compression in transverse section (lower field). The clear manifestation of the orientation compression (see Section 3.4) in the transverse direction is the "necking" effect (Section 2.10). A structure in which the prevailing number of the AM bonds are at the θ-temperature (position 4 in Figure 5.7) requires minimum consumption of energy for fracture to occur; therefore, such a structure would have the best machinability.

It is discussed in Section 9.9 that the state of a material, inherent or acquired in technological processes (Section 9.6), can be changed using force, thermal, or other fields. As an example, consider how we can reduce energy consumption in the cutting process and improve machinability by controlling the state of a material in force fields. Let cylindrical workpiece 1 of a ductile

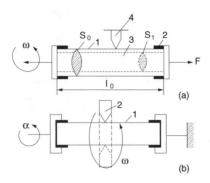

FIGURE 9.35
Machining of metals in the stress–strain state induced by (a) tension and (b) torsion of the workpiece.

material (in Figure 5.7, its MB factor is shown by position 4 and its macroscopic behavior is characterized by diagram oab_2 in Figure 2.22) be stretched grips 2 in the axial direction by force **F** (Figure 9.34) at the elastic stage (in region oa of Figure 2.22). Before loading, its length was l_0 and cross section was s_0, while after loading the length becomes l_1 and cross section s_1.

Because of the three-dimensional nature of the Ω-potential (Equation 2.96) and the chaotic character of orientation of the AM bonds formed during solidification, the value of the initial work for resistance to fracture is of equal probability in any direction of workpiece 1 on a flat section of the three-dimensional diagram (Figure 9.15); it is determined by figure 1, close to a circumference in shape. In diagram "$\sigma - \varepsilon$" (Figure 2.22), this state is identified with the origin of coordinates (point 0).

Elastic deformation in tension (Figure 9.35a) changes the sizes and shape of the workpiece (from position 1 to position 3). Elongation in the axial direction and shortening in the transverse direction take place due to the occurrence of the orientation processes in the AM set N of the bonds (Section 3.4). They lead to the formation of asymmetry in the diagram of the resistance work (Figure 9.15), which is transformed from figure 1 into ellipse 2. In the "$\sigma - \varepsilon$" diagram (Figure 2.22), the state of a material will no longer be characterized by point 0. In the axial direction (along axis 90 to 90 in Figure 9.15), it is identified with point G_2, and in the transverse direction (along axis 0 to 0 in Figure 9.15) with point G_3.

It follows from Section 9.9 that a tensile stressed–deformed state leads to redistribution of both potential and kinetic parts of the overall internal energy (Equation 9.36). Its manifestation is in the fact that microenergy distribution in the axial direction (e.g., position 7 in Figure 9.4) becomes more flattened (taking position 6), whereas in the transverse direction it may take shape 8 or become even more pointed, while moving toward the compresson phase. Experience of measuring longitudinal ($-\varepsilon$) and transverse ($+\varepsilon$) strains (Figure 2.22), as well as variation in microhardness (see Section 3.9) during loading, confirms this statement. So, impact of external loads leads to formation of energy anisotropy in the workpiece structure; its elasticity grows in a transverse direction (in Figure 2.22, characterized by plot $G_3 0 G_2 a$, rather than plot $0 G_2 a$), whereas the work of fracture in cutting decreases. (Compare shape and sizes of the diagram in Figure 9.15 along axis 0 to 0 in positions 1 and 2.)

It is proposed that the effect described should be used to decrease energy consumption in cutting.[16] The point of the proposal (see Figure 9.35) is that a tensile prestress is first induced in workpiece 1 by applying force **F** through grips 2; then it is fixed and the billet brought into rotation at angular velocity ω and cut by cutting tool 4. The level of the preliminary stress for each material is selected experimentally and controlled by relative strains or by hardness indicators.

A similar effect is produced also by preliminary static torsion of workpiece 1 to angle α (Figure 9.34b). In this case, the cutting force is generated by rotating, not workpiece 1, but tool 2 at angular velocity ω.

To improve the efficiency of application of external force field, a local (within the machining zone) triaxial state of stress can be used. As shown by Astakhov[356] (figure 4.8 in his book), the degree of triaxiality has a very strong effect on the strain to fracture of different metals and alloys. Because the state of stress in the machining zone is mainly determined by the tool geometry, this geometry should be selected so that the strain at fracture is minimal.

The efficiency of the process of adjustment of the state and, therefore, the operation of cutting can be fundamentally increased by using thermal (Figure 9.29), magnetic (Figure 9.26), and electric (Figure 3.2) fields and, especially, their combinations. For example, application of the local thermal fields through the preheating or cryogenic cooling of the workpiece make it possible to machine special high alloys that cannot be machined at normal conditions.

References

1. Paton, B.E., New trends in improvement of strength and residual life of welded structures (in Russian), *Avtomaticheskaya Svarka (Automatic Welding)*, 9–10, 3, 2000.
2. Felbeck, D.K. and Atkins, A.G., *Strength and Fracture of Engineering Solids*, 2nd ed., Prentice-Hall, Englewood Cliffs, NJ, 1996.
3. Troshenko, B.T., Ed., *Resistance of Engineering Materials to Deformation and Fracture (Handbook)* (in Russian), Naukova Dumka, Kiev, 1993, V. 1–2.
4. Hertzberg, R.W., *Deformation and Fracture Mechanics of Engineering Materials*, 4th ed., John Wiley & Sons, New York, 1996.
5. Second International Conference on the Strength of Materials (ICSMA-II), Fundamental aspects of the strength of crystalline materials, Prague, Czech Republic, August 1997.
6. Pollock, D.D., *Physical Properties of Materials for Engineers*, 2nd ed., CRC Press, Boca Raton, FL, 1993.
7. Smith, C.O., *The Science of Engineering Materials*, 3rd ed., Prentice-Hall, Englewood Cliffs, NJ, 1986.
8. Klyuev, V.V., Nondestructive testing and diagnostics of safety (in Russian), *Zavodskaya Laboratoriy: Diagnostics of Materials*, 1, 16, 1998.
9. Sorin, Ya.M., *Physical Nature of Reliability* (in Russian), Standartizdat, Moscow, 1969.
10. Regel, V.R., Slutsker, A.M., and Tomashevsky, E.I., *Kinetic Nature of Strength of Solids* (in Russian), Nauka, Moscow, 1974.
11. Hudson, J.B., *Thermodynamics of Materials: A Classical and Statistical Synthesis*, John Wiley & Sons, New York, 1996.
12. Komarovsky, A.A., Control of the Stress–Strain State of Materials and Structures (in Russian), Vysshaya Shkola, Kiev, 1966.
13. Nishida, S.-I., *Failure Analysis in Engineering Application*, Butterworth-Heinemann, Oxford, 1992.
14. Komarovsky, A.A., Chornobyl: cover. How much did it serve? (in Russian), Kyiv, *Vistnyk Chornobylya*, 3–4, 1998.
15. Galyas, A.A. and Poluyanovsky, S.A. *Principles of Thermomechanical Fracture of Rocks* (in Russian), Naukova Dumka, Kiev, 1972.
16. Komarovsky, A.A. and Astakhov, V.P., Implementation of the latest breakthrough in the physics of materials in metal cutting: analysis and preliminary results, *Manuf. Sci. Eng.*, 10, 319, 1999.
17. Komarovsky, A.A., Wreck and catastrophes are not fatal inevitability (in Russian), Kyiv, *Weekly Mirror*, July, 1997.
18. Komarovsky, A.A., *State-of-Art and Prospects of Development of Strength Science* (Scientific forecasts) (in Russian), Naukovedeniye, Kyiv, 2001, 7–12.
19. Lampman, S.R., Ed., *Fatigue and Fracture*, ASM Handbook, Vol. 19, ASM International, Materials Park, OH, 1996.

20. Levich, V.G., *Course of Theoretical Physics* (in Russian), Phyzmatgiz, Moskow, 1962.
21. Komarovsky, A.A., Physics of strength, reliability and durability (in Russian), *Tyazholoye Mashinostroyeniye (Truck Manufacturing)*, 9, 24, 2000.
22. Libovits, G., Ed., *Fracture (Modern Concepts)* (in Russian), Mir, Moscow, 1976, V. 1–6.
23. Landau, L.D. and Lifshits, L.M., *Statistical Physics* (in Russian), Nauka, Moscow, 1964.
24. Komarovsky, A.A., Physics of deformation and fracture (in Russian), *Prikladnaya Phyzika (Applied Physics)*, 1, 88, 2001.
25. Landau, L.D. and Lifshits, E.M., *Course of Theoretical Physics (Mechanics, Electrodynamics)* (in Russian), Nauka, Moscow, 1969.
26. Bogoroditsky, N.P. et al., *Theory of Dielectrics* (in Russian), Energiya, Moscow, 1965.
27. Blokhintsev, D.I., *Principles of Quantum Mechanics* (in Russian), Vysshaya Shkola, Moscow, 1962.
28. Landau, L.D. and Lifshits, E.M., *Quantum Mechanics. Part I* (in Russian), Gostekhizdat, Moscow, 1948.
29. Kiffer, I.I., *Testing of Ferromagnetic Materials* (in Russian), Energoizdat, Moscow, 1965.
30. Frenkel, Ya.I., *Introduction to the Theory of Metals* (in Russian), Nauka, Moscow, 1972.
31. Landau, L.D., Akhiezer, A.I., and Lifshits, E.M., *Course of General Physics. Mechanics and Molecular Physics* (in Russian), Nauka, Moscow, 1969.
32. Reif, F., *Statistical Physics* (in Russian), Nauka, Moscow, 1972.
33. Zhurkov, S.N., Dilaton mechanism of strength of solids, *FTT*, 25, 3119, 1983.
34. Pettifor, D.G., *Bonding and Structure of Molecules and Solids*, Oxford Science Publications, Clarendon Press, Oxford, 1995.
35. Kuzmenko, V.A., *Development of Ideas about the Deformation Nature of Materials* (in Russian), UkrNIINTI, Kiev, 1968.
36. Finkel, V.M., *Physics of Fracture (Crack Growth in Solids)* (in Russian), Metallurgiya, Moscow, 1970.
37. Greshnikov, V.A. and Drobot, Yu.V., *Acoustic Emission* (in Russian), Standartizdat, Moscow, 1978.
38. Novodvorsky, I., Electrons from fracture: on the electron and X-ray radiation in fracture of solids (in Russian), *Izvestiya*, Moscow, 1984, N160.
39. Kusov, A.A. and Vettegren, V.I., Calculation of durability of loaded chain of atoms in unharmonic approximation, *FTT*, 22, 3350, 1980.
40. Rzhevsky, V.V. and Novak, G.Ya., *Principles of Physics of Rock. Part II. Thermo- and Electrodynamics of Rock* (in Russian), Moscow Institute of Radioelectronics and Mining Electromechanics, Moscow, 1964.
41. Chalmers, B., *Theory of Solidification* (in Russian), Metallurgiya, Moscow, 1968.
42. Lifanov, I.S. and Sherstniov, N.G., *Metrology, Means and Methods of Quality Testing in Building* (in Russian), Stroyizdat, Moscow, 1979.
43. Silvestrov, A.V. and Shagimordinov, R.M., Brittle fracture of steel structures and ways of its prevention (in Russian), *Problemy Prochnosti (Problems of Strength)*, 5, 88, 1972.
44. Ekobori, T., *Physics and Mechanics of Fracture and Strength of Solids* (in Russian), Metallurgiya, Moscow, 1971.
45. Mack Lin, D., *Mechanical Properties of Metals* (in Russian), Metallurgiya, Moscow, 1965.

46. Bridgman, P.W., *Fracturing of Metals*, ASM, Metals Park, OH, 1948.

47. Fridel, Zh., *Dislocations* (in Russian), Mir, Moscow, 1967.

48. Berg, O.Ya., *Physical Principles of Strength Theory of Concrete and Reinforced Concrete* (in Russian), Stroyizdat, Moscow, 1962.

49. Nozdriov, V.F. and Senkevich, A.A., *Course of Statistical Physics* (in Russian), Vysshaya Shkola, Moscow, 1969.

50. Terletsky, Ya.P., *Statistical Physics* (in Russian), Vysshaya Shkola, Moscow, 1966.

51. Samoylovich, A.G., *Thermodynamics and Statistical Physics* (in Russian), Gostekhizdat, Moscow, 1956.

52. Leontovich, M.A., *Introduction to Thermodynamics* (in Russian), Tekhizdat, Moscow, 1952.

53. Komarovsky, A.A., Diagnostics of structural materials (in Russian), *Tekhnicheskaya Diagnostika i Nerazrushayushchiy Kontrol*, 3, 8, 1999.

54. Komarovsky, A.A., Diagnostics of stress–strain state (in Russian), *Control. Diagnostika*, 2, 22, 2000.

55. Komarovsky, A.A., *Fracture Mechanism of Concrete and Advanced Methods for Ensuring Its Durability* (in Russian), Znaniye, Kiev, 1986.

56. Frenkel, S.Ya., Polymers. Problems, prospects, forecasts (in Russian), in *Physics of Today and Tomorrow (Scientific Forecasts)*, Nauka, Leningrad, 1973, 176.

57. Iveronov, B.I. and Revkovich, G.P., *Theory of X-Ray Dissipation* (in Russian), Izd-vo MGU, Moscow, 1978.

58. Bezukhov, N.I., *Fundamentals of Theory of Elasticity, Plasticity and Creep* (in Russian), Vysshaya Shkola, Moscow, 1968.

59. Bogachiov, I.N., Weinstein, A.A., and Volkov, S.D., *Introduction to Statistical Science of Metals* (in Russian), Metallurgiya, Moscow, 1972.

60. Umansky, Ya.S., *Physical Principles of Science of Metals* (in Russian), Metallurgiya, Moscow, 1965.

61. Haasen, P., *Physical Metallurgy*, 3rd ed., Cambridge University Press, Cambridge, 1996.

62. Regel, B.R. and Slutsker, A.I., Kinetic nature of strength (in Russian), in *Physics of Today and Tomorrow*, Nauka, Leningrad, 1973, 90.

63. Pozdnyakov, O.F. and Regel, V.R., Kinetics of cracks nucleation (in Russian), *Solid State Physics*, 10, 3669, 1968.

64. Zhurkov, S.N., Kuksenko, V.S., and Slutsker, A.I., Thermal domains and crystal instability (in Russian), *Phys. Solid*, 11, 296, 1969.

65. Ashby, M.F., On the engineering properties of materials. Overview N 80. *Acta Metall.*, 37, 5, 1273, 1989.

66. Tsarnekki, E.G., Stesi, D.T., and Zimmerman, D.K., Refractory metals in space engineering (in Russian), in *Properties of Refractory Metals and Alloys*, Metallurgiya, Moscow, 1968, 277.

67. Lement, B.S. and Perlmuter, I., Mechanical properties of alloys based on refractory metals (in Russian), in *Properties and Treatment of Refractory Metals and Alloys*, Izdatelstvo Inostr. Liter., Moscow, 1961, 52.

68. Khimushin, F.F., *Heat Resistant Steels and Alloys* (in Russian), Metallurgiya, Moscow, 1964.

69. Luzhnikov, L.P., Ed., *Materials in Machine-Building. Nonferrous Metals and Alloys* (in Russian), Mashinostroyeniye, Moscow, V. 1, 1967.

70. Mogilevsky, E.P., Ed., *Materials in Machine Building. Structural Steel* (in Russian), Mashinostroyeniye, Moscow, V. 11, 1967.

71. Khimushin, F.F., Ed., *Materials in Machine Building. Special Steels and Alloys* (in Russian), Mashinostroyeniye, Moscow, V. III, 1968.
72. Popov, V.A., Silvestrovich, S.I., and Sheidlin, I.Yu., Ed., *Materials in Machine Building. Nonmetallic Materials* (in Russian), Mashinostroyeniye, Moscow, V. IV, 1968.
73. Kozlov, P.M., *Use of Polymer Materials in Loaded Structures* (in Russian), Khimiya, Moscow, 1966.
74. Brener, S., *Factors Affecting the Strength of Filamentary Crystals. Fiber Composite Materials* (in Russian), Metallurgiya, Moscow, 1967, 24.
75. Honeycomb, R., in *Plastic Deformation of Metals*, Lyubov, B.Ya., Ed. (in Russian; trans. from English), Mir, Moscow, 1972.
76. Gerasimov, E.P., Martynov, V.M., and Sassa, V.S., *Heat Resistant Concretes for Electric Furnaces* (in Russian), Energiya, Moscow, 1969.
77. Garnier, V., Etude de la tenacite et de comportement de quatre aciers sous differentes conditions de temperature et de vitesse de deformation, These, Institut National Polytechnique de Lorraine, France, 1984, 116.
78. Ryzhova, S.I., Study of operation of bent reinforced concrete components at uniform heating up to 250°C, Ph.D. thesis (in Russian), NIIZhB, Moscow, 1967.
79. Bartenev, G.M., *Structure and Mechanical Properties of Inorganic Glasses* (in Russian), Stroyizdat, Moscow, 1996.
80. Hertzfeld, G.-M., Physics of melting, *Phys. Rev.*, 46, 995, 1934.
81. Weiner, I.H. and Boley, B.A., Thermal stresses formed on solidification, *Mech. Phys. Solids*, 11, 145, 1963.
82. Murgatryed, J., Two-phase states of solids, *Nature*, 154, 51, 1944.
83. Fersman, A.E., *Geochemistry* (in Russian), ONTI Press, Moscow, 1937.
84. Frenkel, Ya.I., *Kinetic Theory of Fluids* (in Russian), AN SSSR Press, Moscow, 1946.
85. Osipov, K.A., *Problems of Theory of Heat Resistance of Metals and Alloys* (in Russian), AN SSSR Press, Moscow, 1960.
86. Frenkel, Ya.I., *Statistical Physics* (in Russian), AN SSSR Press, Moscow, 1948.
87. Born, M. and Gopper-Mayer, M., *Theory of Solids* (in Russian), ONTI Press, Moscow, 1938.
88. Fermi, E., *Molecules and Crystals* (in Russian), IL Press, Moscow, 1947.
89. Kitaygorodsky, A.I., *Molecular Crystals* (in Russian), Nauka, Moscow, 1971.
90. Rid, V.T., *Dislocations in Crystals* (in Russian), Metallurgizdat, Moscow, 1957.
91. Gurevich, V.B. and Minorsky, V.P., *Analytical Geometry* (in Russian), Phyzmatgiz, Moscow, 1959.
92. Vorobiov, A.A. and Borobiov, V.A., *Electrical Break-Down and Fracture of Solid Dielectrics* (in Russian), Vysshaya Shkola, Moscow, 1966.
93. Berri, D.P., Fracture of Glassy Polymer, in *Fracture*, Libovits, G., Ed. (in Russian), Mir, Moscow, V. 7, 218, 1973.
94. Sammal, O.B., *Stresses in Concrete and Prediction of Service Life of Concrete and Reinforced Concrete Structures and Constructions* (in Russian), Valgus, Tallinn, 1980.
95. Dondik, I.G., *Mechanical Tests of Metals* (in Russian), Izdatelstvo AN USSR, Kyiv, 1962.
96. Komarovsky, A.A., Scale effect: origins, forms of appearance and dangerous consequences (in Russian), *Tekhnicheskaya Diagnostika i Nerazrushayushchiy Kontrol*, 4, 7, 2001.
97. Alexandrov, A.P. and Zhurkov, S.N., *Phenomenon of Brittle Fracture* (in Russian), ONTI Press, Moscow, 1933.

98. Gul, V.E., *Structure and Strength of Polymers* (in Russian), Khimiya, Moscow, 1971, 344.

99. Peras, A. and Dauknis, V., *Strength of Refractory Ceramics and Methods of Its Study* (in Russian), Mokslas, Vilnius, 1977.

100. Pisarenko, G.S. and Troshchenko, V.T., *Statistical Theories of Strength and Their Application for Metal Ceramic Materials* (in Ukrainian), AN USSR Press, Kiev, 1961.

101. Krol, I.S. et al., Study of scale effect manifestation during compression testing of concrete cubic samples. Investigations in the field of mechanical measurements (in Russian), *Proc. VNIIFTRI*, Moscow, 8(38), 206, 1971.

102. Krol, I.S., Krasnovsky, R.O., and Markov, A.I., On the origins of scale effect in concrete under compression (in Russian), *Izvestiya VUZOV, Stroitelstvo i Arkhitektura*, Novosibirsk, 3, 155, 1975.

103. Ziobron, V., Results of investigation of the scale effect in reinforcing bars (in Russian), in *Ingenernye Konstruktsii*, LISI Press, Leningrad, 1969, 79.

104. Chechulin, B.B., *Scale Effect and Statistical Nature of Strength of Metals* (in Russian), Metallurgizdat, Moscow, 1963.

105. Pisarenko, G.S. and Lebedev, A.A., *Resistance of Materials to Deformation and Fracture in the Complex Stressed State* (in Russian), Naukova Dumka, Kiev, 1969.

106. Kontorova, T.A. and Frenkel, Ya.I., Statistical theory of strength of real crystals (in Russian), *J. Tech. Phys.*, 11(3), 17, 1941.

107. Kontorova, T.A., One of the applications of statistical theory of the scale effect (in Russian), *J. Tech. Phys.*, 13(6), 13, 1943.

108. Fudzin, T. and Dzako, M., *Fracture Mechanics of Composite Materials* (in Russian), Mir, Moscow, 1982.

109. Troshchenko, V.T., *Fatigue and Inelasticity of Metals* (in Russian), Naukova Dumka, Kiev, 1971.

110. Sosnovsky, L.A., Influence of length of samples on critical tensile stresses of steel (in Russian), *Problemy Prochnosti*, 4, 30, 1976.

111. Komarovsky, A.A., Radchenko, L.N., and Chuiko, P.A., Variability of strength characteristics of conventional and high-strength concretes (in Russian), *Stroitelnye Materialy i Konstruktsii*, 1, 17, 1972.

112. Berezovik, S.V., Korzun, S.I., and Bich, P.M., Experimental study of strength of the concrete in bending tension (in Russian), *Proc. Instit. Civil Eng.*, Minsk, 84, 1979.

113. Sytnik, V.I. and Ivanov, Yu.A., Experimental studies of strength and deformability of high-strength concretes (in Russian), *High-Strength Concretes*, Kiev, 54, 1967.

114. Vorobiov, A.A. and Zavadskaya, E.K., *Electric Strength of Solid Dielectrics* (in Russian), GITL Press, Moscow, 1956.

115. Komarovsky, A.A., Measuring of indirect indices of hardness during strength testing of concrete by nondestructive methods (in Russian), *Stroitelnye Materialy i Konstrukcii*, 1, 35, 1976.

116. Golikov, A.E., Influence of moulding technology of high-strength concretes on their physical–mechanical properties (in Russian), *Beton i Zhelezobeton*, 9, 17, 1967.

117. Golikov, A.E., Study of physical–mechanical properties of high-strength concretes for transport structures allowing for the effect of some technological factors, Ph.D. thesis (in Russian), SRI of Concrete, Moscow, 1968.

118. Pisanko, G.N., Study of strength and strain properties of high-strength concretes, Ph.D. thesis (in Russian), SRI of Civil Engineering, Moscow, 1960.
119. Pinus, B.I., Study of properties of high-strength concretes in view of their operating in prestressed structures (in Russian), Ph.D. thesis, SRI of Railway Transport, Novosibirsk, 1968.
120. Saltykov, S.A., *Stereometric Metallography* (in Russian), Metallurgiya, Moscow, 1970.
121. Scillard, R., Review of durability of prestressed reinforced concrete structures in USA, Canada, countries of basin of the Pacific Ocean and the Far East, transl. from English, No. 1747, TsINIS Press, Moscow, 1969.
122. Komarovsky, A.A., Physical principles of adaptation of materials and structures to service conditions (in Russian), *Tyazholoye Mashinostroyeniye*, 5, 7, 2001.
123. Komarovsky, A.A., Popovich, G.A., and Radchenko, L.N., Method of determination of strength and volume structural heterogeneity of building materials (in Russian), Author's cert. of USSR No. 359593, *Bull. Inv.*, 35, 117, 1972.
124. Komarovsky, A.A., Method of determination of solidified concrete composition (in Russian), Author's cert. No. 566182, *Bull. Izobr.*, 27, 128, 1977.
125. Komarovsky, A.A. and Milov, V.A., Complex elastic–plastic method for nondestructive testing of strength of concrete (in Russian), *Stroitelnye Materialy i Konstruktsii*, 6, 29, 1972.
126. Komarovsky, A.A., Radchenko, L.N., and Petrov, A.N., Efficiency of methods of nondestructive testing of strength of light concretes (in Russian), *Stroitelnye Materialy i Konstruktsii*, 3, 20, 1972.
127. Komarovsky, A.A. and Novgorodsky, M.A., On accuracy, sensitivity and resolution of methods for nondestructive testing of concrete (in Russian), in *Application of Nondestructive Methods to Test Quality of the Concrete, Reinforced Concrete Structures and Parts*, Stroyizdat, Moscow, 1973, 35.
128. Komarovsky, A.A., On the problem of estimation of reliability of reinforced concrete structures by instrumentation methods (in Russian), in *Problems of Reliability of Reinforced Concrete Structures*, KyESE, Kuibyshev, 1975, 74.
129. Komarovsky, A.A., Physical principles of prediction of reliability of structures by instrumentation methods (in Russian), in *Experimental Investigations of Engineering Structures*, Budivelnik, Kiev, 1981, 95.
130. Berdichevsky, G.I. et al., *Reference Book on Testing of Strength of Concrete in Structures Using Mechanical Devices* (in Russian), Stroyizdat, Moscow, 1972.
131. Novgorodsky, M.A., Komarovsky, A.A., and Toporovsky, Yu.N., Study of sensitivity of methods for nondestructive testing of strength of heavy concrete (in Russian), *Abstracts on R&D in VUZ USSR*, 13, 53, 1978.
132. Komarovsky, A.A., Metrological support of nondestructive testing of strength of concrete, in *Experimental Investigations of Engineering Structures* (in Russian), Budivelnik, Kiev, 1981, 106.
133. Chonkvetadze, V.A. and Chikovani, T.D., Application of the methods of mathematical statistics for quality control of concrete (in Russian), *Beton i Zhelezobeton*, 6, 17, 1966.
134. Leonovich, M.F., On the nature of change of the variation coefficient of compressive strain of concrete (in Russian), *Beton i Zhelezobeton*, 2, 43, 1980.
135. Krasnyi, I.M., Estimation of accuracy of testing of concrete by the scatter indices of twin samples (in Russian), in *Technology and Properties of Heavy Concrete*, SRI of Concrete, Moscow, 1971.

136. Ratz, A.G., Statistical control of strength of concrete at the reinforced concrete plants (in Russian), *Beton i Zhelezobeton*, 10, 17, 1968.

137. Mints, Sh.I., Spread of strength indices of concrete of different buildings (in Russian), in *Investigations of Conventional and Prestressed Reinforced Concrete Structures*, Stroyizdat, Moscow, 1949, 236.

138. Diosov, A.E. and Malinovsky, A.G., Statistical quality control of concrete in manufacturing of prefabricated parts and structures (in Russian), in *Technology and Properties of Heavy Concrete*, Stroyizdat, Moscow, 1971, 159.

139. Dmitriev, V.P., On technology factor and classes of strength homogeneity of concrete (in Russian), in *Reinforced Concrete and Reinforced Concrete Structures*, Knizhnoye Izdatelstvo, Sverdlovsk, 1967, 358.

140. Ruch, X., Der Einfluss der Festigkeit bei der Betonkontrolle, *Der Bauingenieur*, 10, 11, 1962.

141. Conrad, F., Zur Gutebeurteilung des Betons nach statistischen Werten, *Baustoffindustrie*, 4, 13, 1964.

142. Pisanko, G.P. and Golikov, A.A., Strength and deformability of high-strength concretes based on super high-early-strength cement (in Russian), *Beton i Zhelezobeton*, 7, 22, 1966.

143. Skatynsky, V.I., Chuiko, P.A., and Komarovsky, A.A., Study of physical–chemical characteristics of high-strength concretes (in Russian), *Stroitelnye Materialy i Konstruktsii*, 5, 13, 1971.

144. Vasiliev, P.M., Plastic compressive properties of concrete and their influence on service of some components of concrete and reinforced concrete structures (in Russian), Ph.D. thesis, Institute of Civil Engineering, Leningrad, 1951.

145. Markov, A.I., Estimation of strength of brittle bodies on the basis of parameters of their structure and fracture behavior by an example of concretes (in Russian), in *Investigations in the Field of Mechanical Measurements: Trans. VNIIFTRI*, Moscow, 8(38), 125, 1971.

146. Komarovsky, A.A., On the time dependence of statistical parameters of strength of concrete (in Russian), in *Issue: Problems of Reliability of Reinforced Concrete Structures*, Kuibyshev, 1972, 37.

147. Komarovsky, A.A., Semiglazova, O.M., and Spesivtsev, V.P., Nondestructive testing of strength of concrete with allowance for the time factor (in Russian), *Transportnoye Stroitelstvo*, 3, 24, 1977.

148. Novgorodsky, M.A., Komarovsky, A.A., and Lemechko, V.A., Influence of time factor on main parameters of strength distribution of concrete (in Russian), *Izvestiya VUZOV, Stroitelstvo i Arkhitektura*, Moscow, 7, 66, 1982.

149. Gansen, T., *Creep and Stress Relaxation in Concrete* (in Russian), Stroyizdat, Moscow, 1963.

150. Gnutov, I.A. and Osipov, A.D., Testing of concrete after 30 years service of a structure (in Russian), *Gidrotekhnicheskoye Stroitelstvo*, 3, 10, 1973.

151. Sosnovsky, L.A., *Tribofatics: Problems and Prospects* (in Russian), Bel. NIIZhT Press, Gomel, 1989.

152. Blanter, M.E., *Procedure of Study of Metals and Experimental Data Treatment* (in Russian), Metallurgizdat, Moscow, 1952.

153. Ivanitsky, G.A., Stereology (in Russian), *Nauka i Zhizn*, 9, 65, 1972.

154. Microstructure of cement mortars and petrographic method for their quality control (in Russian), *Trans. NIISMI, NII Building Materials*, Kiev, 1964.

155. Nevill, M., *Properties of Concrete* (in Russian), Stroyizdat, Moscow, 1972.

156. Volf, I.V. et al., Determination of strength of concrete structures by the method of pulling out of bars (in Russian) *Beton i Zhelezobeton*, 10, 17, 1973.

157. Chuiko, P.A., Study of strength of high-strength concretes in structures by non-destructive methods (in Russian), Ph.D. thesis, NII Building Structures, Kiev, 1973.

158. Kublin, I.Ya., Study of concrete in complex compressive and tensile loading (in Russian), Ph.D. thesis, Riga Civil Engineering Institute, Riga, 1961.

159. Nogin, S.I., Shtaltovny, V.A., and Posvizhsky, E.G., Boundaries of microcracking of concrete under compression (in Russian), *Beton i Zhelezobeton*, 3, 16, 1980.

160. Pochtovik, G.Ya. and Shkolnik, I.E., *Determination of a Low Limit of Microcracking by the Ultrasonic Test Results. Issue Nondestructive Testing of Materials* (in Russian), Stroyizdat, Moscow, 1971, 71.

161. GOST 21243-85. Concretes. Determination of Strength by the Method of Pulling Out with Cleavage (in Russian), Standartizdat, Moscow, 1985.

162. Kuznetsov, V.D., *Surface Energy of Solids* (in Russian), Energia, Moscow, 1954.

163. Komarovsky, A.A., Nondestructive testing of strength and homogeneity of concrete based on super high-frequency electromagnetic field (in Russian), in *Trans. All-Union Conf. "Integration of Nondestructive Testing Methods,"* Issue, Stroyizdat, Moscow, 1973, 135.

164. Komarovsky, A.A., Spesivtsev, V.P., and Zhariy, A.E., Control of reinforced concrete structures by super high-frequency radio waves (in Russian), *Transportnoye Stroitelstvo*, 10, 52, 1976.

165. Matveev, V.I. and Paveliev, V.A., Microwaves propagation in building materials. *Issue Commercial Application of Radiointroscopy.* Tsniitei Priborostroyeniya, Moscow, 1967, 18.

166. Rudakov, V.N., Belyanin, A.N., and Zelenkov, A.L., On Macroscopic Structural Anisotropy of Industrial Products (in Russian), *Defektoskopiya, Moscow*, 5, 19, 1966.

167. Odinets, L.L., Optical method for measurement of dielectric permittivity and dielectric losses of solid dielectrics in centimeter range (in Russian), *J. Tech. Phys.*, 19, 1, 107, 1949.

168. Sloushch, V.G., Application of electromagnetic waves of the super high-frequency range for nondestructive testing of refractory products (in Russian), *Defektoskopiya*, 6, 33, 1968.

169. Klevtsov, V.A., Main trends in improvement of methods for evaluation of the state of load-carrying reinforced structures in their restoration (in Russian), *Promyshlennoye Stroitelstvo*, 8, 15, 1984.

170. Byrin, V.N., Application of acoustic emission for diagnostics of industrial objects (in Russian), *Izmereniye, Control, Avtomatizatsiya*, 3, 7, 1977.

171. Novikov, P.V. and Weinberg, V.E., On physical nature of acoustic emission in deformation of metallic materials (in Russian), *Problemy Prochnosti*, 12, 65, 1977.

172. Duncan, H.L., Acoustic emission, Issue ASTM, STP-505, 101, 1972.

173. Hsu, T.T. et al., Acoustic emission, *Acoustic J.*, 60, 15, 1963.

174. Borodin, Yu.P., Gulevsky, I.V., and Nikolaichev, A.N., Acoustic emission — a new method of nondestructive testing (in Russian), Review No. 516, TsAGI Press, Moscow, 1977.

175. Ivanov, V.I., *Methods and Equipment of Control Using Acoustic Emission* (in Russian), Mashinostroyeniye, Moscow, 1980.

176. Greshnikov, V.A. and Drobot, Yu.V., *Acoustic Emission* (in Russian), Gosstandart, Moscow, 1976.

177. Ckoblo, A.B. and Zhigun, A.P., Some aspects of hardness measurement using acoustic emission (in Russian), *Zavodskaya Laboratoriya*, 2, 6, 1980.

178. Hatton, R.H. and Ordt, R.N., *Acoustic Emission Issue Methods of Nondestructive Testing,* Sharp, R.S., Ed., Mir, Moscow, 1972, 27.

179. Acoustic Emission. Materials symposium presented at the December committee week of American Society for Testing and Materials, Bal Harbor, FL, December 7–8, 1971.

180. Nogin, S.I. and Shlyaktsu, M.I., *Apparatus SAKEM-1 for Measurement of Acoustic Emission* (in Russian), NIISK Gosstroya of the USSR Press, Moscow, 1976.

181. Jons, R. and Fekeoaru, I., *Nondestructive Testing of Concrete* (in Russian), Stroyizdat, Moscow, 1974.

182. Kvirikadze, O.N., Influence of loading rate on deformability and tensile strength of concrete (in Russian), *Beton i Zhelezobeton,* 1, 9, 1962.

183. Nikalau, V., Influence of loading rate on strength of concrete (in Russian), *Beton i Zhelezobeton,* 3, 17, 1959.

184. Verker, M., Influence of loading rate and other factors on strength and elastic properties of rock (in Russian), Transl. from Spanish, VNIIFTRI Press, Moscow, 1959, 54.

185. Nemets, I., Dependence between material strength and nature of variations of loading with time (in Russian), *Express-Information "Detali mashin,"* 23, 17, 1962.

186. Sheikin, A.E. and Nikolaeva, V.G., On elastic–plastic tensile properties of concrete (in Russian), *Beton i Zhelezobeton,* 9, 15, 1959.

187. Bazhenov, Yu.M., *High-Strength Fine-Grained Concrete for Reinforced Cement Structures* (in Russian), Gosstroyizdat, Moscow, 1963.

188. Grushko, I.M., Glushchenko, N.F., and Ilyin, A.G., *Structure and Strength of Road Cement Concrete* (in Russian), Khisi Press, Kharkov, 1965.

189. Krol, I.S. and Krasnovsky, R.O., Some problems of the procedure used for determination of tensile characteristics of concrete in axial compression (in Russian), in *Issue Experimental Studies of Engineering Structures,* Nauka, Moscow, 1973, 42.

190. Nilender, Yu.A., Mechanical properties of concrete and reinforced concrete (in Russian), in *Reference Book for Designer of Industrial Structures,* ONTI Press, Moscow, Vol. IV, 1935, 47.

191. Ostashov, I.A., Dependence of deformation of materials upon the time and rate of loading (in Russian), *Sci. Lett. Acad. Building Architec. Ukr. SSR,* AN USSR Press, Kiev, 1953.

192. Voblikov, V.S. and Tedder, R.I., Influence of loading rate on cleavage of rock (in Russian), in *Issue Study of Physical–Mechanical Properties and Method of Explosion Destruction of Rock,* Nauka, Moscow, 1976, 166.

193. Kuntysh, M.F. and Tedder, R.I., Influence of loading rate on compression strength of rock (in Russian), *Study of Physical–Mechanical Properties and Method of Explosion Destruction of Rock,* Nauka, Moscow, 1970, 60.

194. GOST 10180-78, Concretes. Methods for Determination of Compressive and Tensile Strength, Standartizdat, Moscow, 1978.

195. GOST 310.4-76, Cements. Methods for Determination of Bending and Tensile Strength, Standartizdat, Moscow, 1976.

196. GOST 18105-82, Concretes. Homogeneity and Strength Testing, Standartizdat, Moscow, 1982.

197. GOST 21217-85, Concretes. Testing and Estimation of Strength and Homogeneity by Nondestructive Methods, Standartizdat, Moscow, 1985.

198. GOST 12004-86, Reinforcing Steel. Methods of Tensile Tests, Standartizdat, Moscow, 1986.

199. GOST 8829-77, Prefabricated Reinforced Concrete Structures and Parts. Methods of Testing and Estimation of Strength, Rigidity and Crack Resistance, Standartizdat, Moscow, 1977.

200. Rozhkov, A.I., Relationship between structural changes and deformations in concrete under long-time loading (in Russian), *Izvestiya VUZOV Stroitelstvo i Arkhitektura*, 5, 14, 1970.

201. Stepanov, G.V., *Elastic–Plastic Deformation of Materials in Pulsating Loads* (in Russian), Naukova Dumka, Kiev, 1979.

202. Mayer, M.A. and Mur, L.E., Ed., *Shock Waves and High Speed Deformation Phenomena* (in Russian), Metallurgiya, Moscow, 1984.

203. Gvozdev, A.A., State-of-art and problems of theory of reinforced concrete (in Russian), *Beton i Zhelezobeton*, 2, 3, 1955.

204. Gnedenko, B.V., Statistical methods in theory of reliability (in Russian), Reliability of hoisting vehicles, *Trans. VNIIPTTMash*, 1(96), 134, 1970.

205. Bolotin, V.V., *Application of the Methods of Probability and Reliability Theories for Design of Structures* (in Russian), Stroyizdat, Moscow, 1971.

206. Gnedenko, B.V., Some problems of reliability theory as the subject of investigation and teaching (in Russian), *Reliability and Durability of Machines and Equipment*, GOSSTANDARD, Moscow, 1972, 87.

207. Melomedov, I.M., *Physical Principles of Reliability* (in Russian), Energiya, Leningrad, 1970.

208. Sedyakin, N.M., On one physical principle of reliability theory (in Russian), *Izv. AN SSSR "Technicheskaya Kibernetika,"* 3, 18, 1966.

209. Komarovsky, A.A., Equation of state and durability of solids (in Russian), *Ogneupory i Tekhnicheskaya Keramika*, 6, 7, 2001.

210. Komarovsky, A.A., Physics of failures of structural materials (in Russian), *Tekhnicheskaya Diagnostika i Nerazrushayushchiy Kontrol*, 1, 3, 2001.

211. Betekhtin, V.I., Time and temperature dependence of strength of solids (in Russian), *Experimental Investigations of Engineering Structures*, Nauka, Moscow, 1979, 10.

212. Komarovsky, A.A., Prediction of residual life and durability (in Russian), *Tekhnicheskaya Diagnostika i Nerazrushayushchiy Kontrol*, 3, 3, 2000.

213. Komarovsky, A.A., Prediction of life of materials and structures (in Russian) *Kontrol. Diagnostika*, 12, 8, 2000.

214. Medeksha, G.G. and Zhitkyavichus, V.P., Study of long-time cyclic strength in soft asymmetrical and hard loading (in Russian), *Problemy Prochnosty*, 6, 40, 1978.

215. Komarovsky, A.A., Fatigue of materials: physical nature, prediction, prevention and relief (in Russian), *Tekhnicheskaya Diagnostika i Nerazrushayushchiy Kontrol*, 1, 35, 2000.

216. Strizhalo, V.A. and Kalashnik, M.V., Cyclic creep of structural alloys and its relationship with parameters of acoustic emission (in Russian), *Tekhnicheskaya Diagnostika i Nerazrushayushchiy Kontrol*, 9, 58, 2000.

217. Brinkman, C.R. and Korth, G.E., Low cycle fatigue and hold time comparisons of irradiated and unirradiated type 316 stainless steel (in Russian), *Met. Trans.*, 3, 792, 1974.

218. Selton, R.P., High strain fatigue of 20 Cr/25 Ni NB steel at 1025 K. Part II. Total endurances (in Russian), *Mater. Sci. Eng.*, 19(1), 31, 1975.

219. Rulkov, A.I., Complex of means for nondestructive testing of strength of concrete in CMM-2 structures (in Russian), VEM-3, EI NII Stroitelstva Gosstroya ESSR, Tallinn, 1978.

220. Sammal, O.Yu. and Rulkov, A.A., VSM-4 microprocessor sclerometer for determination of strength of concrete (in Russian), *Investigations in the Field of Building*, Valgus, Tallinn, 1981.
221. Zhurkov, S.N., Kinetic nature of strength (in Russian), *Inorg. Mater.*, 3, 13, 1967.
222. Oding, I.A. et al. *Theory of Creep and Long-Time Strength of Metals* (in Russian), Metallurgizdat, Moscow, 1959.
223. Dlin, A.M., *Mathematical Statistics in Engineering* (in Russian), Nauka, Moscow, 1951.
224. Garofalo, F., *Fundamentals of Creep and Creep-Rupture in Metals*, MacMillan, New York, 1965.
225. Mechovsy, J.J. and Passoja, D.E., *Fractal Aspect of Materials*, MRS, Pittsburg, PA, 1985.
226. Simitsy, X., *Fractography Handbook*, Chubu Keiei Kaihatsy Center, Nagoya, Japan, 1985.
227. Brostom, W. and Corneliussen, R.D., *Fracture of Plastics*, Hanser, New York, 1986.
228. Elizavetin, M.A., *Improving Reliability of Machines* (in Russian), Mashinostroyeniye, Moscow, 1973.
229. Ostreikovsky, V.A., *Multifactorial Reliability Testing* (in Russian), Energiya, Moscow, 1978.
230. Newkerk, D.B., General theory, mechanism and kinetics (in Russian), in *Ageing Alloys*, Transl. from English, Zakharova, M.I., Ed., Metallurgizdat, Moscow, 1962, 74.
231. Parshin, A.M., *Structure, Strength and Radiation Damage of Corrosion-Resistant Steels and Alloys* (in Russian), Metallurgiya, Chelyabinsk, 1988.
232. Skryabina, N.E. and Spivak, L.V., Mechanical instability as a result of quasi-liquid state of alloys of metal-hydrogen system (in Russian), Hydrogen treatment of materials. Proc. 3rd Int. Conf. "BOM-2001," Donetsk-Mariupol, May 14–18, Part 1, 132, 2001.
233. Stanyukovich, A.V., *Brittleness and Plasticity of Refractory Materials* (in Russian), Metallurgiya, Moscow, 1967.
234. Gorynin, I.V. et al., Influence of neutron irradiation on radiation strengthening and embrittlement of titanium alloy and titanium-zirconium system (in Russian), *Radiation Defects in Metallic Crystals*, Issue Nauka Kaz. SSR, Alma-Ata, 123, 1978.
235. Pisarenko, G.S. and Kiselevsky, V.N., *Strength and Plasticity of Materials in Radiation Flows* (in Russian), Naukova Dumka, Kiev, 1979.
236. Grounes, M., Review of Swedish work on irradiation effects in canning and core support materials. Effects of irradiation on structural metals, Philadelphia, PA, 1967, 200.
237. Irwin, J.E. and Bement, A.L., Nature of radiation damage to engineering properties of various stainless alloys. Effect of radiation on structural metals, Philadelphia, PA, 1967, 278.
238. Gusev, A.V., Paper N 28/P339a Materials 3rd Int. Conf. Application of Nuclear Energy for Peace Purposes, Geneva, 1964, 178.
239. Ibragimov, Sh.Sh., Effect of radiation on properties of structural materials (in Russian), *Proc. Symp. CMEA Countries*, FEI Press, Obninsk, 1967, 44.
240. Bohm, H. and Ranck, N., The development of closed-loop, servo-hydraulic test system for direct stress monotonic and cyclic crack propagation studies under biaxial loading, *J. Nucl. Mater.*, 23, 184, 1969.
241. Oding, I.A., Ivanova, V.S., and Burdunsky, V.V., *Theory of Creep and Long-Term Strength of Metals* (in Russian), Metallurgizdat, Moscow, 1959.

242. Fridman, Ya.B., *Mechanical Properties of Metals* (in Russian), Oborongiz, Moscow, 1952.

243. Birger, I.A. and Shorr, B.F., Ed., *Thermal Strength of Machine Parts* (in Russian), Mashinostroyeniye, Moscow, 1975.

244. Ischenko, I.I., Ed., *Study and Development of Materials Used for Thermonuclear Fusion Reactor* (in Russian), Nauka, Moscow, 1981.

245. Kiselevsky, V.N., *Change of Mechanical Properties of Steels and Alloys under Radiation Exposure* (in Russian), Naukova Dumka, Kiev, 1971.

246. Meshchansky, N.A., *Density and Strength of Concretes* (in Russian), Gosstroyizdat, Moscow, 1961.

247. Effect of humidity of concrete on its compressive strength, editorial (in Russian), *Techn. Inf. Industry of Prefabricated Reinforced Concrete*, 7, 27, 1969.

248. Koifman, M.I. and Ilnitskaya, B.I., Effect of humidity on tensile and compressive strength of rock (in Russian), in *Investigation of Physical–Chemical Properties of Rocks*, Nauka, Moscow, 1970, 44.

249. Popovich, G.A. and Nikolko, A.P., Effect of humidity and age of concrete on strength indices in its testing by mechanical devices (in Russian), in *Methods of Nondestructive Testing of Reinforced Concrete*, Budivelnyk, Kiev, 1972, 100.

250. Bazhenov, Yu.M., Effect of moisture content on strength of concrete at different loading rates (in Russian), *Beton i Zhelezobeton*, 12, 18, 1966.

251. Elissev, V.I., Influence of nonuniform distribution of moisture content of sections of reinforced concrete members (in Russian), *Inzhenernye Konstrukcii*, 37, 18, 1969.

252. Komarovsky, A.A. and Radchenko, L.N., On effect of moisture content of concrete on reliability and durability of prefabricated reinforced concrete structures (in Russian), *Prob. Reliability Build.*, Issue Sverdlovsk, 1972, 92.

253. Buzhevich, G.A. and Kornev, P.A., *Expanded-Clay Lightweight Concrete* (in Russian), Stroyizdat, Moscow, 1963.

254. Gul, B.E., *Structure and Strength of Polymers* (in Russian), Khimiya, Moscow, 1971.

255. Chentemirov, M.G., Prospects of development and application of light-weight concrete structures (in Russian), *Beton i Zhelezobeton*, 3, 3, 1971.

256. Afanasiev, V.G., Principles of classification of integral systems (in Russian), *Voprosy Philosofii*, 5, 31, 1963.

257. Petrushenko, L.A., *Principles of Feedback Control* (in Russian), Mysl, Moscow, 1967.

258. Makhutov, N.A. and Romanov, A.N., Eds., *Strength of Structures in Low Cycle Loading* (in Russian), Nauka, Moscow, 1983.

259. Popov, N.T. and Marfin, N.I., *Choice of Rational Structures of Village Power Transmission Line Supports* (in Russian), UkrNIINTI Press, Kiev, 1969.

260. Kuznetsov, V.D., *Physics of Solids* (in Russian), Krasnoye Znamya, Tomsk, Vol. 2, 1941.

261. Zhurkov, S.N., Levina, B.Ya., and Sanfirova, T.P., Fluctation of properties of solids (in Russian), *Solid State Phys.*, 2(6), 1040, 1960.

262. Kudryavtsev, I.P., *Texture in Metals and Alloys* (in Russian), Metallurgiya, Moscow, 1965.

263. Shmidt, V., Introduction to testing methods (in Russian), Issue *Behavior of Steel under Cyclic Loads*, Metallurgiya, Moscow, 441, 1983.

264. Babich, E.M., Koshlai, V.A., and Pogorelyak, A.P., Strength of concretes after short-time nonmultiple repeated compressive load (in Russian), in *Problems of Reliability of Reinforced Concrete Structures*, Technika, Kuibyshev, 1975, 12.

265. Akhverdov, I.N., *Principles of Physics of Concrete* (in Russian), Stroyizdat, Moscow, 1981.

266. Sannikov, Yu.D., On the problem of influence of microfractures on energy dissipation in glass-reinforced plastic (in Russian), *Ingenernye Konstrukcii*, LISI Press, Leningrad, 1969, 123.

267. Krasilnikov, K.G. and Skablevskaya, I.N., Physical–chemical nature of moist deformation of cement stone (in Russian), in *Creep and Shrinkage of Concrete*, NIIZhB Gosstroya SSSR, Moscow, 1969, 119.

268. Gusenkov, A.P. and Sheiderovich, R.M., Properties of cyclic deformation diagrams at elevated temperatures (in Russian), in *Deformation and Fracture Resistance at Low Cycle Loading*, Nauka, Moscow, 1967, 64.

269. Murakami, Y., Ed., *Stress Intensity Factors*, Pergamon Press, Oxford, 1987.

270. Tada, H., Paris, P.C., and Irwin, G.R., *The Stress Analysis of Cracks Handbook*, Del Research, Hellertown, PA, 1973.

271. Peterson, R.E., *Stress Concentration Design Factors*, John Wiley & Sons, New York, 1974.

272. Nair, N.V., *Fracture Mechanics: Microstructure and Micromechanisms*, ASM, Metals Park, OH, 1987.

273. Young, W.C., *Roak's Formulas for Stress and Strain*, 6th ed., McGraw-Hill, New York, 1989.

274. Shkolnik, L.M., *Cracks Growing Rate and Life of Metal* (in Russian), Metallurgiya, Moscow, 1973.

275. Cherepanov, G.P., *Mechanics of Brittle Fracture* (in Russian), Nauka, Moscow, 1974.

276. Neiber, G., *Stress Concentration* (in Russian), Gostekhizdat, Moscow, 1947.

277. Gnedenko, B.V., On some problems of reliability theory as the object of investigation and teaching (in Russian), *Reliability and Durability of Machines and Equipment*, Issue, Standartizdat, Moscow, 1972, 87.

278. Kamke, E., *Reference Book of Conventional Differential Equations* (in Russian), Nauka, Moscow, 1976.

279. Smirnov, N.V. and Dunin-Barkovsky, I.V., *Course of Theory of Probability and Mathematical Statistics* (in Russian), Nauka, Moscow, 1969.

280. Sytnik, N.I., Theoretical prerequisites and principles of technology for manufacturing of high-strength concrete (in Russian), in *High-Strength Concrete*, Budivelnyk, Kiev, 1967, 6.

281. Weibule, W.A., Statistical theory of the strength of materials, *Proc. R. Swedish Inst. Eng. Res., Stockholm*, 5(151), 17, 1939.

282. Kontorova, T.A., Statistical theory of strength (in Russian), *J. Tech. Phys.*, 10(11), 7, 1940.

283. Komarovsky, A.A., Scale effect and level of technogenic safety of engineering objects (in Russian), *Tyazholoye Mashinostroyeniye*, 10, 23, 2001.

284. Lermit, R., *Technological Problems of Concrete* (in Russian), Stroyizdat, Moscow, 1959.

285. Rumshinsky, L.Z., *Elements of Probability Theory* (in Russian), Nauka, Moscow, 1966.

286. Freudenthal, A.M., *The Inelastic Behavior of Engineering Materials and Structures*, John Wiley & Sons, New York, 1950.

287. Krol, I.S., Empiric representation of compression diagram of concrete (taken from literature) (in Russian), *Trans. VNIIFTRI*, Moscow, 8(38), 306, 1971.

288. Rzhevsky, V.V. and Protasov, B.I., *Electrical Destruction of Rock* (in Russian), Nedra, Moscow, 1972.

289. Bunin, M.V., Grushko, A.G., and Ilyin, I.M., *Structure and Mechanical Properties of Road Cement Concretes* (in Russian), Building Industry Press, Kharkov, 1968.

290. Krylov, N.A., Kalashnikov, V.A., and Polishchuk, S.M., *Radio Engineering Methods of Quality Testing of Concrete* (in Russian), Stroyizdat, Leningrad, 1966.
291. Gvozdev, A.A. and Baikov, V.N., On the problem of behavior of reinforced concrete structures at the stage close to fracture (in Russian), *Beton i Zhelezobeton*, 9, 22, 1977.
292. Chogovadze, D.V., Study of deformation process and tensile fracture mechanism of cement stone (in Russian), Ph.D. thesis, Building Institute, Tbilisi, 1968.
293. Vaganov, A.I., *Expanded-Clay Light-Weight Concrete* (in Russian), Stroyizdat, Leningrad, 1954.
294. Indenbom, V.L. and Orlov, N.A., Problems of fracture in physics of strength (in Russian), *Problemy Prochnosti*, 12, 21, 1970.
295. Ivanova, V.S. and Terentiev, V.F., *Fatigue Nature of Metals* (in Russian), Metallurgiya, Moscow, 1965.
296. Komarovsky, A.A., Periodic law of variations in state of solids (in Russian), *Ogneupory i Tekhnicheskaya Keramika*, 7, 7, 2001.
297. Kotsanda, S., *Fatigue Fracture of Metals* (in Russian), Metallurgiya, Moscow, 1976.
298. Forrest, P., *Fatigue of Metals* (in Russian), Mashinostroyeniye, Moscow, 1986.
299. Kuslitsky, A.V. et al., Nature of fatigue (in Russian), *Stal*, 2, 151, 1965.
300. Serensen, S.V. et al., *Strength at Low-Cycle Loading. Principles of Design and Testing Methods* (in Russian), Nauka, Moscow, 1975.
301. Troshenko, V.T., *Strength of Metals under Alternating Loads* (in Russian), Naukova Dumka, Kiev, 1978.
302. Gudkov, A.A. and Slivsky, Yu.I., *Methods for Measurement of Hardness of Metals and Alloys* (in Russian), Metallurgiya, Kiev, 1982.
303. Narisova, I., *Strength of Polymer Materials* (in Russian), Khimiya, Moscow, 1987.
304. Sosnovsky, L.A. et al., Complex experimental investigations of properties of 08Kh18N12T steel with regard to long-time service (in Russian), *Vestsi AN BSSR, Ser. PHYZ. Energ.* Navuk, Minsk, 1, 43, 1991.
305. Akchurin, M.M. and Regel, V.R., Study of peculiarities of deformation structure formed under concentrated load on crystal (review) (in Russian), *Zavodskaya Laboratoriya*, 5, 17, 1999.
306. Shcherbakov, V.I. and Titov, A.N., Study of kinetics of cracks by polarization-optical method (in Russian), *Zavodskaya Laboratoriya*, 7, 44, 1998.
307. Milov, V.A., Theory of shock sclerometers (in Russian), *Izvestiya VUZOV, Priborostroyeniye*, 8(8), 26, 1970.
308. Dieter, G.E., *Mechanical Metallurgy*, 3rd ed., McGraw-Hill, New York, 1986.
309. Johnson, W. and Mellor, P.B., *Engineering Plasticity*, John Wiley & Sons, New York, 1983.
310. Honeycombe, R.W.K., *The Plastic Deformation of Metals*, 2nd ed., Edward Arnold, London, 1984.
311. Ravichandran, K.S. and Vasudevan, A.K., Fracture resistance of structural alloys, in *AASM Handbook 7*, Vol. 19, *Fatigue and Fracture*, ASM International, Metals Park, OH, 1996, 381.
312. Hertzberg, R.W., *Deformation and Fracture Mechanics of Engineering Materials*, 3rd ed., John Wiley & Sons, New York, 1989.
313. Pagh, H.L.D., Ed., *The Mechanical Behavior of Materials under Pressure*, Elsevier, New York, 1970.
314. Collins, J.A., *Failure of Materials in Mechanical Design*, John Wiley & Sons, New York, 1981.

315. Chernyshev, V.N. and Sysoev, E.V., Contactless adaptive method of nondestructive testing of thermal–physical properties of materials (in Russian), *Kontrol i Diagnostika*, 2, 31, 2000.

316. Paton, B.E. and Nedoseka, A.Ya., New approaches to estimation of the state of welded structures and determination of their residual life (in Russian), *Tekhnicheskaya Diagnostika i Nerazrushayushchiy Kontrol*, 1, 8, 2000.

317. Paton, B.E. and Nedoseka, A.Ya., New approach to estimation of the state of welded structures (in Russian), *Avtomaticheskaya Svarka*, 9/10, 97, 2000.

318. Knysh, V.V., Determination of cyclic durability of structural members in fatigue crack arresting (in Russian), *Avtomaticheskaya Svarka*, 9/10, 73, 2000.

319. Lanin, V.S. and Egorov, A.G., Fracture of ductile–brittle bodies under combined action of thermal and mechanical loads (in Russian), *Physika i Khimiya Obrabotki Materialov*, 2, 78, 1999.

320. Pokhil, Y.A. et al., Effects of low temperatures and magnetic fields on the structure and physical and mechanical properties of structural materials (in Russian), EMRS Symposium D: Materials under Extreme Conditions, St. Petersburg, October 17–21, 1993, 37.

321. Ermolaev, B.I., Thermal shock resistance of refractory metals, EMRS Symposium D: Materials under Extreme Conditions, St. Petersburg, October 17–21, 1993, 49.

322. Dzigo, Sh. et al., Heat treatment and surfacing using light beam heating (in Russian), *Avtomaticheskaya Svarka*, 9/10, 171, 2000.

323. Kompan, Ya., Magnetically impelled melting of high quality titanium (in Russian), Annual Report of STCU, Kyiv, 1999, 30.

324. Golovin, Yu.I. and Morgunov, R.B., Influence of magnetic field on structure-sensitive properties of real diamagnetic crystals (in Russian), *Materialovedeniye*, 3, 2, 2000.

325. Fujita, Yu. et al., Development of twin-wire TIG welding process with electromagnetic control of the weld pool (in Russian), *Avromaticheskaya Svarka*, 9/10, 152, 2000.

326. Desnenko, V.A. and Peckarskaya, V.I., Plastic deformation of metals in electric field. EMRS, 1993, Symposium, 17–21 October, St. Peterburg, 59, 1993.

327. Lakhtin, Yu.M. and Leontieva, V.P., *Materials Science* (in Russian), Mashinostroyeniye, Moscow, 1972.

328. Schetky, L.M., Shape-memory alloys, *Sci. Am.*, 5(241), 74–92, 1979.

329. Ivanova, V.S. and Gordienko, L.K., *New Trends in Improving Metal Strength* (in Russian), Nauka, Moscow, 1964.

330. Stefanov, B.V., *Technology of Concrete and Reinforced Concrete Products* (in Russian), Budivelnik, Kiev, 1972.

331. Komarovsky, A.A., Behavior of solids in external fields and aggressive environments (in Russian), *Khimicheskaya Phyzika*, 6, 23, 2001.

332. Komarovsky, A.A., Prediction of reliability and durability (in Russian), *Tyazholoye Mashinostroyeniye*, 12, 16, 2000.

333. Komarovsky, A.A., Adaptation of materials and structures to service conditions (in Russian), *Prikladnaya Phyzika*, 5, 13, 2001.

334. Parton, V.Z., Fracture mechanics (in Russian), *Nauka i Zhizn*, 12, 37, 1974.

335. Pronikov, A.S., Content and fundamental trends of reliability and durability science of machines (in Russian), in *Reliability and Durability of Machines and Equipment*, Standartizdat, Moscow, 1972.

336. Maklyaev, P.G. and Fridman, Ya.B., *Anisotropy of Mechanical Properties of Materials* (in Russian), Metallurgiya, Moscow, 1969.

337. Zhukov, A.A. and Sherman, A.D., Eds., *Materials in Machine-Building (Cast Iron)* (in Russian), Mashinostroyeniye, Moscow, V. IV, 1969.

338. Berner, R. and Kronmuller, G., Plastic deformation of single crystals, Orlov, A.N., Ed., Mir, Moscow, 1969.

339. Troshenko, V.T., *Deformation and Fracture of Metals in High Cycle Loading* (in Russian), Naukova Dumka, Kiev, 1981.

340. Tomson, E.R. and Lempe, F.D., Eutectic refractory alloys manufactured by the method of directed solidification. *Composite Materials with Metallic Matrix* (in Russian), Mashinostroyeniye, Moscow, 1978, 110.

341. Komarovsky, A.A., Control of physical–chemical properties of materials (in Russian), *Ogneupory i Tekhnicheskaya Keramika*, 3, 8, 2001.

342. Komarovsky, A.A., Chernobyl: "Shelter." How long will it serve? Newspaper *Delovaya Ukraina*, 84 (534), November 1997.

343. Komarovsky, A.A., "Shelter": Avoiding errors of the past, Newspaper *Delovaya Ukraina*, 87 (537), December 1997.

344. Komarovsky, A.A., "Shelter": What happened and why? Newspaper *Vestnik Chernobylya*, 3–4, March 1998.

345. Kartashov, K.N. and Ushakov, N.A., Ways of improving the technical level of industrial engineering (in Russian), *Bulleten Stroitelnoi Tekhniki*, 11, 41, 1970.

346. Shilkrut, D.I., Theory of propagation of real microcracks in solids during deformation (in Russian), Papers of AN SSSR, Moscow, 122(1), 17, 1958.

347. Sedyakin, N.M. and Kabanov, G.A., On substantiation of physical principle of reliability (in Russian), in *Issue Engineering Means of Automatics*, Nauka, Moscow, 26, 1971.

348. Gusenkov, A.P. and Kotov, P.I., *Long-Time and Nonisothermal Low-Cycle Strength of Structural Elements* (in Russian), Mashinostroyeniye, Moscow, 1988.

349. Krachmalev, V.I., Krilov, B.S., and Treschevskiy, A.N., Vibration-damping structural materials (in Russian), EMRS-1993 Fall Meeting, Symposium D, October 17–21, St. Petersburg, 1993, 48.

350. Ermolaev, B.I., Thermal shock resistance of refractory metals (in Russian), EMRS-1993, Symposium D, October 17–21, St. Petersburg, 1993, 49.

351. Komarovsky, A.A., Intelligent building materials and structures. Symposium Current Trends in Building, Mimer Sinan University, May 5–8, Istanbul, Turkey, 1995.

352. 7th Construction Industry International Symposium, Santiago, Chile, Sept. 17–20, 1995, 3–5.

353. Komarovsky, A.A., The method how to prevent fracture of the materials and structures at subzero temperature in cold regions. 8th Int. Cold Regions Eng. Specialty Conf., Fairbanks, AK, August 7–9, 1996, 5.

354. Komarovsky, A.A., Intelligent materials and structures, Presentation on TV of Ukraine, Kiev, June 1996.

355. Vashchenko, V.N., *Centers of Deep-Focus Earthquakes* (in Russian), Nauchnaya Kniga, Kyiv, 1995.

356. Astakhov, V.P., *Metal Cutting Mechanics*, CRC Press, Boca Raton, FL, 1998.

357. Trent, E.M. and Wright, P.K., *Metal Cutting*, 4th ed., Butterworth-Heinemann, Boston, 2000.

358. Altintas, Y., *Manufacturing Automation: Metal Cutting Mechanics, Machine Tool Vibrations, and CNC Design*, Cambridge University Press, Cambridge, 2000.

359. Childs, T.H.C. et al., *Metal Machining. Theory and Application*, Edward Arnold, London, 2000.

Index

A

Acceleration, tension with, 64
Acceleration well, 33–34, 39–40, 37, 122, 137
Acceleration zone, 94
Acoustic emissions, 290–294
  durability, design of materials for, 350
  dynamic effect in tension and
      compression, 312
  fracture physics, 426–427
  interatomic bond failure and, 413
  stress-strain state diagnostics, 494, 495
Activation energy, initial, 332
Active media, solids in, 363–411
  aging, 364–379
    strain, 376–379
    thermal, 370–375
  defect healing and damaged structure
      restoration, 406–412
  durability of unstable structures, 399–405
  hydrogen embrittlement, 380–383
  moisture induced softening of porous
      materials, 390–399
  radiation damage, 383–390
    embrittlement, 387–388
    long-term strength, 389–390
    strengthening and softening, 384–387
Adaptation of materials to service conditions,
      see Service conditions,
      adapatation of materials and
      structures to
Adiabatic deformability, 578
Aging, 186, 364–379
  control of stress-strain state, 553
  durability of unstable structures, 404, 405
  fatigue prevention and relief, 489–490
  strain, 376–379
  thermal, 370–375
Age of concrete
  durability of unstable structures, 404, 405
  moisture-induced changes, 396, 398
Aggregate states of materials, 112–113,
      156–165

Aggressive environments
  Debye temperature and mechanical
      properties, 54
  energy levels of materials, 123–124
Allotropic transformations, aging, 379
Alloys, see also Metals and alloys; Steel and
      steel alloys
  aging processes, 368, 373
  improvement of material properties, 510
  periodic law of variations in state, 155
  stress-strain diagrams in tension at
      various temperatures, 149
  temperature dependence of strength, 144
Alumina cements, temperature effects on
      strength, 151–152
Aluminum and alloys, 159
  cold treatment, 542–543
  complex stressed states, 274
  durability equation, 330
  durability tests, 348
Ampere forces, 12
Anchor removal from concrete, see Concrete,
      anchor removal from
Angular deformation, 60
Angular momentum, isolated atoms, 2, 3
Angular velocity, stable condition, 45
Anharmonicity parameter, 158
Anisotropy
  adaptation of materials for service, 513
  atomic mechanisms, 21
  CD ratio, 539
  control of properties, 511
  dielectrics, 192
  of electric properties, 278
  external field effects, stages of, 211
  formation of, 531–539
  hardness tests of wrought metals, 543
  polyatomic molecule, 109
  service effects, 273
Annealing, 488, 543
Antimony, 159
Antiparallel spin, atomic, 123
Atomic masses, see Mendeleev periodic law

607

DATE DUE			